Fundamental Biomedical Technologies

Volume 8

Series Editor
Mauro Ferrari, Ph.D. The University of Texas, Houston, TX, USA

More information about this series at http://www.springer.com/series/7045

Aleš Prokop • Volkmar Weissig
Editors

Intracellular Delivery III

Market Entry Barriers of Nanomedicines

Springer

Editors
Aleš Prokop
Biomolecular and Chemical Engineering
Vanderbilt University Biomolecular and Chemical Engineering
Nashville, TN, USA

Volkmar Weissig
Pharmaceutical Sciences
Midwestern University Pharmaceutical Sciences
GLENDALE, AZ, USA

Intracellular Delivery I: 978-94-007-1247-8
Intracellular Delivery II: ISBN 978-94-017-8895-3

ISSN 1559-7083
Fundamental Biomedical Technologies
ISBN 978-3-319-43523-7 ISBN 978-3-319-43525-1 (eBook)
DOI 10.1007/978-3-319-43525-1

Library of Congress Control Number: 2014939942

Printed on acid-free paper

This Springer imprint is published by Springer Nature
The registered company is Springer International Publishing AG
The registered company address is: Gewerbestrasse 11, 6330 Cham, Switzerland

Contents

Part I Introductory Chapters

1 **Overview of Present Problems Facing Commercialization of Nanomedicines** 3
Aleš Prokop and Volkmar Weissig

2 **Precision Drugs and Cell-Specific Drug Delivery** 37
Karel Petrak

Part II EPR Effect and ECM Modification

3 **Extracellular Matrix Degrading Enzymes for Nanocarrier-Based Anticancer Therapy** 49
Pablo Scodeller

4 **Nanocarrier-Based Anticancer Therapies with the Focus on Strategies for Targeting the Tumor Microenvironment** 67
Fransisca Leonard and Biana Godin

Part III How to Extend the Circulation Time of Nanovehicles

5 **A New Approach to Decrease the RES Uptake of Nanodrugs by Pre-administration with Intralipid® Resulting in a Reduction of Toxic Side Effects** 125
Li Liu and Chien Ho

Part IV Differences Between In Vivo Status in Men and Mice

6 **Authentic Vascular and Stromal Structure in Animal Disease Model for Nanomedicine** 149
Hiroshi Nishihara and Mitsunobu R. Kano

Part V Cell-Specific Targeting

7 Ligand-targeted Particulate Nanomedicines Undergoing Clinical Evaluation: Current Status 163
Roy van der Meel, Laurens J.C. Vehmeijer, Robbert Jan Kok, Gert Storm, and Ethlinn V.B. van Gaal

8 Anti-angiogenic Therapy by Targeting the Tumor Vasculature with Liposomes 201
Yu Sakurai and Hideyoshi Harashima

9 Accessing Mitochondrial Targets Using NanoCargos 229
Ru Wen, Afoma C. Umeano, and Shanta Dhar

10 Redox-Responsive Nano-Delivery Systems for Cancer Therapy 255
Amit Singh, Thanh-Huyen Tran, and Mansoor M. Amiji

Part VI Improved Imaging

11 Nano-emulsions for Drug Delivery and Biomedical Imaging 273
Nicolas Anton, François Hallouard, Mohamed F. Attia, and Thierry F. Vandamme

12 The Tumor Microenvironment in Nanoparticle Delivery and the Role of Imaging to Navigate Roadblocks and Pathways 301
Dmitri Artemov and Zaver M. Bhujwalla

13 Microscopic Mass Spectrometry for the Precise Design of Drug Delivery Systems 323
Yasuhiro Matsumura and Masahiro Yasunaga

Part VII Quantitative PK Treatment, Systems Biology and Drug Discovery

14 Pharmacokinetics and Pharmacodynamics of Nano-Drug Delivery Systems 341
Amit Ranjan Maity and David Stepensky

15 PBPK Modelling of Intracellular Drug Delivery Through Active and Passive Transport Processes 363
Lars Kuepfer, Christoph Niederalt, Thomas Wendl, Jan-Frederik Schlender, Michael Block, Thomas Eissing, and Donato Teutonico

16 Exploiting Nanocarriers for Combination Cancer Therapy 375
Yi Wen Kong, Erik C. Dreaden, Paula T. Hammond, and Michael B. Yaffe

Part VIII Market Situation and Commercialization of Nanotechnology

17 The Commercialization of Medical Nanotechnology for Medical Applications 405
David W. Hobson

Index 451

Contributors

Mansoor M. Amiji Department of Pharmaceutical Sciences, School of Pharmacy, Bouve College of Health Sciences, Northeastern University, Boston, MA, USA

Nicolas Anton CNRS UMR 7199, Laboratoire de Conception et Application de Molécules Bioactives, équipe de Pharmacie Biogalénique, University of Strasbourg, Illkirch Cedex, France

Dmitri Artemov Division of Cancer Imaging Research, Department of Radiology, The Johns Hopkins University School of Medicine, Baltimore, MD, USA

Mohamed F. Attia CNRS UMR 7199, Laboratoire de Conception et Application de Molécules Bioactives, équipe de Pharmacie Biogalénique, University of Strasbourg, Illkirch Cedex, France

National Research Center, Cairo, Egypt

Zaver M. Bhujwalla Division of Cancer Imaging Research, Department of Radiology, The Johns Hopkins University School of Medicine, Baltimore, MD, USA

Michael Block Complex Systems Modelling, Bayer Technology Service GmbH, Leverkusen, Germany

Shanta Dhar NanoTherapeutics Research Laboratory, Department of Biochemistry and Molecular Biology, Sylvester Comprehensive Cancer Center, Miller School of Medicine, University of Miami, Miami, FL, USA

Erik C. Dreaden Koch Institute for Integrative Cancer Research, Massachusetts Institute of Technology, Cambridge, MA, USA

Department of Chemical Engineering, Massachusetts Institute of Technology, Cambridge, MA, USA

Thomas Eissing Complex Systems Modelling, Bayer Technology Service GmbH, Leverkusen, Germany

Biana Godin Department of Nanomedicine, Houston Methodist Research Institute, Houston, TX, USA

François Hallouard DC2N Inserm U982, Laboratoire de Pharmacie Galénique Biopharmacie, Faculté de Médicine & Pharmacie, Université de Rouen, Rouen, France

Service de Pharmacie, Centre Hospitalier Asselin Hédelin, Yvetot, France

Paula T. Hammond Koch Institute for Integrative Cancer Research, Massachusetts Institute of Technology, Cambridge, MA, USA

Department of Chemical Engineering, Massachusetts Institute of Technology, Cambridge, MA, USA

Institute for Soldier Nanotechnologies, Massachusetts Institute of Technology, Cambridge, MA, USA

Hideyoshi Harashima Faculty of Pharmaceutical Sciences, Hokkaido University, Sapporo, Japan

Chien Ho Department of Biological Sciences, Carnegie Mellon University, Pittsburgh, PA, USA

David W. Hobson LoneStar PharmTox LLC, Bergheim, TX, USA

Mitsunobu R. Kano Department of Pharmaceutical Biomedicine, Graduate School of Medicine, Dentistry, and Pharmaceutical Sciences, Okayama University, Okayama, Japan

Robbert Jan Kok Department of Pharmaceutics, Utrecht Institute for Pharmaceutical Sciences (UIPS), Utrecht University, Utrecht, The Netherlands

Yi Wen Kong Koch Institute for Integrative Cancer Research, Massachusetts Institute of Technology, Cambridge, MA, USA

Department of Biology, Massachusetts Institute of Technology, Cambridge, MA, USA

Lars Kuepfer Complex Systems Modelling, Bayer Technology Service GmbH, Leverkusen, Germany

Fransisca Leonard Department of Nanomedicine, Houston Methodist Research Institute, Houston, TX, USA

Li Liu Department of Biological Sciences, Carnegie Mellon University, Pittsburgh, PA, USA

Amit Ranjan Maity Department of Clinical Biochemistry and Pharmacology, Faculty of Health Sciences, Ben-Gurion University of the Negev, Beer-Sheva, Israel

Yasuhiro Matsumura Division of Developmental Therapeutics, Exploratory Oncology Research & Clinical Trial Center, National Cancer Center, Kashiwa, Japan

Christoph Niederalt Complex Systems Modelling, Bayer Technology Service GmbH, Leverkusen, Germany

Hiroshi Nishihara Department of Translational Pathology, Hokkaido University Graduate School of Medicine, Sapporo, Japan

Translational Research Laboratory, Hokkaido University Hospital, Sapporo, Japan

Karel Petrak NangioTx Inc, Houston, TX, USA

Aleš Prokop Chemical and Biomolecular Engineering, Vanderbilt University, Nashville, TN, USA

Pharmaceutical Sciences, Midwestern University, Glendale, AZ, USA

Yu Sakurai Faculty of Pharmaceutical Sciences, Hokkaido University, Sapporo, Japan

Jan-Frederik Schlender Complex Systems Modelling, Bayer Technology Service GmbH, Leverkusen, Germany

Pablo Scodeller Laboratory of Cancer Biology, Institute of Biomedicine, Center of excellence for translational medicine, University of Tartu, Tartu, Estonia

Amit Singh Department of Pharmaceutical Sciences, School of Pharmacy, Bouve College of Health Sciences, Northeastern University, Boston, MA, USA

David Stepensky Department of Clinical Biochemistry and Pharmacology, Faculty of Health Sciences, Ben-Gurion University of the Negev, Beer-Sheva, Israel

Gert Storm Department of Pharmaceutics, Utrecht Institute for Pharmaceutical Sciences (UIPS), Utrecht University, Utrecht, The Netherlands

Department of Targeted Therapeutics, University of Twente, Enschede, The Netherlands

Donato Teutonico Complex Systems Modelling, Bayer Technology Service GmbH, Leverkusen, Germany

Thanh-Huyen Tran Department of Pharmaceutical Sciences, School of Pharmacy, Bouve College of Health Sciences, Northeastern University, Boston, MA, USA

Afoma C. Umeano NanoTherapeutics Research Laboratory, Department of Biochemistry and Molecular Biology, Sylvester Comprehensive Cancer Center, Miller School of Medicine, University of Miami, Miami, FL, USA

Roy van der Meel Department of Pharmaceutics, Utrecht Institute for Pharmaceutical Sciences (UIPS), Utrecht University, Utrecht, The Netherlands

Department of Clinical Chemistry & Haematology, University Medical Center Utrecht, Utrecht, The Netherlands

Department of Biochemistry & Molecular Biology, University of British Columbia, Vancouver, BC, Canada

Ethlinn V.B. van Gaal Department of Pharmaceutics, Utrecht Institute for Pharmaceutical Sciences (UIPS), Utrecht University, Utrecht, The Netherlands

Cristal Therapeutics, Maastricht, The Netherlands

OctoPlus, Leiden, The Netherlands

Thierry F. Vandamme CNRS UMR 7199, Laboratoire de Conception et Application de Molécules Bioactives, équipe de Pharmacie Biogalénique, University of Strasbourg, Illkirch Cedex, France

Laurens J.C. Vehmeijer Department of Pharmaceutics, Utrecht Institute for Pharmaceutical Sciences (UIPS), Utrecht University, Utrecht, The Netherlands

Volkmar Weissig Chemical and Biomolecular Engineering, Vanderbilt University, Nashville, TN, USA

Pharmaceutical Sciences, Midwestern University, Glendale, AZ, USA

Ru Wen NanoTherapeutics Research Laboratory, Department of Biochemistry and Molecular Biology, Sylvester Comprehensive Cancer Center, Miller School of Medicine, University of Miami, Miami, FL, USA

Thomas Wendl Complex Systems Modelling, Bayer Technology Service GmbH, Leverkusen, Germany

Michael B. Yaffe Koch Institute for Integrative Cancer Research, Massachusetts Institute of Technology, Cambridge, MA, USA

Department of Biology, Massachusetts Institute of Technology, Cambridge, MA, USA

Department of Biological Engineering, Massachusetts Institute of Technology, Cambridge, MA, USA

Divisions of Acute Care Surgery, Trauma, and Critical Care and Surgical Oncology, Department of Surgery, Beth Israel Deaconess Medical Center, Harvard Medical School, Boston, MA, USA

Masahiro Yasunaga Division of Developmental Therapeutics, Exploratory Oncology Research & Clinical Trial Center, National Cancer Center, Kashiwa, Japan

Introduction to Volume III

As a continuation of the previous two volumes, this third volume concentrates on commercial aspects. Unfortunately, obtaining data from the industry has proven to be almost impossible. Thus, we tried collecting manuscripts with emphasis on some preclinical and clinical applications.

Our introductory chapter (Prokop-Weissig) introduces problems associated with the translation of research from the bench to the clinic and subsequently market. This chapter was written by us to serve as an introduction to the entire volume. We failed to get some topics covered, e.g., on patenting as well as on the situation of funding in the USA.

The editors would like to acknowledge the effort of some individuals to peer-review manuscripts submitted by other authors as well as by outside reviewers. These are Fyllos Stylianopoulos, Ales Prokop, Shanta Dhar, Mansoor Amiji, Gerard D'Souza, Biana Godin, Vladimir Torchillin, Hideoyoshi Harashima, Pablo Scodeller, Volkmar Weissig, Nicolas Anton, Lars Kuepfer, Karel Petrak, Sjoerd Hak and

Ales Prokop
Volkmar Weissig

Part I
Introductory Chapters

Chapter 1
Overview of Present Problems Facing Commercialization of Nanomedicines

Aleš Prokop and Volkmar Weissig

Abstract A critical review is attempted to assess the status of nanomedicine entry onto the market. The emergence of new potential therapeutic entities such as DNA and RNA fragments requires that these new "drugs" will need to be delivered in a cell-and organelle-specific manner. Although efforts have been made over the last 50 years or so to develop such delivery technology, no effective and above all clinically approved protocol for cell-specific drug delivery in humans exists as yet. Various particles, macromolecules, liposomes and most recently "nanomaterials" have been said to "show promise" but none of these promises have so far been "reduced" to human clinical practice.

The focus of this volume is on cancer indication since the majority of published research relates to this application; within that, we focus on solid tumors (solid malignancies). Our aim is to critically evaluate whether nanomaterials, both non-targeted and targeted to specific cells, could be of therapeutic benefit in clinical practice. The emphasis of this volume will be on pharmacokinetics (PK) and pharmacodynamics (PD) in animal and human studies.

Apart from the case of exquisitely specific antibody-based drugs, the development of target-specific drug–carrier delivery systems has not yet been broadly successful at the clinical level. It can be argued that drugs generated using the conventional means of drug development (i.e., relying on facile biodistribution and activity after (preferably) oral administration) are not suitable for a target-specific delivery and would not benefit from such delivery even when a seemingly perfect delivery system is available. Therefore, successful development of site-selective drug delivery systems will need to include not only the development of suitable carriers, but also the development of drug entities that meet the required PK/PD profile.

A. Prokop (✉) • V. Weissig
Chemical and Biomolecular Engineering, Vanderbilt University, Nashville, TN, USA

Pharmaceutical Sciences, Midwestern University, Glendale, AZ, USA
e-mail: ales.prokop@vanderbilt.edu

A. Prokop, V. Weissig (eds.), *Intracellular Delivery III*, Fundamental Biomedical Technologies 8, DOI 10.1007/978-3-319-43525-1_1

In general, human clinical studies are approved only after the expected benefits of targeting have been shown in pre-clinical, *in vivo* animal studies first. Therefore, quantitative data on biodistribution of targeted and non-targeted nanoparticles should be generated as the first step. This should be followed by determining whether an increased presence of nanoparticles in tumors also results in increased concentration of the *free drug* within the tumor space. Any "promise" for reproducing similar data in human clinical studies should be supported by relevant scaling from the animal model used to humans.

For too long now, the same or similar approaches have been used by researchers without success. We believe that new fundamentally different approaches are needed to make cell-specific drug delivery clinical reality. In this volume we want to focus on (a) how nanoparticles could be redesigned from the material-science point of view (for example, redesigning nanoparticles for long-circulating properties, passive (EPR) and active targeting concept); and (b) on the design and properties of drugs that would benefit from cell-specific targeting (examining why active targeting of drug carrier does not necessarily result in drug accumulation in tumor). Further, we will draw attention to (c) the manner pre-clinical animal data should be translated to humans using appropriate scaling, in particular with reference to the differences between mice and men in terms of differing vascular morphology and immunological background.

Successful development of cell-specific drug-delivery systems requires that reliable quantitative pharmacokinetic/pharmacodynamic (PK/PD) data are collected both in animal and human studies. This volume will include (d) information on improved body imaging technologies and on enabling quantitative tools available.

Finally, we address (e) the issue of diminishing academic funding of animal studies and of (f) the current dismal market and proprietary situation in the area of site-specific drug delivery.

Keywords Nanomedicine • Market • Enhanced permeability effect • Targeted delivery • Extracellular matrix components • Imaging • Patenting

Acronyms

μCT	microcomputed tomography
ACA	anticancer agent (functionalized oligomer with attached targeting motif)
Ad-p53	Human Adenovirus Type5 (dE1/E3) expressing Tumor Protein P53 (P53) under a CMV promoter
ADMET	absorption, distribution, metabolism, and excretion – toxicity in pharmacokinetics
AuNC-CS-TPP	Chitosan-coated gold nanocluster – triphenylphosphonium
AuNP-TPP	Triphenylphosphonium gold nanoparticles
BITES	bispecific T-cell engagers

CAFs	cancer associated fibroblasts
CAGR	compound annual growth rate
CBER	Center for Biologics Evaluation and Research
CD3ε	anti-human scFv monoclonal antibody
CT	computed tomography
CTC	circulating tumor cells
DCE-CT	dynamic contrast enhanced computed tomography
DDD	drug discovery and development
DOX	doxorubicin
ECM	extracellular cell matrix
EPR	enhanced permeability retention effect R –endoplasmic reticulum
FA	Folic acid
FMT 3D	fluorescence molecular tomography
FRET	Fluorescence Resonance Energy Transfer
GFP	Green Fluorescence Protein
HA	hyaluronic acid
HPMA	N-(2-hydroxypropyl) methacrylamide
HTS	high throughput screening
HYAL	hyaluronidase
IFP	interstitial fluid pressure
mRNA	messenger RNA
MALDI-IMS	Matrix-assisted laser desorption imaging – ionization mass spectrometry
MHC I	Multihistocompatibility complex I
MHC II	Multihistocompatibility complex II
MRI	magnetic resonance imaging
MSP	mononuclear phagocyte system
NP	nanoparticle
NIRF	near infrared fluorophore
OI	optical imaging
OMICS	a field of study in biology ending in -omics, such as genomics, proteomics or metabolomics
PD	pharmacodynamics
PE-PEG-TPP	phosphatidylethanolamine polyethylene glycol triphenyl phosphonium
PL-TPP	phospholipid triphenyl phosphonium
PEG	polyethylene glycol
PEI-TPP	polyethylene imine triphenyl phosphonium
PET	positron emission tomography
PLGA-PEG-TPP	poly(lactic-co-glycolic acid)- block – polyethylene glycol) triphenylphosphonium
PIT	photo-immunotherapy
PK	pharmacokinetics
PMN	polymorphonuclear leukocyte

RES	reticuloendothelial system
SB	systems biology
SiNP	silica based nanoparticle
TPGS1000-TPP	tocopherol polyethylene glycol 1000 succinate triphenylphosphonium
STPP	stearyl triphenyl phosphonium
SUPR	super enhanced permeability effect
QSAR	quantitative structure activity relationship
T (see Fig. 1) or Tox	toxicology
TAMs	tumor-associated macrophages
TPP	triphenylphosphonium
TSAS	tumor-specific antigen
VW	Volkmar Weissig

1.1 Introduction

It should be stated upfront that the emphasis of this chapter (and Volume) will be on pharmacokinetics (PK) in animal and human studies if available. The focus is on cancer market since it is the most important; as the cancer interest, no doubt, is the fastest growing component of the US market. The majority of literature concerns with this application. We also stress the emphasis on solid tumors (solid malignancies). The controversy of this field is whether targeted (and non-targeted) nanoparticles are of any benefit in clinical practice and how to push towards the market.

The topic of this volume was in part inspired by a statement by Petrák (2005):

> Future efforts will need to be directed to solve, in practical terms, the following fundamental issues:
>
> - The drug–carrier system (including the drug to be delivered) must avoid nonspecific interactions in the vascular compartment (RES).
> - The system should retain its ability to accumulate at the target site(s) (defined in terms of unique anatomical, physiological or disease conditions) and be in a form capable of acting on its pharmacological activity target.
> - Drugs need to be selected, or rather designed, to have the pharmacokinetic properties compatible with the demands of target-selective drug delivery (especially drug retention at the site of delivery and its ability to access its site of molecular action).

According to a report published by BCC Research, the market value of the worldwide nanomedicine industry was $72.8 billion in 2011. The market is estimated to grow at a CAGR of 12.5 % to reach $130.9 billion by the fiscal year 2016. The market for anti-cancer products was valued at $28 billion in the fiscal year 2011 and is anticipated to reach $46.7 billion by the fiscal year 2016. As indicated by Petrák (personal communication) we should note that 10 % rate of inflation would take the market value from $28 to $46 billion in 5 years (i.e., between 2011 and 2016), hence such numbers effectively mean no growth. We don't' know what the market value is today (i.e., in 2016).

Tremendous efforts are underway worldwide, at the bench and in preclinical research, in order to make the big promise of the nano-revolution a reality. However, there is a low number of trials, which reflects neither the massive investments made in the field of nanomedicine nor the general hype associated with the term "nano". This is supported by finding of Weissig in two papers (Weissig et al. 2014; Weissig and Guzman-Villanueva 2015). We believe (BCC Report above) the true promise of nanoscience for drug development still has to materialize.

Soluble cancer drugs present a different kind of problem. When a soluble cancer drug is injected into the patient, it quickly distributes into all the body tissues so that only a small fraction of the drug actually reaches the tumor. Most of the drug enters normal tissues where it kills normal dividing cells, causing the serious side-effects associated with chemotherapy. Another disadvantage to soluble drugs is their rapid elimination from the body through the kidneys. Nanotechnology may offer a solution.

Unfortunately, nanotechnology promises anything but *miniscule effects*, but most of these visions are hypothetical at this point. Most of the present nanotechnologies may come into fruition in 10–20 years. Accordingly, most pitfalls of molecular manufacturing have not yet been explored, because the benefits remain the dominant focus of researchers. We intend to discuss new approaches that would help to realize "dream" of nanomedicine to help the mankind. It is the insufficient innovation which results in a high failure rate in clinic.

We then question the conventional wisdom of definition of nanomedine. We partly redefine nanoparticulate (NP) delivery vehicles, i.e. as to include term "intracellular uptake", although for some applications (i.e. imaging, there is no need to uptake, just to stay firmly at the site). This is in contrast to previously defined nanomedicine or nanopharmaceutical: According to Rivera et al. (2010) nanopharmaceuticals are defined as "pharmaceuticals engineered on the nanoscale, i.e., pharmaceuticals in which the nanomaterial plays a pivotal therapeutic role or adds additional functionality to the previous compound". And, according to the " Medical Standing Committee of the ESF, nanomedicines result from "the science and technology of diagnosing, treating, and preventing disease and traumatic injury, of relieving pain, and of preserving and improving human health, using molecular tools and molecular knowledge of the human body" (ESF–European Science Foundation 2004). The stress on the internalization is important regardless of specific route out of many possible mechanisms. The rapidly proliferating cross-disciplinary area of endocytosis of nanomedicines is at the interface of biology and material science and may bring the next wave of significant technological breakthrough.

As we will see from the assessment discussed below, we have also serious doubts about both targeted and non-targeted carriers, especially in terms of benefits they provide. To resolve all above problems, *we suggest below the following coverage in this volume*. (Interestingly, recently, some other authors considered similar or some additional impediments to drug delivery – in form of nanomedicine. For example, Blanco et al. (2015) highlighted innovative designs, such as the use of nontraditional nanoparticle geometries for improved vascular dynamics, endosomal escape, and multidrug resistance to overcome clinical translations problems. Likewise,

Stylianopoulos and Jain (2015) and separately Bertrand et al. (2014) reviewed our understanding of therapeutic cancer treatments via nanomedicinal approaches, the latter with help of condensed mathematical formulations, discussing most of the impediments as above with a hope that some of the presently developed nanomedicines might become future's blockblusters).

1.2 Extracellular Matrix Manipulation

We would encourage to develop sufficient cell biology understanding of basic principles of cellular uptake and internalization (Hillaireau and Couvreur 2009) and relate them to material science as a basis for nanoparticle (NP) redesign. There is plethora of possible new designs, based on extracellular matrix molecules. Examples are those based on collagen, hyaluronan, and other ECM components.

One recent example features a balance in size: there is a possibility to choose a nanoparticle that is small enough to escape the leaky blood vessels that surround tumors but large enough to avoid rapid clearance from the blood stream via the kidneys. Balancing these two requirements usually results in using nanoparticles that are indeed small enough to accumulate in the vicinity of tumors, but that are really too large to penetrate deeply enough into tumors to have the maximum therapeutic effect. Jain et al. (2015) developed multilayered, or multistage, nanoparticles that partially dissolve once they accumulate around tumors, leaving behind a payload of nanoparticles a mere one-tenth the size of the original delivery vehicle. The remaining 10-nm-diameter nanoparticles, loaded with anticancer drugs, can then diffuse deeply into a tumor's dense interior. The key to the new nanoparticles is a gelatin material that can serve as a substrate for enzymes that are produced at high levels by tumors. Cancer cells use these enzymes to dissolve the extracellular matrix that surrounds organs, enabling these malignant cells to escape into the bloodstream and colonize sites distant from the primary tumor. The researchers took advantage of this enzyme by embedding tiny nanoparticles within the gelatin core of the larger nanoparticles that they designed to be injected into the blood stream.

For this set of experiments, the investigators loaded 100-nm the gelatin nanoparticles with 10-nm quantum dots. While quantum dots are not likely to be used to deliver drugs to tumors, these nanobeacons produce bright optical signals that can be easily monitored as they are released from the larger nanoparticles. Initial experiments using tumors growing in culture showed that the gelatin-degrading enzymes indeed released quantum dots which were able to diffuse farther and more efficiently than the 100 nm particles into the tumors. Subsequent experiments in tumor-bearing mice confirmed these *in vitro* findings, and as a result, the investigators are now planning to repeat these experiments using drug loaded 10-nm particles in place of the quantum dots they used in this study.

Another approach to facilitate the access of nanoparticles deep into tumors is to disrupt a tumor's ability to form the dense extracellular matrix, made of the protein collagen, which keeps nanoparticles in the outer regions of a tumor. Jain et al. used the widely used high-blood pressure medication Losartan to inhibit collagen synthesis. Human clinical studies have shown that Losartan reduces the incidence of cardiac and renal fibrosis by reducing the synthesis of one particular form of collagen (type I). They reasoned that this same inhibitory effect might lead to easier passage of nanoparticles into the deep recesses of a tumor. Consequently, they observed this effect at doses of the drug that were small enough to leave blood pressure unaffected. Their tests showed that Doxil, the first approved nanoparticulate anticancer agent, was more effective at treating dense, fibrotic tumors, such as pancreatic tumors, growing in mice. They also noted in that because long-term Losartan therapy has proven safe in humans, and because many anticancer agents raise blood pressure, administering Losartan with nanoparticles has the strong possibility of benefitting cancer patients.

In addition, many solid tumors develop extensive fibroses, a result of what is termed the desmoplastic reaction. Desmoplasia leads to a significant increase in the production of extracellular matrix (ECM) proteins, as well as extensive proliferation of myofibroblast-like cells. The result is the formation of a dense and fibrous connective tissue that is composed of multiple ECM components, including collagen types I, III, and IV; fibronectin; laminin; hyaluronan (HA); and the glycoprotein osteonectin [also known as secreted protein, acidic and rich in cysteine (SPARC)]. This fibroinflammatory component of the tumor (sometimes called stroma) contributes to an increase in tumor interstitial fluid pressure, blocking perfusion of anticancer therapies to the tumor. Targeting the components of the stromal compartment, in conjunction with cytotoxic agents directed against tumor cells, is gaining attraction as a potential approach to treating patients and overcoming chemoresistance. Hyaluronidase (HYAL), may have the potential to increase penetration of drugs through the stromal compartment and ultimately into tumor cells. With the clinical availability of recombinant HYAL, prospects for targeting HA in the treatment of cancer are improved. Scodeller et al. (2013) have developed new nanoparticle, employing HYAL immobilized on 250 nm silica nanoparticles (SiNP) maintaining specific activity of the enzyme. They noted that tumor volume reduction with SiNP-immobilized HYAL was significantly enhanced compared to non-immobilized HYAL control. In support of the above, prior the Scodeller paper, Whatcott et al. (2011) reported on employment of HYAL itself to facilitate anticancer drug delivery (not in a nano-form). Increased levels of one ECM component—namely, hyaluronan—leads to reduced elasticity of tumor tissue and increased interstitial fluid pressure. Multiple initial reports showed that the addition of hyaluronidase (HYAL) to chemotherapeutic regimens could greatly improve efficacy. Unfortunately, the bovine HYAL used in those studies is limited therapeutically by immunologic responses to treatment. Newly developed recombinant human HYAL has recently been introduced into clinical trials.

1.3 Extending the Blood Nanoparticle Circulation

We next encourage to modify NPs for long circulating properties in order to minimize the MPS uptake (liver, spleen, bone marrow); there are ways of shielding NPs with pre-treatments and surface modifications (stealth particles and liposomes, employing a variety of PEG length and MW, affecting the receptor availability), and charge reversal (negative NP). Pretreatments include e.g. Intralipid, a nutritional supplement approved by FDA. For example, Liu et al. (2013) employed Intralipid for shielding of MSP (more on update in this Volume). Even better results were obtained by Rodriguez et al. (2013) who developed an immune-shielding technology ("self-peptide" shielding or "active stealth strategy"). Foreign particles and cells are cleared from the body by phagocytes that must also recognize and avoid clearance of "self" cells. The membrane protein CD47 is reportedly a "marker of self" in mice that impedes phagocytosis of self by signaling through the phagocyte receptor CD172a. Minimal "Self" peptides were computationally designed from human CD47 and then synthesized and attached to 160 nm nanobeads – virus-size particles – for intravenous injection into NSG (non-obese diabetic – NOD – severe combined immunodefficient IL2rγ^{null}) mice that express a CD172a variant compatible with hCD47. Self peptides delay macrophage-mediated clearance of nanoparticles, which promotes persistent circulation that delay clearance by the liver and spleen and enhances drug delivery to tumors. Recent report states that some of the above ideas are currently being put to the test in the clinic with initial results from anti-cancer clinical trials hopefully to be reported (Sosale et al. 2015).

Another strategy to prolong the circulation is to attach a biomimetic coating derived from membranes isolated from leukocytes (Parodi et al 2013) or from red blood cells (Hu et al. 2011). Likewise, clinical outcomes are not known.

1.4 Passive and Active Targeting

We will discuss drawbacks of passive (EPR) and active targeting now. Often discussed issue is how nanomedicines access the disease site? Here do we recognize passive or active targeting?

Passive targeting of tumors by nanoparticles takes advantage of their endothelial cell lining. The rapid vascularization of solid tumors results in leaky, defective endothelial cells and impaired lymphatic drainage. Nanoparticles ranging from 10 to 100 nm in size then begin to accumulate within tumors because of their ineffective lymphatic drainage. This results in a phenomenon known as the enhanced permeability and retention effect (EPR). Although accumulation in solid tumors is observed, the cellular uptake by neoplastic cells and the subsequent intracellular drug release have been questioned. The EPR effect is purely size- and geometry-dependent mode of action, however. On the other hand, ligand-targeted NPs (active targeting) may prove beneficial in increasing drug exposure due to increased target cell uptake and target tissue retention compared to ligand-lacking NP. Ligand-targeted approaches are crucial for molecules that need to localize intracellularly for therapeutic activity but are not capable of crossing cellular membranes, such as nucleic acids.

Targeting antibodies or ligands are often selected because of their high specificity and high affinities towards overexpressed antigens on target cells and their ability to trigger receptor-mediated endocytosis after binding. However, the targeting ligands or antibodies do not influence the biodistribution of the nanoparticle: biodistribution still depends on passive targeting to tissues with a leaky vasculature, and targeting ligands only triggers internalization after extravasation into the tumor. The bio-distribution of targeted and non-targeted NPs is often similar. The targeted nanoparticles will show benefits compared to their non-targeted counterparts if they can freely and directly access their targets (PEG chains often impair the ligand-cell interaction). Crommelin and Florence (2013) marked the drug targeting concept as intrinsically biased which has perhaps contributed to the hype surrounding drug targeting. The ideal scenario is if the drug delivery system can reach only one target organ while sparing all other organs. This is rarely the case as most delivery systems reach other organs often at even higher concentrations than the tissue/organ of interest thus undermining the aim of drug targeting. There are many issues that need to be addressed in drug targeting such as the undesirable side effects that may result due to uptake by off-target organs e.g., liver (in case of transferrin-based targeted systems) and kidney (in case of folate-based targeted systems). Thus a critical assessment is needed. They also defined the following: the first-line targeting that can be achieved at the organ level; second-line targeting can be achieved at the cell level, and the third-order targeting at the organelle level. Toxicity is also an important criterion: toxicity, or rather the lack of is very important; one cannot even do human clinical studies if preclinical toxicity is not acceptable (i.e., a high benefit/risk ratio). The critical factors of NP systems for success in clinical application with regard to complement activation and hypersensitivity reactions in particular against polyethylene glycol – PEG (Lehner et al. 2015) are to be assessed.

Very recently, Kobayashi et al. (2014) have developed a "super-enhanced permeability and retention effect" (SUPR) concept induced by photo-immunotherapy (PIT). Photo-immunotherapy (PIT) is a newly developed therapy involving the injection of a conjugate composed of an armed monoclonal antibody and a near infrared phthalocyanine dye. Because it damages cells immediately adjacent to the tumor vasculature, PIT results in marked increases in vascular permeability leading to 12- to 25-fold enhanced nanoparticle delivery into cancer tissue in animals. Again, this effect should be confirmed in humans.

Very small nanodelivery objects, micellar formulations, feature exceptions in tumor accumulation because of their small size. Micelle carriers selectively accumulate in tumor tissue owing to the EPR effect and directly reach the cancer cells in order to attack them. Alternatively, the formulation spontaneously disintegrates while it is retained within the tumor tissue. Disintegrated ACA-bound unimers with the attached drug-payload or released ACA immediately reach and enter cancer cells to kill the cancer cells (Matsumura 2014; also Osada et al. 2009). ACA is an optimized block copolymer that is typically functionalized with 6-aminocaproic acid, polyaspartate, polyglutamate, etc. and a drug payload.

Very recent, Etrych et al. (2014) reported on micellar DOX formulation (no size given, possibly very small), without targeting, showed selectivity toward solid tumors in mice: (i) drug accumulation in tumors driven by enhanced permeability and retention (EPR) effect, which results in almost 100 times higher concentration of drug in the solid tumor than in normal tissue, (ii) pH-dependent release of drug from polymer-drug conjugate, which releases free drug more efficiently at a lower pH in tumors. This effect might be explained on the basis of hydrophobicity of the carrier (Maeda 2015). The size is thus also an issue. Eventually, a drug concentration at the site would be of help to know. Advanced cytotoxicity has been observed.

A considerable critique of classical EPR was recently brought up by Nichols and Bae (2014). They noted that clinical outcomes from nano-sized drug delivery systems, however, have indicated that EPR is not as reliable as previously thought. Drug carriers generally fail to provide superior efficacy to free drug systems when tested in clinical trials. A closer look reveals that EPR-dependent drug delivery is complicated by high tumor interstitial fluid pressure (IFP), irregular vascular distribution, poor blood flow inside tumors and perhaps absence of lymphatics in experimental models. Furthermore, the animal tumor models used to study EPR differ from clinical tumors in several key aspects that seem to make EPR more pronounced than in human patients. Khawar et al. (2015) also reviewed the evidence that supports a statement that considerable barriers of tumors via various mechanisms exist, which results in imperfect or inefficient EPR and/or targeting effect. Barua and Mitragotri (2014) review focuses on the current understanding of penetration of NPs through biological barriers. Emphasis is placed on transport barriers.

As a way towards a progress in this area, Wong et al. (2015) presented a mathematical model that provides a quantitative framework to guide preclinical trials of new chemotherapeutic delivery vehicles and ultimately to develop design rules that can increase targeting efficiency and decrease unwanted side effects in normal tissue. Likewise, Stapleton devised a linear mixed effect model and verified it on animals *in vivo*. The intra-tumoral relationship between the tumor microcirculation, elevated IFP, and accumulation of liposomes was investigated through experiments. This was accomplished by evaluation of the tumor microcirculation using dynamic contrast enhanced computed tomography (DCE-CT) and measurement of tumor IFP using a novel image-guided robotic needle placement system connected to the micro-CT scanner. Results have important implications for guiding drug delivery using image-based approaches. Elucidating the factors mediating heterogeneous intra-tumoral delivery of nanoparticles can substantially enhance their use in diagnostic and therapeutic applications. As the tumor imaging by means of CT represents a robust and noinvasive way of imaging, it could be potentially used in a clinical setting.

Very recently, a major development was reported in terms of predicting EPR variability to improve the clinical applications of nanomedicines. Miller et al. (2015) reported that a 30-nm magnetic NP (MNP) in clinical use could predict colocalization of TNPs (therapeutic nanoparticles) by magnetic resonance imaging

(MRI). A central question in nanomedicine is whether imaging could be used to identify patients with higher predisposition to TNP accumulation and, in turn, efficacy of understanding how to best exploit EPR effects for clinical applications, how to design better TNPs, and how to alter key physiologic parameters to maximize distributions to and within tumors. Heterogeneous tumor vascularization is a recognized clinical feature that can be detected using various angiography modalities. To progress further clinically, more human-representative disease models, as patient-derived xenografts, genetically engineered autochthonous mouse models, and larger animals, should be used to study EPR effects in metastatic lesions, as suggested by the authors.

1.5 Differences Between Man and Mice

The predictive value of animal models for a given clinical condition is getting increasingly attention. Models in both large and small animal species have value for pharmacology and toxicology, including the first evaluation of adverse side effects and pharmacological efficacy of innovative disease intervention strategies, as well as the selection in a given therapeutic discovery program. Also, such models are helpful in elucidating pathways in physiological or pathological processes. But, progress in the field has made it increasingly clear that animal models have their limitations regarding translational value (van der Meer et al. 2015).

Still not resolved is a fundamental question why cancer trials in men often fail while performing well in mice. The problem is what is the differences between mice (and other animals) and men when moving to upper scale at translational medicine. Perhaps, it is the vascular and immune systems which are different but no real clues are available.

It looks like that a way out is the identification of molecular abnormalities that are not only critical for the life of cancer—but not normal cells—but are also the dominant or the only molecular abnormality within the tumor. Is it possible to identify a common denominator across all of these abnormalities that is not only critical for the survival of cancer cells but is also not present in normal cells? And to push the envelope even further, is it possible that this common denominator is also present in cancer stem cells, so that if targeted, tumor relapse would be limited as well (Kinnaird 2015).

To make it even more complicated and add more doubts, clinical outcomes from nano-sized drug delivery systems, however, have indicated that EPR is not as reliable as previously thought (Nichols and Bae 2014). The number of publications citing EPR has increased exponentially; this flourish of creativity has largely failed to translate into new clinical therapies. Drug carriers generally fail to provide superior efficacy to free drug systems when tested in clinical trials. A closer look reveals that EPR-dependent drug delivery is complicated by high tumor interstitial fluid pressure (IFP), irregular vascular distribution, and poor blood flow inside tumors, typical for mice (see a subchapter above; Nichols and Bae 2014).

Furthermore, the animal tumor models (often very fresh tumors) used to study EPR differ from clinical tumors in several key aspects that seem to make EPR more pronounced than in human patients. The architecture of tumors in humans is very different with irregular distribution of the vascular bed leaving unperfused areas, and the difficulty of penetrating the irregular extracellular matrix (ECM) found in the tumors. Exceptions seem to be with liposomal systems, as the increase was suggested to be due to the longer circulation time of the smaller liposomes (also micelles), mostly because the smaller particles were able to more easily avoid uptake by the mononuclear phagocyte system (MPS). Thus, tumor (human) angiogenesis is prone to leave large regions devoid of active blood vessels within the tumor. These void spaces can present a formidable distance barrier to diffusion, with some particles requiring hours or even days to diffuse 200 μm.

Several key differences between human and murine tumors may account for the disparity between preclinical and clinical trial results of most drug carriers the models with which EPR was discovered and validated may not be sufficiently representative of clinical tumors to justify our confidence. The collagenase treatment to reduce IFP and improve intratumoral transport as well as normalizing tumor vasculature to provide more reliable drug access proved to be of help (Jain 2013). Concluding, after 30 years, EPR-based cancer therapeutics have left us in only a marginally better position than we would have been without. Few new formulations have been made available for clinical use, and among those, negligible gains in the drug's therapeutic index have proven the norm rather than the exception.

Besides, the significance of a non-vascular microenvironment which consists of non-cancer cells such as cancer associated fibroblasts (CAFs) or tumor-associated macrophages (TAMs) has been recognized as a major factor for cancer promotion. These stromal components could be possible players to modulate pharmacokinetics not only for small molecules but also for nanodrugs, because the stromal cells are usually located between the blood vessels and the pharmacological targets such as cancer cells. However, the pathophysiological approach for non-vascular stromal cells in nanomedicine has been insufficiently discussed (Nishihara 2014).

Zschaler et al. (2014) discusses basic differences in the organization of immune responses and the differences among cells of the innate immune system, as well as the critical reflection of disease models. He also assessed problems in translation of disease data from mouse to man. Furthermore, mouse models tend to form relatively few spontaneous metastases or form metastases with divergent tissue specificity. He concludes that these differences in organization of the immune system greatly disturb direct translation of data from murine disease models to human pathologies. Thus, it is not surprising that in many cases promising therapeutic principles found in mice models do not work in humans.

1.6 Cell-Specific Targeting

The extracellular targets are typically employed for imaging of cells and tissues (in terms of nanomedicine of their vehicles). We should now review the development of cell-specific targeting systems relevant for tumors and circulating tumor cells (TCT). As stated by Petrák (2012), a systematic approach which should be adopted, starting whether the drug under the consideration could benefit from site-selective targeting based on pharmacokinetic and pharmacodynamic behavior, and physiological properties of the intended target site. In our view, target validation requires understanding target biology in the context of disease, and the chemistry of the drug's ability to modulate the target.

Generally, it is believed that the introduction of targeting ligands does not enhance nanoparticle accumulation into tumors as compared to non-targeted ones, but shows higher efficacy by enhancing internalization into tumor cells (Wang and Tanou 2010). Likewise, van der Meel et al. (2013) state that it is believed that ligand-lacking and ligand-targeted PNM have comparable pharmacokinetic parameters, biodistribution and tumor targeting profiles. They continue that PK and *biodistribution data* of ligand-lacking versus ligand-modified PNM are scarce and more importantly: *impossible in the clinic setting*. The number of ligand-targeted PNM tested for *in vivo* target cell internalization is too limited to provide conclusive evidence. How improved efficacy is related to the presence of a targeting ligand cannot be resolved because the comparison of PK and distribution ligand-modified versus ligand-lacking PNM is often not included. Such comparisons are scarce and *clinical trials are not designed to compare ligand-lacking and ligand-targeted PNM.* In the majority of the cases described, there is insufficient literature. The efficacy and safety of ligand-targeted PNM has been shown in animals, but the evidence for the added delivery value of target ligand-coupling to nanomedicines in humans remains to be established.

As stated by Dawidczyk et al. (2014), assessing the efficiency of a targeting ligand in increasing tumor accumulation is complicated by the different control experiments used in these studies. The contribution of passive targeting is assessed by measuring tumor accumulation of the delivery system without attachment of the targeting ligand, with attachment of a non-specific ligand, pre-injection with a blocking molecule or treatment, or with a xenograft formed from a cell line that did not express the target molecule.

As far as circulating cancer cells, recent advances in the fields of nanotechnology and microfluidics have led to the development of several devices for *in vivo* targeting of CTC during transit in the circulation (Immuno-magnetic separation, *In vivo* photo-acoustic flow cytometry (Faltas 2012).

It can be concluded that in the case of many pathologies, ligand-targeted nanomedicines are mostly subjected to the same physiological localization as ligand-lacking nanomedicines and therefore have comparable biodistribution and accumulation profiles. Of course, there are exceptions.

One approach could be to look at the "most promising" (out of thousands of publications that over the years used this description to promote their research work), examine these critically to identify likely reasons for failures, and to provide "hindsight" comments and explanation, and possibly the way forward to correct what was wrong.

1.7 Subcellular Organelle-Targeted Drug Delivery

Over the last decade, subcellular drug targeting, i.e. the specific delivery of biologically active molecules to and into cell organelles has emerged as the "new frontier" (Lim 2007; D'Souza and Weissig 2009) or "third level" (Sakhrani and Padh 2013) of drug targeting for, in hindsight, quite obvious reasons. First it is very well known that essentially all subcellular organelles can be either involved in the etiology and/or subsequently pathogenesis of human diseases and second literally all drugs except those which act on cell membrane receptors exert their pharmacological action on molecular targets on the subcellular level, either in the cytosol or at or inside cell organelles. This also applies to the therapy of infectious diseases as long as the pathogen is harbored inside human cells. Table 1.1 gives a brief (but by no means comprehensive) overview over diseases associated with cell organelles.

Looking back it appears there has been a widely held assumption that mediating cell cytosolic internalization might be sufficient to ensure the interaction of the biologically active molecule with its sub-cellular target based on simple diffusion of the drug molecule throughout the cytosol. One of us (VW) when trying to submit a paper about mitochondria-specific nanocarriers to a prestigious journal was even challenged by an anonymous reviewer with the question as to why to deliver a drug

Table 1.1 Major diseases associated with specific cell organelles

Cell organelle	Disease
Nucleus	All monogenic diseases
	Cancer
Nuclear envelope	Diseases of skeletal and cardiac muscle, neurons and tendons
	Progeria
	Emery-Dreyfuss Muscular Dystrophy
Mitochondria	Large variety of Neuromuscular and neurodegenerative diseases
	Cancer
Lysosomes	Lysosomal storage diseases
Endoplasmic reticulum (ER)	Large number of diseases caused by misfolded proteins and ER stress
Golgi Apparatus	Congenital disorders of glycosylation
Plasma Membrane	Cystic Fibrosis
	Diabetes
	Familial Hypercholesterolemia

to mitochondria in the first place, there were so many mitochondria inside a cell that the drug would eventually get to interact with them anyway. However, increasing evidence has accumulated showing this general assumption to be faulty for two major reasons. First, the cytosol is a highly crowded environment. At least three factors slow down or even prevent the free diffusion of solutes within the cytosol, which are first its fluid-phase viscosity, second collisional interaction due to macromolecular crowding and third binding to subcellular components (Lukacs 2000; Seksek 1997). In this context, "binding" includes specific molecular interactions like substrate/inhibitor enzyme binding as well as the trapping of for example weakly basic molecules inside acidic cellular vesicles (Weissig 2005). Naturally, such an environment which is hostile towards any free diffusion also impedes the intracellular distribution of nanocarriers.

Second, it has been well established that the intracellular disposition of any low-molecular weight molecule is (in addition to the above indicated cytosolic barriers) dictated by their physico-chemical properties. Charge and amphiphilic index, conjugated bond number, log P and pKa value, molecular weight as well as the size of the largest conjugated fragment are all parameters determining the intracellular distribution of low-molecular weight compounds (Horobin 2001). Using the above listed molecular parameters a QSAR model was developed which allowed establishing synthetic guidelines for the chemical modification of molecules not displaying any intrinsic affinity towards mitochondria in order to make them highly mitochondria-specific (Horobin 2007).

Though it is now possible to either modify existing drug molecules or to design entirely new molecules in order to ensure their mitochondrial disposition once they have entered the cytosol the big challenge remains with respect to their biological activity. How to make sure a drug which for example acts on mitochondria, but does not possess any intrinsic mitochondrial affinity, remains its biological activity after it has been chemically modified in order to render it mitochondriotropic? This challenge potentially applies to any drug which is supposed to act on any cell organelle. Just to clarify in this context the use of terms, a drug which has a molecular target at a certain organelle does not per se also targets that organelle, i.e. "moves" to this organelle. Having a target at an organelle and actively targeting that organelle are two different concepts.

For overcoming this challenge, i.e. rendering a drug molecule organelle-specific while preserving its potential pharmacological activity site-specific nanocarriers appear as the most promising alternative. Once a nanocarrier has been designed which possesses a high specifity for a particulate organelle, it potentially could be used to deliver any biologically active molecule to that organelle completely side-stepping the need for altering the chemical structure of that molecule.

The use of nanocarriers for the delivery of biologically active molecules was arguably proposed for the very first time in 1971. Based on their earlier work about the encapsulation of amyloglucosidase into artificial phospholipid vesicles, back then simply called liposomes since the term nanocarrier came into vogue only after the launch of the National nanotechnology Initiative in 2000, Gregory Gregoriadis and Brenda Ryman proposed these lipid vesicles as carriers of enzymes or drugs as a new approach to the treatment of storage diseases (Gregoriadis and Ryman 1971). These authors succeeded in entrapping *Aspergillus niger* amyloglucosidase with an efficiency of between 4 and 10 % together with ^{131}I-labelled albumin into liposomes composed of phosphatidylcholine, cholesterol and dicetyl phosphate. However, following i.v. injection into rats it was found that the large majority of all injected liposomes were eliminated from the circulation within 10 min. Most of the radioactivity was recovered in liver and to some extent in the spleen. This rapid clearance of i.v. injected liposomes via organs of the RES system, as firstly described by Gregoriadis and Ryman in 1971, proved to become the major and almost insurmountable hurdle for getting liposome-based drugs into the clinic. It took almost 20 years to overcome this barrier towards the clinical application of liposomes. In 1990 it was demonstrated that grafting polyethylene glycol (PEG) chains onto the surface of liposomes significantly prolonged the circulation time of liposomes (Klibanov et al. 1990). Such flexible long PEG chains literally shield liposomes from interacting with opsonins thereby significantly slowing down their clearance from circulation via the RES system (Blume and Cevc 1993). In addition, what is extremely important from the pharmaceutical manufacturer's view of point, these PEG chains also minimize vesicle interactions thereby vesicle aggregation, which in turn dramatically improves the stability of such liposomal formulations (Needham et al 1992). Only 5 years later, in 1995, the FDA (USA) approved the first injectable liposomal drug, Doxil, which are doxorubicin-loaded liposomes the surface of which is protected by a layer of PEG chains (Barenholz 2012). Though, it should be noted that Doxil does not possess any subcellular, organelle-targeting specificity.

Back to Gregoriadis and Ryman in 1971, after subcellular fractionation of the liver following i.v. injection of radiolabeled and enzyme loaded liposomes into rats, these authors recovered most of the liposomes and their content in the mitochondrial-lysosomal fraction as well as the cytosol of both, liver parenchymal and Kupffer cells. In the author's words, liposomes indeed revealed themselves as "a promising vehicle for the direction of enzymes and drugs to the liver and spleen" (Gregoriadis and Ryman 1971). Yet, fast forward, 45 years later, no liposome-based enzyme replacement therapy has been approved for clinical use. Though since passing the US orphan drug act in 1983 about ten enzyme replacement therapies have entered the clinic (Mechler et al. 2015) none of these therapies is utilizing any nanocarrier or liposome-based platform. In most cases, recombinant enzymes injected intravenously are targeted to the lysosomes of affected cells by interactions with cell-surface receptors that recognize carbohydrate moieties such as mannose and mannose 6-phosphate on the enzymes (Grubb et al. 2010). A data base search for ongoing or completed clinical trials (Web Ref. a) yielded 523 hits using the search term "lysosomal storage disease" and 1642 hits using "liposomes", but the combination of both search terms did not turn out any results. It remains to be seen whether the early promise of using liposomes for lysosomal enzyme replacement therapy will ever materialize.

Since lysosomal storage diseases are monogenic diseases a completely different alternative approach would be gene therapy, which leads us to the overarching and without a doubt main target inside a cell, the cell nucleus. Early attempts towards manipulating the cellular genome were reported already in 1971, when transgene expression in human cells was demonstrated by Merril et al (1971) and when Munyon et al. (1971) restored the ability to synthesize thymidine kinase in thymidine kinase-deficient mouse cells by infection with ultraviolet-irradiated herpes simplex virus (Munyon et al. 1971). Right after the publication of those two papers in 1971, Friedman and Roblin (1972) expressed in 1972 their concern about the premature character of any attempts directed at gene therapy in human patients and proposed that a sustained effort be made to formulate a complete set of ethic and scientific criteria to guide the development and clinical application of gene therapy techniques. Following a 10-year moratorium asking for the halt of all gene therapy trials involving human patients, the first transfer of genes into humans was approved by the NIH (USA) in January 1989. One year later Rosenberg et al. demonstrated the feasibility and safety of using retroviral gene transduction for human gene therapy by successfully introducing the gene coding for resistance to neomycin into human Tumor Infiltrating Lymphocytes before their infusion into patients (Rosenberg et al. 1990). The 1990s then saw an explosion of academic and commercial efforts towards gene therapy. By the turn of the century, worldwide over 500 clinical trials have been conducted and as of today a total of 2210 clinical trials are listed in the Gene Therapy Clinical Trials Worldwide Database (Web Ref. b). Numerous nanocarrier-based strategies involving a large variety of viral, non-viral and hybrid vectors for the delivery of nucleic acid into the cell nucleus have been developed. Currently, 115 clinical trials are based on cationic lipidic nanocarriers and about 1500 trials are utilizing a variety of viral vectors (Web Ref. b). The large number of clinical trials also reflects a shift of paradigm in gene therapy. While only 9.5 % of all clinical trials are aimed at curing monogenic diseases, 64 % explore a large diversity of anticancer therapies (Web Ref b.).

However, and quite surprisingly, despite the enormous amount of brain power and money spent over the last quarter of a century on getting gene therapy protocols into the clinic, the Center for Biologics Evaluation and Research (CBER) at the FDA (USA) has not yet approved any human gene therapy product for sale (Web Ref c.). World-wide, so far only three gene therapy-based therapies were approved by the corresponding authorities. On October 16, 2003, Shenzhen SiBiono GenTech (Shenzhen, China), obtained a drug license from the State Food and Drug Administration of China (SFDA; Beijing, China) for its recombinant Ad-p53 gene therapy for head and neck squamous cell carcinoma. Sold under the brand name Gendicine, the world's first commercial gene therapy uses an adenoviral vector (Pearson et al. 2004). Russia registered in 2011 Neovasculgen as a gene therapy drug for the treatment of peripheral artery disease (Nikulnikov 2015) and in 2012 Glybera, a therapy for a rare inherited disorder was approved for clinical use in Europe (Yla-Herttuala 2012).

The reasons for this so far very modest success of translational research in gene therapy are manifold and far beyond the scope of this chapter. Nevertheless, it appears to us that one of the major concerns which were expressed by the Panel to Assess the NIH Investment in Research on Gene Therapy already back in 1995 still has not yet been properly addressed: "Significant problems remain in all basic aspects of gene therapy. Major difficulties at the basic level include shortcomings in all current gene transfer vectors and an inadequate understanding of the biological interaction of these vectors with the host" (Web Ref d). The "inadequate understanding" of how a nucleic acid containing nanocarrier interacts with its host applies to all levels of drug delivery, from the systemic level down to the fate of the nanocarrier inside a cell. A Science paper back in 2003 (Savic et al. 2003) claimed that "Micellar nanocontainers distribute to defined cytoplasmic organelles", although the author's triple-labeling confocal microscopy-based data clearly showed a rather random intracellular distribution of the studied block copolymer micelles. As can be clearly seen from the published confocal fluorescence images (Savic et al. 2003), these nanocontainers were found to localize in the cytosol and to partly co-localize with lysosomes, the Golgi apparatus, the endoplasmic reticulum as well as with mitochondria. Considering that in these experiments no organelle-specific ligands have been used in order to render these nanocontainers organelle-specific this outcome was to be expected.

A serendipitous observation in the laboratory made during the late 1990s opened the way towards the design and development of organelle-targeted pharmaceutical nanocarriers. Dequalinium chloride (DQA), a bola-amphiphilic cationic molecule with a delocalized charge center and known since 1987 to accumulate inside mitochondria (Weiss et al. 1987) was found to be able of self-assembling into liposome-like vesicles named DQAsomes (DeQAlinium-based lipoSOMES) at the time of their discovery in 1998 (Weissig et al. 1998a, b), which was a few years before the term "nano" came into vogue. A combination of a quantitative structure-activity relationship (QSAR) model with a Fick-Nernst-Planck physico-chemical model was used to analyze mechanistic aspects of DQA's affinity for energized mitochondria (Horobin et al. 2007). The same theoretical approach has been tried a few years later to predict the intracellular distribution of nanocontainers but only with limited success (Horobin et al. 2010).

DQA's strong affinity for mitochondria combined with its unique feature of being able to form cationic liposome-like vesicles prompted its exploration as the very first mitochondria-specific transfection vector (summarized in (Weissig et al. 2011a, b). The proof-of-concept was provided in 2011 when Lyrawati et al. (Lyrawati et al. 2011) demonstrated the capacity of DQAsomes to deliver an artificial mini-mitochondrial genome construct encoding Green Fluorescence Protein to the mitochondrial compartment of a mouse macrophage cell line which resulted (though with very low efficiency) in the expression of GFP mRNA and protein.

DQAsomes have also been extensively tested as a pharmaceutical nanocarrier for the delivery of low-molecular weight compounds (summarized in Weissig et al. 2011a). For example, the incorporation of paclitaxel into DQAsomes yields a stable colloidal solution of the drug with an average size distribution between 400 and 600 nm and increases the solubility of the drug in comparison to free paclitaxel by a factor of roughly 3000 thereby presenting a much needed alternative to Cremophor-based formulations of the highly insoluble paclitaxel (Cheng et al. 2005). Although DQAsomes have been sufficiently covered by a series of patents (Weissig et al. 2000, 2001; 2003) and commercialization attempts have been made at the beginning of the 2000s, DQA's toxicity eventually proved to become an insurmountable hurdle. To overcome DQA's toxicity issues and to build upon the FDA approval of liposome-based nanomedicines (Doxil – Barenholz 2012), mitochondria-targeted liposomes have been developed as an alternative to DQAsomes. It was demonstrated in 2005 that the surface modification of liposomes with triphenylphosphonium (TPP) cations renders these vesicles highly mitochondriotropic (Boddapati et al. 2005) and follow-up studies in vitro and in vivo showed the capacity of such mitochondria-targeted liposomes to significantly increase the efficiency of pro-apoptotic anticancer drugs known to act on mitochondria (Boddapati et al. 2008; Patel et al. 2010)). Unfortunately, the commercial development of TPP-bearing liposomes was made impossible by the decision of a U.S. patent officer who decided that the design of mitochondriotropic liposomes was "obvious" due to the earlier description of DQAsomes, which is quite astonishing considering the chemical and structural differences between liposomes and DQAsomes.

Nevertheless, the demonstration from 2005 that conjugating TPP cations to the surface of nanovesicles (i.e., liposomes) renders them mitochondriotropic has launched the development of a large variety of TPP –bearing nanocarriers as can be seen in Table 1.2. To what extent the design of all of the listed TPP-bearing nanocontainers will be considered by the US Trade and Patent Office as "obvious" in light of the granted DQAsome patents remains to be seen.

In summary, the principal way towards the design of organelle-targeted pharmaceutical nanocarriers has been shown via the development of mitochondria-targeted carriers. Either the nanocontainer is entirely (or partly?) composed of molecules possessing an inherent affinity towards a specific organelle (like DQAsomes) or the nanocarrier bears organelle-specific ligands on its surface (like STPP liposomes). Whether any of such organelle-targeted pharmaceutical nanocarriers will eventually enter the clinic remains to be seen.

A special case of intracellular targeting is represented by targeting solid tumors which are not easily accessible as most tumor antigens are intracellular proteins. Such targeting with antibodies is often followed by a related immunoreaction response. Tumor antigens are those antigens that are presented by MHC I or MHC II molecules on the surface of tumor cells. Antigens found only on such cells are called tumor-specific antigens (TSAs) and resulting from a tumor-specific mutation. An oncoantigen is a surface or soluble tumor antigen that supports tumor growth and have a causal role in the promotion of carcinogenesis and cannot be easily downmodulated or negatively selected by precancerous lesions under the pressure of a specific immune attack. A major problem of cancer immunotherapy is thus the selection of tumor cell variants that escape immune recognition. Novel strategies will be required to identify new oncoantigens amenable to targeting for human applications (Cavallo et al. 2007). One promising strategy is to recruit a patient's T cells to tumor cells using fusion proteins called bispecific T-cell engagers (BiTEs), which consist of two single-chain antibody fragments, one specific to CD3ε on the T cell and the other specific to an integral membrane protein on the tumor cell. Dao et al. (2015) demonstrated that such access to intracellular antigens accessible to this class of reagents for the first time is possible. Another way is to employ nano-

Table 1.2 Pharmaceutical Nanocarriers rendered mitochondriotropic via conjugation to TPP cations

Nanocarrier	Indication/Therapeutic Cargo	TPP conjugate	Reference
Liposome	None	Stearyl- triphenylphosphonium (STPP)	Boddapati et al. (2005)
Copolymer conjugate	None	(N-(2- hydroxypropyl) methacrylamide) triphenylphosphonium (HPMA-TPP)	Callahan J, Kopecek (2006)
Liposome	Cancer/Ceramide	Stearyl- triphenylphosphonium (STPP)	Boddapati et al. (2008)
Liposome	Cancer/Sclareol	Stearyl- triphenylphosphonium (STPP)	Patel et al. (2010)
Liposome	Cancer/Paclitaxel	Phosphatidylethanolamine – polyethylene glycol- triphenylphosphonium (PE-PEG-TPP)	Biswas et al. (2012a, b)
Liposome	Cancer/Doxorubicin	Stearyl- triphenylphosphonium (STPP) & Folic acid (FA) as co-ligand	Malhi et al. (2012)
Micelles	Various pathologies/Coenzyme Q10	Poly(ethylene glycol)-polycaprolactone-triphenylphosphonium (PEG-PCL-TPP)	Sharma et al. (2012)
Dendrimers	None	5 poly(amidoamine) dendrimer-triphenylphosphonium (PAMAM-TPP)	Biswas et al. (2012a, b)
Nanoparticles	Cancer; Alzheimer; Obesity/lonidamine, α-tocopherol succinate; 2,4-dinitrophenol	Poly(lactic-co-glycolic acid)- block – polyethylene glycol) triphenylphosphonium (PLGA-PEG-TPP)	Marrache and Dhar (2012)
Liposomes	Cancer/Paclitaxel	Tocopherol polyethylene glycol 1000 succinate triphenylphosphonium (TPGS1000-TPP)	Zhou et al. (2013)
Nanoparticles	Cancer/Doxorubicin	Poly (ethylene imine) hyperbranched polymer – triphenylphosphonium (PEI-TPP)	Theodossiou et al. (2013)

Nanoparticles	Atherosclerosis/Quantum dots for imaging	High-density lipoprotein – phospholipid triphenylphosphonium (HDL-TPP)	Marrache and Dhar (2013)
Nanoclusters	Various pathologies/fluorophore for imaging	Chitosan-coated gold nanocluster – triphenylphosphonium (AuNC-CS-TPP)	Zhuang et al. (2014)
Liposomes	None	Synthetic phospholipid – triphenylphosphonium (PL-TPP)	Benien et al. (2015)
Liposomes	None	Egg-phosphatidyl ethanolamine-triphenylphosphonium (PE-TPP)	Guzman-Villanueva et al. (2015)
Dendrimers	None	4 poly(amidoamine) dendrimer-triphenylphosphonium (PAMAM-TPP)	Bielski et al. (2015)
Nanoparticles	Cancer/5-aminolevulinic acid	Gold nanoparticles- triphenylphosphonium (AuNP-TPP)	Yang et al. (2015)
Micelles	Various pathologies/Tetraphenylethene; boronate (FRET-based ratiometric detection of hydrogen peroxide)	Polymere-triphenylphosphonium (TPP)	Qiao et al. (2015)

medicinal approach to allow for intracellular access and targeting. Hong and Zeng (2013) demonstrated this approach for several pathogens. The cancer application remain to be demonstrated, if any.

1.8 Towards the Improved Imaging Technique in Humans

We will shortly review the developments in suitable imaging technique for human body/organ assessment of targeted sites to extend animal PK and PD studies to human body incl. molecular imaging targets. New developments are emerging in noninvasive imaging modalities for monitoring biodistribution, such as applications of molecular imaging to oncology, including molecular-genetic imaging, imaging the tumor microenvironment for precision medicine and theranostics, tracking cellular and immune therapies in cancer cells and vaccines *in vivo*, PET molecular imaging with suitable, single photon emission computed tomography (SPECT), and optical imaging (OI). Nanoimaging or molecular imaging via fabrication of noninvasive *in vivo* analytic nanotools with improved sensitivity and resolution for molecular imaging and for studying pathologic processes in vivo (main benefits include early detection of disease as well as monitoring of various disease stages) is becoming reality.

Anton and Vandamme (2014) stated that new possibilities for molecular imaging are offered by the increased half-life of nanosystems in blood stream, as well as by the specific accumulation in organ of lesions through passive or active targeting mechanisms. Thus, heavy metal or iodinated-loaded nanoparticles are excellent absorbers of X-rays and can offer excellent improvement in medical diagnosis and X-ray imaging by means of computed tomography (CT). CT offers cheaper format compared to MRI. Also, the CT modality has an important place in the future theranostics; however, it means that the nanoparticulate contrast agent must have a reservoir structure, e.g. like the lipid-based or polymeric nano-carriers. The potential industrial scaling up is also reviewed. Recent developments in imaging in humans are presented in this Volume.

OI has become extremely popular in recent years, and it is about to surpass nuclear medicine- and MR-based imaging techniques in the drug delivery field. The combination of FMT (3D fluorescence molecular tomography) with microcomputed tomography (μCT) for molecular imaging purposes, using N-(2-hydroxypropyl) methacrylamide (HPMA)-based and NIRF-labeled polymeric nanocarriers enables the noninvasive assessment of nanomedicine biodistribution. (Kunjachan et al. 2013) established a hybrid imaging protocol for noninvasively visualizing and quantifying the accumulation of near-infrared fluorophore-labeled nanomedicines in tissues other than superficial tumors. To this end, HPMA-based polymeric drug carriers were labeled with Dy750, their biodistribution and tumor accumulation were analyzed using FMT, and the resulting data sets were fused with anatomical μCT data sets in which several different physiologically relevant organs were pre-segmented. Their findings convincingly demonstrate that combining anatomical μCT with molecular FMT facilitates the noninvasive assessment of nanomedicine biodistribution.

Matsumura (2014) developed a new technique for imaging, denoted as MALDI-IMS. In this paper, they first report describing the precise distribution of a micelle-drug distribution by IMS and on successful visualization and quantification of the distribution of a non-radiolabeled and non-chemically modified drug in various frozen tissue slices microscopically. They have successfully visualized different anticancer agents, and demonstrated that the MALDI-IMS technique can be applied to clinical biopsy specimens or surgically resected tissues after neo-adjuvant chemotherapy. Likewise, details on this procedure are presented in this Volume.

1.9 New Quantitative Tools to Explore Pharmacokinetics

Shortly, we encourage to employ new quantitative tools of PK study to link molecular mechanistic (Intracellular) understanding (i.e. signaling) with standard PK methods; extend to both membrane- and perfusion-limited scenarios (Jones and Rowland-Yeo 2013) and to account for scales by means of multiscale systems biology simulation (from molecular mechanisms to up to human body) (Krauss et al. 2012); extend to a generalized PK (Pilari and Huisinga 2010); and employ new systems biology for designing of drug *combination therapy* (at least two drugs, i.e., dual blockage; e.g., Lammers et al. 2010) via search by employing an interrogation of models of signaling pathways (also called network targeting. Yildirim et al. 2007). The latter efforts are absolutely needed because the body is learning that cancer is able to hijack (by-pass) multiple escape mechanisms (of signaling pathways) to avoid destruction by the immune system. Our working definition of Systems Biology (SB) is a "quantitative, postgenomic, postproteomic, dynamic, *multiscale* physiology" (Wikswo et al. 2006). In addition, our adopted definition of Systems Biology is a two-directional, integrative, top-down (mechanism-based) and bottom-up (hypothesis-driven), often hybrid, methodology to systematically study complex biological/biomedical experimental data. These data can span the spectrum of experimental experience from in vitro molecular biology to whole body levels. The characterization of these data includes both qualitative and (preferably) quantitative tools. The impact of this characterization helps us evaluate a functional organization that is dynamic, defining within its scope relationships and interactions among various system components and between living systems and their interaction with the environment. In a clear departure from the mechanism-based reductionist approach, SB will embrace both arms of scientific dogma, reductionist and holistic, by employing a well-defined middle ground. This scope and structure of this middle ground may shift in time as more SB global tools are developed. This effort will allow us to obtain a deeper understanding of disease mechanisms (including the upper level interactions), and derive benefits for expedited drug discovery and improved drug safety and efficacy.

All drugs somehow interfere with signal transduction, receptor signaling and biochemical equilibrium. Additionally, for many drugs we know, and for most we suspect, that they interact with more than one target (pleiotropy). So, there will be simultaneous changes in several biochemical signals, and there will be feedback reactions from the perturbed pathways. In most cases, the net result will not be linearly deductible from single effects and Systems Biology tools must be applied. In addition, this phenomenon is a source of off-target effects. For drug combinations, this is even more complicated. Thus, a clinically relevant 'target' might consist not of a single biochemical entity, but of simultaneous interference of a number of receptors (pathways, enzymes and so on). And, only the net clinical effect will be beneficial. Such target definition, not derived from a single direct chemical interaction will require input from systems biology (Imming et al. 2006).

Prokop and Michelson (2012) defined Steps in Systems Biology/Computational SB:

- Define a preliminary, data-driven (prior knowledge and data mining), hypothesis of a complex process (note that subsequent iterations will define this step further).
- Collect global, dynamic OMICs data with a multitude of different technologies for healthy and disease phenotypes over a range of environmental and genetic changes (from perturbed experiments)
- Analyze them and confront with available information (mining)
- Identify a network model (by exploring the parameter space) that quantitatively recapitulates prior observations and predicts behavior in new environments, and validate it with data collected: reduce model into a correlation or reaction network
- Simulate and interrogate data *in silico* and discover new targets
- Repeat the cycle (return to data collection in order to improve experimental conditions, etc.) to distinguish between competing model hypotheses; repeat the above steps to refine the model over successive iterations and resolve the model inconsistencies and revisions in the topology and in regulatory circuits

The Fig. 1.1 is strictly an academic flowchart. For a real-life scenario, see below Flowchart (Cookbook); called Systems Biology paradigm by Michelson in terms of practical application of systems approach and SB. Although developed for ordinary drugs, it also applies for nanomedicinal systems. The wide range of emerging nanotechnology systems for targeted delivery further increases the need for reliable *in silico* predictions. Several computational approaches at different scales in the design of traditional oral drug delivery systems are available. A multiscale framework for integrating continuum, stochastic, and computational chemistry models has been proposed and a several successful case studies are available (Haddish-Berhane et al. 2007).

Prokop and Michelson then specifies the following steps. "Identifying and characterizing the patient subpopulation identification and its segmentation is a first step. The key is to use either deductive logic (from data mining and/or statistical analysis) or inductive logic (hypothesizing as to the extent and impact of a pathway),

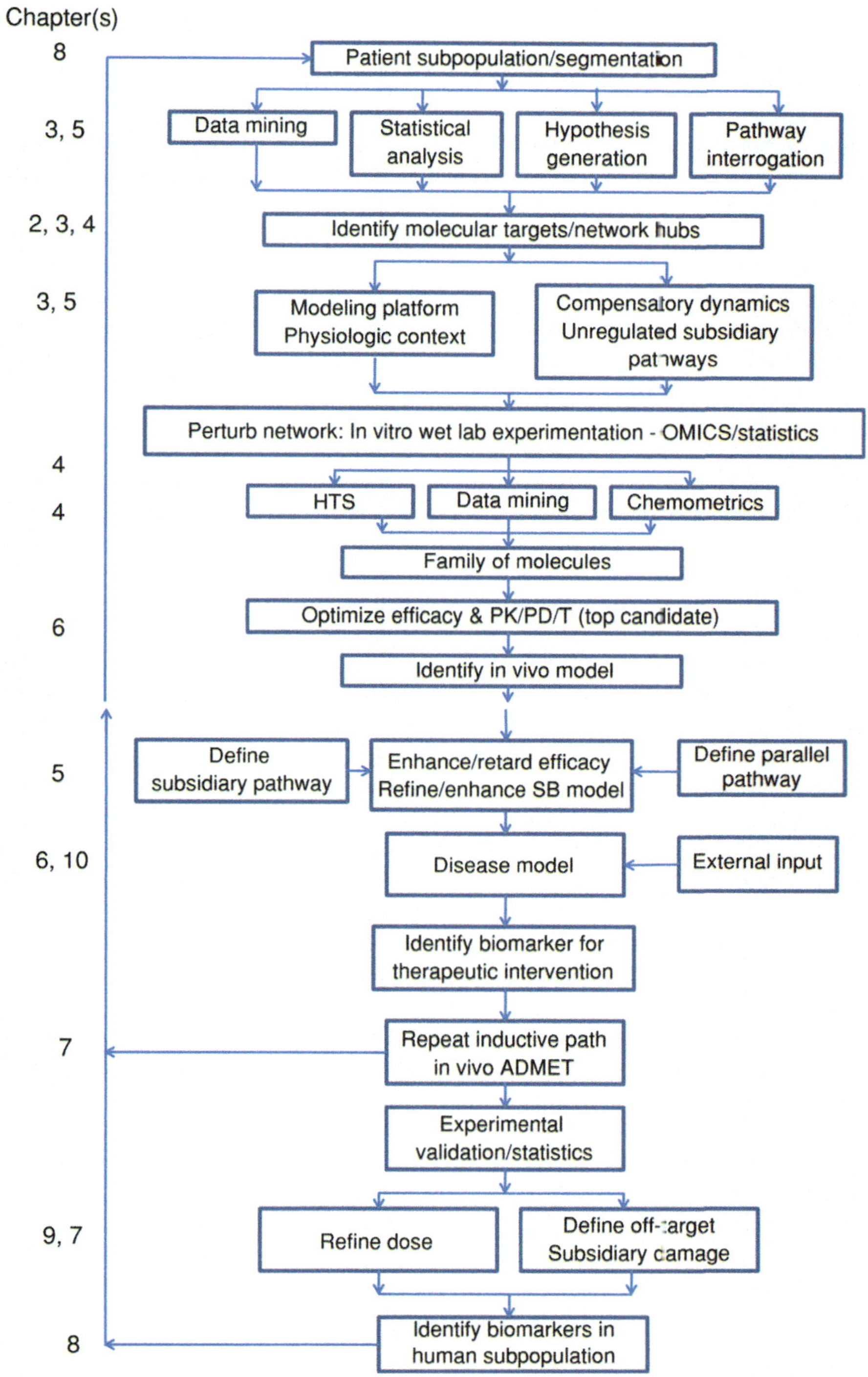

Fig. 1.1 Systems Biology paradigm by Michelson and Prokop, step-by-step approach. This could be used both qualitatively and quantitatively. Note that PK/PD and toxicology is central to this approach. The chapter numbers refer to the original paper (Courtesy of Springer)

identify a candidate molecular target for a particular kind of intervention. Then assure yourself of its potential viability both with a modeling platform of the physiological context (as complete as possible) that will identify possible (hypothetical) compensatory dynamics (e.g., unregulation of subsidiary pathways, etc.). Then, using a highly focused wet lab experimentation strategy, explore the hypothetical landscape (at any level that is feasible, but typically in vitro). Then, when you think you understand your target "well enough", see if there is any molecule that will manipulate it to therapeutic advantage. This can be done using on HTS machinery and robots, *in silico* data mining and chemometrics, or both simultaneously. Once one has a family of molecules (post screening—whether *in silico* or HTS), the leads are typically "optimized" by medicinal chemistry to insure "optimal" efficacy (i.e., are we better—more potent, less toxic—than our competitor?) and that exhibit "optimal" PK/PD/Tox profiles to "insure" one is minimizing the chance of failure in Clinical Development. This is typically an iterative process with close interactions between the in vivo biologists and the medicinal chemists, but the output of this "conversation" is typically the top candidate compound and a few back-ups (just in case). Once one has good reason to believe that there are a family of molecules (it is NEVER just one) that might "work" in a more completely physiological environment (i.e. in vivo), find an appropriate in vivo model (typically a lower species like rodent) that you can make "sick" and then make better. Then, to close the loop to SB even tighter, refine/enhance your model to uncover subsidiary dynamics that might indicate subsidiary or parallel pathways that could either enhance or, more likely, retard the efficacy of your candidate molecule. Then, use the model to develop, refine, and posit more complete testable hypotheses (feedback back to induction again), and design the right experiment in the right way to yield the most informative insight as to how the biology and chemistry are interacting. The work product of this effort should yield some level of insight into the dynamics underlying the disease process and its response to external manipulation. That knowledge should, in turn, yield insights into the existence and accessibility of any informative biomarkers one might use to design/optimize a clinical trial and to eventually identify a subpopulation that will likely form the best subject cohort for therapeutic intervention. Now, do it again for ADMET. This loop of scientific induction/deduction and focused hypothesis generation and experimental validation forms the Holy Grail of DDD."

This Volume presents some details on Bayer's systems biology approach to PK/PD and as well on the combination therapy methodology, based on systems biology.

1.10 How can Academia Survive With Diminishing Funding and How it Affects Nanomedicine Patenting?

We then discuss rapidly diminishing funding for PK studies in academia and limited proprietary effort on this side. As stated "NIH's ability in US to support vital research at more than 2500 universities and organizations across the nation is

Table 1.3 Recent NIH nano funding

Year 2015	Year 2016	Year 2017 projected
364.0	382.0	382.0

reeling from a decline in funding that threatens our health, our economy, and our standing as the world leader in biomedical innovation. After 10 years of essentially flat budgets eroded by the effects of inflation, and now precipitously worsened by the impact of sequestration (an automatic, across-the-board 5.5 % cut in NIH support), NIH's purchasing power has been cut by almost 25 % compared to a decade ago." (Rockey and Collins 2013, see the Web Ref e.). Furthermore, as reported by Kwame Boadi in 2014 (Web Ref f.), "A new nationwide survey by the American Society for Clinical Oncology found that three out of four cancer researchers said the current federal funding situation is negatively affecting their ability to conduct research. The survey reports 28 % of the cancer researchers have decided to participate in fewer federally-funded clinical trials; 27 % have postponed the launch of a clinical trial; and 23 % have had to limit patient enrollment on a clinical trial." And, again by Boadi, "These funding figures, however, obscure a critical factor that affects America's capacity to continue leading the world in biomedical research—inflation. The cost of conducting biomedical research is increasing faster than the cost of other goods and services in the economy. As a result, although much was made of the partial rollback of sequestration in FY 2014 and FY 2015 under the Murray-Ryan budget deal, this partial relief ultimately will do little to stem the trend in recent years of rapidly diminishing purchasing power for biomedical research funding dollars. The decisions of the past decade, however, have begun to threaten America's pre-eminence in biomedical research. As NIH budgets began stagnating after 2003 and declining in 2010, the costs of conducting biomedical research continued to rapidly increase. As such, the lost purchasing power that NIH has experienced over the past decade cannot be undone by simply repealing sequestration. Rather, in order to secure America's position as the global leader in biomedical research for the foreseeable future, Congress must pursue significant new investments above and beyond repealing sequestration." To demonstrate in numbers, we present the NIH funding in 2015/2016 years and expected expenditures for 2016 (in dollar in millions) (Table 1.3).

From the same source, funding is rather flat since 2001 year until now, indicating a decrease in real dollars (Reference g.). Within the NIH funding, however, a special attention is paid to funding in oncology. Potential benefits are being explored by NCI through support of multi-disciplinary research under the umbrella of the NCI Alliance for Nanotechnology in Cancer.

In terms of patenting, Big Pharma business model, which relies on a few blockbusters to generate enormous profits, is clearly not valid anymore. Patent expiration on several blockbusters in recent years is already altering the drug landscape. Additionally, market forces are dictating that new R&D approaches be developed and implemented so that drug companies can continue to discover and fill the pipeline with novel compounds (or develop reformulations of older compounds). Some of these strategies revolve around nanotechnology and miniaturization. Nananomedicine is already having an enormous impact in this regard. However, the burgeoning number of new nanomedicine patent applications filed at the PTO, coupled with the continued issuance of surprisingly broad patents, is creating a chaotic patent landscape where competing players are unsure as to the validity and enforceability of numerous issued patents. If this trend continues, it could stifle competition and limit access to some inventions (Bawa 2007).

1.11 Analysis of Present Market Push

We then will look at new methods for market identification (i.e. via cancer biomarkers, computational approach) – and assess why market is lagging on delivery of promises. As stated by Petrák (personal communication) a suggested approach would be to (a) select applications (e.g., diagnostics, therapy, etc.), (b) select diseases, (c) determine the "unmet medical need", (d) review/list applications of "nanomedicine" that has been shown (in scientific terms) to have an impact (or at least a potential, in cancer area), and finally (e) based on the above, discuss the "market push".

Why promises of "nanomedicine" solving problems are not met? It could be also that the science focus is somewhat distorted: investigators do not appreciate the importance of clinical trials. Further, a contributing factor is the dismal funding situation in academia and Pharma emphasis on block-busters. Petrák notes (personal communication) that "effective cell-specific drug delivery should lead actually to treat the disease (as compared to the currently prevailing trend of treating symptoms, and/or attenuating disease). Current drug block-busters tend to be administered to chronic conditions resulting in repeated sales. It needs to be considered how a drug therapy that actually cures the disease would be priced (this is not subject of this Volume, however)".

Zamboni (Lucas et al. 2016) emphasize the importance of preclinical animal models in understanding and evaluating the interaction between nanomedicines and the immune system, including the MPS, which ultimately influences PK (clearance and distribution) and from the body by the liver and kidneys. Nanoparticles are engulfed and cleared by the monocytes, macrophages, and dendritic cells that make up the MPS. Because the function of the MPS varies greatly across patients, the ability of physicians to predict the proper dosage for nanodrugs is more difficult than with conventional pharmaceuticals. A probe for characterizing an interaction between nanomedicines and MPS was suggested by Bill Zamboni. The authors also stress that an understanding of the biomolecular interactions of nanomedicines and that the MPS is critical for optimal development of nanomedicines and better prediction of efficacy such therapies.

Web Citations and References

Web Citations

a. https://clinicaltrials.gov/. Accessed 23 Nov 2015
b. www.wiley.com/egacy/willeychi/genmed/clinical/. Accessed 15 Dec 2015.
c. http://www.researchgate.net/publication/237620188_Report_and_Recommendations_of_thPanel_to_Assess_the_NIH_Investment_in_Research_on_Gene_Therapy. Accessed 5 Dec 2015
d. http://www.researchgate.net/publication/237620188_Report_and_Recommendations_of_the_Panel_to_Assess_the_NIH_Investment_in_Research_on_Gene_Therapy. Accessed 5 Dec 2015
e. http://nexus.od.nih.gov/all/2013/09/24/one-nation-in-support-of-biomedical-research/
f. https://www.americanprogress.org/issues/economy/report/2014/03/25/86369/erosion-of-funding-for-the-national-institutes-of-health-threatens-u-s-leadership-in-biomedical-research/
g. http://www.nano.gov/about-nni/what/funding

References

Anton N, Vandamme TF (2014) Nanotechnology for computed tomography: a real potential recently disclosed. Pharm Res 31:20–34

Barenholz Y (2012) Doxil(R) – the first FDA-approved nano-drug: lessons learned. J Control Release 160:117–134

Barua S, Mitragotri S (2014) Challenges associated with penetration of nanoparticles across cell and tissue barriers: a review of current status and future prospects. Nano Today 9:223–243

Bawa R (2007) Patents and nanomedicine. Nanomedicine 2(3):351–374

BCC Report: Nanotechnology in medical applications. The global market, Jan 19, 2012, 187 pages. BCC Research LLC, Wellesley

Benien P, Solomon MA, Nguyen P, Sheehan EM, Mehanna AS, D'Souza GG (2015) Hydrophobized triphenyl phosphonium derivatives for the preparation of mitochondriotropic liposomes: choice of hydrophobic anchor influences cytotoxicity but not mitochondriotropic effect. J Liposome Res 26:1–7, March

Bertrand N, Wu J, Xu X, Kamaly N, Farokhzad OC (2014) Cancer nanotechnology: the impact of passive and active targeting in the era of modern cancer biology. Advan Drug Deliv Rev 66(1):2–26

Bielski ER, Zhong Q, Brown M, da Rocha SR (2015) Effect of the conjugation density of triphenylphosphonium cation on the mitochondrial targeting of poly(amidoamine) dendrimers. Mol Pharm 12:3043–3053

Biswas S, Dodwadkar NS, Deshpande PP, Torchilin VP (2012a) Liposomes loaded with paclitaxel and modified with novel triphenylphosphonium-PEG-PE conjugate possess low toxicity, target mitochondria and demonstrate enhanced antitumor effects in vitro and in vivo. J Control Release 159:393–402

Biswas S, Dodwadkar NS, Piroyan A, Torchilin VP (2012b) Surface conjugation of triphenylphosphonium to target poly (amidoamine) dendrimers to mitochondria. Biomaterials 33:4773–4782

Blanco E, Shen H, Ferrari M (2015) Principles of nanoparticle design for overcoming biological barriers to drug delivery. Nat Biotechnol 33(9):941–951

Blume G, Cevc G (1993) Molecular mechanism of the lipid vesicle longevity in vivo. Biochim Biophys Acta 1146:157–168

Boddapati SV, Tongcharoensirikul P, Hanson RN, D'Souza GG, Torchilin VP, Weissig V (2005) Mitochondriotropic liposomes. J Liposome Res 15:49–58
Boddapati SV, D'Souza GG, Erdogan S, Torchilin VP, Weissig V (2008) Organelle-targeted nanocarriers: specific delivery of liposomal ceramide to mitochondria enhances its cytotoxicity in vitro and in vivo. Nano Lett 8:2559–2563
Callahan J, Kopecek J (2006) Semitelechelic HPMA copolymers functionalized with triphenylphosphonium as drug carriers for membrane transduction and mitochondrial localization. Biomacromolecules 7:2347–2356
Cavallo F, Calogero RA, Forni G (2007) Are oncoantigens suitable targets for anti-tumour therapy? Nat Rev Cancer 7(9):707–713
Chauhan VP, Stylianopoulos T, Boucher Y, Jain RK (2011) Delivery of molecular and nanoscale medicine to tumors: transport barriers and strategies. Annu Rev Chem Biomol Eng 2:281–298
Cheng SM, Pabba S, Torchilin VP, Fowle W, Kimpfler A, Schubert R, Weissig V (2005) Towards mitochondria-specific delivery of apoptosis-inducing agents: DQAsomal incorporated paclitaxel. J Drug Delivery Sci Technol 15:81–86
Crommelin DJ, Florence AT (2013) Towards more effective advanced drug delivery systems. Int J Pharm 454(1):496–511
Dao T, Pankov D, Scott A, Korontsvit T, Zakhaleva V, Xu Y, Xiang J, Yan S, Direito de Dawidczyk CM, Russell LM, Searson PC (2014) Nanomedicines or cancer therapy: state-of-the-art and limitations to pre-clinical studies that hinder future development. Front Chem 2:69, 25 Aug. eCollection
Dao T, Pankov D, Scott A, Korontsvit T, Zakhaleva V, Xu Y, Xiang J, Yan S, de Morais Guerreiro MD, Veomett N, Dubrovsky L, Curcio M, Doubrovina E, Ponomarev V, Liu C, O'Reilly RJ, Scheinberg DA (2015) Therapeutic bispecific T-cell engager antibody targeting the intracellular oncoprotein WT1. Nat Biotechnol 33(10):1079–1086
D'Souza GG, Weissig V (2009) Subcellular targeting: a new frontier for drug-loaded pharmaceutical nanocarriers and the concept of the magic bullet. Expert Opin Drug Deliv 6:1135–1148
Dawidczyk CM, Russell LM, Searson PC (2014) Nanomedicines for cancer therapy: state-of-the-art and limitations to pre-clinical studies that hinder future developments. Front Chem 2:69
Eaton MAW, Levy L, Fontaine OMA (2015) Delivering nanomedicines to patients: a practical guide. Nanomed Nanotechnol Biol Med 11:983–992
Etrych T, Chytil P, Koziolova E, Širova M, Hoffman S, Muller T, Mader K, Řihova B, Ulbrich K (2014) Polymer-drug conjugates for targeted tumor therapy. Anticancer Res 34:5761–6258 (extended abstract)
European Science Foundation (2004) Nanomedicine – an ESF–European Medical Research Councils (EMRC) forward look report. France ESF, Strasbourg
Faltas B (2012) Cornering metastases: therapeutic targeting of circulating tumor cells and stem cells. Front Oncol 3(2):68, Epub 3 Jul 2012
Friedmann T, Roblin R (1972) Gene therapy for human genetic disease? Science 175:949–955
Gregoriadis G, Ryman BE (1971) Liposomes as carriers of enzymes or drugs: a new approach to the treatment of storage diseases. Biochem J 124:58P
Groenink L, Folkerts G, Henk-Jan Schuurman H-J (2015) European Journal of Pharmacology, Special issue on translational value of animal models: Introduction. Eur J Pharmacol 759:1–2
Grubb JH, Vogler C, Sly WS (2010) New strategies for enzyme replacement therapy for lysosomal storage diseases. Rejuvenation Res 13:229–236
Guzman-Villanueva D, Mendiola MM, Nguyen HX, Weissig V (2015) Influence of triphenylphosphonium (TPP) cation hydrophobization with phospholipids on cellualr toxicity and mitochondrial selectivity. SOJ Pharm Pharm Sci 2:1–9
Haddish-Berhane N, Rickus JL, Haghighi K (2007) The role of multiscale computational approaches for rational design of conventional and nanoparticle oral drug delivery systems. Int J Nanomed 2(3):315–331

Hillaireau H, Couvreur P (2009) Nanocarriers' entry into the cell: relevance to drug delivery. Cell Mol Life Sci 66(17):2873–2896

Hong CW, Zeng Q (2013) Tapping the treasure of intracellular oncotargets with munotherapy. FEBS Lett 588(2014):350–355

Horobin RW (2001) Uptake, distribution, and accumulation of dyes and fluorescent probes within living cells. A structure-activity modelling approach. Adv Colour Sci Technol 4:101–107

Horobin RW (2010) Can QSAR models describing small-molecule xenobiotics give useful tips for predicting uptake and localization of nanoparticles in living cells? And if not, why not? In: Weissig V, D'Souza GGM (eds) Organelle-specific pharmaceutical nanotechnology, vol 2010. Wiley, Hoboken, pp 193–206

Horobin RW, Trapp S, Weissig V (2007) Mitochondriotropics: a review of their mode of action, and their applications for drug and DNA delivery to mammalian mitochondria. J Control Release 121:125–136

Hu CM, Zhang L, Aryal S, Cheung C, Fang RH, Zhang L (2011) Erythrocyte membrane-camouflaged polymeric nanoparticles as a biomimetic delivery platform. Proc Natl Acad Sci U S A 108:10980–10985

Imming P, Sinning C, Meyer A (2006) Drugs, their targets and the nature and number of drug targets. Nat Rev Drug Discov 5(10):821–834 Erratum in Nat Rev Drug Discov 6(2):126 (2007)

Jain RK (2013) Normalizing tumor microenvironment to treat cancer: bench to bedside to biomarkers. J Clin Oncol 31(17):2205–2208

Jain R, Stylianopoulos T, Economides EA, Baish JW, Fukumura D, Jain RK (2015) Towards optimal design of cancer nanomedicines: multi-stage nanoparticles for the treatment of solid tumors. Ann Biomed Eng 11 Feb 2015. [Epub ahead of print]

Jones HM, Rowland-Yeo K (2013) Basic concepts in physiologically based pharmacokinetic modeling in drug discovery and development. Pharmacometrics Syst Pharmacol 2:e63

Khawar IA, Kim JH, Kuh H-J (2015) Improving drug delivery to solid tumors: priming the tumor microenvironment. J control Release 201:78–89

Kinnaird A, Michelakis ED (2015) Metabolic modulation of cancer: a new frontier with great translational potential. J Mol Med 93:127–142

Klibanov AL, Maruyama K, Torchilin VP, Huang L (1990) Amphipathic polyethyleneglycols effectively prolong the circulation time of liposomes. FEBS Lett 268:235–237

Kobayashi H, Watanabe R, Choyke PL (2014) Improving conventional enhanced permeability and retention (EPR) effects; what is the appropriate target? Theranostics 4(1):81–89

Krauss M, Schaller S, Borchers S, Findeisen R, Lippert J, Kuepfer L (2012) Integrating cellular metabolism into a multiscale whole-body model. PLOS Comput Biol 8(10):e1002750, 1 Oct 2012

Kunjachan S, Gremse F, Thee B, Koczera P, Pola R, Pechar M, Etrych T, Ulbrich K, Storm G, Kiessling F, Lammers T (2013) Noninvasive optical imaging of nanomedicine biodistribution. ACS Nano 7(1):252–262

Lammers T, Storm G, Kiessling F (2010) Nanomedicine formulations for combination therapies. Nano Rev 1:5705–5708

Lehner R, Wang X, Marsch S, Hunziker P (2015) Intelligent nanomaterials for medicine: carrier platforms and targeting strategies in the context of clinical application. Nanomed: Nanotechnol Biol Med 9:742–757

Lim CS (2007) Organelle-specific targeting in drug delivery and design. Adv Drug Deliv Rev 59:697

Liu L, Hitchens TK, Ye Q, Wu Y, Barbe B, Prior DE, Li WF, Yeh F-C, Foley LM, Bain DJ, Ho C (2013) Decreased reticuloendothelial system clearance and increased blood half-life and immune cell labeling for nano- and micron-sized superparamagnetic iron-oxide particles upon pre-treatment with Intralipid. Biochim Biophys Acta 1830(6):3447–3453

Lucas AT, Madden AJ, Zamboni WC (2016) Challenges in preclinical to clinical translation for anticancer carrier-mediated agents. WIREs Nanomed Nanobiotechnol 8:642–653. doi:10.1002/wnan.1394

Lukacs GL, Haggie P, Seksek O, Lechardeur D, Freedman N, Verkman AS (2000) Size-dependent DNA mobility in cytoplasm and nucleus. J Biol Chem 275:1625–1629

Lyrawati D, Trounson A, Cram D (2011) Expression of GFP in the mitochondrial compartment using DQAsome-mediated delivery of an artificial mini-mitochondrial genome. Pharm Res 28:2848–2862

Maeda H (2015) Toward a full understanding of the EPR effect in primary and metastatic tumors as well as issues related to its heterogeneity. Adv Drug Deliv Rev 91:3–6

Malhi SS, Budhiraja A, Arora S, Chaudhari KR, Nepali K, Kumar R, Sohi H, Murthy RS (2012) Intracellular delivery of redox cycler-doxorubicin to the mitochondria of cancer cell by folate receptor targeted mitocancerotropic liposomes. Int J Pharm 432:63–74

Marrache S, Dhar S (2012) Engineering of blended nanoparticle platform for delivery of mitochondria-acting therapeutics. Proc Natl Acad Sci U S A 109:16288–16293

Marrache S, Dhar S (2013) Biodegradable synthetic high-density lipoprotein nanoparticles for atherosclerosis. Proc Natl Acad Sci U S A 110:9445–9450

Matsumura Y (2014) The drug discovery by nanomedicine and its clinical experience. Jpn J Clin Oncol 44(6):515–525

Mechler K, Mountford WK, Hoffmann GF, Ries M (2015) Pressure for drug development in lysosomal storage disorders – a quantitative analysis thirty years beyond the US orphan drug act. Orphanet J Rare Dis 10:46

Merril CR, Geier MR, Petricciani JC (1971) Bacterial virus gene expression in human cells. Nature 233:398–400

Miller MA, Gadde S, Pfirschke C, Engblom C, Sprachman MM, Kohler RH, Yang KS, Laughney AM, Wojtkiewicz G, Kamaly N, Bhonagiri S, Pittet MJ, Farokhzad C, Weissleder R (2015) Predicting therapeutic nanomedicine efficacy using a companion magnetic resonance imaging nanoparticle. Sci Transl Med 7(314):314ra183. 18 Nov 2015

Munyon W, Kraiselburd E, Davis D, Mann J (1971) Transfer of thymidine kinase to thymidine kinaseless L cells by infection with ultraviolet-irradiated herpes simplex virus. J Virol 7:813–820

Needham D, McIntosh TJ, Lasic DD (1992) Repulsive interactions and mechanical stability of polymer-grafted lipid membranes. Biochim Biophys Acta 1108:40–48

Nichols JW, Bae YH (2014) EPR: evidence and fallacy. J Control Release 28(190):451–464

Nikulnikov PI (2015) Experience of application of the gene inductor of vascular endothelium growth factor in treatment of patients, suffering atherosclerotic Ischemia of the lower extremities tissues. Klin Khir 2015:41–43

Nishihara H (2014) Human pathological basis of blood vessels and stromal tissue for nanotechnology. Adv Drug Deliv Rev 74:19–27

Osada K, Christie RJ, Kataoka K (2009) Polymeric micelles from poly(ethyleneglycol)–poly(amino acid) block copolymer for drug and gene delivery. J R Soc Interface 6:S325–S339

Parodi A, Quattrocchi N, van de Ven AL, Chiappini C, Evangelopoulos M, Martinez JO, Brown BS, Khaled SZ, Yazdi IK, Enzo MV, Isenhart L, Ferrari M, Tasciotti E (2013) Synthetic nanoparticles functionalized with biomimetic leukocyte membranes possess cell-like functions. Nat Nanotechnol 8:61–68

Patel NR, Hatziantoniou S, Georgopoulos A, Demetzos C, Torchilin VP, Weissig V, D'Souza GG (2010) Mitochondria-targeted liposomes improve the apoptotic and cytotoxic action of Sclareol. J Liposome Res 20:244–249

Pearson S, Jia H, Kandachi K (2004) China approves first gene therapy. Nat Biotechnol 22:3–4

Petrák K (2005) Essential properties of drug-targeting delivery systems. Drug Discov Today 10(23/24):1667–1673

Petrák K (2012) Targeted drug delivery – Quo vadis? Drug Dev Res 73(2):59–65

Petrak K, Hasirci V (2006) Targeting drug-delivery systems: promises, promises, and more promises…Let's change the paradigm. J Biomater Sci Polym Ed 17(11):1207–1208

Pilari S, Huisinga W (2010) Lumping of physiologically-based pharmacokinetic models and a mechanistic derivation of classical compartmental models. J Pharmacokinet Pharmacodyn 37:365–405

Prokop A, Michelson S (2012) Systems biology in biotech and pharma. A changing paradigm, Springer briefs in pharmaceutical science and drug development. Springer, Dordrecht, p 127

Qiao J, Liu Z, Tian Y, Wu M, Niu Z (2015) Multifunctional self-assembled polymeric nanoprobes for FRET-based ratiometric detection of mitochondrial H2O2 in living cells. Chem Commun (Camb) 51:3641–3644

Rivera Gil P, Hühn D, del Mercato LL, Sasse D, Parak WJ (2010) Nanophar¬macy: inorganic nanoscale devices as vectors and active compounds. Pharmacol Res 62(2):115–125

Rodriguez PL, Harada T, Christian DA, Pantano DA, Tsai RK, Discher DF (2013) Minimal "self" peptides that inhibit phagocytic clearance and enhance delivery of nanoparticles. Science 339:971–975

Rosenberg SA, Aebersold P, Cornetta K, Kasid A, Morgan RA, Moen R, Karson EM, Lotze MT, Yang JC, Topalian SL et al (1990) Gene transfer into humans – immunotherapy of patients with advanced melanoma, using tumor-infiltrating lymphocytes modified by retroviral gene transduction. N Engl J Med 323:570–578

Sakhrani NM, Padh H (2013) Organelle targeting: third level of drug targeting. Drug Des Devel Ther 7:585–599

Savic R, Luo L, Eisenberg A, Maysinger D (2003) Micellar nanocontainers distribute to defined cytoplasmic organelles. Science 300:615–618

Scodeller P, Catalano PN, Salguero N, Duran H, Wolosiuk A, Soler-Illia GJ (2013) Hyaluronan degrading silica nanoparticles for skin cancer therapy. Nanoscale 5(20):9690–9698

Scodeller P (2014) Hyaluronidase and other extracellular matrix degrading enzymes for cancer therapy: new uses and nano-formulations. J Carcinog Mutagen 5:4–8

Seksek O, Biwersi J, Verkman AS (1997) Translational diffusion of macromolecule-sized solutes in cytoplasm and nucleus. J Cell Biol 138:131–142

Sharma A, Soliman GM, Al-Hajaj N, Sharma R, Maysinger D, Kakkar A (2012) Design and evaluation of multifunctional nanocarriers for selective delivery of coenzyme Q10 to mitochondria. Biomacromolecules 13:239–252

Sosale NG, Spinler KR, Alvey C, Discher DE (2015) Macrophage engulfment of a cell or nanoparticle is regulated by unavoidable opsonization, a species-specific 'marker of self' CD47, and target physical properties. Curr Opin Immunol 35:107–112

Stapleton S, Milosevic MD, Tannock IF, Allen C, Jaffray DA (2015) The intra-tumoral relationship between microcirculation, interstitial fluid pressure and liposome accumulation. J Control Release 211:163–170

Stylianopoulos T, Jain RK (2015) Design considerations for nanotherapeutics in oncology. Nanomed 11(8):1893–1907

Theodossiou TA, Sideratou Z, Katsarou ME, Tsiourvas D (2013) Mitochondrial delivery of doxorubicin by triphenylphosphonium-functionalized hyperbranched nanocarriers results in rapid and severe cytotoxicity. Pharm Res 30:2832–2842

Thiel C, Schneckener M, Krauss M, Ghallab A, Hoffmann U, Kanacher T, Zellmer S, Gebhard R, Hengstler JG, Kuepfer L (2015) A Systematic evaluation of the use of physiologically based pharmacokinetic modeling for cross-species extrapolation. J Pharm Sci 104:191–206

van der Meel R, Vehmeijer LJC, Kok RJ, Storm G, van Gaal EVB (2013) Ligand-targeted particulate nanomedicines undergoing clinical evaluation: current status. Advan Drug Deliv Rev 65(10):1284–1298

Wang M, Thanou M (2010) Targeting nanoparticles to cancer. Pharmacol Res 62:90–99

Weiss MJ, Wong JR, Ha CS, Bleday R, Salem RR, Steele GD Jr, Chen LB (1987) Dequalinium, a topical antimicrobial agent, displays anticarcinoma activity based on selective mitochondrial accumulation. Proc Natl Acad Sci U S A 84:5444–5448

Weissig V (2005) Targeted drug delivery to mammalian mitochondria in living cells. Expert Opin Drug Deliv 2:89–102

Weissig V (2011a) From serendipity to mitochondria-targeted nanocarriers. Pharm Res 28:2657–2668
Weissig V (2011b) Mitochondrial delivery of biologically active molecules. Pharm Res 28:2633–2638
Weissig V (2015) DQAsomes as the prototype of mitochondria-targeted pharmaceutical nanocarriers: preparation, characterization, and use. Methods Mol Biol 265:1–11
Weissig V, Guzman-Villanueva D (2015) Nanopharmaceuticals (part 2): products in the pipeline. Int J Nanomed 10:1245–1257
Weissig V, Lasch J, Erdos G, Meyer HW, Rowe TC, Hughes J (1998a) DQAsomes: a novel potential drug and gene delivery system made from Dequalinium. Pharm Res 15:334–337
Weissig V, Moegel HJ, Wahab M, Lasch J (1998b) Computer simulations of DQAsomes. Proc Int Symp Control Release Bioact Mater 25:312
Weissig V, Hughes J, Rowe T, Lasch J (2000) Materials and methods for intracellular delivery of biologically active molecules, US Patent No. 6,090,619
Weissig V, Hughes, J, Rowe T, Lasch J (2001) Materials and methods for intracellular delivery of biologically active molecules, US Patent No. 6,171,863 B1
Weissig V, Hughes J, Rowe T, Lasch J (2003) Materials and methods for intracellular delivery of biologically active molecules, US Patent No. 6,627,618 B2
Weissig V, Pettinger TK, Murdock N (2014) Nanopharmaceuticals (part 1): products on the market. Int J Nanomedicine 9:4357–4373
Whatcott CJ, Han H, Posner RG, Hostetter G, Von Hoff DD (2011) Targeting the tumor microenvironment in cancer: why hyaluronidase deserves a second look. Cancer Discov 1:291–296
Wikswo JP, Prokop A, Baudenbacher FJ, Cliffel D, Csukas B, Velkovsky M (2006) The engineering challenges of BioMEMS: the integration of microfluidics, micro- and nano-devices, models, and external control for systems biology. IEE Proc Nanobiotechnol 153(4):81–101
Wong AD, Ye M, Ulmschneider MB, Searson PC (2015) Quantitative analysis of the enhanced permeation and retention (EPR) Effect. PLoS ONE 10(5):e0123461. doi:10.1371/journal
Yang Y, Gao N, Hu Y, Jia C, Chou T, Du H, Wang H (2015) Gold nanoparticle-enhanced photodynamic therapy: effects of surface charge and mitochondrial targeting. Ther Deliv 6:307–321
Yildirim MA, K-Il G, Cusick ME, Barabasi A-L, Vidal M (2007) Drug target network. Nat Biotechnol 25:1119–1126
Yla-Herttuala S (2012) Endgame: glybera finally recommended for approval as the first gene therapy drug in the European Union. Mol Ther 20:1831–1832
Zhou J, Zhao WY, Ma X, Ju RJ, Li XY, Li N, Sun MG, Shi JF, Zhang CX, Lu WL (2013) The anticancer efficacy of paclitaxel liposomes modified with mitochondrial targeting conjugate in resistant lung cancer. Biomaterials 34:3626–3638
Zhuang Q, Jia H, Du L, Li Y, Chen Z, Huang S, Liu Y (2014) Targeted surface-functionalized gold nanoclusters for mitochondrial imaging. Biosens Bioelectron 55:76–82
Zschaler J, Schlorke D, Arnhold J (2014) Differences in innate immune response between man and mouse. Crit Rev Immunol 34(5):433–454

Chapter 2
Precision Drugs and Cell-Specific Drug Delivery

Karel Petrak

Abstract Precision medicine is an approach to prevent and treat disease that takes into account people's individual variations in genes, environment, and lifestyle. The current Precision Medicine Initiative of the US Government is to: "generate the scientific evidence needed to move the concept of precision medicine into clinical practice". In the first approximation, precision medicine may provide a more accurate diagnosis of the disease, but may not have the means to offer an improved therapy.

The aim of drug targeting is to generate pharmacologically effective drug concentration at the site of disease while keeping a very low/minimal drug concentration in the rest of the body, away from the site of disease. Targeted drugs are thus "precision drugs" needed to bring precision medicine into the clinical practice.

Focusing on the area of cancer therapy, this review examines the essential requirements that must be met for tumor-cell-targeted drug-delivery systems to work. It examines the progress to date and draws conclusions to offer an optimal paradigm for future drug-delivery systems' development.

Keywords Precision drugs • Cell specific • Drug delivery • Antibody • Cancer • Tumor

Abbreviations

Abs	Antibodies
ADC	Antibody-Drug Conjugate
AML	Acute Myeloid Leukemia
CD	Cluster of Differentiation
IC_{50}	The half maximal inhibitory concentration

K. Petrak (✉)
NangioTx Inc, 2368 Rice Blvd, Suite A, 77005-2600 Houston, TX, USA
e-mail: petrak@nangiotx.com

A. Prokop, V. Weissig (eds.), *Intracellular Delivery III*, Fundamental Biomedical Technologies 8, DOI 10.1007/978-3-319-43525-1_2

PMI	Precision Medicine Initiative
TAA	Tumor-Associated Antigen
TSA	Tumor-Specific Antigen
US	United States
VEGF	Vascular Endothelial Growth Factor
VEGFR	Vascular Endothelial Growth Factor Receptor

2.1 Introduction

By way of introduction, the advent of "precision medicine initiative (PMI)", an effort to treat and prevent disease based on individual differences in genetics, environment and lifestyle, is now 1-year old. Ambitious goals of PMI are to: "transform the way research is done, prevention is practiced, treatments are developed and cures are found". At a White House celebration of the PMI anniversary, five reasons were given why these goals will be met (Giusti 2016):

- Patients Are Ready,
- Its A-Team Leadership,
- It's Powered by Cross-Sector Partnerships,
- It Has Bipartisan Support, and
- Obama Has Every Reason to Get It Done.

Putting this new knowledge into clinical practice will require that corresponding "precision drugs" are developed. It may be "just politics" but such a broad rationale and support for developing new therapies is essential.

2.2 Precision Drugs

This review adopts the view expressed in other recent reviews of site-specific drug delivery. It has been argued that: "drugs to be delivered must exhibit the pharmacokinetic properties matching the essential requirements of cell specific-drug delivery" (Boddy et al. 1989; Petrak 2005), that too many promises had been made about efficacy of new drug-delivery system (Petrak 2013) and that calling old materials (such as polymers, particles, liposomes, etc.) new names (typically adding "nano" at the start of the word) does not solve the drug-delivery issues (Petrak 2006). For too many decades now, the field has been "turning the handle" and using old approaches without getting anywhere near of creating the "magic bullet" of Paul Ehrlich (Ehrlich 1954; Petrak 2012, 2015b). It was argued (Petrak 2015a) that "… research in this field needs do adopt a new paradigm that centers on "self-targeting carriers" such as antibodies combined with "high-potency" drugs specifically selected or developed *de novo* that fully meet the specific pharmacokinetic requirements of targeted drug delivery". It is clear that old approaches using non-specific vehicles such as polymers, particles and liposomes have not worked, and for very good reasons are unlikely to be made to work. Let us therefore examine how well the new concepts might be working, focusing on the field of cancer therapy. The term "targeting" has been used not only for physical delivery of drugs to specific body locations such as cell types etc., but also when influencing (e.g., by inhibiting) mechanism by which disease processes operate. This review focuses on the former use of the term.

2.3 Antibodies

I was not by any means the first one to argue for antibodies (Abs) to be used for delivering anticancer drugs to tumor targets (Rowland et al. 1993; Panowksi et al. 2014).

Decades ago, Rowland reported that mice could be cured of cancer. But the effective delivery of anticancer drugs to tumors in humans has not yet become a reality for a number of reasons: some being due to irrelevance of mouse model to predict outcome in humans, and another being associated with consequences of scaling therapy from mice to men (Fujimori et al. 1990).

When considering cell-specific drug targeting using Abs, key elements are the presence of tumor antigens, availability of antibodies to such tumor antigens, uptake of antibody-drug construct by tumor cells, the choice of drug to be delivered, its linkage to the antibody and the payload, and the release and site-residence of the drug.

2.4 Tumor Antigens

Ideally, structures uniquely associated with a given tumor should be identified and viable option to which antibody-drug complexes could be targeted. In this context, there are two groups of antigens expressed on tumor cells: tumor-associated antigens (TAAs) and tumor-specific antigens (TSAs). Both TSAs and TAAs are mostly portions of intracellular molecules expressed on the cell surface as part of the major histocompatibility complex.

2.4.1 Tumor-Associated Antigens

Many such antigens have been described in the literature. For example, haematopoietic differentiation antigens associated with cluster of differentiation (CD) groupings (Van den Eynde and Scott 1998), growth factors (Hudis 2007; Schoeberl et al. 2009; Cañadas et al. 2010; Scartozzi et al. 2010), vascular endothelial growth factor

(VEGF), VEGF receptor (VEGFR), and integrins (Schliemann and Neri 2010) and many others (Deckert 2009; Zarour et al. 2003).

Tumor-associated antigens are not exclusively expressed by tumor cells but are also found on normal tissues and when shed are found in circulation. Blood-born cancer cell antigens are relatively easier to target. Antigens associated with solid tumors present much harder locations to access. TAAs may be useful in imaging and in diagnostics but offer little, if any, opportunity as targets for tumor-specific delivery of drugs. Significant barriers are presented by the heterogeneity of target antigen expression in the tumor (Van den Eynde and Scott 1998), and by the distribution and pharmacokinetics of antibodies (Pillay et al. 2011) as influenced by tumor blood supply and interstitial pressure. To overcome this, a direct injection of antibody-drug conjugates into the tumor may eventually be the best option (Van der Jeught et al. 2015).

2.4.2 *Tumor-Specific Antigens*

According to the National Cancer Institute, "tumor-specific antigen is a protein or other molecule that is unique to cancer cells or is much more abundant in them. These molecules are usually found in the plasma (outer) membrane, and they are thought to be potential targets for immunotherapy or other types of anticancer treatment" (NCI Dictionary of Cancer Terms 2015). The issue here lies with the term "much more abundant". As more sensitive methods of detection are being developed, "tumor-specific" structures are being found to be also present in non-neoplastic tissues, albeit in small amounts. However, when one compares the volume of tumor tissues with that of the rest of the body, the overall mass balance may show any "specificity" to be of little use to enable effective tumor-specific drug targeting.

It also needs to be kept in mind that tumors and their environment are by no means static. As discussed by Vesely and Schreiber (2013), surveillance by the immune system, chronic inflammation, immune-selection of poorly immunogenic variants, and suppression of antitumor immunity (together referred to as cancer immune-editing) is a dynamic process; composition of tumor-specific antigens as determined at any one specific time is likely to change as the disease progresses. There are multiple other mechanisms that can make a treatment with an antibody-drug complex ineffective – in addition to the above-mentioned tumor heterogeneity of targeted-antigen expression, physical properties and pharmacokinetics of antibodies and intra-tumoral microenvironment (including vascularity and interstitial pressure) have an impact on access to, and uniform distribution of the antibody-drug in the tumor (Pillay et al. 2011; Beck et al. 2010). Issues associated with the drug dose, drug release at the site and its pharmacokinetics will be discussed below.

New methods of identifying tumor-specific antigens are becoming available (Barrett et al. 2015; Barrett et al. 2013).

Some antibodies that specifically bind tumor-surface antigens can act as effective drugs. Antibodies lacking a direct therapeutic effect have been instead applied as antibody-drug conjugates (ADCs) to deliver cytotoxic drugs to tumors, with some success. Brentuximab vedotin (Adcetris®) and ado-trastuzumab emtansine (Kadcyla®) have been approved by the US Food and Drug Administration. FDA-approval documents for the two products (FDA Drug Approval Package ADCETRIS (brentuximab vedotin) 2011; FDA Drug Approval Package Kadcyla 2013) do present mechanistic schemes for how the putative drug delivery to target cancer cell is to take place, however, no quantitative data substantiating that this actually happens have been presented. The usually required safety, toxicity and efficacy test needed for drug approval do not address the issue of drug targeting. The therapeutic window for most reported ADCs remains narrow, no doubt at least in part being associated with an unfavorable distribution of the antibody target as mentioned above. Efforts aimed at further improvements to enhance the therapeutic potential of these constructs have been thoroughly discussed by Panowksi et al. (2014) that addressed ADC heterogeneity and reviewed various site-specific conjugation strategies in current use and highlighted critical factors to be considered when designing next-generation ADC therapeutics.

2.5 Uptake by Cells & Drug Release

A useful tumor antigen must be localized to the cell-surface to allow ADC binding. Further, it is important that tumor antigen is able to be internalized following antibody binding. Internalization typically occurs via receptor-mediated endocytosis, followed by antibody-drug degradation in the lysosome that need to results in optimal free and active-drug release and effective therapeutic action. It is not always the case that cell-surface antigens are endocytosed nor at a desired rate. It is further desirable that antibody does not recycle to the cell surface and that the intracellular processing required for effective drug release occur at the required rate (see later under drug requirements).

As discussed elsewhere (Boddy et al. 1989; Petrak 2005, 2013, 2015a), it is essential that after its release, the drug must remain at the target in its active form long enough locally to reach its pharmacologically effective concentration. Many conventional small-molecular-size drugs do not satisfy this condition. However, many anticancer drugs, because of their relatively low solubility, do. Poor water solubility of many anticancer agents (such as paclitaxel and camptothecin) presents difficulty when used for direct parenteral administration. Consequently, various formulation strategies using drug-carrier systems (such as polymeric micelles) have been explored to overcome their poor solubility (Garcia-Carbonero and Supko 2002; Singla et al. 2002; Mu et al. 2005). In this context, antibodies can serve as drug carriers offering exquisite target specificity. However, it is critical that drug-antibody linkages and their degradation at the target site are such that the drug is released from its antibody without increasing is aqueous solubility.

Further, linker needs to be made with consideration of the target antigen to be used (Polson et al. 2009) as the mechanism available for drug release may depend on the mode of antibody uptake by the cells. Ideally, the drug should only be released inside the target cells to maximize drug efficacy and minimize its toxicity. In practice, optimal linker selection will ultimately "depend on experimentally determining the optimal combination of the correct linker, the target antigen and desired payload" (Panowksi et al. 2014).

2.6 Tumor Environment & Drug Access

It has well been recognized that angiogenesis plays an important role in cancer development and progression. The vasculature of solid tumors is fundamentally different from that of normal vasculature (Siemann 2011). The blood supply to the normal tissues of the body is maintained by an orderly and efficient vascular network, whereas aggressive growth of the neoplastic-cell population and associated over-expression of pro-angiogenic factors in tumors leads to the development of disorganized blood-vessel networks that are fundamentally different from normal vasculature. This creates an obstacle to the efficient distribution of anticancer-therapy drugs to tumor cells (Huang et al. 2015).

So called "vascular normalization theory" tries to explain additional antitumor effects of inhibitors of vascular endothelial growth factor (VEGF) signaling when the inhibitors are combined with chemotherapeutics. However, vascular normalization is time and dose dependent and thus difficult to implement clinically. An alternative strategy of "vascular promotion therapy" (Siemann 2011) that can increase chemotherapeutics delivery and intracellular uptake of the drug by increasing tumor blood vessel density and blood flow which leads to reduced cancer growth and metastasis is also being investigated clinically. Hence, both apparently contradictory approaches, pro- and anti-angiogenesis, may be applicable in influencing tumor environment.

Penetration by antibodies of tumor mass, with antibodies being of a rather large-molecular size, is very much influenced and often prevented evenly and completely to reach tumor cancer cells. Combining an antibody-drug complex administration with "vascular normalization" may well be a successful approach provided an optimal timing of the therapy can be identified using the "precision medicine" approach.

2.7 Drug Payload

How many molecules of cytotoxic drugs must an antibody bring to cancer cells in order to exert therapeutic effect? Let us make the following approximation, and take cell volume as 1 $\mu m^3 = 1 \times 10^{-15}$ L, and IC_{50} of a cytotoxic drug (e.g., deoxorubicin) as 1 μM (keeping in mind that this IC value, typically determined in an *in vitro* experiment, represent the concentration of the drug that needs to be in the external environment of the cell to kill 50 % of the cells). Under "optimal" (but unrealistic) static conditions (i.e., all drug being released from the antibody-drug conjugate and all remaining close to the cell surface), a minimum of 600 drug molecules would need to be brought to each tumor cell. Considering the intercellular volume of the tumor and aiming to kill close to 100 % of all tumor cells, we can make a conservative estimate that some 6000 drug molecules per cell need to be delivered to tumors. As stated by Panowksi et al. (2014), the percent of an injected antibody that localizes to a solid tumor is very small (0.003–0.08 % injected dose per gram of tumor). For this reason, and given the above "ball-park" calculation, it is necessary to use cytotoxic compounds of sub-nanomolar potency, such as auristatin (free drug IC_{50}: 10^{-11}–10^{-9} M) or maytansinoid (free drug IC_{50}: 10^{-11}–10^{-9} M). The potency of antibody-drug conjugates may increase with increasing drug loading, however, the hydrophobicity of the construct also increases which leads to an increased plasma clearance (Lyon et al. 2015). Using hydrophilic spacers between the drug and the antibody (Lyon et al. 2015) is unlikely to improve the overall behavior of the antibody-drug complex.

2.8 A Word of Caution

NEWS RELEASE

For Immediate Release: June 21, 2010

FDA: Pfizer Voluntarily Withdraws Cancer Treatment Mylotarg from U.S. Market

"Pfizer Inc. today announced the voluntary withdrawal from the U.S. market of the drug Mylotarg (gemtuzumab ozogamicin) for patients with acute myeloid leukemia (AML), a bone marrow cancer."…"Mylotarg was approved in May 2000 under the FDA's accelerated approval program."… "Under accelerated approval, the company is required to conduct additional clinical trials after approval to confirm the drug's benefit."…

"A confirmatory, post approval clinical trial was begun by Wyeth (now Pfizer) in 2004. The trial was designed to determine whether adding Mylotarg to standard chemotherapy demonstrated an improvement in clinical benefit (survival time) to AML patients. The trial was stopped early when no improvement in clinical benefit was observed, and after a greater number of deaths occurred in the group of patients who received Mylotarg compared with those receiving chemotherapy alone."…"Mylotarg was granted an accelerated approval to allow patient access to what was believed to be a promising new treatment for a devastating form of cancer," said Richard Pazdur, M.D., director, Office of Oncology Drug Products, part of FDA's Center for Drug Evaluation and Research." (Jefferson 2010).

"My Cancer Genome" website (Abramson 2016) offers a partial list of 70 "targeted cancer therapies" defined as "drugs designed to interfere with specific molecules necessary for tumor growth and progression". On examining the efficacy and side effects of these medications one is likely to conclude that "targeting" here is more an intent rather than an outcome. It would be prudent that approval of any new drug or therapy that claims cell-specific drug delivery requires data showing that a "substantial" fraction (more than 50 %?) of the administered dose actually reaches its cancer target.

2.9 Conclusions

It is nice to see that some (Scott et al. 2012) consider that "…cancer therapy has achieved considerable success in recent years…", and brim with optimism that "Antibody–drug conjugates are powerful new treatment options…"; however, even these authors still recognize that "…promise of antibody therapeutics in cancer is dependent on having a better understanding…, on applying innovative approaches to target and antibody selection…" This review shows that there are many obstacles to be overcome on the way of developing effective targeting antibody-drug conjugates:

1. Cancer-cell specific antigens for selected indication
2. Corresponding antigen-specific antibody with the required binding strength to the target
3. Drug that has adequate potency and matching pharmacokinetic behavior
4. Drug-antibody linkage enabling timely release of the drug in its required form
5. Antibody-drug complex with minimal non-specific interaction *in vivo*
6. Antibody-drug complex stability away from the target site (i.e., no or minimal release of the drug away from the target to minimize toxicity and side effects)
7. Antibody-drug complex payload sufficient to generate the required pharmacological concentration of the drug at the intra-cellular target site
8. Adequate access to the tumor cells, achieved by
   - Direct intra-tumoral injection or
   - Appropriate timing of systemic administration to coincide with vascular normalization.

But above all, new paradigms of cell-specific targeting need to be discovered and implemented.

References

Abramson R (2016) Overview of targeted therapies for cancer. *My cancer genome* https://www.mycancergenome.org/content/molecular-medicine/overview-of-targeted-therapies-for-cancer/

Barrett CL, Schwab RB, Jung H, Crain B, Goff DJ, Jamieson CH, Thistlethwaite PA, Harismendy O, Carson DA, Frazer KA (2013) Transcriptome sequencing of tumor subpopulations reveals a spectrum of therapeutic options for squamous cell lung cancer. PLoS One 8(3):e58714. doi:10.1371/journal.pone.0058714, Epub 20 Mar 2013

Barrett CL, DeBoever C, Jepsen K, Saenz CC, Carson DA, Frazer KA (2015) Systematic transcriptome analysis reveals tumor-specific isoforms for ovarian cancer diagnosis and therapy. Proc Natl Acad Sci U S A 112(23):E3050–E3057. doi:10.1073/pnas.1508057112, Epub 26 May 2015

Beck A, Wurch T, Bailly C, Corvaia N (2010) Strategies and challenges for the next generation of therapeutic antibodies. Nat Rev Immunol 10:345–352

Boddy A, Aaron L, Petrak K (1989) Efficiency of drug targeting: steady-state considerations using a three-compartment model. Pharm Res 6(5):367–372

Cañadas I, Rojo F, Arumí-Uría M, Rovira A, Albanell J, Arriola E (2010) C-MET as a new therapeutic target for the development of novel anticancer drugs. Clin Transl Oncol 12:253–260

Deckert PM (2009) Current constructs and targets in clinical development for antibody-based cancer therapy. Curr Drug Targets 10:158–175

FDA Drug Approval Package (2011) ADCETRIS (brentuximab vedotin). Seattle Genetics, Inc., Application No.: 125399 http://www.accessdata.fda.gov/drugsatfda_docs/nda/2011/125399_adcetris_toc.cfm

FDA Drug Approval Package (2013) Kadcyla (ado-trastuzumab emtansine) Injection, Genentech, Inc., Application No.: 125427 http://www.accessdata.fda.gov/drugsatfda_docs/nda/2013/125427Orig1s000TOC.cfm

Ehrlich P (1954) The partial function of cells. Int Arch Allergy Appl Immunol 5:67–86

Fujimori K, Covell DG, Fletcher JE, Weinstein JN (1990) A modeling analysis of monoclonal antibody percolation through tumors: a binding-site barrier. J Nucl Med 31(7):1191–1198

Garcia-Carbonero R, Supko JG (2002) Current perspectives on the clinical experience, pharmacology, and continued development of the camptothecins. Clin Cancer Res 8:641–661

Giusti K (2016) http://www.forbes.com/sites/matthewherper/2016/02/29/5-reasons -to-be-optimistic-about-the-precision-medicine-initiative/#208621091eda

Huang D, Lan H, Liu F, Wang S, Chen X, Jin K, Mou X (2015) Anti-angiogenesis or pro-angiogenesis for cancer treatment: focus on drug distribution. Int J Clin Exp Med 15;8(6):8369–8376. eCollection

Hudis CA (2007) Trastuzumab-mechanism of action and use in clinical practice. N Engl J Med 357:39–51

Jefferson E (2010) FDA: Pfizer voluntarily withdraws cancer treatment mylotarg from U.S. Market, http://www.fda.gov/NewsEvents/Newsroom/PressAnnouncements/ucm216448.htm

Lyon RP, Bovee TD, Doronina SO, Burke PJ, Hunter JH, Neff-LaFord HD, Jonas M, Anderson ME, Setter JR, Senter PD (2015) Reducing hydrophobicity of homogeneous antibody-drug conjugates improves pharmacokinetics and therapeutic index. Nat Biotechnol 33:733–735. doi:10.1038/nbt

Mu L, Elbayoumi TA, Karkala A, Erdogan S, Dabholkar RD, Levchenko TS, Mongayt DA, Torchilin VP (2005). Microencapsulation of poorly soluble anti-cancer drugs into micelles made of polyethylene glycol-phosphatidyl ethanolamine (Peg-Pe) conjugates. In: 15th international symposium on microencapsulation, Parma (Italy), 18–21 September 2005

NCI Dictionary of Cancer Terms (2015) http://www.cancer.gov/publications/dictionaries/cancer-terms?cdrid=44628

Panowksi S, Bhakta S, Raab H, Polakis P, Junutula JR (2014) Site-specific antibody drug conjugates for cancer therapy. MAbs 6(1):34–45. Published online 1 Nov 2013. doi: 10.4161/mabs.27022

Petrak K (2005) Essential properties of drug-targeting delivery systems. Drug Discov Today 10:1667–1673

Petrak K (2006) Nanotechnology and site-targeted drug delivery. J Biomater Sci in Special Issue on Nanobiomaterials (Vasif Hasirci, Karel Petrak (eds)), 17(11), 1209–1221

Petrak K (2012) Targeted drug delivery—Quo Vadis? Drug Dev Res 73(2) 59–65. Article first published online: 29 Dec 2011. DOI: 10.1002/ddr.20492

Petrak K (2013) Targeting drug-delivery systems: promises, promises, and more promises. Let's change the paradigm. In: Valerio V (ed) Recent advances in drug delivery research., pp 167–180

Petrak K (2015a) Precision medicine and site-specific drug delivery. Adv Cancer Res 3(3), No 3:26, 1–4

Petrak K (2015b) The "Magic Bullets" for drug delivery? Int J Drug Dev Res 7:26–28

Pillay V, Gan HK, Scott AM (2011) Antibodies in oncology. Nat Biotechnol 28:518–529

Polson AG, Calemine-Fenaux J, Chan P, Chang W, Christensen E, Clark S, de Sauvage FJ, Eaton D, Elkins K, Elliott JM (2009) Antibody-drug conjugates for the treatment of non-Hodgkin's lymphoma: target and linker-drug selection. Cancer Res 69:2358–2364. doi:10.1158/0008-5472.CAN-08-2250

Rowland AJ, Pietersz GA, McKenzie IFC (1993) Preclinical investigation of the antitumor effects of anti-CD19-idarubicin immunoconjugates. Cancer Immunol Immunother 37(3):195–202

Scartozzi M, Bianconi M, Maccaroni E, Giampieri R, Berardi R, Cascinu S (2010) Dalotuzumab, a recombinant humanized mAb targeted against IGFR1 for the treatment of cancer. Curr Opin Mol Ther 12:361–371

Schliemann C, Neri D (2010) Antibody-based vascular tumor targeting. Recent Results Cancer Res 180:201–216

Schoeberl B, Pace EA, Fitzgerald JB, Harms BD, Xu L, Nie L, Linggi B, Kalra A, Paragas V, Bukhalid R, Grantcharova V, Kohli N, West KA, Leszczyniecka M, Feldhaus MJ, Kudla AJ, Nielsen UB (2009) Therapeutically targeting ErbB3: a key node in ligand-induced activation of the ErbB receptor–PI3K axis. Sci Signal 2:ra31

Scott AM, Wolchok JD, Old LJ (2012) Antibody therapy of cancer. Nat Rev Cancer 12:278–287

Siemann DW (2011) The unique characteristics of tumor vasculature and preclinical evidence for its selective disruption by tumor-vascular disrupting agents. Cancer Treat Rev 37(1):63–74

Singla AK, Garg A, Aggarwal D (2002) Paclitaxel and its formulation. Int J Pharm 235:179–192

Van den Eynde BJ, Scott AM (1998) In: Roitt DPJ, Roitt IM (eds) Encyclopedia of immunology. Academic, London, pp 2424–2431

Van der Jeught K, Bialkowski L, Daszkiewicz L, Broos K, Goyvaerts C, Renmans D, Van Lint S, Heirman C, Thielemans K, Breckpot K (2015) Targeting the tumor microenvironment to enhance antitumor immune responses. Oncotarget, 6(3):1359–1381, 30 Jan

Vesely MD, Schreiber RD (2013) Cancer immunoediting: antigens, mechanisms, and implications to cancer immunotherapy. Ann N Y Acad Sci 1284:1–5. doi:10.1111/nyas.12105

Zarour HM, DeLeo A, Olivera J, Finn OJ, Storkus WJ (2003) Categories of tumor antigens. In: Holland-Frei cancer medicine, 6th edn. BC Decker Inc, Hamilton

Part II
EPR Effect and ECM Modification

Chapter 3
Extracellular Matrix Degrading Enzymes for Nanocarrier-Based Anticancer Therapy

Pablo Scodeller

Abstract Perfusion is seriously hampered in solid tumors that overexpress extracellular matrix components like hyaluronic acid, collagen and chondroitin sulfate. Tumor penetration of circulating material is most affected for nanocarriers because of their size and lower diffusion. Due to this, the aid of enzymes that degrade the matrix becomes essential when applying nanocarrier based anticancer therapies like oncolytic viruses or nano-formulated drugs like Abraxane or Doxil.

In the case of hyaluronic acid and chondroitin sulfate, these biomacromolecules bind water very tightly and as a consequence of this, tumors that abound in them have an elevated resistance to a contacting fluid. In the case of collagen, the dense molecular network represents a physical barrier for any nanocarrier and also compresses the tumor vessels obstructing perfusion. Drug penetration and therapeutic index increase when anticancer therapies are combined with enzymes that specifically degrade the over-expressed matrix component.

This chapter will cover the use of matrix degrading enzymes for enhancing the delivery of nanocarriers, including oncolytic viruses and synthetic nanoparticles. Ongoing clinical trials will be discussed.

Keywords Extracellular matrix • Chemoadjuvant • Hyaluronidase • Collagenase • Chondroitinase • Oncolytic virus • Nanocarrier • Polymersome • Abraxane • Doxil

Abbreviations

HA	hyaluronic acid
Hyal	hyaluronidase
IFP	interstitial fluid pressure
GFP	green fluorescent protein
ECM	extracellular matrix

P. Scodeller (✉)
Laboratory of Cancer Biology, Institute of Biomedicine, Center of excellence for translational medicine, University of Tartu, Ravila 14b, Tartu, Estonia
e-mail: pablo.david.scodeller@ut.ee; pscodeller@sbpdiscovery.org

A. Prokop, V. Weissig (eds.), *Intracellular Delivery III*, Fundamental Biomedical Technologies 8, DOI 10.1007/978-3-319-43525-1_3

b-HABP	biotynilated hyaluronic acid binding protein
CSPG	chondroitin sulfate proteoglycan
PDA, PDAC	Pancreatic ductal adenocarcinoma
FDA	Food and drug administration
OV	Oncolytic virus
Chase	Chondroitinase

3.1 Matrix in Tumors

Matrix components like hyaluronic acid (HA), collagen, and chondroitin sulfate are overexpressed in many solid cancers, altering the perfusion of tumors with the anti-cancer drug containing solutions and correlating with a poor prognosis.

The overexpression of HA, chondroitin sulfate, or collagen results in a gel-like extracellular matrix which resists the penetration of external fluid and drugs or macromolecules and specially nanocarriers dissolved therein (see Fig. 3.1), elevating the interstitial fluid pressure (IFP) (Chauhan et al. 2013),(Fadnes et al. 1977),(Heldin et al. 2004),(PP et al. 2012) or the solid stress of the tumor, always decreasing the perfused vessel fraction (defined as the number of perfused vessels over number of CD31 positive vessels) (Chauhan et al. 2013).

HA is overexpressed in cancers of the pancreas, lung, ovary, melanoma, osteosarcoma, mastocitoma, fibrosarcoma, and neuroendocrine hepatic carcinomas; being the most excessive case of HA overproduction in pancreatic ductal adenocarcinoma (PDAC) (PP et al. 2012). Collagen is aberrantly expressed in pancreatic ductal carcinoma (Armstrong et al. 2004), breast tumors (Plodinec et al. 2012),(Chauhan et al. 2013), sarcomas and hepatomas (Nugent and Jain 1984). Chondroitin sulfate and chondroitin sulfate proteoglycans (CSPGs) are overexpressed in glioblastomas (Dmitrieva et al. 2011) and breast tumors (Ricciardelli et al. 2002; Suwiwat et al. 2004).

Among all the stroma barriers, the most tackled one has been HA, with the use of its degrading enzyme hyaluronidase (Hyal), which is close to reaching the clinic for the oncology market (see http://meetinglibrary.asco.org/content/160151-173.) thanks to its ability to potentiate Abraxane® and also Gemcitabine therapy. The reasons for Hyal being the most studied ECM enzyme in this field are that, unlike collagen, the turnover rate of HA is very rapid, which means that HA degradation that may occur in healthy tissue and joints is more tolerable. A *pegylated* recombinant version of a human Hyal (Hylenex from Halozyme) has repeatedly been used in clinical trials without raising immunogenicity concerns. Also, only one enzyme is needed to degrade HA, whereas in the case of collagen, different enzymes are needed to degrade the different types of collagen.

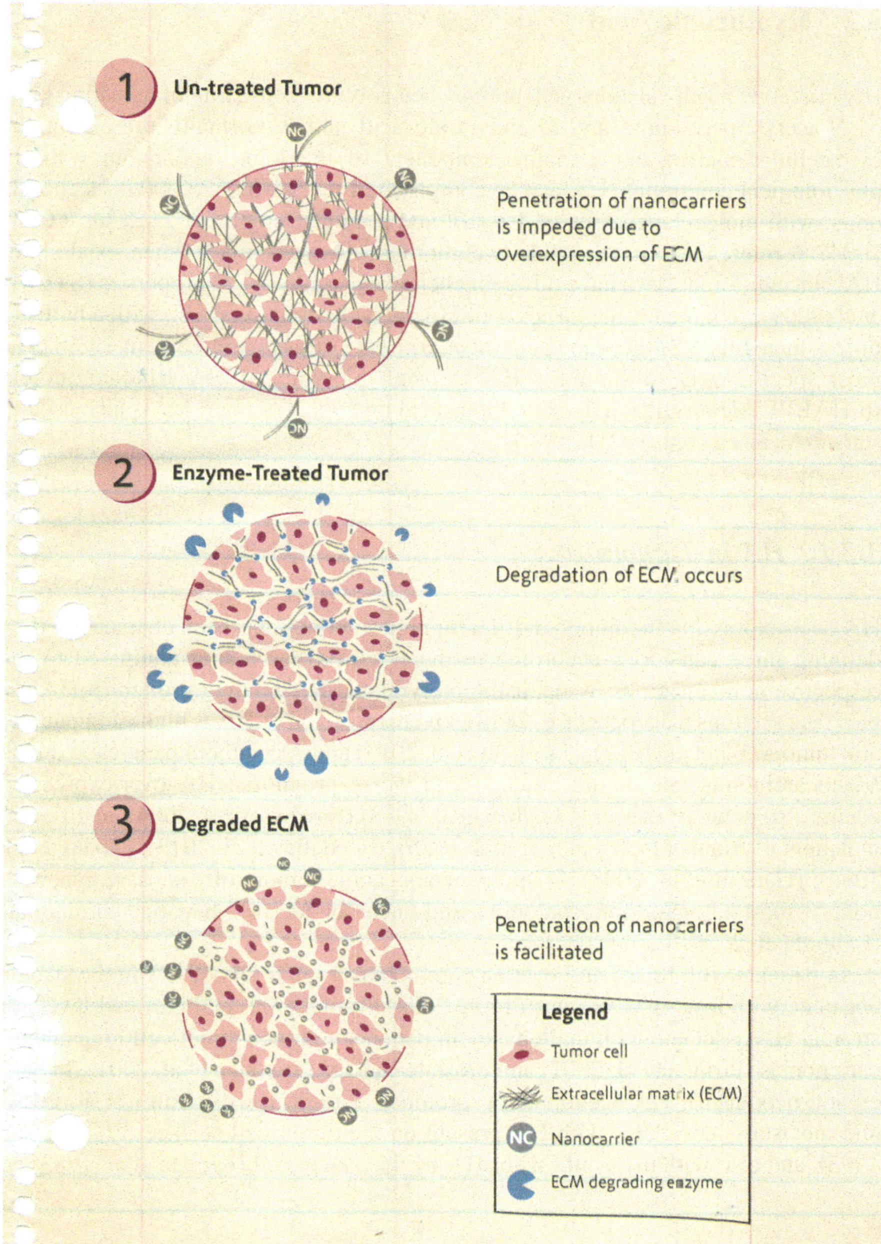

Fig. 3.1 Schematics representing the use of ECM degrading enzymes to facilitate the delivery of therapeutic nanocarriers to tumor cells

3.2 Hyaluronic Acid

HA is a large, negatively charged, unbranched polymer consisting of repeating units of N-acetylglucosamine and D-glucuronic acid and is normally present in the extracellular matrix and a major component of the joints and ovum wall. In physiological conditions, HA can be composed of up to 2000–25,000 disaccharide units, with molecular masses of 10^6–10^7 and polymer lengths of 2–25 μm (Toole 2004). A reference to its molecular weight must always be made when speaking of HA, since different sizes have different chemical and biological properties (STERN et al. 2006). In humans, the turnover of HA in the body is very fast, with a half-life of less than 2 days in the skin (Laurent et al. 1991), (Fraser and Laurent 1989). The most specific way to stain for HA is using biotynilated HA Binding protein (b-HABP). Alcian Blue, a dye that binds acid mucopolysaccharides, is also used to stain HA (Slsca et al. 1971).

3.2.1 HA in Tumors

HA over-expression in tumors of different mouse models has been characterized by defining three categories according to the HA over-expression compared to the respective non-cancerous tissue, using biotynilated HA binding protein (b-HABP) on tissue sections (Jacobetz et al. 2013). Overexpression of HA is highest in pancreatic tumors (PP et al. 2012), (Jacobetz et al. 2013) and specially in pancreatic ductal adenocarcinoma, one of the cancers with worse prognosis. HA over-expression occurs also in breast cancer (Chauhan et al. 2013),(Suwiwat et al. 2004), in different melanoma variants (Turley and Tretiak 1985), (Scodeller et al. 2013), (Zhang et al. 1995), (Ganesh et al. 2008), in spontaneous canine tumors of osteosarcoma, adenoma, mastocitoma, fibrosarcoma, and neuroendocrine hepatic carcinomas (Laborda et al. 2014).

In patients with breast and ovarian carcinomas, high levels of HA in the stroma correlate with low survival rates (Toole 2004),(Toole 2009) and elevated levels of stromal HA associate with malignancy in patients with non-small-cell lung adenocarcinomas (Pirinen et al. 2001) and prostate cancer (Lipponen et al. 2001). In pancreatic neoplasms, HA accumulation promotes tumor growth (Kultti et al. 2014) and metastasis correlates with HA present on the tumor cell surface (Zhang et al. 1995), and also with the synthesis of HA by the cancer cell (Kimata et al. 1983).

3.2.2 Physico-Chemical Properties of Hyaluronic Acid

High molecular weight HA solutions in water show, by fast field cycling NMR relaxometry, three different hydration layers surrounding the molecule, irrespective of HA concentration. The first hydration layer consists of strongly restrained water molecules with the slowest motion. At low concentrations of HA in water, the amido group is suggested to participate mainly in intra-molecular water bridges, and at high concentrations of HA, inter-molecular water bridges become more important, explaining the rheological properties of highly concentrated aqueous solutions of HA (Průšová et al. 2010).

Ellipsometric thicknesses of HA modified substrates incubated with different proteins are smaller than for the non HA modified substrate, demonstrating the anti-fouling properties of HA (Bauer et al. 2013).

It is interesting to note that some *Streptococci* surround themselves with a capsule of HA which is considered a virulence factor for these bacteria (Bisno et al. 2003), (Wessels et al. 1991).

Metastasis correlates with HA present on the tumor cell surface (Zhang et al. 1995), and also with the synthesis of HA by the cancer cell (Kimata et al. 1983).

Altogether, this evidence points to the fact that, besides making a stiff gel-like matrix that compresses tumor vessels and obstructs perfusion, HA stealths cancer cells from antibodies or immune cells.

3.2.3 Hyaluronidase

HA is cleaved by enzymes known as hyaluronidases. In humans, there are six Hyal genes, which encode enzymes that have different properties and cellular locations (Toole 2004). In mammals, Hyal is present in the head of the sperm, to degrade the ovum wall (Swyer 1947).

Subcutaneous use of Hyal, derived from crude extracts of ovine or bovine testicular tissue, was approved by the Food and drug administration (FDA) in 1948 for hypodermoclysis (Dunn et al. 2010), as an adjunct in subcutaneous urography and for treatment of vitreous hemorrhage (Hyaluronidase Monograph (Amphadase, Hydase, Vitrase, Hylenex) Final, in: National PBM Drug Monograph, VHA PharmacyBenefits Management Service and the Medical Advisory Panel, Washington, DC, 2008). This bovine Hyal is homologous to PH20 Hyal, the one found in human sperm.

Hyal is a glycosylated protein and removal of the N-linked glycans with N-glycosidase results in complete inactivation of enzymatic activity (Li et al. 2002). Isoelectric focusing of this enzyme reveals six isoforms with isoelectric points ranging between 5.1 and 6 (Li et al. 2002). Each oligosaccharide derived by enzymatic treatment of HA contains a double bond at the non-reducing end uronate and this allows for its detection using absorbance at 230 nm (Sakai et al. 2007). Enzymatic digestion of HA has been characterized by Nuclear Magnetic Resonance (NMR), Raman spectroscopy, and Infrared and UV-vis spectroscopy (Alkrad et al. 2003). In vitro, Hyal (purified from stonefish venom) degrades up to the tetra- hexa-, octa- and deca-saccharides as major end products, but not disaccharides, as characterized by HPLC and NMR (Sugahara et al. 1992).

The standard enzymatic assay for Hyal is a turbidimetric assay. Under the conditions of the assay, high molecular weight HA forms cloudy precipitates when incubated with bovine serum albumin, while HA fragments produced by the enzymatic cleavage do not.

Hyal may be specifically inhibited with apigenin (Liu et al. 1996).

However, bovine derived Hyal is a crude extract in which less than 1 % in weight is Hyal. To tackle this problem and the issue of immunogenicity in humans when using a bovine derived extract, a recombinant human Hyal (rHuPH20) named HYLENEX® was introduced in the market by Halozyme Inc. Hylenex is produced by genetically engineered Chinese Hamster Ovary (CHO) cells containing a gene encoding for a soluble fragment of human Hyal (PH20) and its use was approved by FDA in 2005 for certain uses outside oncology (http://www.drugs.com/history/hylenex.html).

3.2.4 Hyaluronidase as an Adjuvant

Administration of hyal to cancerous tissues lowers the IFP within the tumors (Heldin et al. 2004), (Whatcott et al. 2011) in a non-linear and concentration-dependent manner (Brekken and de Lange Davies 1998). Hyal treatment leads to vascular decompression in pancreatic tumors, as has been demonstrated by measuring mean vessel area in tissue sections stained with CD31 before and after hyal treatment (Manuel et al. 2015).

Hyal is being studied for cancer treatments, as an adjuvant for cytotoxic chemotherapy (Baumgartner et al. 1998). It acts by degrading HA, lowering the interstitial fluid pressure (IFP) and thus enhancing drug penetration (Stern 2008). Pretreatment with bovine testis derived Hyal enhances the penetration and activity of the oncolytic drug Adriamycin in breast cancer models, in vitro and in vivo (Beckenlehner et al. 1992), and produces selective Melphalan enrichment in malignant melanomas implanted in nude mice (Muckenschnabel et al. 1996). Hyal also favors the cytotoxicity of anticancer drugs in vitro (St Croix et al. 1998).

Hyal acts also as an adjuvant for macromolecules: it has been used as adjuvant in the delivery of 125I-labeled monoclonal antibody against an osteosarcoma-associated antigen (Brekken et al. 2000) and to help the delivery of monoclonal antibody therapy using Trastuzumab or Cetuximab (Singha et al. 2014).

3.2.5 Hyaluronidase for Nanocarrier Based Anti-Cancer Therapy

Hyal has been successfully used to favor the dispersion of oncolytic viruses. Pretreatment with intratumoral injections of the enzyme largely enhanced the penetration of green fluorescent protein (GFP)-expressing oncolytic virus, in prostate and melanoma models in mice (Ganesh et al. 2008), which was accompanied by a significant therapeutic index increase in both models.

A modified version of the oncolytic human adenovirus ICOVIR5, engineered to express PH20 Hyal (denoted as ICOVIR 17) enhanced the spread of the virus within the tumor respect to the non Hyal expressing virus, and concomitantly enhanced suppression of tumor growth (Guedan et al. 2010). In a more recent study by the same group (Laborda et al. 2014), a version of ICOVIR17 presenting the RGD motif on its surface was constructed. The tripeptide RGD motif binds to αv integrins, which are overexpressed in tumor endothelia (Ruoslahti 2002). This new system proved very effective in treating different spontaneous tumors in dogs (Laborda et al. 2014). Currently, the same group is using Hyal expression to improve the distribution of VCN-01 (Martínez-Vélez et al. 2015), a replication competent adenovirus specifically engineered to replicate in tumors with a defective retinoblastoma protein pathway.

Hyal also augments tumor penetration of synthetic nanocarriers like liposomal doxorubicin, improving the efficacy of the treatment (Eikenes et al. 2005).

The pretreatment with Hyal has even shown to favor the spread of microsized objects as is the case with *Salmonella* bacteria used for cancer therapy. Bacterium *Salmonella typhimurium* (ST), which is naturally tropic for the hypoxic tumor environment, and a strain carrying a shRNA expression plasmid against the immunosuppressive molecule indoleamine 2,3-dioxygenase 1 (shIDO) have been described (Blache et al. 2012). When systemically delivered into mice, shIDO silences host IDO expression and leads to high intratumoral cell death that is associated with significant tumor infiltration by polymorphonuclear neutrophils. However, when combined with HA depletion, this treatment led to complete regression of the pancreatic tumors (Manuel et al. 2015). Mice treated with the combination of pegylated PH20 Hyal, PEGPH20, and shIDO-ST showed a significant decrease in tumor burden with 100% of mice responding and approximately 30–60% of mice with no discernible evidence of tumors 2 weeks after treatment and extending to 8 weeks following tumor implantation.

3.2.6 *Immobilized Hyaluronidase for Nanocarrier Based Anti-Cancer Therapy*

In 2013 the first nano-formulation of Hyal was described (Scodeller et al. 2013). Melanomas implanted in nude mice were pretreated with equal enzymatic activities of free or immobilized Hyal prior to chemotherapy application (local injections of Carboplatin), and tumor volume after the treatment study was significantly smaller than when using the free enzyme as chemoadjuvant. The Hyal functionalized nanocarrier system alone, without any chemotherapy, had no effect on tumor growth. The tumor tissue degrading properties of these particles were shown by analyzing tumors with Field Emission-Scanning Electron Microscopy (FE-SEM) after just one application of the Hyal-modified nanoparticles (see Fig. 3.2).

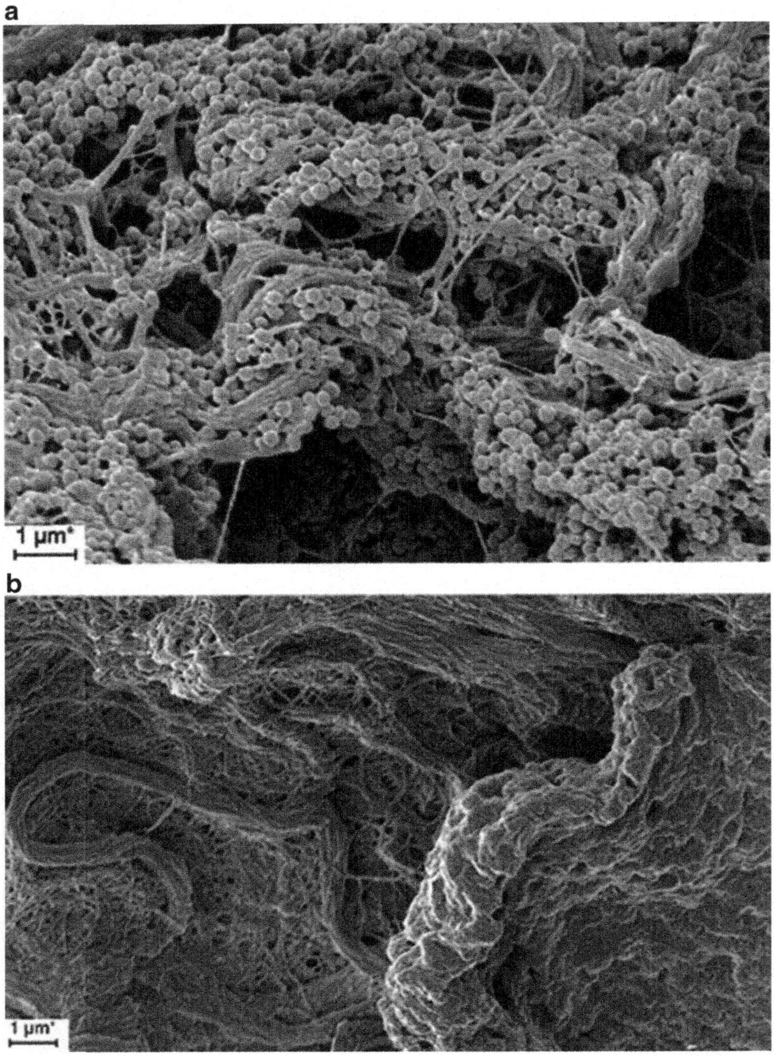

Fig. 3.2 Surface of A375 melanoma tumors in mice as seen with Field Electron Scanning Electron Microscopy (FE- SEM). (**a**) Two hours after one injection of Hyal-modified particles. (**b**) Two hours after one injection of PBS. Reprinted with permission of reference (Scodeller et al. 2013)

To functionalize these particles with Hyal the authors used the layer-by-layer (LBL) self-assembly technique (Decher 1997) which consists of the alternate adsorption of polycations and polyanions from water solutions, where each layer deposition is accompanied by charge reversal as a result of charge overcompensation (Caruso et al. 1999). The LBL technique is appealing when functionalizing planar (Scodeller et al. 2010) and colloidal surfaces (Scodeller et al. 2008) with enzymes because enzymatic activity is maintained and multiple layers may be deposited to yield a larger catalytic response. The particles used in this study were 250 nm sized silica spheres, of low cost and easy to synthesize; similar particles proved to be well tolerated in mice (Botella et al. 2011). In this nano-formulated Hyal, multivalent interactions with the target tissue is the proposed mechanism to explain the enhanced adjuvant effect respect to free Hyal (ligand-receptor affinity scales exponentially with number of valences).

The finding of an enhanced chemoadjuvant effect of Hyal upon its immobilization on nanoparticles has recently been confirmed by another research group (Zhou et al. 2016). In that work, the recombinant human hyaluronidase, rHuPH20, was covalently coupled to the surface of poly(ethylene glycol) (PEG) derivatized, doxorubicin loaded nanoparticles ("PEG-PH20-NP"). The presence of PEG enhanced the half-life of rHuPH20 from 1.4 min to almost 9 h when it was on the *pegylated* nanoparticles, and this allowed the authors to systemically administer the particles. The penetration of PEG-PH20-NP in ECM mimicking gels, was significantly higher than PEG-NP plus equal units of the free enzyme. The researchers also observed a significantly larger decrease in the amount of HA in cellular cultures of 4T1 (a mouse metastatic breast cancer cell line) after treating with PEG-PH20-NP. In animal experiments using this breast cancer model, the PEG-PH20-NP treated group showed the highest tumor penetration, tumor reduction and survival.

3.2.7 *Safety and Clinical Trials with Human Recombinant Hyaluronidase*

Recombinant human Hyal from Halozyme Inc. has shown to not stimulate an acute inflammatory response (Rosengren et al. 2015), (Huang et al. 2014) and has proven safe also in the administration to healthy volunteers of Ondansetron®(Dychter et al. 2014), a medication used to prevent nausea and vomiting caused by cancer chemotherapy, radiation therapy, or surgery. Phase II clinical trials are under way to treat pancreatic ductal carcinoma patients with a combination of *pegylated* recombinant human Hyal, PEGPH20, and Gemcitabine or Abraxane® (an approved nanoformulation of Paclitaxel). The latest interim results from this clinical trial (Hingorani, SR et al. Interim results of a randomized phase II study of PEGPH20 added to nab-paclitaxel/gemcitabine in patients with stage IV previously untreated pancreatic cancer. J Clin Oncol 34, 2016 (suppl 4S; abstr 439)) show that in patients with high HA content in pancreas, PEGPH20 pretreatment increases progression free survival (mPFS) by 114 % respect to the chemotherapy treatment alone, from 4.3 to 9.2 months. Patients with high HA represented 41 % of total patients and there was no response to Hyal pretreatment in patients with low HA. These positive results have encouraged the initiation of a global phase III trial beginning in Q1 2016 (NCT01839487).

Toxicity and biodistribution preclinical studies with oncolytic adenovirus VCN-01 in immunodeficient mice and hamster models of melanoma and pancreatic adenocarcinoma have recently been published (Rodriguez-Garcia et al. 2015). The promising outcome of these studies has motivated the group to begin conducting two phase I clinical studies: one in pancreatic cancer by endoscopic ultrasound-guided intratumoral injection and another one targeting different tumor types by intravenous administration (NCT02045602 and NCT02045589).

3.3 Collagen and Collagenase

Collagen is the most abundant protein in animals. It is found in connective tissue such as tendons, cartilage, the organic matrix of bone, and the cornea of the eye. The collagen helix is a unique secondary structure: it is left-handed and has three amino acid residues per turn. Collagen is coiled: three separate polypeptides, called alpha chains are super-twisted about each other (Nelson and Cox 2008). The amino acid sequence in collagen is generally a repeating tripeptide unit, Gly-X-Y where X is often Pro, and Y is often 4-Hyp (4-hydroxyproline). Only Gly residues can be accommodated at the very tight junctions between the individual alpha chains. The Pro and 4-Hyp residues allow for the sharp twisting of the collagen helix. The tight wrapping of the alpha chains in the collagen triple helix provides tensile strength greater than that of a steel wire of the same cross section. Collagen fibrils are supramolecular assemblies consisting of triple-helical collagen molecules (sometimes referred to as tropocollagen molecules) of approximately 300 nm of length and 1.5 nm in diameter, associated in a variety of ways to provide different degrees of tensile strength. Fibrillar collagens are type I, II, III, V, XI. In humans, fibrillar collagen is cleaved by matrix metalloproteinases MMP-1, MMP-8 and MMP-13, and type IV collagen is cleaved by MMP-2 and MMP-9.

3.3.1 Collagen in Cancer

In human breast cancer collagen type IV is over-expressed as evidenced by immunohistochemical staining of a large number of patient samples (Ioachim et al. 2002). In that same study higher collagen levels correlated with lower survival.

Type I collagen, a fibrillar collagen, is over-expressed in pancreatic cancer (Armstrong et al. 2004),(Shintani et al. 2006), up to threefold compared with normal pancreas. It promotes malignant phenotypes, cell migration and maximal proliferation in pancreatic cancer cell lines (Grzesiak and Bouvet 2006). In a mouse model of pancreatic cancer, increased metastasis was associated with an up-regulation of N-Cadherin triggered by collagen I. Collagen I is also over-expressed in breast cancer, being higher at the rim of the tumors, and the high levels of tumor collagen correlate with mechanical stiffening of the tissue, as measured with nano-indentation (Plodinec et al. 2012).

Collagen type V is also involved in promoting the malignant phenotype of pancreatic ductal adenocarcinoma (Berchtold et al. 2015). Collagen VI is overexpressed in ovarian cancer patients and increases their resistance to oncolytic drugs like Cisplatin (Sherman-Baust et al. 2003).

3.3.2 Collagenase for Nanocarrier Based Anti-Cancer Therapy

Penetration of macromolecules and nanoparticles in tissues abundant in collagen may be increased by combining them with collagenase. Intraperitoneal pretreatment with collagenase increased the tumor penetration of an antibody (IgG, 150 kDa, approximately 6 nm in diameter) in ovarian cancers in mice, detected by radioisotope labeling of the antibody (Choi 2006). Intraperitoneal treatment during 7 days with Losartan in mice reduced collagen content of breast tumors, and concomitantly increased penetration of Evans Blue in tumors and significantly improved the therapeutic efficacy of Doxil (Cun et al. 2015), (Diop-Frimpong et al. 2011), an FDA approved nano-formulation of doxorubicine.

Collagenase can specially facilitate the penetration of nanoparticles, in tissues presenting high collagen content. In a three-dimensional, multicellular spheroid tumor model, collagenase from *Clostridium histolyticum* (which cleaves human collagen types I, II, and III) resulted in a significant decrease in collagen content. This collagenase treatment increased the penetration of 20 nm and 40 nm fluorescent nanoparticles by as much as 7 and 12-fold, respectively, and threefold for 100 nm particles. Incubation of the same spheroids with 100 nm collagenase-coated nanoparticles resulted in a significant increase (almost fourfold) in spheroid core fluorescence compared with albumin-coated nanoparticles (Goodman et al. 2007).

Collagenase has been used to enhance oncolytic nanocarriers-based cancer therapy. It significantly elevated the antitumor activity of an oncolytic adenovirus in U87 and U251 glioblastoma subcutaneous xenografts in mice when compared with the viral therapy alone (Kuriyama et al. 2001). A one-time intratumoral injection of bacterial collagenase was given 24 h prior to starting the treatment with the virus. A potential side effect of using proteases and virus inoculation would be invasion of tumor cells into adjacent areas and the initiation of metastatic tumor spread, however, no evidence of an increased potential for metastatic tumor spread was detected in that study.

In another example, collagenase of bacterial origin was employed to deplete fibrillar collagen in melanoma, which proved to be a major impediment for the transport of a herpes simplex virus (HSV) (McKee 2006) (120 nm in diameter) and led to improved oncolytic viral therapy. Melanomas were grown in the dorsal skin windows in SCID mice and imaged with live multiphoton imaging. Collagen was monitored with second harmonic generation imaging and GFP-expressing viral particles were tracked with green fluorescence. Local injections of virus showed that the viral particles distributed primarily within collagen-free areas of the tumor with limited penetration into collagen-rich regions. The area of viral infection was localized to 15 % of the entire tumor mass, corresponding to the site of injection. Even 11 days after the initial injection, viral vectors could not penetrate sufficiently to infect the entire tumor mass. However, when collagenase was co-injected along with the virus, it resulted in a nearly 3-fold increase in the area of viral distribution compared with HSV particles alone. A blue dye–conjugated 2MDa dextran co-injected with HSV was found to penetrate into collagen-rich regions, where viral particles were excluded, showing that large nanocarriers are most affected in their spreading. Most importantly, tumor sizes were significantly smaller when collagenase was used in combination with the HSV virus.

The velocity of 15 nm super-paramagnetic iron oxide nanoparticles travelling through an *in vitro* model of ECM, under the influence of a magnetic field, was dramatically increased when they were derivatized with microbial collagenase (Kuhn et al. 2006). Also, the enzyme stability was maintained for up to 5 days when it was immobilized on the nanoparticle surface.

3.4 Chondroitin Sulfate Proteoglycans (CSPGs)

CSPGs are composed of a core protein onto which chondroitin sulfate is covalently attached.

Chondroitin Sulfate (CS) is an anionic linear polysaccharide which consists of alternating disaccharide units of D-glucuronic acid and N-Acetyl-D-galactosamine (→4GlcAβ1 → 3GalNAcβ1→) and can be modified by sulfate groups in different positions. Sulfate groups occur mainly at C4 and/or C6 of N-acetylgalactosamine and/or C2 of glucuronic acid. Chondroitin sulfate is attached to serine residues of the protein cores via a tetrasaccharide linkage consisting of xylose, two galactose molecules and glucuronic acid. The biosynthesis of CS chains on core proteins results in the formation of certain CSPGs, such as aggrecan (the major CSPG of cartilage), versican (the common CSPG of non-cartilaginous connective tissues), decorin and biglycan (Asimakopoulou et al. 2008).

3.4.1 CSPGs in Cancer

Chondroitin sulfate proteoglycans (CSPGs), such as versican, accumulate in tumor stroma and play a key role in tumor growth and invasion. The high expression of CSPGs in fast growing tissues and cells is correlated with chondroitin sulfate (CS) chains and the sulfation pattern (Asimakopoulou et al. 2008). In patients with node negative primary breast cancer, high levels of versican and tenascin in the peritumoral stromal tissue were found to be predictors for risk and relapse, independent of the tumor size, suggesting that these CSPGs play a primary role in breast cancer spread for patients with node negative disease (Ricciardelli et al. 2002), (Suwiwat et al. 2004).

3.4.2 *Chondroitinase in Nanocarrier Based Anticancer Therapy*

Chondroitinase (Chase) was expressed in an oncolytic, GFP-expressing virus (OV-Chase) and tested in-vivo in two mouse models (Gli36DEGFR and U87DΔEGFR) of glioblastoma (Dmitrieva et al. 2011). Chase greatly enhanced the penetration of the viral particles in in-vitro grown glioma spheroids, in subcutaneous and intracranial tumors in mice, compared with the same virus without the enzyme. The authors also showed that the enzyme only affects the spread of the virus and not its replication or infectivity.

The oncolytic virus armed with Chase drastically improved the efficacy against glioma models in vivo. Gli36DEGFR and U87DΔEGFR cells were intracranially implanted in mice, and 10 days after tumor cells implantation, mice were treated with OV-Chase, OV alone, or PBS, intratumorally for the subcutaneous tumors and stereotactically for the intracranial tumors.

The use of the oncolytic virus alone had no effect either on survival or on subcutaneous tumor growth, compared with PBS control, but when the Chase-expressing virus was used, it largely prolonged the survival and diminished the volume of the subcutaneous implanted tumors, showing that CSPGs are a total impediment to the spread of the virus. Two of the mice treated with OV-Chase lived more than 80 days, and showed no evidence of intracranial tumor by histologic examination.

3.5 Conclusions and Outlook

Extracellular matrix degrading enzymes greatly enhance the dispersion of nanocarriers within many tumors and improve the outcome of treatments. The benefit is for nanocarriers in general, whether they are oncolytic virus nanocarriers or nano-formulated drugs such as Abraxane or Doxil.

An extracellular matrix degrading enzyme, Hyal, is close to receiving approval for a synthetic nanocarrier (Abraxane) based anticancer therapy, and is in earlier clinical studies as part of an oncolytic virus nanocarrier.

The depletion of collagen and CSPGs to aid nano-cancer therapies has shown exciting results in mice models. However the path to bedside applications for these enzymes in oncology is still far. To decrease immunogenicity in patients, recombinant human versions of collagenases like MMP-1 and MMP-8 should be engineered. These human collagenases have shown promise as helpers in nanocarrier based anticancer therapy (Mok et al. 2007).

As opposed to HA, collagen does not have only one enzyme that degrades it, instead, different enzymes degrade different types of collagen. This will require that each case be analyzed to decide which collagenase to treat with.

Moreover, the fact that the half-life of collagen is extremely long (over 100 years for cartilage collagen and 15 for skin collagen according to recent calculations (Verzijl et al. 2000)), might complicate the setting for a clinical translation, due to unwanted off-target degradation. To reduce this off target effect in healthy tissue, the enzyme should be directed to the site of lesion. To this end, enzymes may be constructed fused with short tumor homing peptides (Järvinen and Ruoslahti 2010; Järvinen 2012),(Curnis et al. 2000).

A dramatic impact of Chase has been reported in oncoviral therapy for glioblastoma. This is a recent and encouraging study for a promising field still unexplored: the use of Chase to aid nanocarrier cancer therapy.

Many tumors do not overexpress one but multiple matrix elements. Pancreatic tumors over-express HA and collagen, and breast tumors upregulate collagen and different CSPGs. In the future, this fact should be considered and a multiple targeting of stroma components contemplated. This could be achieved with a cocktail of enzymes fused with peptides that target the tumor stroma, or through the use of tumor targeted porous silicon nanoparticles (Joo et al. 2015) that can load proteins inside their pores (Wu et al. 2015).

As for the chemotherapy to be used, besides oncolytic viruses and the familiar nanoformulations of drugs such as Abraxane and Doxil, in the future, delivery systems that only release their cargo inside the cell will improve the therapeutic index. These include pH-sensitive polymeric particles such as polymersomes, proficient endosomal escapers that are already proving superior to Abraxane (Simon-Gracia et al. 2016).

Acknowledgements The author would like to Patricia Scodeller for the graphic design, and Eduardo Scodeller, Matilde Bonder, and Lorena Simon-Gracia for their careful proofreading of this chapter.

References

Alkrad JA, Mrestani Y, Stroehl D et al (2003) Characterization of enzymatically digested hyaluronic acid using NMR, Raman, IR, and UV-vis spectroscopies. J Pharm Biomed Anal 31:545–550. doi:10.1016/S0731-7085(02)00682-9

Armstrong T, Packham G, Murphy LB et al (2004) Type I collagen promotes the malignant phenotype of pancreatic ductal adenocarcinoma. Clin Cancer Res 10:7427–7437. doi:10.1158/1078-0432.CCR-03-0825

Asimakopoulou AP, Theocharis AD, Tzanakakis GN, Karamanos NK (2008) The biological role of chondroitin sulfate in cancer and chondroitin-based anticancer agents. In Vivo (Brooklyn) 22:385–390

Bauer S, Arpa-Sancet MP, Finlay JA et al (2013) Adhesion of marine fouling organisms on hydrophilic and amphiphilic polysaccharides. Langmuir 29:4039–4047. doi:10.1021/la3038022

Baumgartner G, Gomar-Höss C, Sakr L et al (1998) The impact of extracellular matrix on the chemoresistance of solid tumors – experimental and clinical results of hyaluronidase as additive to cytostatic chemotherapy. Cancer Lett 131:85–99. doi:10.1016/S0304-3835(98)00204-3
Beckenlehner K, Bannke S, Spruss T et al (1992) Hyaluronidase enhances the activity of adriamycin in breast cancer models in vitro and in vivo. J Cancer Res Clin Oncol 118:591–596. doi:10.1007/BF01211802
Berchtold S, Grünwald B, Krüger A et al (2015) Collagen type V promotes the malignant phenotype of pancreatic ductal adenocarcinoma. Cancer Lett 356:721–732. doi:10.1016/j.canlet.2014.10.020
Bisno AL, Brito MO, Collins CM (2003) Molecular basis of group A streptococcal virulence. Lancet Infect Dis 3:191–200. doi:10.1016/S1473-3099(03)00576-0
Blache CA, Manuel ER, Kaltcheva TI et al (2012) Systemic delivery of Salmonella typhimurium transformed with IDO shRNA enhances intratumoral vector colonization and suppresses tumor growth. Cancer Res 72:6447–6456. doi:10.1158/0008-5472.CAN-12-0193
Botella P, Abasolo I, Fernández Y et al (2011) Surface-modified silica nanoparticles for tumor-targeted delivery of camptothecin and its biological evaluation. J Control Release 156:246–257. doi:10.1016/j.jconrel.2011.06.039
Brekken C, de Lange DC (1998) Hyaluronidase reduces the interstitial fluid pressure in solid tumours in a non-linear concentration-dependent manner. Cancer Lett 131:65–70
Brekken C, Hjelstuen MH, Bruland ØS, de Lange DC (2000) Hyaluronidase-induced periodic modulation of the interstitial fluid pressure increases selective antibody uptake in human osteosarcoma xenografts. Anticancer Res 20:3513–3519
Caruso F, Lichtenfeld H, Donath E, Möhwald H (1999) Investigation of electrostatic interactions in polyelectrolyte multilayer films: binding of anionic fluorescent probes to layers assembled onto colloids. Macromolecules 32:2317–2328. doi:10.1021/ma980674i
Chauhan VP, Martin JD, Liu H et al (2013) Angiotensin inhibition enhances drug delivery and potentiates chemotherapy by decompressing tumour blood vessels. Nat Commun 4:2516. doi:10.1038/ncomms3516
Choi J (2006) Intraperitoneal immunotherapy for metastatic ovarian carcinoma: resistance of intratumoral collagen to antibody penetration. Clin Cancer Res 12:1906–1912. doi:10.1158/1078-0432.CCR-05-2141
Cun X, Ruan S, Chen J et al (2015) Acta Biomaterialia A dual strategy to improve the penetration and treatment of breast cancer by combining shrinking nanoparticles with collagen depletion by losartan. ACTA Biomater 31:186. doi:10.1016/j.actbio.2015.12.002
Curnis F, Sacchi A, Borgna L et al (2000) Enhancement of tumor necrosis factor alpha antitumor immunotherapeutic properties by targeted delivery to aminopeptidase N (CD13). Nat Biotechnol 18:1185–1190. doi:10.1038/81183
Decher G (1997) Fuzzy nanoassemblies: toward layered polymeric multicomposites. Science 277:1232–1237. doi:10.1126/science.277.5330.1232
Diop-Frimpong B, Chauhan VP, Krane S et al (2011) Losartan inhibits collagen I synthesis and improves the distribution and efficacy of nanotherapeutics in tumors. Proc Natl Acad Sci U S A 108:2909–2914. doi:10.1073/pnas.1018892108
Dmitrieva N, Yu L, Viapiano M et al (2011) Chondroitinase ABC I-mediated enhancement of oncolytic virus spread and antitumor efficacy. Clin Cancer Res 17:1362–1372. doi:10.1158/1078-0432.CCR-10-2213
Dunn AL, Heavner JE, Racz G, Day M (2010) Hyaluronidase: a review of approved formulations, indications and off-label use in chronic pain management. Expert Opin Biol Ther 10:127–131. doi:10.1517/14712590903490382
Dychter SS, Harrigan R, Bahn JD et al (2014) Tolerability and pharmacokinetic properties of ondansetron administered subcutaneously with recombinant human hyaluronidase in minipigs and healthy volunteers. Clin Ther 36:211–224. doi:10.1016/j.clinthera.2013.12.013
Eikenes L, Tari M, Tufto I et al (2005) Hyaluronidase induces a transcapillary pressure gradient and improves the distribution and uptake of liposomal doxorubicin (Caelyx) in human osteosarcoma xenografts. Br J Cancer 93:81–88. doi:10.1038/sj.bjc.6602626

Fadnes HO, Reed RK, Aukland K (1977) Interstitial fluid pressure in rats measured with a modified wick technique. Microvasc Res 14:27–36. doi:10.1016/0026-2862(77)90138-8
Fraser JR, Laurent TC (1989) Turnover and metabolism of hyaluronan. Ciba Found Symp 143:41–49, 281–285
Ganesh S, Gonzalez-Edick M, Gibbons D et al (2008) Intratumoral coadministration of hyaluronidase enzyme and oncolytic adenoviruses enhances virus potency in metastatic tumor models. Clin Cancer Res 14:3933–3941. doi:10.1158/1078-0432.CCR-07-4732
Goodman TT, Olive PL, Pun SH (2007) Increased nanoparticle penetration in collagenase-treated multicellular spheroids. Int J Nanomedicine 2:265–274
Grzesiak JJ, Bouvet M (2006) The alpha2beta1 integrin mediates the malignant phenotype on type I collagen in pancreatic cancer cell lines. Br J Cancer 94:1311–1319. doi:10.1038/sj.bjc.6603088
Guedan S, Rojas JJ, Gros A et al (2010) Hyaluronidase expression by an oncolytic adenovirus enhances its intratumoral spread and suppresses tumor growth. Mol Ther 18:1275–1283. doi:10.1038/mt.2010.79
Heldin C-H, Rubin K, Pietras K, Östman A (2004) High interstitial fluid pressure — an obstacle in cancer therapy. Nat Rev Cancer 4:806–813. doi:10.1038/nrc1456
Huang Z, Zhao C, Chen Y et al (2014) Recombinant human hyaluronidase PH20 does Not stimulate an acute inflammatory response and inhibits lipopolysaccharide-induced neutrophil recruitment in the air pouch model of inflammation. J Immunol 192:5285–5295. doi:10.4049/jimmunol.1303060
Ioachim E, Charchanti A, Briasoulis E et al (2002) Immunohistochemical expression of extracellular matrix components tenascin, fibronectin, collagen type IV and laminin in breast cancer: their prognostic value and role in tumour invasion and progression. Eur J Cancer 38:2362–2370. doi:10.1016/S0959-8049(02)00210-1
Jacobetz MA, Chan DS, Neesse A et al (2013) Hyaluronan impairs vascular function and drug delivery in a mouse model of pancreatic cancer. Gut 62:112–120. doi:10.1136/gutjnl-2012-302529
Järvinen TAH (2012) Design of target-seeking antifibrotic compounds. Methods Enzymol 509:243–261. doi:10.1016/B978-0-12-391858-1.00013-7
Järvinen TAH, Ruoslahti E (2010) Target-seeking antifibrotic compound enhances wound healing and suppresses scar formation in mice. Proc Natl Acad Sci U S A 107:21671–21676. doi:10.1073/pnas.1016233107
Joo J, Liu X, Kotamraju VR et al (2015) Gated luminescence imaging of silicon nanoparticles. ACS Nano 9:6233–6241. doi:10.1021/acsnano.5b01594
Kimata K, Honma Y, Okayama M et al (1983) Increased synthesis of hyaluronic acid by mouse mammary carcinoma cell variants with high metastatic potential. Cancer Res 43:1347–1354
Kuhn SJ, Finch SK, Hallahan DE, Giorgio TD (2006) Proteolytic surface functionalization enhances in vitro magnetic nanoparticle mobility through extracellular matrix. Nano Lett 6:306–312. doi:10.1021/nl052241g
Kultti A, Zhao C, Singha NC et al (2014) Accumulation of extracellular hyaluronan by hyaluronan synthase 3 promotes tumor growth and modulates the pancreatic cancer microenvironment. BioMed Res Int 2014:817613. doi:10.1155/2014/817613
Kuriyama N, Kuriyama H, Julin CM et al (2001) Protease pretreatment increases the efficacy of adenovirus-mediated gene therapy for the treatment of an experimental glioblastoma model advances in brief protease pretreatment increases the efficacy of adenovirus-mediated gene therapy for the treatment o. 1805–1809.
Laborda E, Puig-Saus C, Rodriguez-García A et al (2014) A pRb-responsive, RGD-modified, and hyaluronidase-armed canine oncolytic adenovirus for application in veterinary oncology. Mol Ther 22:986–998. doi:10.1038/mt.2014.7
Laurent UB, Dahl LB, Reed RK (1991) Catabolism of hyaluronan in rabbit skin takes place locally, in lymph nodes and liver. Exp Physiol 76:695–703
Li M-W, Yudin AI, Robertson KR et al (2002) Importance of glycosylation and disulfide bonds in hyaluronidase activity of macaque sperm surface PH-20. J Androl 23:211–219

Lipponen P, Aaltomaa S, Tammi R et al (2001) High stromal hyaluronan level is associated with poor differentiation and metastasis in prostate cancer. Eur J Cancer 37:849–856

Liu D, Pearlman E, Diaconu E et al (1996) Expression of hyaluronidase by tumor cells induces angiogenesis in vivo. Proc Natl Acad Sci U S A 93:7832–7837. doi:10.1073/pnas.93.15.7832

Manuel ER, Chen J, D'Apuzzo M et al (2015) Salmonella-based therapy targeting indoleamine 2,3-Dioxygenase coupled with enzymatic depletion of tumor hyaluronan induces complete regression of aggressive pancreatic tumors. Cancer Immunol Res 3:1096–1107. doi:10.1158/2326-6066.CIR-14-0214

Martínez-Vélez N, Xipell E, Vera B, Acanda de la Rocha A, Zalacain M, Marrodan L, Gonzalez-Huarriz M, Toledo G, Cascallo M, Alemany R, Patiño A, Alonso AM (2016) The oncolytic adenovirus VCN-01 as therapeutic approach against pediatric osteosarcoma. Clin Cancer Res 22(9):2217–2225. doi:10.1158/1078-0432.CCR-15-1899.

McKee TD (2006) Degradation of fibrillar collagen in a human melanoma xenograft improves the efficacy of an oncolytic herpes simplex virus vector. Cancer Res 66:2509–2513. doi:10.1158/0008-5472.CAN-05-2242

Mok W, Boucher Y, Jain RK (2007) Matrix metalloproteinases-1 and −8 improve the distribution and efficacy of an oncolytic virus. Cancer Res 67:10664–10668. doi:10.1158/0008-5472.CAN-07-3107

Muckenschnabel I, Bernhardt G, Spruss T, Buschauer A (1996) Hyaluronidase pretreatment produces selective melphalan enrichment in malignant melanoma implanted in nude mice. Cancer Chemother Pharmacol 38:88–94

Nelson DL, Cox MM (2008) Lehninger principles of biochemistry, 5th edn., pp 1–1294

Nugent LJ, Jain RK (1984) Extravascular diffusion in normal and neoplastic tissues. Cancer Res 44:238–244

Pirinen R, Tammi R, Tammi M et al (2001) Prognostic value of hyaluronan expression in non-small-cell lung cancer: increased stromal expression indicates unfavorable outcome in patients with adenocarcinoma. Int J Cancer 95:12–17

Plodinec M, Loparic M, Monnier CA et al (2012) The nanomechanical signature of breast cancer. Nat Nanotechnol 7:757–765. doi:10.1038/nnano.2012.167

Provenzano PP, Cuevas C, Chang AE et al (2012) Enzymatic targeting of the stroma ablates physical barriers to treatment of pancreatic ductal adenocarcinoma. Cancer Cell 21:418–429

Průšová A, Conte P, Kučerík J, Alonzo G (2010) Dynamics of hyaluronan aqueous solutions as assessed by fast field cycling NMR relaxometry. Anal Bioanal Chem 397:3023–3028. doi:10.1007/s00216-010-3855-9

Ricciardelli C, Brooks JH, Suwiwat S et al (2002) Regulation of stromal versican expression by breast cancer cells and importance to relapse-free survival in patients with node-negative primary breast cancer. Clin Cancer Res 8:1054–1060

Rodriguez-Garcia A, Gimenez-Alejandre M, Rojas JJ et al (2015) Safety and efficacy of VCN-01, an oncolytic adenovirus combining fiber HSG-binding domain replacement with RGD and hyaluronidase expression. Clin Cancer Res 21:1406–1418. doi:10.1158/1078-0432.CCR-14-2213

Rosengren S, Dychter SS, Printz MA et al (2015) Clinical immunogenicity of rHuPH20, a hyaluronidase enabling subcutaneous drug administration. AAPS J 17:1144–1156. doi:10.1208/s12248-015-9782-0

Ruoslahti E (2002) Specialization of tumour vasculature. Nat Rev Cancer 2:83–90. doi:10.1038/nrc724

Sakai S, Hirano K, Toyoda H et al (2007) Matrix assisted laser desorption ionization-time of flight mass spectrometry analysis of hyaluronan oligosaccharides. Anal Chim Acta 593:207–213. doi:10.1016/j.aca.2007.05.005

Scodeller P, Flexer V, Szamocki R et al (2008) Wired-enzyme core-shell Au nanoparticle biosensor. J Am Chem Soc 130:12690–12697. doi:10.1021/ja802318f

Scodeller P, Carballo R, Szamocki R et al (2010) Layer-by-layer self assembled osmium polymer mediated laccase oxygen cathodes for biofuel cells: the role of hydrogen peroxide. J Am Chem Soc 132:11132–11140

Scodeller P, Catalano PN, Salguero N et al (2013) Hyaluronan degrading silica nanoparticles for skin cancer therapy. Nanoscale 5:9690–9698. doi:10.1039/c3nr02787b

Sherman-Baust CA, Weeraratna AT, Rangel LBA et al (2003) Remodeling of the extracellular matrix through overexpression of collagen VI contributes to cisplatin resistance in ovarian cancer cells. Cancer Cell 3:377–386. doi:10.1016/S1535-6108(03)00058-8

Shintani Y, Hollingsworth MA, Wheelock MJ, Johnson KR (2006) Collagen I promotes metastasis in pancreatic cancer by activating c-Jun NH2-terminal kinase 1 and up-regulating N-cadherin expression. Cancer Res 66:11745–11753. doi:10.1158/0008-5472.CAN-06-2322

Simon-Gracia L, Hunt H, Scodeller PD et al (2016) Paclitaxel-loaded polymersomes for enhanced intraperitoneal chemotherapy. Mol Cancer Ther 15:670–680. doi:10.1158/1535-7163.MCT-15-0713-T

Singha NC, Nekoroski T, Zhao C et al (2014) Tumor-associated hyaluronan limits efficacy of monoclonal antibody therapy. Mol Cancer Ther 14:523–532. doi:10.1158/1535-7163.MCT-14-0580

Slsca RF, Gkamt HHLAI, Tholqard JC (1971) Histochemical demonstration of acid mucopolysaccharide production by tissue cultured epithelial-like cells *. 181:173–181

St Croix B, Man S, Kerbel RS (1998) Reversal of intrinsic and acquired forms of drug resistance by hyaluronidase treatment of solid tumors. Cancer Lett 131:35–44, doi: S0304-3835(98)00199-2 [pii]

Stern R (2008) Hyaluronidases in cancer biology. Semin Cancer Biol 18:275–280. doi:10.1016/j.semcancer.2008.03.017

Stern R, Asari A, Sugahara K (2006) Hyaluronan fragments: an information-rich system. Eur J Cell Biol 85:699–715. doi:10.1016/j.ejcb.2006.05.009

Sugahara K, Yamada S, Sugiura M et al (1992) Identification of the reaction products of the purified hyaluronidase from stonefish (Synanceja horrida) venom. Biochem J 283(Pt 1):99–104

Suwiwat S, Ricciardelli C, Tammi R et al (2004) Expression of extracellular matrix components versican, chondroitin sulfate, tenascin, and hyaluronan, and their association with disease outcome in node-negative breast cancer. Clin Cancer Res 10:2491–2498

Swyer G (1947) The hyaluronidase content of semen. Biochem J 41:409

Toole BP (2004) Hyaluronan: from extracellular glue to pericellular cue. Nat Rev Cancer 4:528–539. doi:10.1038/nrc1391

Toole BP (2009) Hyaluronan-CD44 interactions in cancer: paradoxes and possibilities. Clin Cancer Res 15:7462–7468. doi:10.1158/1078-0432.CCR-09-0479

Turley EA, Tretiak M (1985) Glycosaminoglycan production by murine melanoma variants in vivo and in vitro. Cancer Res 45:5098–5105

Verzijl N, DeGroot J, Thorpe SR et al (2000) Effect of collagen turnover on the accumulation of advanced glycation end products. J Biol Chem 275:39027–39031. doi:10.1074/jbc.M006700200

Wessels MR, Moses AE, Goldberg JB, DiCesare TJ (1991) Hyaluronic acid capsule is a virulence factor for mucoid group A streptococci. Proc Natl Acad Sci U S A 88:8317–8321. doi:10.1073/pnas.88.19.8317

Whatcott CJ, Han H, Posner RG et al (2011) Targeting the tumor microenvironment in cancer: why hyaluronidase deserves a second look. Cancer Discov 1:291–296. doi:10.1158/2159-8290.CD-11-0136

Wu C-C, Hu Y, Miller M et al (2015) Protection and delivery of anthelmintic protein Cry5B to nematodes using mesoporous silicon particles. ACS Nano 9:6158–6167. doi:10.1021/acsnano.5b01426

Zhang L, Underhill CB, Chen L (1995) Hyaluronan on the surface of tumor cells is correlated with metastatic behavior hyaluronan on the surface of tumor cells is correlated with metastatic behavior. Cancer Res 55:428–433

Zhou H, Fan Z, Deng J et al (2016) Hyaluronidase embedded in nanocarrier PEG shell for enhanced tumor penetration and highly efficient antitumor efficacy. Nano Lett 16:3268. doi:10.1021/acs.nanolett.6b00820, acs.nanolett.6b00820

Chapter 4
Nanocarrier-Based Anticancer Therapies with the Focus on Strategies for Targeting the Tumor Microenvironment

Fransisca Leonard and Biana Godin

Abstract In the past couple of decades, advanced knowledge of cancer biology has revealed the predominant role of tumor microenvironment in tumor growth and progression. Different components in tumor microenvironment release sets of biological signals, which contribute to cell growth and survival, epithelial-mesenchymal transition during metastases, angiogenesis, and development of drug resistance. The new understanding about the cancer cell-stroma interactions has been utilized to design various nanocarriers targeting the tumor environment for drug delivery. This can be achieved passively by using differential physiological features in the tumor microenvironment (e.g. different blood flow patterns or fenestrations in the tumor endothelium) or actively by recognizing specific molecular signatures on the cells in the tumor microenvironment. This chapter will discuss the biology of the tumor microenvironment and identify various cells and extracellular components that can be exploited to enhance cancer therapies and efficiently target nanotherapies to tumor loci.

Keywords Nanocarriers • Tumor microenvironment • Anticancer therapy • Endothelial cells • Tumor-associated macrophages • Cancer-associated fibroblast

Abbreviations

NP	Nanoparticle
EPR	Enhanced Permeation and Retention
ECM	Extracellular matrix
TAM	Tumor-associated macrophage
CAF	Cancer-associated fibroblasts

F. Leonard • B. Godin (✉)
Department of Nanomedicine, Houston Methodist Research Institute,
6670 Bertner Ave., R8-213 Houston, TX, USA
e-mail: bgodin@houstonmethodist.org; bianagodinv@gmail.com

A. Prokop, V. Weissig (eds.), *Intracellular Delivery III*, Fundamental Biomedical Technologies 8, DOI 10.1007/978-3-319-43525-1_4

IIC	Infiltrating immune cells
CTL	Cytotoxic T-lymphocytes
TIL	Tumor-infiltrating lymphocytes
CAA	Cancer-associated adipocytes
ADF	Adipocyte-derived fibroblasts
HIF	Hypoxia-inducible factor
VEGF	Vascular endothelial growth factor
PDGF	Platelet-derived growth factor
TNF-α	Tumor necrosis factor alpha
MMP	Matrix metalloproteinase
EGF	Endothelial growth factor
TGF-β	Transforming growth factor beta
FGF	Fibroblast growth factor
PlGF	Placental growth factor
SDF	Stromal cell-derived factor
HGH	Hepatocyte growth factor
EGF	Epidermal growth factor
IGF	Insulin-like growth factor
GM-CSF	Granulocyte macrophage colony-stimulating factor
M-CSF	Monocyte-colony stimulating factor
CSF-1	Colony stimulating factor-1
ICAM-1	Intracellular adhesion molecule-1
FABP	Fatty acid-binding protein
IL	Interleukin
IFN-γ	Interferon gamma
CCL	Chemokines (C-C motif) ligand
CXCL	C-X-C motif chemokine
CD	Cluster of differentiation
MIP	Macrophage inhibitory protein
PGE2	Prostaglandin E2
ZA	Zoledronic acid
HRG	Histidine-rich glycoprotein
STAT	Signal transducers and activators of transcription
TRAIL	TNF-related apoptosis-inducing ligand
NF-KB	Nuclear factor kappa b
CTLA-4	Cytotoxic T-lymphocyte-associated protein
SIRPα	Signal regulatory protein alpha
ACT	Adoptive cell therapy
ROS	Reactive oxygen species
NSCLC	Non-small cell lung cancer
PD-1	Programmed death-1
PD-L1	Programmed death ligand 1
MAPK	Mitogen-activated protein kinase
α-SMA	α-smooth musche actin
MPS	Mononuclear phagocytic system

PEG	Poly(ethylene)glycol
PAMAM	Polyamidoamine
PEI	Polyethylenimine
PLGA	Poly(lactic-co-glycolic acid)

4.1 Introduction: Tumor Microenvironment

In the past, cancer research has been focused on cancer biology, including identifying and understanding the role of oncogenes and tumor suppressor genes involved in activation and loss of function contributing to the progression and transformation of the cells. In the recent decades, along with new discoveries and gained knowledge about the complexity of cancer, there is a rising recognition of the importance of the tumor microenvironment (TME). The interplay between the deregulation in the physiology of tumor cells and the cells of TME is vital for tumor initiation, growth and progression. This knowledge drives novel and more efficient therapeutics approaches to achieve improved and safe therapies, which will localize more specifically to the tumor milieu. These efforts require collaborative interaction of multidisciplinary teams, including clinicians, biological and material scientists, and biomedical engineers. Increased efficacy in therapeutics can potentially be achieved by designing drugs or drug carriers that target abnormalities in TME passively, including increased vascularization and blood vessel leakiness, the low pH, hypoxia condition, as well as with active targeting via binding to receptors overexpressed on the cancer cells and on the cells of the TME.

The non-neoplastic cells of the TME facilitate tumor survival and development by providing growth factors, cytokines, remodeling the vascular network and restructuring the extracellular matrices. They are also important factors in forming metastatic niches and, thus, defining the sites of tumor metastasis (Quail and Joyce 2013). TME composition varies between different tumor types, different sites of the same tumor (e.g. primary and metastatic) and, in solid tumors, the non-neoplastic cells can constitute up to 90 % of tumor mass (De Kruijf et al. 2011; Downey et al. 2014; Yan et al. 2005). Below, we briefly introduce the major cells and non-cellular elements in the TME (Fig. 4.1) including their crosstalk signaling network to control TME (Fig. 4.2).

4.1.1 Angiogenic Blood Vessels

In order to support the growing cell mass and to grow beyond the diffusion-limited maximal mass (2 mm), most of the solid tumors are characterized by neovascularization, a *de novo* formation of the blood vessels to provide the nutrients (Brannon-Peppas and Blanchette 2012). The resulting new vasculature, so-called tumor-associated vessels, is irregular, chaotic and inherently unstable (Di Tomaso et al. 2005). These blood vessels are mainly responsible for the transport of

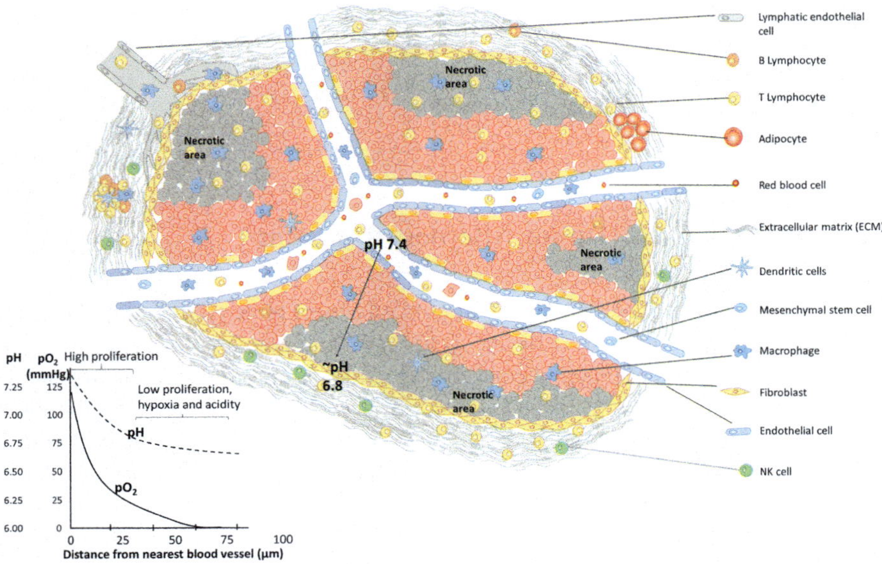

Fig. 4.1 Schematic presentation of the components in the tumor microenvironment, representing the main cell elements and the levels of oxygen. Oxygen availability decreases and acidity increases along with the distance to the blood vessel

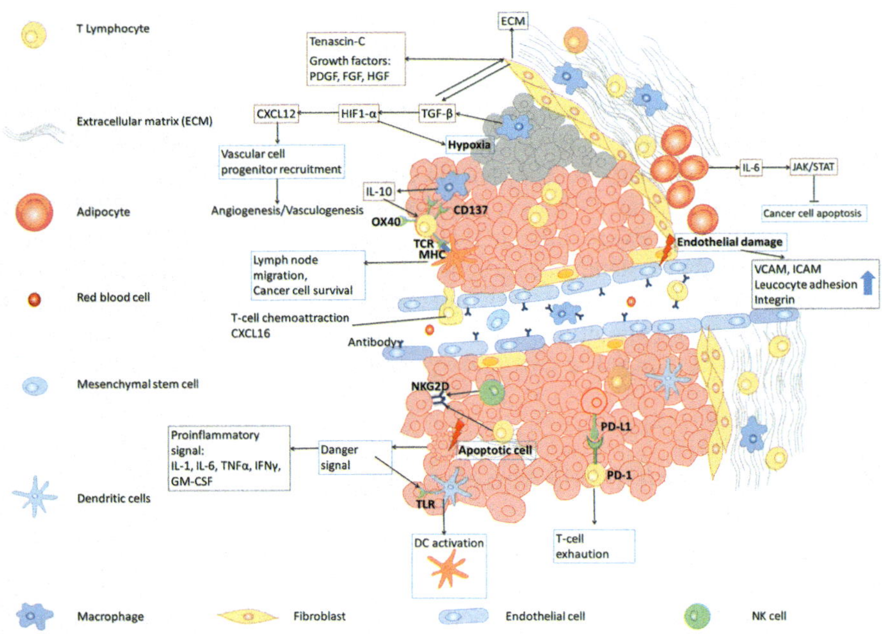

Fig. 4.2 Schematic of tumorigenic microenvironment and signaling network between cellular components and their effect on processes occuring in TME to support cancer cell growth

nutrients and oxygen to the tumor cells, and other physiological functions of these cells are being investigated. Tumor neovascularization involves tube-forming endothelial cells and their supporting pericytes that comprise angiogenic vasculature. Induction of an angiogenic switch causes the hyperproliferation of neoplasia and tumor. This involves upregulation of hypoxia-inducible factor (HIF), vascular endothelial growth factor (VEGF), platelet-derived growth factor (PDGF), tumor necrosis factor (TNF)-α, integrin αvβ3, and matrix metalloproteinases (MMPs) (Hanahan and Coussens 2012). Several drugs target the regulation of angiogenesis mechanism, thus impairing the growth of the angiogenic vessels and the tumor mass, as a result of restriction of the transport of blood-borne mitogenic growth factor. When the vessels grow in an uncontrolled fashion, the formed vasculature is different from a normal one in terms of density of the endothelial cells and the molecular signatures, characterizing dividing cells. The endothelial cells in the tumor tissue do not form a tight monolayer, producing fenestrations of 100–700 nm, which enable the passage of macromolecules and nanotherapeutics across the impaired endothelium. The entities are further retained there based on the differences in the osmotic pressure and pour lymphatic drainage between the tumor and healthy tissue. This Enhanced Permeation and Retention (EPR) effect first described by Matsumura and Maeda (Matsumura and Maeda 1986) as well as the irregular and sometimes oscillating blood flow which changes the margination patterns of the objects as compared to the normal vasculature are responsible for the passive targeting of nanotherapeutics to the tumor tissue (Decuzzi and Ferrari 2006). These mechanisms will be discussed in more details in the following sections. The majority of the clinically used nanotherapeutics account for EPR effect.

4.1.2 *Lymphatic System and Infiltrating Immune Cells*

Lymphatic system plays a critical role in maintaining tissue fluid homeostasis, regulating immune response, and is the site for lymph node metastasis (Alitalo 2011). Although the lymphatic vessels on the tumor periphery and beyond are functional and hyperplastic, they are absent within the tumor margins in humans and other species (Leu et al. 2000). This is most likely due to the compression of lymphatic vessels caused by the high osmotic and hydraulic pressure characterizing many tumors due to the fluid retention and the growing tumor. The lack of functional lymphatic vessels within the tumor causes poor drainage, a contributing factor of the EPR effect.

Solid tumors contain infiltrates of diverse leukocyte subsets including myeloid and lymphoid-lineage cells. These cells supply mitogenic growth factors which promote proliferation, and also increase the production of epidermal growth factor (EGF), transforming growth factor (TGF)-β, TNF-α, fibroblast growth factor(FGF), interleukins, histamine, heparin, as well as proteolytic enzymes such as metallo-, serin- and cysteine-proteases to cleave and modify extracellular matrix (ECM) (Hanahan and Coussens 2012).

Major actors among the infiltrating immune cells in TME are the tumor-associated macrophages (TAMs). TAMs are alternatively differentiated macrophages called M2 macrophage, which can modulate tissue remodeling and angiogenesis, suppress T cell activity/proliferation and play significant role in tumor survival. Their density has been correlated with poor disease-free survival in cholangiocarcinoma, hepatocellular carcinoma, as well as in several other epithelial cancers (Zhang et al. 2012).

TAMs are characterized by a constitutive high expression of several other growth factors, including VEGF, FGF1 and 2, PDGF, granulocyte macrophage colony-stimulating factor (GM-CSF), insulin-like growth factor (IGF)-1 and TGF-β, and pro-inflammatory mediators, such as interleukin (IL)-1, −6, −8, prostaglandins, interferon (IFN)-γ, TNF-α, Chemokines (C-C motif) ligand (CCL)17 and CCL18, or the interleukins IL-1ra, IL-6, IL-10, and arginase-1. TAMs also express low cluster of differentiation (CD51), carbopeptidase M and IL-12 (Hanahan and Coussens 2012). TAMs express α4 integrins, which can bind to VCAM, activating Ezrin that mediates receptor tyrosine signaling and PI3K/Akt signaling, suppressing cell programmed death. TAMs also protects tumor from chemotherapy-induced cell death via cathepsin protease-dependent pathway (Chen et al. 2011; Shree et al. 2011).

TAMs are recruited from circulating monocytes and regulated by homing proteins, such as monocyte chemotactic protein (MCP)-1/CCL2, CCL3, CCL4, CCL5, CCL8, macrophage inhibitory protein (MIP)-1α, macrophage migration inhibitory factor, as well as the growth factors VEGF and monocyte-colony stimulating factor (M-CSF), expressed by tumor cells or several cellular components of the TME (Mukaida et al. 2014). Arriving in the TME, monocytes are exposed to soluble factors, such as prostaglandin E2 (PGE2), and cytokines, such as IL-2, IL10 and TGF-β1, causing their differentiation into M2 macrophages (Burkholder et al. 2014).

This knowledge can be utilized for cancer therapy to revert M2 polarization to M1 using pro-inflammatory molecules. Drugs such as zoledronic acid (ZA), a clinical drug for bone resorption therapy, and has recently been repurposed to be exploited in cancer immunotherapy since ZA can regulate macrophage functions and reduce cell migration, infiltration, proliferation. ZA boosted the production of type-1 cytokines by TAM in response to immunomodulators such as IL-12 and polyI:C, which can revert M2 phenotype of TAMs to a preferably anti-tumor M1 macrophage, as well as suppressing the expression of MMP-9 by TAM and increasing the proliferation of activated Tγδ lymphocytes in *in vitro* model of prostate cancer cell (Chanmee et al. 2014; Tsagozis et al. 2008). Histidine-rich glycoprotein (HRG) inhibited tumor growth and metastasis by reverting macrophage polarization and inducing vessel normalization by placental growth factor (PlGF) downregulation in tested T241 fibrosarcoma, Panc02 pancreatic tumor and 4T1 breast tumor cell lines (Rolny et al. 2011). Antitumor drug Trabectedin (ET-743), a standard treatment of soft tissue sarcoma and relapsed platinum-sensitive ovarian cancer patients, can kill TAMs via TRAIL receptors-mediated apoptosis (Chanmee et al. 2014), although the action also affects monocyte/macrophage-mediated host defense, which potentially cause adverse effects. A new peptide construct M2pep can bind to and kill M2 macrophages selectively, limiting the side effects (Cieslewicz et al. 2013).

TAMs effects on tumor invasiveness are mostly derived from studies performed in cancers featuring a rich stroma, such as in breast and colon carcinomas. In a mouse model of breast cancer, expression of colony stimulating factor-1 (CSF-1) is highest at the invasive edge of tumor cells, a site highly populated with macrophages. CSF-1 released by breast tumor cells has been associated with recruitment of TAM, and TAMs reciprocate by releasing EGF, which stimulates tumor cells to migrate and metastasize. Similar findings have been observed in a genetic model of colon cancer. Another study also identified the enrichment of macrophages in the tumor periphery of liver metastasis of the breast cancer in mice model (Tanei et al. 2016). In general, CSF-1 is highly upregulated in breast cancer (Lin et al. 2001), correlated with poor prognosis in breast, endometrial, hepatocellular and colorectal cancer (Groblewska et al. 2007; Zhu et al. 2008), but not overexpressed by pancreatic cancer cells (Panni et al. 2013). Inhibition of CSF-1R dimerization using humanized monoclonal antibody RG7155 has been shown to reduce specific macrophage population in tumor tissue in patients with diffuse-type giant cell tumor (Ries et al. 2014). PLX3397, a potent tyrosine kinase inhibitor of CSF-1R also decreased macrophage and lymphocyte infiltration and effective for adoptive cell therapy (ACT) (Mok et al. 2014). Macrophage depletion reached almost fivefold decrease in PLX3397-treated lympho-depleted C57BL/6 mice with established subcutaneous SM1-OVA tumors.

TAMs also secrete cytokines involved in angiogenesis and ECM remodeling, including VEGF, IL-8, TNF-α and MMPs, namely MMP-2 and −9. Additionally, TAMs may promote lymphangiogenesis by VEGF-C secretion, which most likely enhanced by hypoxic conditions present in most desmoplastic tumors (Cadamuro et al. 2013a). Specifically TAMs can regulate angiogenesis via production of VEGF-A and MMP-9, thereby increasing bioavailability of VEGF and activating the angiogenic switch (Deryugina and Quigley 2015). The modification of ECM can be a double edge sword for tumor cells: (1) on one hand it can uncage bioactive mitogenic agents, which normally act as a repair mechanism for tissue damage but misused by neoplastic cells to help sustain their hyperproliferation. (2) Alternatively, the process can also cleave cell-cell or cell-ECM adhesion molecules, disabling the growth and suppressing adhesion complexes needed to maintain the homeostasis (Lu et al. 2011). Infiltrating immune cells (IICs) can also regulate angiogenesis processes by upregulating cytokines such as VEGF, bFGF, TNF-alpha, TGF-beta, PDGF, PlGF, neuropilin-1, various chemokines (C-X-C motif chemokine (CXCL)-12, IL-8/CXCL8), MMPs (MMP-2, −7, −9, −12, −14), serine proteases (urokinase-type plasminogen activator), cysteine cathepsin proteases, DNA-damaging molecules (ROS), histamine, and other bioactive mediators (nitric oxide) (Hanahan and Coussens 2012).

TAMs are reprogrammed cells that have less “antigen-presenting” characteristics which can blunt tumor growth by disrupting angiogenesis and normalize vascular development. This knowledge has been also targeted for cancer therapy, e.g. histidine-rich glycoprotein HRG (a host-produced protein deposited in the stroma) (Rolny et al. 2011) or CSF-1 blockade (DeNardo et al. 2011) can induce TAM reprogramming resulted in vascular normalization and improved responses to chemotherapy. In both cases, anti-tumor activity was enhanced by cytotoxic

T-lymphocytes (CTLs), indicating the complex signaling processes within the tumor microenvironment. It was shown that the increased concentration of tumor-infiltrating lymphocytes (TIL) improves the survival of early stage non-small cell lung cancer (NSCLC) (Djenidi et al. 2015). Moreover, T-cells equipped with T-cell receptor (TCR), play a central role in cell-mediated immunity and can regulate antigen-specific, effector and immune memory responses. However, most of the tumor will not express sufficient antigens to stimulate the host antigen-presenting cells, resulting in inadequate expression of MHC class I- and II-peptide molecules in T-cells. TILs are mostly consist of cytotoxic memory CD8+ T-cells, and can increase the expression of the T-cell survival factors enumerated previously, such as p-Foxo3a, IGF and osteopontin, and cytokines, such as IL-7 and IL-15, and moreover, triggered the stimulation of AKT pathway and Wnt, increasing the T-cell survival in the tumor. Several approaches have been done to utilize these patient-derived T-cells and genetically modify them for immunotherapies targeting tumor cells (Rosenberg et al. 1988). In terms of nanotherapies, T cells can serve as a vital target for nanomedicine-based vaccines. Moreover, the affinity of these circulating cells to the tumor tissue is being explored for delivery of the nanotherapies pre-loaded into the cells to the malignant loci.

Another promising therapy targeting infiltrating immune cells is the anti-programmed death (PD)-L1 antibody. PD-L1 itself are pro-oncogenic in two distinct mechanisms, by inducing apoptosis in T-cells and by enhancing the tumor cell survival cells in hepatocellular carcinoma, where overexpression of PD-L1 corresponds with a poor prognosis and increased mortality. An anti-PD-L1 antibody is currently undergoing phase III clinical trials in the US for several types of cancers, including melanoma and non-small cell lung, colorectal, ovarian, pancreatic, gastric, breast and renal cell cancers (Brahmer et al. 2012).

Another approach to modify TAMs function is by inhibiting macrophages-dependent biological processes. Cycooxygenase-2 inhibitors have been shown to reduce macrophage production of VEGF-C, which subsequently led to suppression of lymphangiogenesis and lymph node metastasis in two murine models engrafted with human gastric carcinoma and human lung adenocarcinoma cell lines (Na et al. 2013).

In the context of this chapter, it is important to point out that macrophages as well as other populations of immune cells are naturally occurring phagocytes, which will uptake the solid particulates promptly. Thus, targeting these cell populations with nanovectors represents one of the important strategies for enabling preferential uptake of nanotherapeutics in the tumors (Sounni and Noel 2013) (Table 4.1).

4.1.3 Cancer-Associated Fibroblasts (CAF)

Fibroblasts are the most abundant cell type and function mostly as connective and structural framework of tissues due to their ECM secretion. Although, only with increasing evidence in the last decade that the interactions between cancer cells and

Table 4.1 Drugs targeting tumor microenvironment currently in preclinical research

Elements in tumor	Strategy	Drug/targeting moiety	Nanocarriers	Reference
Vasculature	Antiangiogenesis	Anti-VEGF monoclonal antibody (Sc-7269)	Radioimmunoconjugate-^{131}I Dextran Magnetic Nanoparticles (DMN)	(Chen et al. 2006)
		TNP-470 (Lodamin)	PEG-PLA	(Satchi-Fainaro et al. 2004)
		Anti-VCAM monoclonal antibody (mAb) (M/K-271)	Immunoliposomes (IL)	(Gosk et al. 2008)
		^{10}Boron/VEGF	Dendrimer	(Backer et al. 2005)
		siRNA/RGD	PEG-PLGA	(Wang et al. 2009)
	Vascular normalization	Anti-VEGF mAb (Bevacizumab)	nanophotoactivatable liposome (nanoPAL)	(Tangutoori et al. 2016)
	Targeting hypoxia	–	Layer-by-layer assembled polyelectrolyte NPs	(Poon et al. 2011)
		–	hypoxia-activated NIR-triggered chitosan NPs,	(Lin et al. 2013)
		Palladium meso-tetraphenylporphyrin/ Herceptin	Polystyrene NPs	(Napp et al. 2011)
	Photothermal therapy	Doxorubicin	Phorphysome nanoliposome	(Lovell et al. 2011)
		Gentamicin	Porphyrin–phospholipid liposomes	(Carter et al. 2014)
		–	Polyethylene glycol (PEG)-protected gold nanorods (NR)	(Maltzahn et al. 2009)
		–	90)Y labeled N-(2-hydroxypropyl) methacrylamide (HPMA) copolymer with gold NR	(Buckway et al. 2014)

(continued)

Table 4.1 (continued)

Elements in tumor	Strategy	Drug/targeting moiety	Nanocarriers	Reference
TAMs	TAM depletion	CSF-1R antibody	–	(Ries et al. 2014)
		Chlondronate	Liposome	(Zeisberger et al. 2006)
		CCR2 siRNA	Liposome	(Leuschner et al. 2011)
	Reduction of TAM and conversion of TAM phenotype	Imatinib	–	(Cavnar et al. 2013)
	Revert TAM polarization and NK cells stimulation	Sorafenib	–	(Sprinzl et al. 2013)
	Restricting TAM biological functions	Cycooxygenase-2 inhibitors	–	(Nakanishi et al. 2011)
	Macrophages as trojan horse for drugs	nAb-paclitaxel	Silicon mesoporous particles (SMP)	(Tanei et al. 2016), (Leonard et al. 2016)
		Gold nanoshell	Gold nanoshell	(Choi et al. 2007)
T-cells	Checkpoint inhibitors: PD-L1, CTLA-4	Ipilimumab (Yervoy®), pembrolizumab (Keytruda®), and nivolumab (Opdivo®)	–	(Barbee et al. 2015)
	Co-stimulatory agonists	CD137, GITR, OX40, and ICOS	–	(Peggs et al. 2009)
	Cancer vaccines	IL-15Sa and IL-21	Liposomal NPs	(Stephan et al. 2010)
		Photothermal therapy	Carbon nanotube	(Wang et al. 2014)
		DPX-0907 (DepoVax™)	Liposome made from PMV from B16-OVA	(van Broekhoven et al. 2004)
		Mannan	Liposome	(Berinstein et al. 2012)
			NPs, microspheres, liposomes	(Brandhonneur et al. 2009; Espuelas et al. 2008; Saraogi et al. 2011)
	T-cell response and proliferation	IL-12	Liposome	(Simpson-Abelson et al. 2009)
		Mannan	PLGA NPs	(Hamdy et al. 2011)

CAF	Reversing CAF-phenotype and CAF effect	TGFβ inhibitors	–	(Groessl et al. 2014)
		Cyclopamine	–	(Walter et al. 2010)
		Dasatinib	–	(Haubeiss et al. 2010)
	Nutrient depletion	N-acetyl-cysteine		(Lisanti et al. 2013)
	Altering host metabolism	Metformin		(Leone et al. 2014)
	Stroma depletion	Docetaxel	Carboxymethylcellulose NPs	(Murakami et al. 2013)
		Cisplatin&RAPA	PLGA NP	(Guo et al. 2014)
		Navitoclax™	Nanoliposome	(Chen et al. 2016)
	Blocking PDGFR	Imatinib	–	(Pietras et al. 2008)
ECM	Improving drug penetration and cell apoptosis	Doxorubicin/HA-targeting enzyme PEGPH20	–	(Provenzano et al. 2012)
	ECM degradation	Trastuzumab	–	(Beyer et al. 2011)
		Lodamin	–	(Adams et al.)
		Doxorubicin/HA	PEG-PCL NP	(Yadav et al. 2008)
		Doxorubicin/HA	PLGA NP	(Hyung et al. 2008)
		Hyaluronan	Liposome	(Mizrahy et al. 2014)
	Inhibition of cell migration	Taxol and Doxorubicin	–	(Harisi et al. 2009)

the stroma have attracted considerable attention as one of the components supporting tumor growth and progression. They are activated during wound healing and fibrosis, the condition present in tumor, and the tumor has been considered as "the wound that does not heal". CAF are recruited via pro-inflammatory chemokines and cytokines released by inflammatory cells and by the tumor cells themselves, with signaling pathways similar to tissue repair/wound healing. CAF can also be recruited from circulating bone marrow-derived mesenchymal cells that co-express both hematopoietic and fibroblast markers. These mesenchymal stem cells can be recruited and differentiated into either myofibroblasts defined by expression of α-smooth musche actin (α-SMA) (Quante et al. 2011), or into adipocytes defined by expression of fatty acid-binding protein (FABP4) (Qian et al. 2010).

In general, normal functioning fibroblasts can limit cancer cell proliferation (Bissell and Hines 2011), the characteristic that is lost after transition to CAF. The reprogramming from normal fibroblasts can be caused by other abberant conditions, such as fibrosis, edema or infections, which can produce proteases or other paracrine factors that disrupt normal epithelial structures (Räsänen and Vaheri 2010). Fibroblasts may become activated in response to cytokines released by cancer cells and inflammatory cells such as PDGF, TGF-β and FGF-2 (Kalluri and Zeisberg 2006b), as well as MCP-1, and PDGF (Marsh et al. 2013). Several microRNA has been found to be downregulated (miR-31 and miR-214) or upregulated (miR-155) in ovarian CAF compare to the normal fibroblasts, suggesting their role for defining CAF function (Mitra et al. 2012).

CAF have diverse characteristics and functional contributions in various organs, but generally possess a higher proliferative index, as compared with fibroblasts in normal tissues (Tlsty and Coussens 2006). They express increased level of cytokines and chemokines such as stromal cell-derived factor (SDF)-1/CXCL12, CXCL14, CCL2, CCL8 (Togo et al. 2013). In colorectal cancer, CAF has been shown to express growth factors such as hepatocyte growth factor (HGH), epidermal growth factor (EGF) family members, variety of FGF, and which in turn activated mitogenic signals such as mitogen-activated protein kinase (MAPK) and phosphatidylinositol 3-kinase (Quante et al. 2013). CAF can also orchestrate metastasis-promoting communication in colorectal cancer by expressing TGF-β (Tommelein et al. 2015). Secretion of IL-11 by TGF-β-stimulated CAF triggers STAT3 signaling in cancer cells (Sanchez-Lopez et al. 2015). CAF are important for neoplastic cells survival via the release of diffusible paracrine factors insulin-like growth factor(IGF)-1 and IGF-2 (Hanahan and Coussens 2012), as well as ECM molecules and ECM-remodeling proteases. As an example, Yes-activating protein (YAP) transcription factor is the main player, allowing CAF to remodel ECM by modulating cytoskeleton and matrix stiffness to support tumorigenesis (Calvo et al. 2013). The presence of CAF is also posed as a physical barrier that can hinder the transport of therapeutics and has been correlated with bad prognosis due to low therapeutic responses (Castells et al. 2012), especially in pancreatic cancer (Dimou et al. 2012).

Adipocytes in TME has also been reported to be able to differentiate to cancer-associated adipocytes (CAA) and further into adipocyte-derived fibroblasts (ADF)

(Bochet et al. 2013). CAA has been reported to reduce the cytotoxicity effect of chemotherapy (trastuzumab) (Duong et al. 2015) and radiation therapy (Bochet et al. 2011). They also confer a radioresistant phenotype to breast cancer cells dependent on adipocyte-derived IL-6. Interleukin (IL)-6 is overexpressed in cholangiocarcinoma (CCA) cells in response to hypoxia, and represents a key determinant of cancer invasiveness. Elevated levels of IL-6 plays a key role in promoting cancer cell proliferation and inhibition of apoptosis, by binding to its receptor (IL-6Rα) and co-receptor gp130 (glycoprotein 130), and further activates the JAK/STAT signaling pathway of the Janus kinases (JAK) and signal transducers and activators of transcription (STATs): STAT1 and STAT3 (Landskron et al. 2014).

Since CAF contribute to the production of the dense cellular and non-cellular (through the synthesis of collagen, hyaluronic acid, elastin, glycosaminoglycans, reticular and elastic fibers, glycoproteins (Kalluri and Zeisberg 2006a)) tumor stroma, these cells are the major players in the inability of the drugs to transport across the tumor tissue. In some tumors fibrous stroma comprises >90 % of the tumor mass (e.g. pancreatic adenocarcinoma) preventing the efficient transport of therapeutics and impairing the pharmacological responses. Strategies that can overcome these obstacles are currently being sought.

4.1.4 Non-cellular Components

ECM consists of a complex mixture of proteins, glycoproteins, proteoglycans, and polysaccharides. These components are produced by fibroblasts and have various functions. Within the tumor, ECM stiffness is increased, and its structure is characterized by a loosely woven morphology and nonplanar orientation (Ruiter et al. 2002). These alterations of biochemical and physical properties of ECM favor the tumor progression and resistance to therapies (Giussani et al. 2015). Together with the cellular stroma elements, ECM forms an additional barrier that hinder the chemotherapy penetration into the tumor lesions (Choi et al. 2013). There is a complex signaling processes from the TME cellular components and ECM, where changes in ECM architectures are signaled from the stromal and tumor cells, and these changes impact the tumor stroma. Several studies have identified upregulation of some factors that affect ECM, such as: proteases (cathepsin F, K, and L, MMP2), protease inhibitors (thrombospondin2), serin peptidase inhibitor clade G (C1 inhibitor), member 1 (SERPING1), cytostatin C and tissue inhibitor of metalloproteinase (TIMP3) in ductal carcinoma in situ (DCIS) myoepithelial cells (Allinen et al. 2004). CAF can also modulate ECM via MMP-1 and MMP-3 production (Cirri and Chiarugi 2011). Angiotensin has been used in ECM remodeling strategy, and its inhibition has been found to enhances drug delivery and potentiates chemotherapy via inhibition of collagen I synthesis and subsequent tumor blood vessel decompression (Chauhan et al. 2013; Diop-Frimpong et al. 2011) (Fig. 4.3).

Hyaluronic acid (HA) is another major ECM component that is important for monocyte/macrophage trafficking, in association with HA receptor, CD44. Various

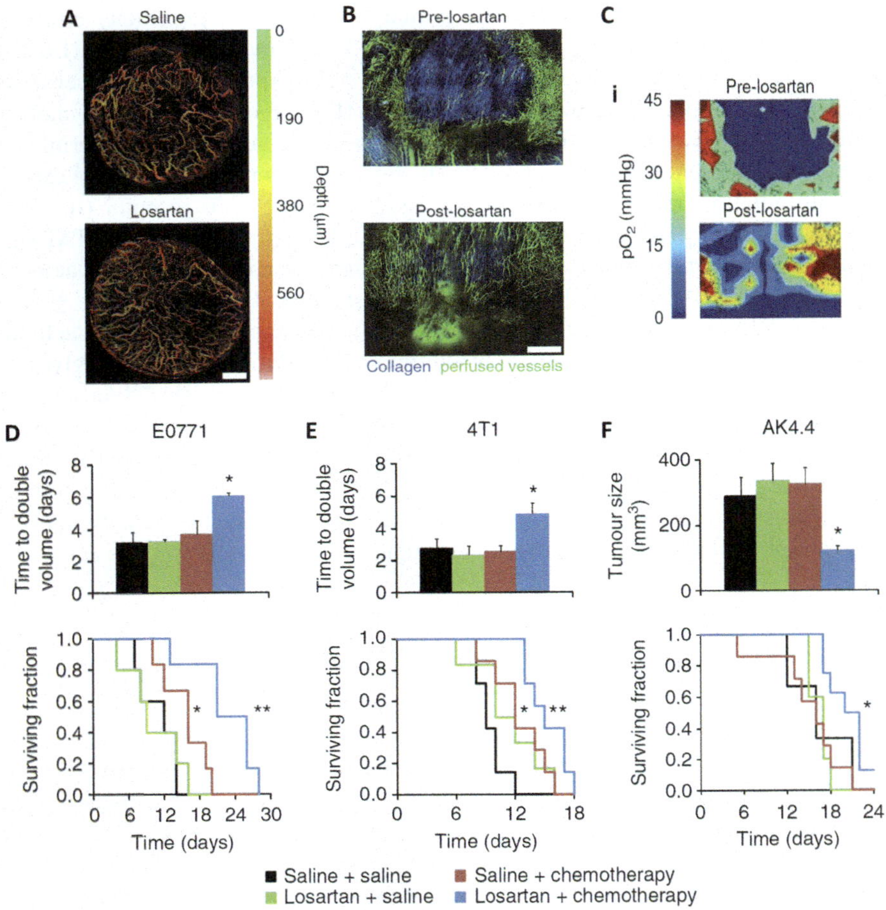

Fig. 4.3 Inhibition of collagen I synthesis by angiotensin. (**a**). Representative images from intravital optical frequency-domain imaging of perfused vessels with losartan treatment. E0771 tumours in control mice have a low density and poor distribution of perfused vessels in three dimensions, whereas losartan-treated mice showed a more even distribution and higher density of perfused vessels. Scale bar, 1 mm. (**b**). Representative images from intravital multiphoton microscopy of perfused tumour vessels (*green*) and collagen (*blue*), showing that losartan increases the density of perfused vessels in an E0771 breast tumor (**c**). Losartan increases oxygenation in some tumours whereas control tumours decrease in oxygen levels. Losartan also appears to result in a more homogenous distribution of well-oxygenated tumour tissue. (**d**-**f**). Quantification of tumour growth rates, based on the time to reach double the initial volume in response to treatment withlosartan or saline control (40 mg kg-1 daily from day 0 onwards) in combination with either the small-molecule chemotherapeutic doxorubicin or saline control (2 mg kg-1 every 3 days from day 1 onwards) (*top* panels) and animal survival for grafted mice following the initiation of treatment with doxorubicin and losartan in monotherapy and in combinatorial therapy (*bottom* panel) in (**d**) E0771 cells graft, (**e**) 4 T1, and (**f**) in AK4.4 tumor graft. Reproduced from (Chauhan et al. 2013) with permission from Nature Communication

HA binding partners, such as inter-α-inhibitor (IαI) heavy chains, TNF-stimulated gene-6 (TNF-6), and proteoglycan versican also play important roles in ECM formation during inflammatory reaction (Chanmee et al. 2014). Versican is one component of the ECM that interacts with cells by binding to non-integrin and integrin receptors and to other ECM components that associate with the cell surface and can influence the ability of cells to proliferate, migrate, adhere, and remodel the ECM (Wight et al. 2014) via the secretion of inflammatory cytokines, such as TNF-α and IL-6 from macrophages through interaction with TLR2 and its co-receptors TLR6 and CD14, which can promote metastasis in a mouse model (Kim et al. 2009).

Tumor ECM also induces macrophages to produce TNF-α and MIP-2 that play role in Toll-like receptor (TLR)-2 and TLR-4-dependant macrophage recruitment and activation. Tumor-derived HA fragments are also able to promote M2 macrophage polarization by a transient early activation of macrophages (Kuang et al. 2007).

4.2 Characteristics to be Considered to Design Nanoparticles Targeting Tumor Microenvironment

Achieving desirable bioavailability of therapeutic agents has been especially challenging in the fields of gene therapy (Wong et al. 2007) and oncology (Godin et al. 2010; Peer et al. 2007; Zamboni 2008). Drug molecule should reach the intended site of action in sufficient concentrations to exert efficient therapeutic action and be protected in the systemic circulation to prevent side-effects (Khalil et al. 2006). The use of nanoscale vectors has been intensively investigated to deliver drugs, genetic material, and imaging contrast agents. Aside from increasing bioavailability and targeting, nanocarriers have been used to enable the in vivo administration of a drug or imaging agent that is chemically unstable, water insoluble, or toxic. Nanocarriers can be made of various materials and modified according to the target moiety (Fig. 4.4). Encapsulation in the nanocarriers was also reported to enhance the efficacy of hard to dissolve hydrophobic drugs (Nie et al. 2007). Typically, tumor therapies are administered intravenously, and the drugs have to overcome several biological barriers to achieve the diseased target tissue (Bally et al. 1999). These barriers span from the systemic ones (e.g. interactions with endothelium, ECM and non-target cells; capture by the mononuclear phagocytic system (MPS), and non-specific biodistribution), to the barriers on the cellular level (e.g. cell/nuclear membrane, lysosomal degradation, efflux pumps). Additionally, particle aggregation and disassembly may be caused by the interactions with blood components, tissue, and immune cells which are specialized in uptake and removal of foreign particles,. These can lead to non-specific organ accumulation and low targeting efficiency and adverse effects (Hunter and Moghimi 2010; Mislick and Baldeschwieler 1996).

Better understanding of the physiological differences between cancerous and normal tissue has opened new opportunities to target cancer microenvironment, allowing better design and development of new nanocarriers targeting tumor. Design

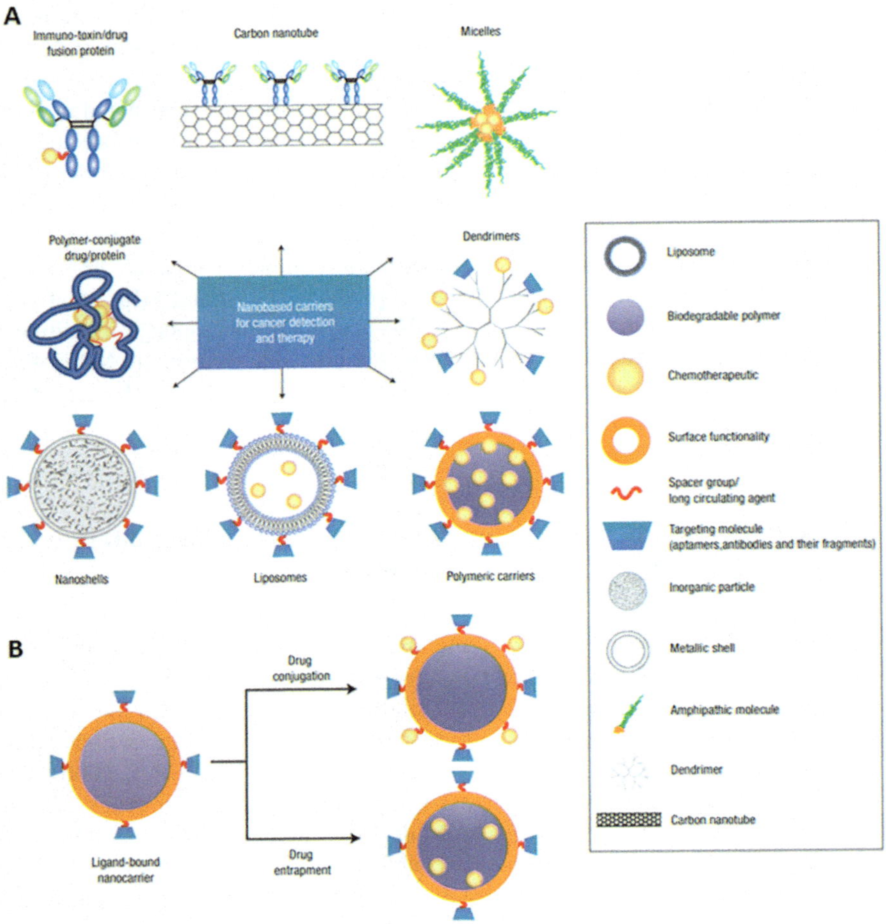

Fig. 4.4 Examples of nanocarriers for targeting cancer. (**a**). A whole range of delivery agents are possible but the main components typically include a nanocarrier, a targeting moiety conjugated to the nanocarrier, and a cargo (such as the desired chemotherapeutic drugs). (**b**). Schematic diagram of the drug conjugation and entrapment processes. The chemotherapeutics could be bound to the nanocarrier, as in the use of polymer–drug conjugates, dendrimers and some particulate carriers, or they could be entrapped inside the nanocarrier. Reproduced from (Peer et al. 2007) with permission from Nature Nanotechnology

of drug nanovectors involves the manipulation of several critical factors, including their chemical functionality and mechanical properties, the degree of coating or encapsulation of the drug to the nanovector, size and shape of the particle, and the surface charge (Table 4.2). The role of spherical particle size and composition in biodistribution (Sun et al. 2005), cellular binding (Cortez et al. 2007), cell entry pathways (Rejman et al. 2004a), cell uptake (Hu et al. 2007; Win and Feng 2005; Zauner et al. 2001), and tumor bed penetration (Dreher et al. 2006a; Goodman et al. 2007) has been extensively studied. It is possible to modulate various physico-

Table 4.2 A comparative summary of organ biodistribution of nanovectors as a function of their size, shape and surface charge (Data from (Blanco et al. 2015; Decuzzi et al. 2010b))

Effect			Lung	Liver	Spleen	Kidney
Size	Spherical	< 20 nm	+	+	+	++++
		20–150 nm	++	++	++	+
		> 150 nm	+++	+++	+++	+
	Discoidal	600 x 400 nm	+	+++	+	++
		1000 x 400 nm	++	++	+	++
		1700 x 400 nm	+++	+	++	+
Shape		Spherical	++	+++	+	+
		Hemispherical	o	+++	++	+
		Cylindrical	+	++++	++	+
		Discoidal	+++	++	++	+
Surface charge		Negative	+	+	+	+
		Neutral	+	+	+	+
		Positive	++	+++	++	+

chemical characteristics of the nanocarriers, including geometry and surface charge, or to add moieties that will target elements overexpressed in the TME to improve their distribution to the tumor tissue and therapeutic efficacy. These manipulations in characteristics affect particles' transport in the bloodstream, extravasation out of the blood vessels, diffusion through the tumor tissue, degree of cellular internalization, and subsequent uptake pathway as well as their long-term fate.

Further in this chapter we will be discussing a few strategies that have been utilized to enable better accumulation of the therapeutics in the tumor microenvironment using nanocarriers.

4.2.1 *Geometry*

Geometry is a very important feature in the nature and accounts for the, size and shape of the nanoparticles. Early nanoparticles are characterized by a spherical geometry, mostly due to the easiness of their fabrication, as during the bottom-up fabrication (self-assembly), the spherical shape is generally the most thermodynamically stable one. Advances in material sciences have enabled fabrication of nanotherapeutics with varying shapes mimicking biological entities. Most of the naturally occurring blood-born particulates and cells (e.g. erythrocytes, platelets, bacteria) possess non-spherical geometry which allows them to efficiently navigate in the blood stream. Moreover, cells are known to change their geometry when their function is changed. An example of this behavior is the circulating monocytes, which change their spherical geometry to an edgy non-spherical one when they are differentiated to macrophages. Traditionally, nanocarriers fabrication process utilize "bottom-up" method, relying on self-assembly of polymers, phospholipids or metallic materials based on hydrophobic and electrostatic interactions. New fabrication processes enable the fabrication of various shapes/sizes of nanocarriers and use "top down" approach or a combination of "bottom-up" and "top down" approach. These methods such as photolithography, particle replication in non-wetting templates (PRINT®), silicone molding, as well as layer-by-layer hydrophobic-hydrophilic block copolymer formation using sacrificial rigid core templates.

4.2.1.1 Size

As it was mentioned above, in the presence of tumor-associated inflammation and angiogenesis, blood vessels in the tumor are characterized by the presence of aberrant proliferating endothelial cells with disorganized branches, irregular capillary bed, enlarged interendothelial gaps, as well as disrupted basement membrane, with enhanced vascular permeability. Size is very important for designing nanoparticles drug delivery system, as they can determine in vivo fate. The size of gap junction between endothelial cells leaky tumor vasculature varies between 100 and 700 nm (in some cases up to 1200 nm) (Hobbs et al. 1998). Small particles (<5 nm) (e.g. some micelles, quantum dots, or carbon-based nanofullerens) will diffuse out in tumor and normal tissue, while not being able to retain in the tumor interstitium (Dreher et al. 2006b). These smaller particles generally undergo a rapid renal clearance, and thus would be less suitable to serve as a sustained release drug delivery system without any further modification. Tumor lymphatic vessels are also compromised and collapsed, leading to poor drainage from tumor tissue (Leu et al. 2000) and causing prolonged resident times of larger nanocarriers in the tumor vasculature. These specific physiological conditions contribute to the EPR effect which affects the retention of particles between 10-500 nm (Maeda et al. 2001; Torchilin 2011). To achieve long-circulating NPs that accumulate in the tumor tissue, a diameter between 30 nm and 200 nm is desired (Jain and Stylianopoulos 2010). Larger NPs (>100 nm) remain trapped in the extracellular matrix between cells and do not extravasate far beyond the blood vessel, while smaller NPs (20 nm) penetrate deeper into the tumor tissue but are not retained beyond 24 h.

EPR effect has also been found to be amplified by chemotherapies that affect VEGF, such as doxorubicin (Minko et al. 2000). However, due the high interstitial pressure within the tumor lesion, the nanocarriers have difficulties penetrating the lesion and have limited distribution only around tumor vicinity (Baxter and Jain 1989). Another emerging problem in defining the optimum size for nanoparticles is that with the successful treatment, the vascular branches are becoming more organized and less leaky. For example, in breast cancer model in mice after VEGFR2 blockade, the 10 nm nanoalbumin bound-paclitaxel (Abraxane®) was more efficient than 100 nm liposomal doxorubicin (Doxil®) (Jain 2013).

In general, particles without modifications are treated as a foreign body and targeted by MPS following opsonization of nanocarriers with serum proteins such as IgM, fibronectin, and C-reactive proteins, which are recognized by MPS as the "capture" signal (Owens and Peppas 2006). Particles larger than 0.5 μm without any modifications can invoke "bacteria-like" responses and will be cleared by MPS within minutes or hours via phagocytosis by macrophages, while smaller particles up to 200 nm can invoke "virus-like" responses (Xiang et al. 2006). This tendency has been exploited to design NPs that can invoke, suppress, or avoid immune response (Dobrovolskaia and McNeil 2007). For tumors located in organ of MPS, such as hepatocarcinoma or liver metastasis of breast and lung tumors, the process can be beneficial for accumulating nanovectors in the liver and unfolding of their therapeutic effects. In other cases, nanocarriers are generally designed to avoid immune system.

Interestingly, the dimensions of the contact point of particles play an important role in determining whether macrophages initiate phagocytosis or simply spread on particles (Champion and Mitragotri 2006). Non-symmetrical particles are internalized faster by macrophages if the point of contact is on the smaller dimension of the particles, this the particle size is sculpturing the ability of phagocytes to complete the process of the particle uptake (Champion and Mitragotri 2006).

Further, non-functionalized systems less than 200 nm in size (Grosse et al. 2005; Rejman et al. 2004b), such as transferrin or low-density lipoprotein receptors for specific binding (Brodsky et al. 2001; Dautryvarsat et al. 1983; Schmid 1997; Takei and Haucke 2001), are typically internalized via clathrin-mediated endocytosis (CME) (Grosse et al. 2005; Rejman et al. 2004b), which is considered kinetically the most effective (Goncalves et al. 2004; Zhou and Huang 1994) endocytic uptake pathway for various particles. Micro-sized particles typically undergo a classic endosome-mediated phagocytosis trafficking to the lysosomes (Koval et al. 1998; Rejman et al. 2004c).

4.2.1.2 Shape

Shape has also been shown to play a critical role in the margination of NPs and can determine the fate of the particles (Gentile et al. 2008; Lee et al. 2009). Particle shape is important in the mechanism of cell entry (Champion et al. 2007; Champion and Mitragotri 2006; Gratton et al. 2008a) and the release rate of the therapeutic cargo (Goldberg et al. 2007; Mader et al. 1997). Rod-shaped micelles have been shown to have circulation lifetime ten times longer than spherical micelles (Geng et al. 2007). In general, particles in the sub-100 nm particle size range show advantage of spheres over rods (Chithrani et al. 2006; Qiu et al. 2010). However in NP over 100 nm size, Gratton *et al* (Gratton et al. 2008a) found that rods were taken up the most, followed by spheres, cylinders, and cubes, and increasing the aspect ratio of nanorods seems to decrease total cell uptake. Nanotubes endocytosis rate was 1000 times higher than for spherical gold NPs, and similar exocytosis kinetics was measured for poly(d,l-lactide-co-glycolide), single-walled nanotubes (SWNT), and Au nanoparticles across distinct cells (Jin et al. 2009a). Further, dynamic modulation of particles geometry can be used to precisely control particle-cell interactions (Caldorera-Moore et al. 2010a, b). Yoo and Mitragotri (2010b) have designed stimulus-responsive polymeric particles by controlling a fine balance between polymer viscosity and interfacial tension (Fig. 4.5). The NP's shape can be tuned based

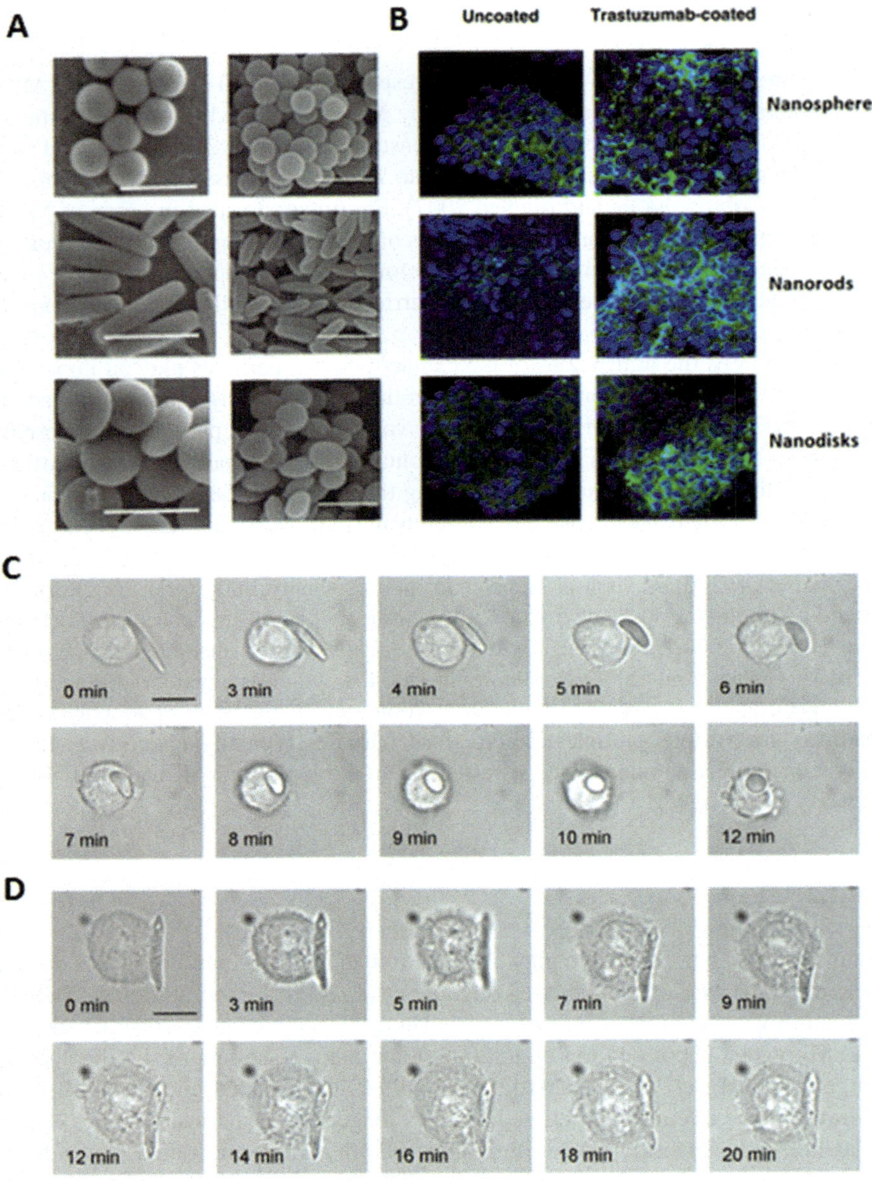

Fig. 4.5 Importance of shape for intracellular uptake of nanocarriers and its modulation to increase macrophage uptake. (**a**). Scanning electron micrographs of particles used in this study and images of their uptake in BT-474 cells. Microspheres, rods, and disks (scale: 2 μm). (*Right*) Nanospheres, rods, and disks. (Scale bar: 500 nm.) (**b**). Confocal micrographs of nanoparticle uptake in BT-474 cells. (*Left*) Nanoparticles without trastuzumab. (*Right*) Nanoparticles with trastuzumab. (**c-d**), Time-lapse video microscopy clips of shape-dependent phagocytosis by macrophage. C. shape-switching PLGA-ester ED (mixture of two PLGAs, AR = 5) was initially attached on a macrophage and not phagocytosed. The macrophage then quickly internalized the particle once shape switched to near-sphere shape. (**d**). Macrophage spread on a PLGA-acid ED [molecular mass 4.1 kDa, T_g(mid) = 27 °C, AR = 5], which do not switch shape at pH 7.4, but could not complete phagocytosis. All particles were opsonized with mouse IgG before the experiments. (Scale bar: 10 μm). Reproduced from (Barua et al. 2013; Yoo and Mitragotri 2010a) with permission from PNAS

on the external stimuli (temperature, pH, or chemicals in the medium), enabling the modulation of phagocytosis of particles that previously were not internalized by the cells.

Discoidal or hemispherical particles have different flow characteristics from spherical particles, and, depending on their size, can accumulate in different organs (Table 4.2). Various shapes of nanocarriers behave differently in the blood flow, leading to different margination dynamics, causing unique distribution patterns *in vivo*. In the study by Decuzzi et al., discoidal particles marginate more towards endothelial cells when compared with hemi-spherical and spherical particles (Decuzzi and Ferrari 2006; Serda et al. 2011). Another study utilizing variation of shape with constant size has shown that particle shape, rather than size, which plays a dominant role for determining the complexity of the local actin structure (Champion et al. 2007; Champion and Mitragotri 2006). The margination increased the potential of nanocarriers to accumulate in various organs before they are cleared in the liver. Additionally, size of discoidal particles has been found to contribute to biodistribution in vivo (Table 4.2). Smaller particles have increased accumulation within the liver, and kidney, and most importantly, in tumor vasculature. Larger particles, however, are mostly enriched in the heart, lung and spleen. Larger size of 1700 nm was inefficient for EPR effect in the TME (Decuzzi et al. 2010), and tend to be sticking to the lung vasculature, which may cause increased risk of embolism. Oblate spheroidal can dramatically affect firm adhesion under flow, as oblate-shaped particles are subjected to torques resulting in tumbling and rotation, in contrary to classical spherical particles (Gavze and Shapiro 1998; Lee et al. 2009). These predictions of particle distribution have been confirmed by both in vitro experiments with silica particles (Gentile et al. 2008) and by in vivo studies with silicon discoidal particles (Godin et al. 2012) and discoidal polystyrene particles targeted to Intercellular Adhesion Molecule 1 (ICAM-1) molecules, that both exhibited higher tumor accumulation/targeting specificity compared to spherical particles (Muro et al. 2008a). Further, after being taken up by the cells, various sizes of ICAM-1 targeted spherical and elliptical shapes polymeric nanoparticles (100 nm to 10 μm) have shown different characteristics according to the size, and particle size above 1 μm caused particles to accumulate and remain longer in pre-lysosomal compartments, while submicron nanovectors were trafficked immediately to lysosomes (Muro et al. 2008b). Discoidal nanovectors were internalized by endothelial cells through a combination of phagocytosis and micropinocytosis mechanisms that are actin-driven and involve extensive membrane reorganization with the formation of pseudopodia and extension of the cell membrane to engulf the particles (Muro et al. 2004; Serda et al. 2009).

Due to the progresses in various fabrication techniques that enable the geometry tuning of the nanocarriers, various size and shape of drug nanocarriers is being explore. The hollow microcapsules composed of polymeric film produced by a combination of top-down and bottom-up fabrication approaches, have shown to possess an extended in vivo circulation time (Merkel et al. 2011), controllable degradation (De Koker et al. 2007), tunable drug release (Kozlovskaya et al. 2009; Liu et al. 2005; Ng et al. 2011), capable of surface functionalization for enhanced cellular association, and function as suitable drug delivery agents to target cells (Cortez et al. 2006; De Cock et al. 2010; Johnston et al. 2012; Palamà et al. 2010; Patel et al. 2012; Pridgen et al. 2007; Sukhorukov et al. 2007; Tao et al. 2011; Vergaro et al. 2011; Wang et al. 2007; Yan et al. 2010). Recently, Alexander et al. reported that the shape of hollow cubical particles can improve the interaction with breast cancer cells and endothelial cells as compared to their spherical counterpart (Alexander et al. 2015). Interestingly in the same work, the elasticity of the particles determined their interactions with the immune cells.

Cell ability to deform seems to be critical for particle phagocytic predisposition, although there has been contradicting findings and further characterization is needed. Macrophage is known to have preference for rigid particles since they resemble bacteria and other pathogens with sturdy cell walls. In one study, soft polyacrylamide particles were unable to stimulate the assembly of actin filaments required for the proper phagosomes formation, as opposed to rigid particles which had same total polymer mass and surface properties (Beningo and Wang 2002). However, particle rigidity can have an opposite effect on opsonization. Rigid liposomes and poly(methyl acrylate) PMA-based core-shell NPs with rigid core are known to decrease complement activation and hence decrease phagocytosis (Allen et al. 1991; Sun et al. 2005).

The effect of rigidity can also be observed in a study with poly(methacrylic acid) capsules, where the spherical capsules were internalized by HeLa cells at a faster rate than rod-shaped capsules with a higher aspect ratio (Shimoni et al. 2012), while in case of sized rigid particles, rod-shaped particles with higher aspect ratio showed a higher internalization rate compared to their spherical counterpart (Gratton et al. 2008b).

NPs deformability can be the key of overcoming physical filtration barriers in the body (Fox et al. 2009; Mitragotri and Lahann 2009). Several studies have been published on particles that resembled red blood cells (RBCs) in size and shape, and have been shown to deform in restricted channels (Haghgooie et al. 2009) or capillaries (Doshi et al. 2009) with smaller than the particle diameter. Unfortunately, their modulus was poor compared to RBCs (Doshi et al. 2009). The advantage of RBC-shaped particles has been explored by Merkel at al, where out of varying size of hydrogel PRINT particles, the ones comparable with RBC in size, shape and deformable was the particles with the longest circulating, and is uniquely suited to avoid clearance (Merkel et al. 2012).

Geng et al. have compared soft spherical assemblies and flexible filaments, and found ten times increased in vivo circulation time for non-spherical filomicelles (Geng et al. 2007). This study was translated to a xenograft mouse model, and delivery of paclitaxel using filomicelles caused filomicelle length-dependent tumor shrinkage. This finding shows an advantageous of long, worm-like structure for a prolonged circulation time, as compared to classical sphere.

4.2.2 Surface Charge

Surface charge of nanocarriers can influence their accumulation in TME (Bazak et al. 2014). Blood vessels have a net negative charge on the luminal side caused by carboxylate sugar and sulfate group and particle uptake is driven by electrostatic interactions (Jin et al. 2009b; Srinivasan and Sawyer 1970; Wang et al. 2010). Negatively charged heparan sulfate proteoglycans and integrins play major role in the cellular binding of positively charged particles, shown by reduced binding of cationic particles to the cell membrane in proteoglycan-deficient mutant cells (Mislick and Baldeschwieler 1996). Clustering of trans-membrane proteins and syndecans at the plasma membrane during binding to cationic particles can stimulate interaction with the actin cytoskeleton and result in the formation of tension fibers, which provides the energy required for internalization (Woods and Couchman 1994). Positively charged particles have been reported to have increased efficacy of imaging, gene transfer, and drug delivery, and are taken up at a faster rate (Slowing et al. 2006; Thorek and Tsourkas 2008). However, they are also associated with strong immune response and increased opsonization and removal from circulation, and needs to be fine-tuned, depending on the strategy and target disease. Cationic NPs exhibited the lowest half-life in the bloodstream and can cause several complications such as hemolysis and platelet aggregation (Albanese et al. 2012).

In some cases, where dendritic cell maturation and vaccine-induced immune responses are sought, the cationic nanocarriers can be advantageous. Positively charged liposome has been shown to have higher accumulation in tumor blood vessels and lower in the spleen compared to the negatively charged particles (Campbell et al. 2002). Positive charge of nanocarrier surface causes swelling and disruption of lysosomes due to buffering of H^+, a process that allows their escape from endosomes and entrance to the cytoplasm, while anionic particles are taken up into the lysosome potentially causing cell death (Fröhlich 2012).

In smaller NPs (2 nm), positive charge can perturb the cell membrane potential and inhibit cell proliferation via Ca^{2+} influx into the cells (Arvizo et al. 2010). In larger NPs (4–20 nm), surface charge can induce reconstruction of lipid bilayers (Wang et al. 2008), where polycationic molecules interact with fluid-phase domains of lipid bilayers (Mecke et al. 2005), inducing the development of plasma membrane pores leading and membrane disruption. Binding of negatively charged NPs to a lipid bilayer can cause local gelation, whereas positively charged NPs induce membrane fluidity.

Additionally, protein corona can be formed via particle (especially of high surface density)interactions with serum proteins such as lipoproteins, immunoglobulin, complement and coagulation factors, acute phase proteins and metal-binding and binding proteins (Cedervall et al. 2007) during circulation in the blood. This can alter particle-cell interaction and thus, protecting of the cells from the damage induced by the highly positive NP surface (Wang et al. 2013).

4.2.3 pH

Average extracellular pH in the tumor is between 5.8 and 7.6, lower than the pH of normal tissue of 7.0–8.0 (Tannock and Rotin 1989) due to increased fermentative metabolism and poor perfusion.

Acidic pH promotes local invasive growth and metastasis, and regulation of pH may play an important role in maintaining the viability of tumor cells. Acidic TME also affects upregulation of significant genes such as MMP-9, acidic sphingomyelinase in mouse B16 melanoma (Kato et al. 2007), platelet-derived endothelial cell growth factor (thymidine phosphorylase) in human breast cancer cells (Griffiths et al. 1997), inducible isoform of nitric oxide synthase (iNOS) in mice macrophages via activation of NF-κB (Bellocq et al. 1998), VEGF-A in glioma and glioblastoma cells (Fukumura et al. 2001; Xu et al. 2002). Acidic and hypoxia condition also regulates IL-8 expression in human pancreatic adenocarcinoma and ovarian carcinoma cells, causing increased tumorigenesis and metastatic potential (Shi et al. 1999). H^+ that diffuses from the TME into the adjacent normal tissues causes tissue remodeling and allows local invasion. Administration of proton pump inhibitors which buffer tumor acidity, can improved CD8+ cell function in vivo and increase the efficacy of immunotherapy. Oral administration of sodium bicarbonate was able to increase peritumoral pH and inhibit tumor growth and local invasion in preclinical models (Estrella et al. 2013). Combination of bicarbonate therapy with anti-CTLA-4, anti-PD1, or adoptive T-cell transfer can improve antitumor responses in multiple models and can lead to the cure (Pilon-Thomas et al. 2016).

To deliver drugs into the TME, nanocarriers need to be stable in neutral pH and can be activated in low pH of TME. This can be achieved in two ways: prodrug, or mostly by designing tunable nanocarriers for loaded drug content release in low pH. In regards to tunable particles, Yang et al developed a tumor acidity-responsive sheddable poly(ethylene)glycol (PEG) layer for delivery of siRNA (Yang et al. 2012), and Meng et al developed pH-Sensitive Polymeric Nanoparticles (Meng et al. 2014). Additionally, another approach with pH-sensitive polymer was developed using computational simulation to follow the particle dynamics (Ding and Ma 2013). Hollow elastic particles made from polymeric materials have also been shown to mimic erythrocyte and able to change size and shape based on environmental pH (Kozlovskaya et al. 2014), while their shape can be manipulated by controlling time, temperature and stirring condition during the fabrication.

4.2.4 Modifications of Nanocarriers

4.2.4.1 NP Modification for Targeting

The pathophysiological changes in TME, such as upregulation of specific receptors has been widely used to direct NP systems into the tumor region by conjugation with targeting antibodies and ligands (Allen 2002). However, antibody-drug conjugate have failed in clinical trials due to the limited number of drug molecule that can be conjugated without compromising their target affinity (Panowksi et al. 2014). The use of NP systems such as liposomes and polymers can potentially increase the number of drugs conjugated to NPs without limiting the targeting moiety (Kamaly et al. 2012). Ideal targeting moiety should be tumor specific and homogenously within all targeted tumor cells, however, in most cases, the cancer cell heterogeneity limits the targeting efficacy (Janku 2014). Furthermore, the choice of ligands depends on whether the drug conjugates destination is to be internalized after binding of ligand to the receptors. Internalization via receptor mediated endocytosis cause an endosome formation, in which acidic pH cause drug release into cytoplasm. Additionally, in choosing the receptor and ligands, it is important that the surface antigen and receptors should not be shed into blood circulation. The most prominent targets for the targeting are folate, transferrin, and lectin receptors.

Higher activity of drug against tumor growth was observed in heparin-folate-Taxol (paclitaxel) tested in mice model (Cho et al. 2008), when compared to the heparin-Taxol conjugates or free drug. Similar system has also been found to inhibit proliferation of P-gp overexpressing cell line (Sun et al. 2011).

Aptamers can be also as a targeting moiety, while some of them possess therapeutic efficacy. Aptamer is based on oligonucleotide molecules that exhibit affinity to the particular tertiary structure of the receptors (Thiviyanathan et al. 2007). Aptamer binding screening a specific receptor are conducted by a quick, efficient and automated selection method, SELEX (Tuerk and Gold 1990). When one or more of the phosphoryl non-bridging oxygen atoms of the aptamer modified with sulfur, they produce monothiophosphate-backbone-modified DNA aptamers (thioaptamers), which displayed tighter binding to the RNase H domain of the HIV Reverse transcriptase and inhibited its activity in vitro. When conjugated to liposome and transfected into HIV infected astrocytoma, the DNA thioaptamers demonstrated > 70 % efficiency (Thiviyanathan et al. 2007). Hybrid thioaptamers consist of a combination of thiophosphate and unaltered phosphate show maximum affinity and specificity for receptors and are nuclease resistive (Serda et al. 2010).

Aptamer conjugation to therapeutics or the surface of particles has been used to target different cancer cell types (Debbage 2009). Small interfering RNA conjugation with anti-prostate specific membrane antigen (anti - PSMA) aptamer has been used to silence genes that are responsible for malignancies, and to slow tumor growth in mice without eliciting side effects associated with cancer therapies (Chu et al. 2006).

In another study docetaxel-encapsulated poly(lactic-co-glycolic acid) NP conjugated with aptamer targeting prostate specific membrane antigens exhibited a considerable increase in in vitro cellular toxicity in comparison to the non- targeted NPs (Farokhzad et al. 2006).

Targeting tumor microenvironment can also be done using phage display peptide libraries that allows the presentation of large peptide and protein libraries on the surface of filamentous phage of single-stranded DNA viruses that infect a number of gram-negative bacteria, leading to the selection of peptides and proteins, includ-

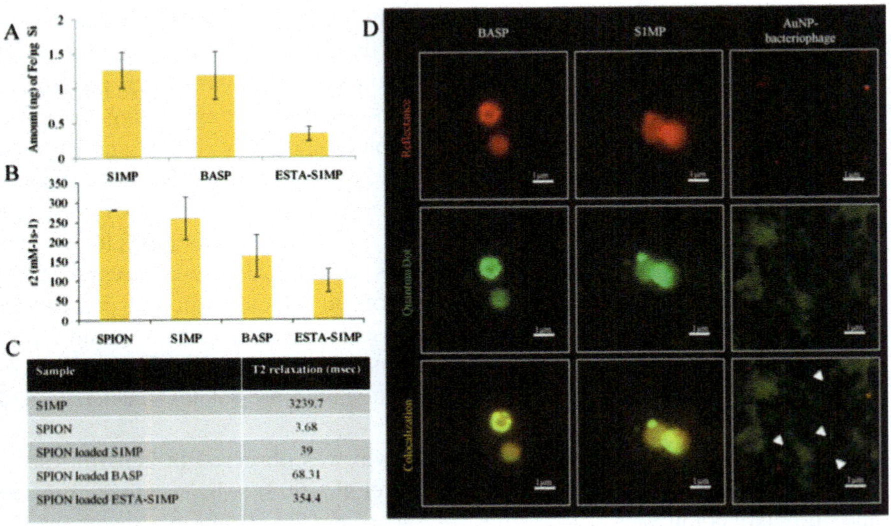

Sample	T2 relaxation (msec)
SIMP	3239.7
SPION	3.68
SPION loaded SIMP	39
SPION loaded BASP	68.31
SPION loaded ESTA-SIMP	354.4

Fig. 4.6 Characterization of SPION loading into stage 1 mesoporous particles (*S1MP*) (**a**) Iron content analysis in S1MP, BASP and ESTA-S1MP following loading of SPION using the wet incipient method as quantified by ICP-OES; (**b**) Comparison of transverse relaxivity, r2 and (**c**) relaxation times, T2, of SPION loaded particles; (**d**) Fluorescence microscopic images of QD (*green*) loaded S1MP and BASP (*red*) showing S1MP colocalization with QD (*yellow*). In a control experiment, free QD nonspecifically bound to AuNP-bacteriophage networks (*arrow heads*) enable the visualization of the bacteriophage fibers (*green*) while the AuNP appear as red dots. Reproduced from (Srinivasan et al. 2013)

ing antibodies (Pasqualini et al. 1995; Pasqualini and Ruoslahti 1996). A Multistage Nanovector based on Bacteriophage Associated Silicon Particles (BASP), obtained by mixing AuNP-bacteriophage networks with porous silicon particles has shown that addition of bacteriophage to the porous particles can increase drug load efficiency and increase the accumulation in the tumor up to threefold when applied intravenously (Srinivasan et al. 2013) (Fig. 4.6).

4.2.4.2 NP Modification to Avoid Mononuclear Phagocyte System MPS Clearance

Another reason for nanovactors modification is to avoid rapid clearance by MPS, and can be achieved by adding PEG or by block copolymerization with hydrophilic and hydrophobic domains to the particles surface(Adams et al. 2003; Harris et al. 2012).

Dunn et al. (1994) showed that PEGylation of 100 nm polystyrene particles can overcome removal and sequestration in the resident Kupffer cells of the liver in rats. The process involves prevention of opsonisation when PEG was conjugated, and drastically increases the blood half-life of all nanomaterials regardless of surface charge. The blood half-life is generally directly correlated to the extension of PEG length, which causes the thickening of the protective layer (Ferrault et al. 2009). These "stealth" long circulating particles improve accumulation in target tissue since they are able to circulate in the systemic blood flow longer, increasing the chance to accumulate in the leaky vasculature. Stealth liposomes are the most clinically used nanoparticles (e.g. Doxyl®). Pristine and PEGylated single-walled carbon nanotubes were found to be toxic *in vitro*, but can persist within liver and spleen macrophages for 4 months without apparent toxicity *in vivo* in rats and rabbits (Schipper et al. 2008). When nanocarriers include a metallic component, such as gold, silver, or iron oxide, they can additionally be activated by triggering with an external energy source, such as radio-frequency (Glazer et al. 2010) or near-infrared (Loo et al. 2004) energy.

During the last decade, CD47 (integrin-associated protein) has been identified to be the marker of self for RBC that prevented elimination from the bloodstream by binding to the inhibitory receptor signal regulatory protein alpha (SIRPα) (Oldenborg et al. 2000). This could be used for designing long circulating nanocarriers delaying macrophage-mediated clearance, and promoting circulation time to enhance dye and drug delivery to tumors (Rodriguez et al. 2013).

4.2.4.3 NP Modification for Intracellular Fate Modulation

To avoid acidic and/or enzymatic degradation in the late endosome or lysosome after NP uptake, it is advantageous to use conjunction with a polymer material, such as polyethylenimine (PEI) (Sonawane et al. 2003), polyamidoamine (PAMAM) dendrimers (Sonawane et al. 2003) or an imidazole-containing polymers (Pack et al. 2000), that can induce endosomal escape of the polyplex into the cytoplasm of the cell before significant degradation occurs. Coating NP with cationic polymers, such as polyethylenimine (PEI) initiates endosomal escape via so-called "proton sponge effect" due to the buffering inside endosomes. This includes additional pumping of protons into the endosome, along with a concurrent influx of chloride ions to neutralize the net charge, causing an increased ionic strength inside the endosome and osmotic swelling, physical rupture of the endosome, and the escape of the vesicle material into the cytosol and, therefore, avoiding the degradative lysosomal trafficking pathway (Behr 1997).

Caveolae-mediated endocytosis maybe an advantageous internalization mechanism for gene transfection strategies as the vesicles that result from the internalization in this pathway do not develop into lysosomes and, therefore, avoid significant degradation (Gabrielson and Pack 2009; Sahay et al. 2010). Conjugation to caveola-targeting TX3.833 antibody was reported to increase drug delivery to the lung up to 172-fold and achieving rapid, localized efficacy (McIntosh et al. 2002). Subsequent publication from the same group reported aminopeptidase P (APP), the target enzyme for TX3.833 antibody, to be >20-fold enriched in the isolated caveolae compared to the luminal endothelial plasma membranes of lung endothelium. The antibody was able to accumulate 446-fold more in the lung than control mIgG. Further coating of 10–15 nm colloidal gold NP with APP antibody was able to induce specific and rapid concentration of NP in the lung tumor microvasculature, and the APP antibody was more superior than non-caveolae-targeting antibody mCD34 (Oh et al. 2007).

NP size and shape also dictates the number of ligands accessible to interact with the target receptor. The curvature of the NP surface determines the avidity of binding of NP-linked ligands to available receptor. The binding is most likely causes the uptake of NP via receptor-mediated endocytosis, caused by a localized decrease in Gibbs free energy that induces the membrane to wrap around the NP to form a closed vesicle structure (Chithrani et al. 2006; Gao et al. 2005; Jiang et al. 2008). The vesicle buds off the membrane and fuses with other vesicles to form endosomes, which are further fused with lysosomes, where degradation occurs. As it was earlier mentioned, uptake of particles is size dependent, and most likely related to the membrane wrapping processes. Small ligand-coated NPs may interact with only one or two cell receptors, while similar NP of ~100 nm size has many more ligand-receptor interactions per particle. To be able to produce enough free energy to drive membrane wrapping, several small-ligand coated NPs must bind to receptors in close proximity one to another. Thermodynamically, a 40–50 nm NP is capable of recruiting and binding enough receptors to successfully produce membrane wrapping. Particles >50 nm can bind to such a large number of receptors that can exhaust receptor on cell surface, limiting the uptake depletion of receptor, and causing redistribution of receptors on the cells surface via diffusion to compensate for local depletion. Mathematical modeling of this phenomenon has shown that NPs 30–50 nm in diameter are the most optimal for ligand-targeted endocytosis, since there is no ligand shortage on the NP, and there is no localized receptor shortage on the cell surface (Yuan et al. 2010).

Additionally, lysosome can cleave targeting ligands from NPs by the protease cathepsin L inside endo-lysosomal vesicles (See et al. 2009), and cause slow particle core structure decomposition (e.g. in quantum dots) by its enzymes (Fischer et al. 2010). Therapeutic payload is intended to act in the cell compartments such as nucleus and cytoplasm and the accumulation in lysosome and endo-lysosome can strongly limit its efficacy, therefore, the escape from endosomes is favorable. If NP is engineered to escape the endo-lysosomal system and enter the cytoplasm, it may interact with intracellular organelles and potentially can affect cell behavior. Cytoplasm also contains nucleases and proteases that can neutralize the active compound. It was shown that plasmid DNA is degraded in the cytoplasm of HeLa and COS cells with a half-life of 50–90 min (Lechardeur et al. 1999). Additionally, the high viscosity of cytoplasm environment can decreases the mobility of macromolecules (Dauty and Verkman 2005; Kao et al. 1993; Lukacs et al. 2000).

Cell function can also be influenced by the NP, it has been observed that 40–50 nm Herceptin-coated gold NPs altered cellular apoptosis by influencing the activation of caspase enzymes (Jiang et al. 2008). Similarly, receptor-specific peptides conjugated to NP surface can improve their ability to induce angiogenesis (Kanaras et al. 2011). These reports highlight the advantage of having the ligand bound to NP as opposed to being free in solution, since they create a region of highly concentrated ligands, which increases avidity and, alters cell signaling.

A concern with NP-ligand complexes is that the ligands are potentially denatured when bound to the engineered surface, thus affecting its binding to the receptor, increasing nonspecific interactions or provoking inflammation. For example, when lysosome is bound to gold nanoparticles, it denatures and interacts with other lysozyme molecules inducing the formation of protein-NP aggregates (Zhang et al. 2009). Fibrinogen can also unfolds when bound to the surface of polyacrylic acid-coated gold NPs, causing binding to the integrin receptor Mac-I and cause an inflammatory response (Deng et al. 2011).

Once arriving in the cytosol, NP can elicit biological responses by disrupting mitochondrial function, thereby provoking production of reactive oxygen species and activating the oxidative stress mediates signaling cascade (AshaRani et al. 2009), which can induce oxidative DNA damage and promote micronuclei formation (Berneburg et al. 2006). Certain types of NPs can induce nuclear DNA damage that leads to gene mutations, cell cycle arrest, cell death, or carcinogenesis. When cells proliferate, NPs that persist in the cytosol during mitosis will be distributed within the daughter cells (Rees et al. 2011), However, the effect of the NPs on daughter generations remains unclear and additional work is still required to fully understand the fate and toxicity of internalized NPs.

4.3 Nanovectors Targeting Tumor Microenvironment

Research in the field of nanotechnology towards cancer therapeutics has been growing exponentially in the recent years. The strategy involves the expectation for better accumulation and transport of therapeutics to the disease tissue based on their pathological conditions. A vast numbers of nanovectors made of different materials with various geometries and surface properties have been explored for clinical applications of cancer chemotherapies. These nanovectors include polymer conjugates (Duncan 2006), lipid based carriers like liposomes (Torchilin 2005) and micelles (Torchilin 2007), porous silicon (Serda et al. 2010) and silica (Rosenholm et al. 2010) based nanovectors, polymeric nanoparticles (Danson et al. 2004), carbon nanotubes (Bianco 2004), dendrimers (Gillies and Frechet 2005), gold nanoparticles (Sokolov et al. 2003), nanowires (Cui et al. 2001) and nanofibers (Venugopal and Ramakrishna 2005). Cancer nanotherapeutics that are commercially available and in the pipeline are summarized in Table 4.3.

4.3.1 Vasculature

Tumor vasculature is characterized by a leaky endothelium with heterogeneous vessel sizes and shapes. They supply cancer cells with nutrients and oxygen needed for the tumor growth. But, due to the vascular leakiness coupled with impaired lymphatic drainage, it also allows preferential accumulation of therapeutic-loaded NP

Table 4.3 Nanotechnology-based Chemotherapeutics approved by Food and Drug Administration and in clinical trials (Data gathered from FDA and clinicaltrial.gov)

Drug name	Platform	Targeting mechanism	Active ingredient	Indication	Year approved/clinical study stage
Doxil® (USA)	liposome		doxorubicin	refractory Kaposi's sarcoma, reccurent breast, ovarian cancer	1995 (USA)/1996 (EU)
DaunoXome®	liposome		daunorubicin citrate	Kaposi's sarcoma	1996 (USA)
Myocet®	liposome		doxorubicin	combinational therapy of reccurent breast, ovarian cancer	2000 (EU/Canada)
Abraxane®	albumin-bound paclitaxel nanoparticles		paclitaxel	metastatic breast cancer	2005 (USA)/2008 (EU)
Genexol-PM®	Micelle		paclitaxel	metastatic breast cancer	2007 (South Korea)/2009 (Serbia/Vietnam/India/Philippines)
ONCO TCS®	liposome		vincristine sulfate	relapsed aggressive non-Hodgkin Lymphoma	2012 (USA)
KADCYLA®	antibody-drug conjugate		mertansine (DM1)	breast cancer	2013 (USA)
Onivyde™	liposome		irinotecan	metastatic pancreatic cancer previously treated with gemcitabine	2015 (USA)
LEP-ETU	liposome		paclitaxel	ovarian cancer	2015 (USA-orphan drug designation)
Lipoplatin™	liposome		cisplatin	pancreatic/head&neck/breast cancer	clinical phase IV
EndoTAG®-1	cationic liposome		paclitaxel	breast cancer/pancreatic cancer	clinical phase III
ThermoDox®	temperature sensitive liposome		doxorubicin	hepatocellular carcinoma	clinical phase III
ADI-PEG 20™	polymer-protein conjugate		arginine deiminase	hepatocellular carcinoma	clinical phase III
Livatag®	polymeric NPs		doxorubicin	hepatocellular carcinoma	clinical phase III
Etirinotecan pegol	polymer-drug conjugate		irinotecan	breast cancer	clinical phase III

OPAXIO®	polymer-drug conjugate		paclitaxel	NSCLC, ovarian, head& neck cancer, gliablastoma multiforme	clinical phase III
Taxopresin®	polymer-drug conjugate		paclitaxel	metastatic malignant melanoma	clinical phase III
Atragen™			tretinoin	acute promyelocytic leukemia, Kaposi's sarcoma	clinical phase II/III
Accurin®	polymeric NPs (PLGA)	prostate-specific membrane antigen (PSMA)	doxetaxel	solid tumor	clinical phase II
SPI-077	liposome		cisplatin	head and neck cancer, lung cancer	clinical phase II
Aurimmune®	PEGylated gold NPS		TNF-alpha	head&neck cancer	clinical phase II
LE-SN38	liposome		SN-38 active metabolite of irinotecan	metastatic colorectal carcinoma	clinical phase I/II
CALAA-01	cyclodextrin-containing polymeric NPs	transferrin	siRNA	solid tumor	clinical phase I
MBP-426	liposome	transferrin	oxaliplatin	solid tumor	clinical phase I
SGT53-01	liposome	transferrin receptor specific-scAb	p53 gene	solid tumor	clinical phase I
INX-0125	liposome		Vinorelbine	Advanced solid tumors	clinical phase I
CPX-1	liposome		Irinotecan HCI: floxuridine	Colorectal cancer	clinical phase II

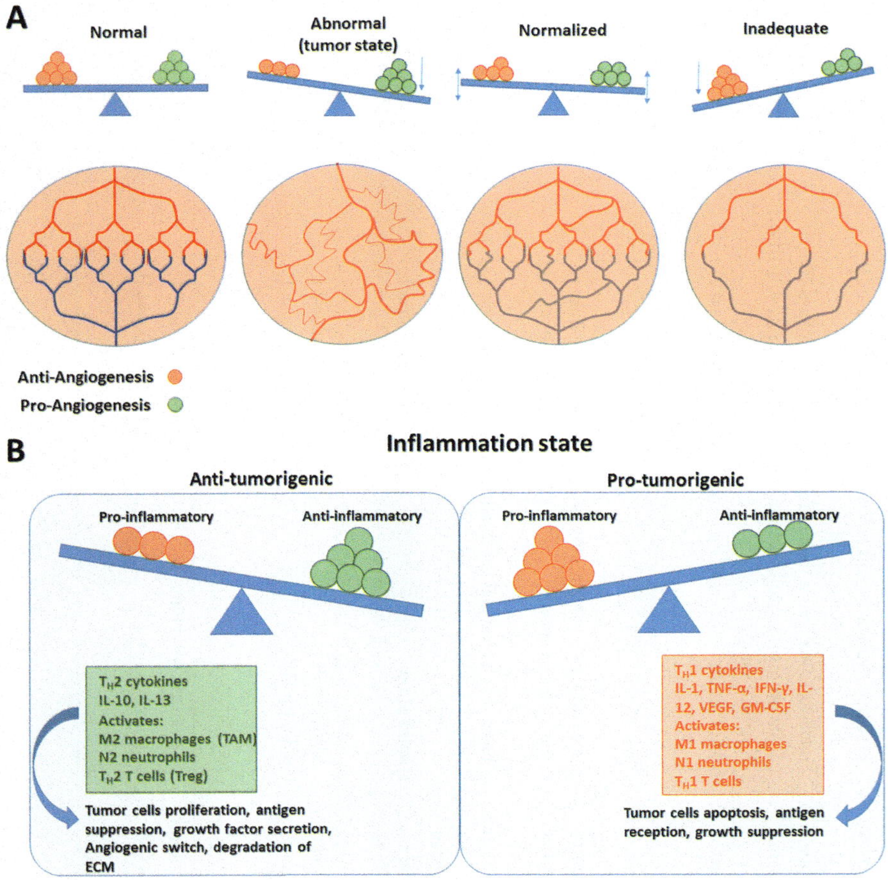

Fig. 4.7 (**a**). Balance of pro- and anti-angiogenic molecules to restore the vascular network to the normal state (figure modified from (Jain 2013)). (**b**). Balance of pro- and anti-inflammatory cytokine in TME to define tumor fate. Pro-tumorigenic TME is skewed to Th2 phenotypes, induce differentiation of macrophage to M2 macrophages, while anti-tumorigenic TME display Th1 phenotype.

in the tumor interstitial space. These characteristics have been used to passively increase the accumulation of nanotherapeutics such as polymeric drug conjugates, polymeric NPs, liposomes, micelles and other particulates and macromolecules. On the other hand, due to the neovascularization process, the newly formed tumor associated blood vessels possess specific molecular signatures which are lacking in the normal vessels. These can be used for targeting nanocarriers to the diseased vasculature. In all case, the process of normalization of vasculature have to be fine-tuned to achieve a controlled vascular repair and not a rapid vessel pruning, which can be contraproductive for the outcome of the therapies (Fig. 4.7a).

Additionally, various growth factor receptors have been identified that can serve as the targets on the tumor vasculature. Targets on the tumor vasculature can also be used for the normalization of the vasculature. As an example, high concentration of a humanized monoclonal antibody directed against the VEGF, bevacizumab (15 mg/kg) has been shown to improve progression-free survival (PFS) in the second-line treatment of patients with human epidermal growth factor receptor 2 (HER2)-negative metastatic breast cancer in phase III breast cancer trial (Brufsky et al. 2011).

Below we summarize a few studies with particles targeting angiogenesis and hypoxia conditions.

4.3.1.1 Angiogenesis

VEGF receptor is overexpressed on the angiogenic endothelial cells. Dextran magnetic NP conjugated to radiolabeled ^{131}I anti-VEGF monoclonal antibodies has tested in hepatic carcinoma mouse model and has been found to significantly increase the retention percentages of radioactivity (% ID/g) in tumors after intratumoral and intravenous injection, reaching 100.96 % ID/g and 90.00 % even at 48 h after, respectively. These values were much higher than retention in normal organs evaluated (Chen et al. 2006). VEGF moiety has also been explored for boron neutron capture therapy (BNCT), a therapy using irradiation and which is lethal only to those cells that bind or internalize ^{10}B in sufficient concentrations. A fifth-generation polyamidoamine dendrimer containing 128 reactive amino groups was reacted with 105–110 decaborate molecules to produce a macromolecule with 1050–1100 boron atoms per dendrimer (VEGF-BD), and then conjugated to thiol groups of VEGF at a 4:1 molar ratio to form VEGF-BD. The systemic treatment of VEGF-BD selectively eliminated only VEGFR-2-overexpressing cells in 4T1 tumor-bearing mice and did not affect the majority of tumor endothelial cells expressing low to moderate levels of VEGFR-2 (Backer et al. 2005).

Another example of nanocarriers reducing drug toxicity can be found with the use of TNP-470, a potent angiogenic inhibitor. When TNP-470 is conjugated with large N-(2-Hydroxypropyl)methacrylamide (HPMA) polymers, the formulation called Caplostatin™ and the system significantly reduces neurotoxicity of TNP-470 by inhibiting hyperpermeability of tumor blood vessels to penetrate across blood–brain barrier when injected intravenously (Satchi-Fainaro et al. 2004). Loading of TNP-470 into PEG-PLA micelles has also been explored and currently in clinical trial under the name Lodamin™. This formulation has been developed as the first oral, broad-spectrum angiogenesis inhibitor. Lodamin™ was absorbed through the intestine, accumulating in the tumor and showed 31 % of inhibition of angiogenesis via *in vivo* analysis with the corneal micropocket angiogenesis assay. Lodamin™ was found to retain TNP-40 potency and significantly increased half-life, selectively accumulated in tumor tissue, blocked angiogenesis, and inhibited primary tumor growth in mouse models of melanoma and lung cancer significantly, with no apparent side effects (Benny et al. 2008).

VCAM is an immunoglobulin-like transmembrane glycoprotein receptor overexpressed on the tumor associated endothelial cells and can mediate immune cell adhesion. Liposome with VCAM-1 antibody (α-VCAM-Ls) selectively target human tumor vasculature *in vivo*, and can provide a way to enhance delivery of drugs to growing tumor vessels and change endothelial functions (Gosk et al. 2008). α-VCAM-Ls specifically bound to activated endothelial cells *in vitro* under static and flow conditions and effectively accumulated in tumor vessels with increasing intensities from 30 min to 24 h in vivo in Colo 677 tumor xenografts, not increasing their accumulation in MPS organs.

Vascular normalization may also convert immunosuppressive TME towards immunostimulatory environment, therefore, improving immunotherapy outcome (Huang et al. 2013). The chance of tumor cells to cross the endothelial barrier is reduced, which in turn significantly inhibits the cancer cell shedding into the circulation, limiting the metastasis potential (Ma and Waxman 2008).

One of the first ligands known to target vasculature was RGD peptide, which recognizes integrin $\alpha v\beta 3$ and $\alpha 5\beta 1$. Drug delivery systems utilizing the ligand have been thoroughly studied *in vitro* and *in vivo* (Marelli et al. 2013). RGD grafting to paclitaxel (PTX)-loaded PEGylated PLGA-based NP *was* demonstrated to increase targeting to tumor vessels as well as the effective retardation of tumor growth and prolonged survival times of treated mice when compared to non-targeted nanoparticles (Danhier et al. 2009).

Binding of RGD to pH sensitive NP loaded with siRNA controling tumor cell proliferation has resulted in receptor-mediated cellular uptake and high gene silencing efficiency in U87 cells and succeeded in increasing adhesion and reducing tumor growth by factor of two, compared to the non-targeted siRNA-NP (Wang et al. 2009).

4.3.1.2 Hypoxia

Tumor hypoxia is caused by the compression of blood vessels, combined with poor lymphatic drainage and overall disorganization of structural features supporting diffusion. Hypoxic regions are also generally associated with high interstitial pressure build up and lack of drug penetration into the region causing physiological resistance to the systemically administered therapeutics. Delivery of drugs into the hypoxic area remains a significant challenge for tumor therapy. Several nanocarriers have been developed for hypoxia environment in the tumor. Layer-by-layer assembled polyelectrolyte NPs can target hypoxic region of the tumor based on the low pH (Poon et al. 2011). The low pH inside tumor lesion triggers the shedding of the outer layer of NPs, exposing charged NP surface can be subsequently taken up by the cells, exhibiting the promising potential of targeting low pH and hypoxia for systemic tumor targeting. Another system with hypoxia-activated NIR-triggered chitosan NPs, consists of a hypoxic sensor, a photoactiveable trigger, and a caged drug has been developed for cancer cell targeting and showed highly preferential release of cancer drug in the hypoxic region of the tumor (Lin et al. 2013). Polystyrene NPs (PS-NPs) doped with oxygen-sensitive NIR-emmisive palladium meso-tetraphenylporphyrin can be selectively activated and emit optical signals when they experience hypoxic condition (Napp et al. 2011). When PS-NPs was functionalized with PEG and monoclonal antibody herceptin, their binding to HER2/neu-overexpressing tumor cells was confirmed *in vitro* and was found within the tumor upon hypoxic condition in tumor-bearing mouse.

4.3.2 Immune and Lymphatic Cells

TAMs are one of the major mediators of tumor cell proliferation and invasion in various carcinomas, and their presence is often correlated to poor prognosis. Strategies for TAM-targeted cancer therapy have been focused on inhibition of macrophage recruitment, re-differentiation of macrophage phenotype from M2 to anti-tumor M1, and suppression of TAMs survival (Chanmee et al. 2014) (Fig. 4.7b). While most of the current nanoparticle systems were designed to avoid uptake by MPS to increase their half-life, new approaches to enhance their uptake by monocytes and macrophages have been developed to target TAM.

Bindarit is an indazolic derivative that has been proven to have anti-inflammatory activity in a variety of inflammatory diseases (e.g. lupus nephritis, arthritis and pancreatitis) (Mora et al. 2012). The therapeutic effects of bindarit have been associated with its ability to selectively interfere with monocyte recruitment and the early inflammatory response. Inhibition of CCL2/MCP-1 with Bindarit has been successful in decreasing macrophage recruitment rate and suppress tumor growth in melanoma-bearing mice injected intraperitoneally with bindarit at a dose of 100 mg/kg twice a day, abolishing vascularization and rendering tumors almost completely necrotic (Gazzaniga et al. 2007). TAM express VEGFR2, and the expression is critical for VEGF-induced recruitment of macrophages into tumors. Various antibodies targeting VEGF and VEGFR2 have been utilized to reduce macrophage infiltration and tumor growth (Dineen et al. 2008). 2C3, a monoclonal antibody to human VEGF that specifically blocks VEGF interaction with VEGFR2 can decrease macrophage infiltration into pancreatic tumors in tumor bearing mice.

To increase targeting in tumor associated endothelial cells and macrophages in stroma of orthotopic human pancreatic cancers, Yokoi et al. developed porous silicon nanocarriers conjugated with Ly6C (Yokoi et al. 2013). Ly6C is a mouse homolog of CD59, a marker that is upregulated in tumor associated endothelial cells as well as infiltrating cells in the pancreatic stroma. Targeted nanocarriers accumulated to tumor associated endothelial cells within 15 min after intravenous injection. At 4 h after administration, $9.8 \pm 2.3\,\%$ of injected dose/g tumor of the Ly6C targeting nanocarriers accumulated in the pancreatic tumors, almost 20 times higher than control, therefore can serve as a basis for improved therapy.

Therapeutic cancer vaccine formulation has been in clinical trial, with DepoVax™ as the first liposomal formulation tested, made from processed HLA-A2 restricted peptides presented by breast, ovarian and prostate cancer cells antigens to create therapeutic cancer vaccine, DPX-0907 (Berinstein et al. 2012), showing the potential of cancer vaccine as immunotherapy for cancer.

Several nanoformulations have been developed that target immune cells. As an example, IL-12 loaded liposome delivered via intratumoral injection caused an enhanced memory T cell response (Simpson-Abelson et al. 2009), but did not enter the peripheral blood circulation, which can help in the safety concerns. This system was tested in *in vivo* tumor xenografts of subcutaneous implantation of non-disrupted pieces of human lung, breast or ovarian tumors into immunodeficient mice and was found to have local and sustained release of IL-12 into the TME, reactivating tumor-associated quiescent effector memory T cells to proliferate, produce and release IFN-γ resulting in the killing of tumor cells in situ. Stealth liposomes made of plasma membrane vesicles (PMV) from the highly metastatic murine melanoma (B16-OVA) and a surrogate tumor antigen (OVA) were found to target DCs in vitro and in vivo, and elicit response dependent on the simultaneous delivery of both antigen and a DC maturation or "danger signal" signal (IFN-γ or lipopolysaccharide) and act as DC vaccines in syngeneic C57BL/6 mice, the preparations stimulated strong B16-OVA-specific cytotoxic T cell responses (van Broekhoven et al. 2004). Cationic liposome-DNA complexes (LDC) increased the antigen-specific CD8(+) T cells and also generated significant antitumor activity against established tumors in tumor-bearing mice, while DNA complexes themselves were not able to achieve the effect (U'Ren et al. 2006).

A lipid nanoparticle loaded with CCR2-silencing short interfering RNA possessed a rapid blood clearance, accumulated in the spleen and bone marrow, and localized to monocytes when administered systemically in mice (Leuschner et al. 2011).

Another promising strategy in cancer therapy is the targeting of macrophages and its utilization as trojan horse to transfer drugs into cancer milieu in sufficient concentration. This strategy use a natural phagocytic activity of macrophages and its preference in taking up submicron to micron-sized particles. Choi et al. reported a system made of gold nanoshell that can be taken up by macrophages, transported into the tumor spheroid, and cause cancer cells to succumb to Au nanoshell-based photoinduced cell death using NIR light (Choi et al. 2007). In a recent study, silicon microparticles loaded with nAb-PTX (Abraxane®) has been utilized for liver and lung metastasis of breast cancer, and the mechanisms include passive targeting towards liver organ, uptake of the system by local liver macrophages, and release of partial drug in the TME, which in turn increased the MCP-1 concentration to attract additional macrophages deeper into the lesions (Leonard et al. 2016; Tanei et al. 2016). Efficiency in this system has been mathematically modelled and verified in mouse model of breast cancer liver metastasis (Tanei et al. 2016) (Fig. 4.8). The increased accumulation of PTX in mouse model of lung and liver metastasis breast cancer caused increased level of apoptosis and inhibition of proliferation, ultimately reducing the volume growth of the tumor and increasing the survival rate of the treated mice.

Mannose-grafted nanocarriers can increase uptake by DCs, PLGA NPs coated with mannan (natural polymannose isolated from S. cerevisiae) have higher binging affinity and produce increased CD4+ and CD8+ T-cell responses compared to unmodified NPs (Hamdy et al. 2011). Mannose receptor have been also targeted for

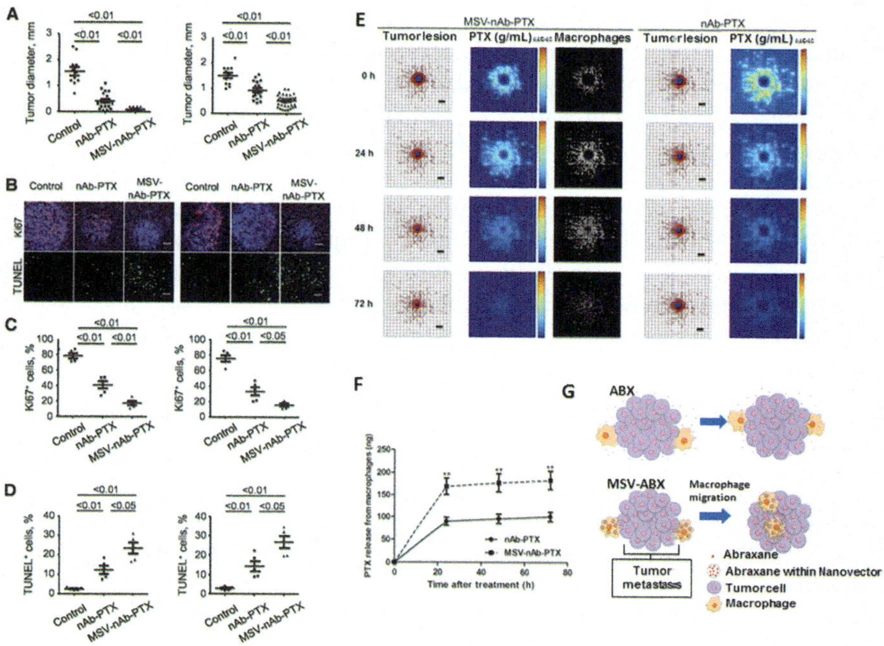

Fig. 4.8 Macrophage as Trojan horse for drug delivery into tumor lesions. (**a-d**) Histologic and immunofluorescent analysis of MSV-nAb-PTX and nAb-PTX effects on the breast (*left*) and lung (*right*) liver metastasis. (**a**), quantification of the tumor diameters in the liver of the mice treated with control (PBS), nAb-PTX, or MSV-nAb-PTX (MSV-nAb-PTX vs. nAb-PTX, $P<0.01$ in both models). Immunohistochemical staining (**b**; scale bar, 100 μm) and quantitative analysis (**c** and ***d***) of proliferating Ki67-positive cancer cells (*pink*) and apoptotic TUNEL-positive cancer cells (*green*). Nuclei were stained with DAPI (*blue*). (**e**), Simulation of breast cancer liver metastasis therapy with MSV-nab-PTX (*left*) and nab-PTX (*right*): Presentation of tumor lesions, concentration of drug (PTX) released from either macrophages or vessels, and macrophages (only in the case of MSV-nab-PTX). Tumor lesions at the indicated times post treatment initiation. In this simulation, the viable tumor tissue (*red*) encloses a hypoxic region (*blue*) without necrosis. The dense capillary network in the liver is modelled by the rectangular grid, with a few irregular sprouts generated through angiogenesis. Size of the lesion is simulated in parallel with drug distribution. PTX represents drug concentration (gm/mL) in the lesion. MSV-nab-PTX are retained in the lesion through the interaction with macrophages while drug is slowly released in the proximity of the tumor cells. Individual macrophages (*white*) are recruited to vicinity of the lesion based on chemoattraction to hypoxic regions. Bar, 200 μm. F, Amount of active compound PTX released from the nAb-PTX and MSV-nAb-PTX pre-treated macrophages *in vitro* for 24-72 h as assessed by LC/MS-MS. G, Schematic of proposed mechanism of transport for MSV-nAb-PTX by macrophages. nAb-PTX loaded MSV were taken up in higher degree in the tumor microenvironment by macrophages compared to free nAb-PTX, further causing accumulation of macrophages deeper into the lesion and leading to higher concentration of active compound PTX and higher tumor cell death. Reproduced from (Leonard et al. 2016; Tanei et al. 2016)

vaccine treatment, such as Mannose-PEG3-NH2-decorated PLGA microsphere (Brandhonneur et al. 2009), mannosylated-gelatin NPs(Saraogi et al. 2011), and liposome loaded with mannosyl lipid derivate (Espuelas et al. 2008). Other target with similar effect is lectin, anti-lectin antibodies and lectin ligands can enhance humoral and cellular responses to induce immunity against melanomas in vivo (Conniot et al. 2014).

There has been a slower progress in the approach of lymphatic system restoration, mostly due to the complexity of lymph flow regulation. Instead of being generated from heart drive, such as blood flow, the lymphatic system is driven by lymphatic contraction and their valves along the lymphatics. In the tumor margin, the valves did not function normally because of hyperplasia (Jain 2013).

4.3.3 CAF

The prostate, breast, bile duct/cholangiocarcinoma (CCA) and pancreas cancers are tumors that rely on its surrounding stroma during development and progression. Targeting CAF with nanotherapies has not been intensively studied to these days. Below are the potential strategies to combine with nanotherapies as well as a summary of current publications of the topic.

Estrogen receptor and HER-2/neu were among the earliest targets investigated, ultimately leading to the widespread use of tamoxifen and trastuzumab (Hue et al. 2004). Other classes of targeted therapies include cyclin-dependent kinase inhibitors, histone deactylase inhibitors, and receptor tyrosine kinase inhibitors (Thurn et al. 2011).

Hepatic stellate cells are known to secrete CXCL12 [or SDF-1], the ligand for CXCR4, a receptor highly expressed in cancer cells (Hong et al. 2009). As a potential therapeutic for CCA, intervention to reduce the extent of CAF, treatment with the specific CXCR4 inhibitor (AMD3100) hampered the CCA-mediated migration of hepatic stellate cells (Cadamuro et al. 2013a). Furthermore, IL-6 secreted by tumor cells within a hypoxic environment has been linked with recruitment of mesenchymal cells and an antibody against IL-6 has resulted in attenuation of mesenchymal cell migration (Kyurkchiev et al. 2014).

Cyclopamine, an inhibitor of the Hedgehog signaling pathway acting on Smoothened, unleashed TRAIL on neoplastic cholangiocytes and induced their apoptosis reversing the effects of PDGF-B released by CAF(Cadamuro et al. 2013a). *In vivo* experiments have found that Cyclopamine can reduce tumor size, but increase its metastatic potential. IPI-926, a novel derivate of Cyclopamine, can potentially improve the efficacy of Gemcitabine in ductal pancreatic cancer, by inhibiting the Hedgehog signaling pathway and therefore depleting the TRS (Olive et al. 2009).

Another study with tyrosine kinase inhibitor Imatinib mesylate was to target the CAF-emitted-PDGF-B that is signaled to bile duct cancer cells. By blocking the PDGFR- expressed by these cancer cells, cholangiocytes, Imatinib resulted in increases of TRAIL and apoptosis of tumoral cholangiocytes. PDGFR- can be interfered directly using Imatinib mesylate or indirectly by inhibiting the downstream effectors of this pathway, in particular the Rho GTPases with NSC23766 (Rac1 inhibitor) and CASIN (Cdc42 inhibitor) and/or JNK signaling with SP600125(Cadamuro et al. 2013b). Reduction in CAF migration was similarly obtained with both approaches. As with Cyclopamine, Imatinib also reduced CCA tumor size and metastasis in rodents. Pro-apoptotic agents, such as Navitoclax, have been shown to act selectively on CAF. By activating Bax, Navitoclax reduced the number of -SMA-positive CAF as well as CCA tumor mass, which is associated with an improved survival in rodents. In recent year, a whole-tumor vaccine expressing fibroblast activation protein has been found to improve antitumor immunity against both tumor cells and CAF (Chen et al. 2015).

MMP inhibition can also potentially modulate cell communications within the tumor reactive stroma. However, so far, MMP inhibition does not seem to improve patient outcome in many epithelial cancers (van Zijl et al. 2011). MMP inhibitors are promising since they play a role in both epithelial-CAF signaling and ECM remodeling. Several broad-spectrum MMP inhibitors (e.g., marimastat and tanomastat/BAY12-9566) have failed in randomized phase III trials in patients with breast and ovarian cancer because they do not prolong overall and/or progression-free survival or exhibit dose-dependent toxicity (Coussens et al. 2002; Hirte et al. 2006; Sparano et al. 2004). Future efforts possibly need focus on the development of highly selective MMP inhibitors (Zhang and Liu 2013).

CAF targeting nanocarriers are being explored for cancer therapy. Docetaxel conjugate nanoparticles, known as Cellax™ has been found to reduce α-smooth muscle actin (α-SMA) content by 82 % and 70 % in 4T1 and MDA-MB-231 orthotopic breast tumor models, respectively (Murakami et al. 2013). After Cellax™ treatment, the tumor perfusion was increased by approximately 70-fold, tumor vascular permeability was enhanced by more than 30 %, tumor matrix was decreased by 2.5-fold, and tumor interstitial fluid pressure was suppressed by approximately threefold in the mice models compared to docetaxel, and nab-paclitaxel groups. Significant tumor stromal depletion occurred within 16 h (~50 % depletion) postinjection, and the α-SMA (Alpha-SMA) stroma population was almost undetectable (~3 %) by 1 week, which ultimatively led to reduction of lung by 7- to 24-fold, whereas native docetaxel and nab-paclitaxel treatments were ineffective. Another approach enables the co-delivery of cisplatin with rapamycin (RAPA), an mTOR inhibitor that may offer notable therapy through antiangiogenic activity (Guo et al. 2014). Traditionally, cisplatin are too hydrophilic to be loaded into PLGA NP, however cisplatin can be made hydrophobic by coating a nanoprecipitate (cores) of the drug with dioleoylphosphatidic acid (DOPA). The presence of resulting DOPA coated cisplatin cores are not only compatible, but also significantly improved the encapsulation of RAPA into PLGA NP, and both can be coencapsulated in PLGA NPs at a molar ratio to promote synergistic antitumor activity. The formulation has been found to induce significant apoptosis on A375-luc human melanoma cells in vitro. In vivo, the system greatly reduced the fibroblast population, and therefore, potentially decreased the level of collagen and inhibited the growth of A375-luc melanoma in a xenograft tumor model. Chen et al. reported a new system with nanoliposomes modified with peptide FH which specifically binds to tenascin C, a protein mainly expressed by CAFs, loaded with Navitoclax (Nav), forming FH-SSL-Nav (Chen et al. 2016). Nav is a targeted high-affinity small molecule, which can inhibit the anti-apoptotic activity of BCL-2 and BCL-XL FH-SSL-Nav. FH-SSL-Nav displayed reduced tumor growth to half of the control by eradicating CAFs in Hep G2 tumor-bearing nude mice model.

4.3.4 ECM

ECM forms a barrier for drug delivery and can affect its efficacy, especially for larger molecules (150,000 MW Dextran) (Jain 1994; Nugent and Jain 1984). In many tumors, the narrow spacing between collagen fibrils hinder particles larger than 60 nm in diameter penetration across collagen matrix to reach the tumor cells. Interestingly, even within the brain tumor, there is 5- to 10-fold difference in diffusion between cranial windows (CW) and dorsal chamber (DC) when the molecule diameters approaching the interfibrillar spacing, which most likely due to higher levels of collagen type I and its organization into fibrils in DC (Pluen et al. 2001). This result emphasize the heterogeneity of ECM matrix and the difficulties to remodel ECM To effectively increase large molecule penetration, which mostly restricted by fibrillar collagen (Brown et al. 2003), standard approaches include treatment with either collagen degrading enzymes such as MMP1 and MMP8 (Choi et al. 2013).

When attached to a nanocarrier, HA can act as protective structural component, targeting moiety, and a targeting coating. HA-conjugated nanocarriers targeting CD44 over-expressing cancer cells to transport drugs to tumor site. HA-PEG-PCL NP to deliver doxorubicin in Ehrlich ascited tumor-bearing mice. HA-decorated PLGA NP enhance doxorubicin uptake by cultured human breast adenocarcinoma cells (Yadav et al. 2008). Liposome loaded with hyaluronidase disrupt the HA matrix in human osteosarcoma solid tumor model (Platt and Szoka 2008). Mizrahy et al. have developed hyaluronan-coated lipid nanoparticles which exhibit target HA receptors (CD44 and CD168) and opposing effects on NCI/ADR-RES (human ovarian adenocarcinoma Adriamycin resistant) cell proliferation mediated by CD44(Mizrahy et al. 2014). The same group also loaded mitomycin C into NP, which enhanced its targeting and antitumor activity as well as increased circulation time to 7-fold and 70-fold longer than nt-LIP and free mitomycin C in 3 tumor models: BALB/c bearing C-26 solid tumors; C57BL/6 bearing B16F10.9 or (separately) D122 lung metastasis. Further, delivery of proteases, such as collagenase, gelatinase, hyaluronidase has potential to be explored since they can aid in breaking down dense ECM barrier and increases transport of NPs and drugs into the tumor (Choi et al. 2012).

4.4 Conclusion and Perspectives

Tumor microenvironment represents intricate ecosystem in which neoplastic cells recruit and co-exist in synergy with their non-neoplastic neighboring cells. Multiple works over the past few decades have validated the importance of TME in the development and progression of the malignant disease. As a result, targeting elements in the tumor microenvironment is evolving as a vital strategy to impede tumor growth. There are major structural and functional changes at the interface between tumor cells and adjacent TME cells and non-cellular elements. These deviations from the normal physiology of vascular, immune and connective tissues/systems can be targeted not only pharmacologically, but also through designing nano-carriers with specific physical/chemical/biological characteristics which are able to localize into the proximity of the tumor cells enabling substantial therapeutic benefits over conventional molecular therapies. The dimensions of nanovectors allow them to interact on both pathologically relevant scales: nanometer scale (proteins, antibodies, sugars, genetic materials) and micrometer scale (cells and extracellular structures), thus enabling to specifically engineer and fine tune delivery systems for cancer and other conditions. Moreover, nanovectors can also allow the design of multi-modal theranostic systems that simultaneously perform a multitude of tasks, spanning from therapy to serving as a contrast agent or a reporter for the disease progression/therapeutic outcomes. The benefits of currently marketed and under clinical and preclinical investigation cancer nanotherapies as compared to their molecular counterparts are: extended circulation time (Table 4.3), increased stability of therapeutics, enhanced drug concentration in the tumor tissue, improved therapeutic index, lesser toxicity with improved stability and multifunctionality. However, nanotherapeutics are not much different from the molecular therapies in terms of the problems arising en route to their regulatory approval and many of them fail during the various phases of the process.

Deeper understanding of cancer cell physiology, tumor microenvironment and its complex signaling processes is vital to develop new cancer nanotherapeutics that will be more specific and render active only in the proximity of the tumor loci. We also envision that nano-modular systems can be designed from the pre-formulated nano-components based on the needs and the physiological features of the specific patient. Additionally, as with the conventional therapies, not all the patients will benefit from all the newly developed nanotherapies, thus, it is very important to understand what are the factors that will allow pre-selction of the patient population and personalization of nanomedicines for the efficient cancer elimination.

References

Adams ML, Lavasanifar A, Kwon GS (2003) Amphiphilic block copolymers for drug delivery. J Pharm Sci 92(7):1343–1355. doi:10.1002/jps.10397

Albanese A, Tang P, Chan C (2012) The effect of nanoparticle size, shape, and surface chemistry on biological systems. Annu Rev Biomed Eng 14:1–16

Alexander JF, Kozlovskaya V, Chen J et al (2015) Cubical shape enhances the interaction of layer-by-layer polymeric particles with breast cancer cells. Adv Healthcare Mater 4(17):2657–2666. doi:10.1002/adhm.201500537

Alitalo K (2011) The lymphatic vasculature in disease. Nat Med 17(11):1371–1380. doi:10.1038/nm.2545

Allen TM (2002) Ligand-targeted therapeutics in anticancer therapy. Nat Rev Cancer 2(10):750–763. doi:10.1038/nrc903

Allen TM, Austin GA, Chonn A et al (1991) Uptake of liposomes by cultured mouse bone marrow macrophages: influence of liposome composition and size. Biochim Biophys Acta 1061(1):56–64

Allinen M, Beroukhim R, Cai L et al (2004) Molecular characterization of the tumor microenvironment in breast cancer. Cancer Cell 6(1):17–32. doi:10.1016/j.ccr.2004.06.010

Arvizo R, Miranda O, Thompson M et al (2010) Effect of nanoparticle surface charge at the plasma membrane and beyond. Nano Lett 10(7):2543–2548

AshaRani PV, Low Kah Mun G, Hande MP et al (2009) Cytotoxicity and genotoxicity of silver nanoparticles in human cells. ACS Nano 3(2):279–290. doi:10.1021/nn800596w

Backer MV, Gaynutdinov TI, Patel V et al (2005) Vascular endothelial growth factor selectively targets boronated dendrimers to tumor vasculature. Mol Cancer Ther 4(9):1423–1429. doi:10.1158/1535-7163.mct-05-0161

Bally MB, Harvie P, Wong FMP et al (1999) Biological barriers to cellular delivery of lipid-based DNA carriers. Adv Drug Deliv Rev 38(3):291–315, doi: http://dx.doi.org/10.1016/S0169-409X(99)00034-4

Barbee MS, Ogunniyi A, Horvat TZ et al (2015) Current status and future directions of the immune checkpoint inhibitors ipilimumab, pembrolizumab, and nivolumab in oncology. Ann Pharmacother 49(8):907–937. doi:10.1177/1060028015586218

Barua S, Yoo J-W, Kolhar P et al (2013) Particle shape enhances specificity of antibody-displaying nanoparticles. Proc Natl Acad Sci 110(9):3270–3275. doi:10.1073/pnas.1216893110

Baxter LT, Jain RK (1989) Transport of fluid and macromolecules in tumors. I. Role of interstitial pressure and convection. Microvasc Res 37(1):77–104. doi:10.1016/0026-2862(89)90074-5

Bazak R, Houri M, Achy SE et al (2014) Passive targeting of nanoparticles to cancer: a comprehensive review of the literature. Mol Clin Oncol 2(6):904–908. doi:10.3892/mco.2014.356

Behr J (1997) The proton sponge: a trick to enter cells the viruses did not exploit. Chimia 51:34–36

Bellocq A, Suberville S, Philippe C et al (1998) Low environmental pH is responsible for the induction of nitric-oxide synthase in macrophages: evidence for involvement of nuclear factor-kB activation. J Biol Chem 273(9):5086–5092. doi:10.1074/jbc.273.9.5086

Beningo KA, Wang YL (2002) Fc-receptor-mediated phagocytosis is regulated by mechanical properties of the target. J Cell Sci 115(Pt 4):849–856

Benny O, Fainaru O, Adini A et al (2008) An orally delivered small-molecule formulation with antiangiogenic and anticancer activity. Nat Biotechnol 26(7):799–807. doi:10.1038/nbt1415

Berinstein NL, Karkada M, Morse MA et al (2012) First-in-man application of a novel therapeutic cancer vaccine formulation with the capacity to induce multi-functional T cell responses in ovarian, breast and prostate cancer patients. J Transl Med 10:156. doi:10.1186/1479-5876-10-156

Berneburg M, Kamenisch Y, Krutmann J et al (2006) 'To repair or not to repair – no longer a question': repair of mitochondrial DNA shielding against age and cancer. Exp Dermatol 15(12):1005–1015. doi:10.1111/j.1600-0625.2006.00508.x

Beyer I, Li Z, Persson J et al (2011) Controlled extracellular matrix degradation in breast cancer tumors improves therapy by Trastuzumab. Mol Ther 19(3):479–489. doi:10.1038/mt.2010.256

Bianco A (2004) Carbon nanotubes for the delivery of therapeutic molecules. Expert Opin Drug Deliv 1(1):57–65. doi:10.1517/17425247.1.1.57

Bissell MJ, Hines WC (2011) Why don't we get more cancer? A proposed role of the microenvironment in restraining cancer progression. Nat Med 17(3):320–329. doi:10.1038/nm.2328

Blanco E, Shen H, Ferrari M (2015) Principles of nanoparticle design for overcoming biological barriers to drug delivery. Nat Biotechnol 33(9):941–951. doi:10.1038/nbt.3330

Bochet L, Meulle A, Imbert S et al (2011) Cancer-associated adipocytes promotes breast tumor radioresistance. Biochem Biophys Res Commun 411(1):102–106. doi:10.1016/j.bbrc.2011.06.101

Bochet L, Lehuédé C, Dauvillier S et al (2013) Adipocyte-derived fibroblasts promote tumor progression and contribute to the desmoplastic reaction in breast cancer. Cancer Res 73(18):5657–5668. doi:10.1158/0008-5472.can-13-0530

Brahmer JR, Tykodi SS, Chow LQM et al (2012) Safety and activity of anti–PD-L1 antibody in patients with advanced cancer. N Engl J Med 366(26):2455–2465. doi:10.1056/NEJMoa1200694

Brandhonneur N, Chevanne F, Vié V et al (2009) Specific and non-specific phagocytosis of ligand-grafted PLGA microspheres by macrophages. Eur J Pharm Sci 36(4–5):474–485, doi: http://dx.doi.org/10.1016/j.ejps.2008.11.013

Brannon-Peppas L, Blanchette JO (2012) Nanoparticle and targeted systems for cancer therapy. Adv Drug Deliv Rev 64(SUPPL):206–212. doi:10.1016/j.addr.2012.09.033

Brodsky F, Chen C, Knuehl C et al (2001) Biological basket weaving: formation and function of clathrin-coated vesicles. Annu Rev Cell Dev Biol 17:517–568

Brown E, McKee T, diTomaso E et al (2003) Dynamic imaging of collagen and its modulation in tumors in vivo using second-harmonic generation. Nat Med 9(6):796–800. doi:10.1038/nm879

Brufsky AM, Hurvitz S, Perez E et al (2011) RIBBON-2: a randomized, double-blind, placebo-controlled, phase III trial evaluating the efficacy and safety of bevacizumab in combination with chemotherapy for second-line treatment of human epidermal growth factor receptor 2–negative metastatic breast cancer. J Clin Oncol. doi:10.1200/jco.2010.34.1255

Buckway B, Frazier N, Gormley AJ et al (2014) Gold nanorod-mediated hyperthermia enhances the efficacy of HPMA copolymer – (90)Y conjugates in treatment of prostate tumors. Nucl Med Biol 41(3):282–289. doi:10.1016/j.nucmedbio.2013.12.002

Burkholder B, Huang RY, Burgess R et al (2014) Tumor-induced perturbations of cytokines and immune cell networks. Biochimica et Biophysica Acta Rev Cancer 1845(2):182–201. doi:10.1016/j.bbcan.2014.01.004

Cadamuro M, Nardo G, Indraccolo S et al (2013a) Platelet-derived growth factor-D and Rho GTPases regulate recruitment of cancer-associated fibroblasts in cholangiocarcinoma. Hepatology 58(3):1042–1053. doi:10.1002/hep.26384

Cadamuro M, Morton SD, Strazzabosco M et al (2013b) Unveiling the role of tumor reactive stroma in cholangiocarcinoma: an opportunity for new therapeutic strategies. Translat Gastrointestinal Cancer 2(3):130–144

Caldorera-Moore M, Guimard N, Shi L et al (2010a) Designer nanoparticles: incorporating size, shape and triggered release into nanoscale drug carriers. Expert Opin Drug Deliv 7:479–495

Caldorera-Moore M, Guimard N, Shi L et al (2010b) Designer nanoparticles: incorporating size, shape and triggered release into nanoscale drug carriers. Expert Opin Drug Deliv 7(4):479–495. doi:10.1517/17425240903579971

Calvo F, Ege N, Grande-Garcia A et al (2013) Mechanotransduction and YAP-dependent matrix remodelling is required for the generation and maintenance of cancer-associated fibroblasts. Nat Cell Biol 15(6):637–646. doi:10.1038/ncb2756

Campbell RB, Fukumura D, Brown EB et al (2002) Cationic charge determines the distribution of liposomes between the vascular and extravascular compartments of tumors. Cancer Res 62(23):6831–6836

Carter KA, Shao S, Hoopes MI et al (2014) Porphyrin–phospholipid liposomes permeabilized by near-infrared light. Nat Commun 5:3546. doi:10.1038/ncomms4546

Castells M, Thibault B, Delord JP et al (2012) Implication of tumor microenvironment in chemoresistance: tumor-associated stromal cells protect tumor cells from cell death. Int J Mol Sci 13(8):9545–9575. doi:10.3390/ijms13089545

Cavnar MJ, Zeng S, Kim TS et al (2013) KIT oncogene inhibition drives intratumoral macrophage M2 polarization. J Exp Med 210(13):2873–2886. doi:10.1084/jem.20130875
Cedervall T, Lynch I, Lindman S et al (2007) Understanding the nanoparticle-protein corona using methods to quantify exchange rates and affinities of proteins for nanoparticles. Proc Natl Acad Sci U S A 104(7):2050–2055
Champion J, Mitragotri S (2006) Role of target geometry in phagocytosis. Proc Natl Acad Sci U S A 103(13):4930–4934
Champion J, Katare Y, Mitragotri S (2007) Particle shape: a new design parameter for micro- and nanoscale drug delivery carriers. J Control Release 121(1):2–9
Chanmee T, Ontong P, Konno K et al (2014) Tumor-associated macrophages as major players in the tumor microenvironment. Cancers 6(3):1670
Chauhan VP, Martin JD, Liu H et al (2013) Angiotensin inhibition enhances drug delivery and potentiates chemotherapy by decompressing tumour blood vessels. Nat Commun 4:2516. doi:10.1038/ncomms3516
Chen J, Wu H, Han D et al (2006) Using anti-VEGF McAb and magnetic nanoparticles as double-targeting vector for the radioimmunotherapy of liver cancer. Cancer Lett 231(2):169–175. doi:10.1016/j.canlet.2005.01.024
Chen Q, Zhang XF, Massagué J (2011) Macrophage binding to receptor VCAM-1 transmits survival signals in breast cancer cells that invade the lungs. Cancer Cell 20(4):538–549. doi:10.1016/j.ccr.2011.08.025
Chen M, Xiang R, Wen Y et al (2015) A whole-cell tumor vaccine modified to express fibroblast activation protein induces antitumor immunity against both tumor cells and cancer-associated fibroblasts. Sci Rep 5:14421, doi: 10.1038/srep14421, http://www.nature.com/articles/srep14421#supplementary-information
Chen B, Wang Z, Sun J et al (2016) A tenascin C targeted nanoliposome with navitoclax for specifically eradicating of cancer-associated fibroblasts. Nanomed Nanotechnol Biol Med 12(1):131–141, doi: http://dx.doi.org/10.1016/j.nano.2015.10.001
Chithrani B, Ghazani A, Chan W (2006) Determining the size and shape dependence of gold nanoparticle uptake into mammalian cells. Nano Lett 6(4):662–668
Cho K, Wang X, Nie S et al (2008) Therapeutic nanoparticles for drug delivery in cancer. Clin Cancer Res 14(5):1310–1316. doi:10.1158/1078-0432.ccr-07-1441
Choi M-R, Stanton-Maxey KJ, Stanley JK et al (2007) A cellular trojan horse for delivery of therapeutic nanoparticles into tumors. Nano Lett 7(12):3759–3765. doi:10.1021/nl072209h
Choi KY, Swierczewska M, Lee S et al (2012) Protease-activated drug development. Theranostics 2(2):156–178. doi:10.7150/thno.4068
Choi IK, Strauss R, Richter M et al (2013) Strategies to increase drug penetration in solid tumors. Front Oncol. doi:10.3389/fonc.2013.00193
Chu TC, Twu KY, Ellington AD et al (2006) Aptamer mediated siRNA delivery. Nucleic Acids Res 34(10):e73, doi: ARTN e73, DOI 10.1093/nar/gkl388
Cieslewicz M, Tang J, Yu JL et al (2013) Targeted delivery of proapoptotic peptides to tumor-associated macrophages improves survival. Proc Natl Acad Sci U S A 110(40):15919–15924. doi:10.1073/pnas.1312197110
Cirri P, Chiarugi P (2011) Cancer associated fibroblasts: the dark side of the coin. Am J Cancer Res 1(4):482–497
Conniot J, Silva JM, Fernandes JG et al (2014) Cancer immunotherapy: nanodelivery approaches for immune cell targeting and tracking. Front Chem 2:105. doi:10.3389/fchem.2014.00105
Cortez C, Tomaskovic-Crook E, Johnston APR et al (2006) Targeting and uptake of multilayered particles to colorectal cancer cells. Adv Mater 18(15):1998–2003. doi:10.1002/adma.200600564
Cortez C, Tomaskovic-Crook E, Johnston A et al (2007) Influence of size, surface, cell line, and kinetic properties on the specific binding of A33 antigen-targeted multilayered particles and capsules to colorectal cancer cells. ACS Nano 1(2):93–102
Coussens LM, Fingleton B, Matrisian LM (2002) Matrix metalloproteinase inhibitors and cancer—trials and tribulations. Science 295(5564):2387–2392. doi:10.1126/science.1067100

Cui Y, Wei Q, Park H et al (2001) Nanowire nanosensors for highly sensitive and selective detection of biological and chemical species. Science 293(5533):1289–1292, 10.1126/science.1062711.293/5533/1289 [pii]

Danhier F, Vroman B, Lecouturier N et al (2009) Targeting of tumor endothelium by RGD-grafted PLGA-nanoparticles loaded with Paclitaxel. J Control Release 140(2):166–173, doi: http://dx.doi.org/10.1016/j.jconrel.2009.08.011

Danson S, Ferry D, Alakhov V et al (2004) Phase I dose escalation and pharmacokinetic study of pluronic polymer-bound doxorubicin (SP1049C) in patients with advanced cancer. Br J Cancer 90(11):2085–2091, 10.1038/sj.bjc.6601856.6601856 [pii]

Dautryvarsat A, Ciechanover A, Lodish H (1983) Ph and the recycling of transferrin during receptor-mediated endocytosis. Proc Natl Acad Sci U S A 80(8):2258–2262

Dauty E, Verkman A (2005) Actin cytoskeleton as the principal determinant of size-dependent DNA mobility in cytoplasm: a new barrier for non-viral gene delivery. J Biol Chem 280(9):7823–7828

De Cock LJ, De Koker S, De Geest BG et al (2010) Polymeric multilayer capsules in drug delivery. Angew Chem Int Ed 49(39):6954–6973. doi:10.1002/anie.200906266

De Koker S, De Geest BG, Cuvelier C et al (2007) In vivo cellular uptake, degradation, and biocompatibility of polyelectrolyte microcapsules. Adv Funct Mater 17(18):3754–3763. doi:10.1002/adfm.200700416

De Kruijf EM, Van Nes JGH, Van De Velde CJH et al (2011) Tumor-stroma ratio in the primary tumor is a prognostic factor in early breast cancer patients, especially in triple-negative carcinoma patients. Breast Cancer Res Treat 125(3):687–696. doi:10.1007/s10549-010-0855-6

Debbage P (2009) Targeted drugs and nanomedicine: present and future. Curr Pharm Des 15(2):153–172

Decuzzi P, Ferrari M (2006) The adhesive strength of non-spherical particles mediated by specific interactions. Biomaterials 27(30):5307–5314

Decuzzi P, Godin B, Tanaka T et al (2010) Size and shape effects in the biodistribution of intravascularly injected particles. J Control Release 141(3):320–327. doi:10.1016/j.jconrel.2009.10.014

DeNardo DG, Brennan DJ, Rexhepaj E et al (2011) Leukocyte complexity predicts breast cancer survival and functionally regulates response to chemotherapy. Cancer Discovery 1(1):54–67. doi:10.1158/2159-8274.cd-10-0028

Deng Z, Liang M, Monteiro M et al (2011) Nanoparticle-induced unfolding of fibrinogen promotes Mac-1 receptor activation and inflammation. Nat Nanotechnol 6(1):39–44

Deryugina EI, Quigley JP (2015) Tumor angiogenesis: MMP-mediated induction of intravasation- and metastasis-sustaining neovasculature. Matrix Biol 44–46:94–112. doi:10.1016/j.matbio.2015.04.004

Di Tomaso E, Capen D, Haskell A et al (2005) Mosaic tumor vessels: cellular basis and ultrastructure of focal regions lacking endothelial cell markers. Cancer Res 65(13):5740–5749. doi:10.1158/0008-5472.can-04-4552

Dimou A, Syrigos KN, Saif MW (2012) Overcoming the stromal barrier: technologies to optimize drug delivery in pancreatic cancer. Ther Adv Med Oncol 4(5):271–279. doi:10.1177/1758834012446008

Dineen SP, Lynn KD, Holloway SE et al (2008) Vascular endothelial growth factor receptor 2 mediates macrophage infiltration into orthotopic pancreatic tumors in mice. Cancer Res 68(11):4340–4346. doi:10.1158/0008-5472.can-07-6705

Ding H-M, Ma Y-Q (2013) Controlling cellular uptake of nanoparticles with pH-sensitive polymers. Sci Rep 3:2804. doi:10.1038/srep02804

Diop-Frimpong B, Chauhan VP, Krane S et al (2011) Losartan inhibits collagen I synthesis and improves the distribution and efficacy of nanotherapeutics in tumors. Proc Natl Acad Sci U S A 108(7):2909–2914. doi:10.1073/pnas.1018892108

Djenidi F, Adam J, Goubar A et al (2015) CD8+CD103+ tumor-infiltrating lymphocytes are tumor-specific tissue-resident memory T cells and a prognostic factor for survival in lung cancer patients. J Immunol 194(7):3475–3486. doi:10.4049/jimmunol.1402711

Dobrovolskaia M, McNeil S (2007) Immunological properties of engineered nanomaterials. Nat Nanotechnol 2:469–478
Doshi N, Zahr A, Bhaskar S et al (2009) Red blood cell-mimicking synthetic biomaterial particles. Proc Natl Acad Sci U S A 106:21495–21499
Downey CL, Simpkins SA, White J et al (2014) The prognostic significance of tumour-stroma ratio in oestrogen receptor-positive breast cancer. Br J Cancer 110(7):1744–1747. doi:10.1038/bjc.2014.69
Dreher M, Liu W, Michelich C et al (2006a) Tumor vascular permeability, accumulation, and penetration of macromolecular drug carriers. J Natl Cancer Inst 98(5):335–344
Dreher MR, Liu W, Michelich CR et al (2006b) Tumor vascular permeability, accumulation, and penetration of macromolecular drug carriers. J Natl Cancer Inst 98(5):335–344. doi:10.1093/jnci/djj070
Duncan R (2006) Polymer conjugates as anticancer nanomedicines. Nat Rev Cancer 6(9):688–701, 10.1038/nrc1958 nrc1958 [pii]
Dunn S, Brindley A, Davis S et al (1994) Polystyrene-poly (ethylene glycol) (PS-PEG2000) particles as model systems for site specific drug delivery. 2. The effect of PEG surface density on the in vitro cell interaction and in vivo biodistribution. Pharm Res 11:1016–1022
Duong MN, Cleret A, Matera EL et al (2015) Adipose cells promote resistance of breast cancer cells to trastuzumab-mediated antibody-dependent cellular cytotoxicity. Breast Cancer Res 17(1):57. doi:10.1186/s13058-015-0569-0
Espuelas S, Thumann C, Heurtault B et al (2008) Influence of ligand valency on the targeting of immature human dendritic cells by mannosylated liposomes. Bioconjug Chem 19(12):2385–2393. doi:10.1021/bc8002524
Estrella V, Chen T, Lloyd M et al (2013) Acidity generated by the tumor microenvironment drives local invasion. Cancer Res 73(5):1524–1535. doi:10.1158/0008-5472.can-12-2796
Farokhzad OC, Cheng J, Teply BA et al (2006) Targeted nanoparticle-aptamer bioconjugates for cancer chemotherapy in vivo. Proc Natl Acad Sci 103(16):6315–6320. doi:10.1073/pnas.0601755103
Fischer H, Hauck T, Gómez-Aristizábal A et al (2010) Exploring primary liver macrophages for studying quantum dot interactions with biological systems. Adv Mater 22(23):2520–2524
Fox M, Szoka F, Frechet J (2009) Soluble polymer carriers for the treatment of cancer: the importance of molecular architecture. Acc Chem Res 42:1141–1151
Fröhlich E (2012) The role of surface charge in cellular uptake and cytotoxicity of medical nanoparticles. Int J Nanomedicine 7:5577–5591. doi:10.2147/ijn.s36111
Fukumura D, Xu L, Chen Y et al (2001) Hypoxia and acidosis independently up-regulate vascular endothelial growth factor transcription in brain tumors in vivo. Cancer Res 61(16):6020–6024
Gabrielson N, Pack D (2009) Efficient polyethylenimine-mediated gene delivery proceeds via a caveolar pathway in HeLa cells. J Control Release 136(1):54–61
Gao H, Shi W, Freund L (2005) Mechanics of receptor-mediated endocytosis. Proc Natl Acad Sci U S A 102(27):9469–9474
Gavze E, Shapiro M (1998) Motion of inertial spheroidal particles in a shear flow near a solid wall with special application to aerosol transport in microgravity. J Fluid Mech 371:59–79
Gazzaniga S, Bravo AI, Guglielmotti A et al (2007) Targeting tumor-associated macrophages and inhibition of MCP-1 reduce angiogenesis and tumor growth in a human melanoma xenograft. J Investig Dermatol 127(8):2031–2041. doi:10.1038/sj.jid.5700827
Geng Y, Dalhaimer P, Cai S et al (2007) Shape effects of filaments versus spherical particles in flow and drug delivery. Nat Nanotechnol 2:249–255
Gentile F, Chiappini C, Fine D et al (2008) The effect of shape on the margination dynamics of non-neutrally buoyant particles in two-dimensional shear flows. J Biomech 41:2312–2318
Gillies ER, Frechet JM (2005) Dendrimers and dendritic polymers in drug delivery. Drug Discov Today 10(1):35–43, 10.1016/S1359-6446(04)03276-3 S1359644604032763 [pii]
Giussani M, Merlino G, Cappelletti V et al (2015) Tumor-extracellular matrix interactions: identification of tools associated with breast cancer progression. Semin Cancer Biol 35:3–10. doi:10.1016/j.semcancer.2015.09.012

Glazer ES, Zhu C, Massey KL et al (2010) Noninvasive radiofrequency field destruction of pancreatic adenocarcinoma xenografts treated with targeted gold nanoparticles. Clin Cancer Res 16(23):5712–5721. doi:10.1158/1078-0432.ccr-10-2055

Godin B, Serda RE, Sakamoto J et al (2010) Nanoparticles for cancer detection and therapy. Wiley-VCH Verlag GmbH & Co, KGaA

Godin B, Chiappini C, Srinivasan S et al (2012) Discoidal porous silicon particles: fabrication and biodistribution in breast cancer bearing mice. Adv Funct Mater 22(20):4225–4235. doi:10.1002/adfm.201200869

Goldberg M, Langer R, Jia X (2007) Nanostructured materials for applications in drug delivery and tissue engineering. J Biomater Sci Polymer Edn 18(3):241–268

Goncalves C, Mennesson E, Fuchs R et al (2004) Macropinocytosis of polyplexes and recycling of plasmid from clathrin-dependent pathway impair the transfection efficiency into human hepatocarcinoma cells. Mol Ther 10:373–385

Goodman T, Olive P, Pun S (2007) Increased nanoparticle penetration in collagenase-treated multicellular spheroids. Int J Nanomedicine 2(2):265–274

Gosk S, Moos T, Gottstein C et al (2008) VCAM-1 directed immunoliposomes selectively target tumor vasculature in vivo. Biochim Biophys Acta Biomembr 1778(4):854–863, doi: http://dx.doi.org/10.1016/j.bbamem.2007.12.021

Gratton S, Ropp P, Pohlhaus P et al (2008a) Effect of particle design on cellular internalization pathways. Proc Natl Acad Sci U S A 105(33):11613–11618

Gratton SEA, Ropp PA, Pohlhaus PD et al (2008b) The effect of particle design on cellular internalization pathways. Proc Natl Acad Sci 105(33):11613–11618. doi:10.1073/pnas.0801763105

Griffiths L, Dachs GU, Bicknell R et al (1997) The influence of oxygen tension and pH on the expression of platelet- derived endothelial cell growth factor/thymidine phosphorylase in human breast tumor cells grown in vitro and in vivo. Cancer Res 57(4):570–572

Groblewska M, Mroczko B, Wereszczyńska-Siemiątkowska U et al (2007) Serum levels of granulocyte colony-stimulating factor (G-CSF) and macrophage colony-stimulating factor (M-CSF) in pancreatic cancer patients. Clin Chem Lab Med 45:30

Groessl M, Slany A, Bileck A et al (2014) Proteome profiling of breast cancer biopsies reveals a wound healing signature of cancer-associated fibroblasts. J Proteome Res 13(11):4773–4782. doi:10.1021/pr500727h

Grosse S, Aron Y, Thevenot G et al (2005) Potocytosis and cellular exit of complexes as cellular pathways for gene delivery by polycations. J Gene Med 7(10):1275–1286

Guo S, Lin CM, Xu Z et al (2014) Co-delivery of cisplatin and rapamycin for enhanced anticancer therapy through synergistic effects and microenvironment modulation. ACS Nano 8(5):4996–5009. doi:10.1021/nn5010815

Haghgooie R, Toner M, Doyle P (2009) Squishy non-spherical hydrogel microparticles. Macromol Rapid Comm 31:128–134

Hamdy S, Haddadi A, Shayeganpour A et al (2011) Activation of antigen-specific T cell-responses by mannan-decorated PLGA nanoparticles. Pharm Res 28(9):2288–2301. doi:10.1007/s11095-011-0459-9

Hanahan D, Coussens L (2012) Accessories to the crime: functions of cells recruited to the tumor microenvironment. Cancer Cell 21(3):309–322. doi:10.1016/j.ccr.2012.02.022

Harisi R, Kenessey I, Olah JN et al (2009) Differential inhibition of single and cluster type tumor cell migration. Anticancer Res 29(8):2981–2985

Harris JM, Martin NE, Modi M (2012) Pegylation. Clin Pharmacokinet 40(7):539–551. doi:10.2165/00003088-200140070-00005

Haubeiss S, Schmid JO, Mürdter TE et al (2010) Dasatinib reverses Cancer-associated Fibroblasts (CAFs) from primary lung carcinomas to a phenotype comparable to that of normal fibroblasts. Mol Cancer 9:168. doi:10.1186/1476-4598-9-168

Hirte H, Vergote IB, Jeffrey JR et al (2006) A phase III randomized trial of BAY 12–9566 (tanomastat) as maintenance therapy in patients with advanced ovarian cancer responsive to primary surgery and paclitaxel/platinum containing chemotherapy: a National Cancer Institute of Canada Clinical Trials Group Study. Gynecol Oncol 102(2):300–308. doi:10.1016/j.ygyno.2005.12.020

Hobbs SK, Monsky WL, Yuan F et al (1998) Regulation of transport pathways in tumor vessels: role of tumor type and microenvironment. Proc Natl Acad Sci U S A 95(8):4607–4612. doi:10.1073/pnas.95.8.4607

Hong F, Tuyama A, Lee TF et al (2009) Hepatic stellate cells express functional CXCR4: role in stromal cell–derived factor-1α–mediated stellate cell activation. Hepatology 49(6):2055–2067. doi:10.1002/hep.22890

Hu Y, Xie J, Tong Y et al (2007) Effect of PEG conformation and particle size on the cellular uptake efficiency of nanoparticles with the HepG2 cells. J Control Release 188:7–17

Huang Y, Goel S, Duda DG et al (2013) Vascular normalization as an emerging strategy to enhance cancer immunotherapy. Cancer Res 73(10):2943–2948. doi:10.1158/0008-5472.can-12-4354

Hue HL, Ruiwen Z, Rex CH et al (2004) Wnt/β-Catenin signaling pathway as novel cancer drug targets. Curr Cancer Drug Targets 4(8):653–671, doi: http://dx.doi.org/10.2174/1568009043332709

Hunter AC, Moghimi SM (2010) Cationic carriers of genetic material and cell death: a mitochondrial tale. Biochim Biophys Acta (BBA) Bioenerget 1797(6–7):1203–1209, doi: http://dx.doi.org/10.1016/j.bbabio.2010.03.026

Hyung W, Ko H, Park J et al (2008) Novel hyaluronic acid (HA) coated drug carriers (HCDCs) for human breast cancer treatment. Biotechnol Bioeng 99(2):442–454. doi:10.1002/bit.21578

Jain RK (1994) Barriers to drug delivery in solid tumors. Sci Am 271(1):58–65

Jain RK (2013) Normalizing tumor microenvironment to treat cancer: bench to bedside to biomarkers. Journal of Clin 31(17):2205–2218

Jain R, Stylianopoulos T (2010) Delivering nanomedicine to solid tumors. Nat Rev Clin Oncol 7(11):653–664

Janku F (2014) Tumor heterogeneity in the clinic: is it a real problem? Ther Adv Med Oncol 6(2):43–51. doi:10.1177/1758834013517414

Jiang W, Kim B, Rutka J et al (2008) Nanoparticle-mediated cellular response is size-dependent. Nat Nanotechnol 3(3):145–150

Jin H, Heller D, Sharma R et al (2009a) Size-dependent cellular uptake and expulsion of single-walled carbon nanotubes: single particle tracking and a generic uptake model for nanoparticles. ACS Nano 3:149–158

Jin H, Heller DA, Sharma R et al (2009b) Size-dependent cellular uptake and expulsion of single-walled carbon nanotubes: single particle tracking and a generic uptake model for nanoparticles. ACS Nano 3(1):149–158. doi:10.1021/nn800532m

Johnston APR, Kamphuis MMJ, Such GK et al (2012) Targeting cancer cells: controlling the binding and internalization of antibody-functionalized capsules. ACS Nano 6(8):6667–6674. doi:10.1021/nn3010476

Kalluri R, Zeisberg M (2006a) Fibroblasts in cancer. Nat Rev Cancer 6(5):392–401. doi:10.1038/nrc1877

Kalluri R, Zeisberg M (2006b) Fibroblasts in cancer. Nat Rev Cancer 6(5):392–401. doi:10.1038/nrc1877

Kamaly N, Xiao Z, Valencia PM et al (2012) Targeted polymeric therapeutic nanoparticles: design, development and clinical translation. Chem Soc Rev 41(7):2971–3010. doi:10.1039/c2cs15344k

Kanaras A, Bartczak D, Sanchez-Elsner T et al (2011) Receptor mediated interactions between colloidal gold nanoparticles and human umbilical vien endothelial cells. Small 7:388–394

Kao H, Abney J, Verkman A (1993) Determinants of the translational mobility of a small solute in cell cytoplasm. J Cell Biol 120(1):175–184

Kato Y, Ozawa S, Tsukuda M et al (2007) Acidic extracellular pH increases calcium influx-triggered phospholipase D activity along with acidic sphingomyelinase activation to induce matrix metalloproteinase-9 expression in mouse metastatic melanoma. FEBS J 274(12):3171–3183. doi:10.1111/j.1742-4658.2007.05848.x

Khalil IA, Kogure K, Akita H et al (2006) Uptake pathways and subsequent intracellular trafficking in nonviral gene delivery. Pharmacol Rev 58(1):32–45. doi:10.1124/pr.58.1.8

Kim S, Takahashi H, Lin WW et al (2009) Carcinoma-produced factors activate myeloid cells through TLR2 to stimulate metastasis. Nature 457(7225):102–106. doi:10.1038/nature07623

Koval M, Preiter K, Adles C et al (1998) Size of IgG-opsonized particles determines macrophage response during internalization. Exp Cell Res 242(1):265–273, 10.1006/excr.1998.4110 S0014-4827(98)94110-6 [pii]

Kozlovskaya VA, Kharlampieva EP, Erel-Unal I et al (2009) Single-component layer-by-layer weak polyelectrolyte films and capsules: loading and release of functional molecules. Polym Sci Ser A 51(6):719–729. doi:10.1134/s0965545x09060170

Kozlovskaya V, Alexander JF, Wang Y et al (2014) Internalization of red blood cell-mimicking hydrogel capsules with pH-triggered shape responses. ACS Nano 8(6):5725–5737. doi:10.1021/nn500512x

Kuang D-M, Wu Y, Chen N et al (2007) Tumor-derived hyaluronan induces formation of immunosuppressive macrophages through transient early activation of monocytes. Blood 110(2):587–595. doi:10.1182/blood-2007-01-068031

Kyurkchiev D, Bochev I, Ivanova-Todorova E et al (2014) Secretion of immunoregulatory cytokines by mesenchymal stem cells. World J Stem Cells 6(5):552–570. doi:10.4252/wjsc.v6.i5.552

Landskron G, De La Fuente M, Thuwajit P et al (2014) Chronic inflammation and cytokines in the tumor microenvironment. J Immunol Res 2014, 149185. doi:10.1155/2014/149185

Lechardeur D, Sohn K, Haardt M et al (1999) Metabolic instability of plasmid DNA in the cytosol: a potential barrier to gene transfer. Gene Ther 6(4):482–497

Lee S-Y, Ferrari M, Decuzzi P (2009) Shaping nano-/micro-particles for enhanced vascular interaction in laminar flows. Nanotechnology 20(49):495101

Leonard F, Curtis LT, Yesantharao P et al (2016) Enhanced performance of macrophage-encapsulated nanoparticle albumin-bound-paclitaxel in hypo-perfused cancer lesions. Nanoscale. doi:10.1039/c5nr07796f

Leone A, Di Gennaro E, Bruzzese F et al (2014) New perspective for an old antidiabetic drug: metformin as anticancer agent. In: Zappia V, Panico S, Russo LG, Bucillon A, Della Ragione F (eds) Advances in nutrition and cancer. Springer, Berlin/Heidelberg, pp 355–376

Leu AJ, Berk DA, Lymboussaki A et al (2000) Absence of functional lymphatics within a murine sarcoma: a molecular and functional evaluation. Cancer Res 60(16):4324–4327

Leuschner F, Dutta P, Gorbatov R et al (2011) Therapeutic siRNA silencing in inflammatory monocytes in mice. Nat Biotechnol 29(11):1005–1010. doi:10.1038/nbt.1989, doi: http://www.nature.com/nbt/journal/v29/n11/abs/nbt.1989.html#supplementary-information

Lin EY, Nguyen AV, Russell RG et al (2001) Colony-stimulating factor 1 promotes progression of mammary tumors to malignancy. J Exp Med 193(6):727–740

Lin Q, Bao C, Yang Y et al (2013) Highly discriminating photorelease of anticancer drugs based on hypoxia activatable phototrigger conjugated chitosan nanoparticles. Adv Mater 25(14):1981–1986. doi:10.1002/adma.201204455

Lisanti MP, Martinez-Outschoorn UE, Sotgia F (2013) Oncogenes induce the cancer-associated fibroblast phenotype: metabolic symbiosis and "fibroblast addiction" are new therapeutic targets for drug discovery. Cell Cycle 12(17):2723–2732. doi:10.4161/cc.25695

Liu X, Gao C, Shen J et al (2005) Multilayer microcapsules as anti-cancer drug delivery vehicle: deposition, sustained release, and in vitro bioactivity. Macromol Biosci 5(12):1209–1219. doi:10.1002/mabi.200500176

Loo C, Lin A, Hirsch L et al (2004) Nanoshell-enabled photonics-based imaging and therapy of cancer. Technol Cancer Res Treat 3(1):33–40

Lovell JF, Jin CS, Huynh E et al (2011) Porphysome nanovesicles generated by porphyrin bilayers for use as multimodal biophotonic contrast agents. Nat Mater 10(4):324–332. doi:10.1038/nmat2986, doi: http://www.nature.com/nmat/journal/v10/n4/abs/nmat2986.html#supplementary-information

Lu P, Takai K, Weaver VM et al (2011). Extracellular matrix degradation and remodeling in development and disease. Cold Spring Harbor Perspectives in Biology, 3(12). doi: 10.1101/cshperspect.a005058

Lukacs G, Haggie P, Seksek O et al (2000) Size-dependent DNA mobility in cytoplasm and nucleus. J Biol Chem 275(3):1625–1629
Ma J, Waxman DJ (2008) Combination of anti-angiogenesis with chemotherapy for more effective cancer treatment. Mol Cancer Ther 7(12):3670–3684. doi:10.1158/1535-7163.mct-08-0715
Mader K, Bacic G, Domb A et al (1997) Noninvasive in vivo monitoring of drug release and polymer erosion from biodegradable polymers by EPR spectroscopy and NMR imaging. J Pharm Sci 86(1):126–134
Maeda H, Sawa T, Konno T (2001) Mechanism of tumor-targeted delivery of macromolecular drugs, including the EPR effect in solid tumor and clinical overview of the prototype polymeric drug SMANCS. J Control Release 74(1–3):47–61. doi:10.1016/s0168-3659(01)00309-1
Maltzahn G v, Park J-H, Agrawal A et al (2009) Computationally guided photothermal tumor therapy using long-circulating gold nanorod antennas. Cancer Res 69(9):3892–3900. doi:10.1158/0008-5472.can-08-4242
Marelli UK, Rechenmacher F, Sobahi TRA et al (2013) Tumor targeting via integrin ligands. Front Oncol 3:222. doi:10.3389/fonc.2013.00222
Marsh T, Pietras K, McAllister SS (2013) Fibroblasts as architects of cancer pathogenesis. Biochim Biophys Acta Mol basis Dis 1832(7):1070–1078. doi:10.1016/j.bbadis.2012.10.013
Matsumura Y, Maeda H (1986) A new concept for macromolecular therapeutics in cancer chemotherapy: mechanism of tumoritropic accumulation of proteins and the antitumor agent smancs. Cancer Res 46(12 I):6387–6392
McIntosh DP, Tan X-Y, Oh P et al (2002) Targeting endothelium and its dynamic caveolae for tissue-specific transcytosis in vivo: a pathway to overcome cell barriers to drug and gene delivery. Proc Natl Acad Sci U S A 99(4):1996–2001. doi:10.1073/pnas.251662398
Mecke A, Lee D-K, Ramamoorthy A et al (2005) Synthetic and natural polycationic polymer nanoparticles interact selectively with fluid-phase domains of DMPC lipid bilayers. Langmuir 21(19):8588–8590. doi:10.1021/la051800w
Meng F, Zhong Y, Cheng R et al (2014) pH-sensitive polymeric nanoparticles for tumor-targeting doxorubicin delivery: concept and recent advances. Nanomedicine 9(3):487–499. doi:10.2217/nnm.13.212
Merkel TJ, Jones SW, Herlihy KP et al (2011) Using mechanobiological mimicry of red blood cells to extend circulation times of hydrogel microparticles. Proc Natl Acad Sci U S A 108(2):586–591, 10.1073/pnas.1010013108 1010013108 [pii]
Merkel TJ, Chen K, Jones SW et al (2012) The effect of particle size on the biodistribution of low-modulus hydrogel PRINT particles. J Control Release 162(1):37–44. doi:10.1016/j.jconrel.2012.06.009
Minko T, Kopečkova P, Pozharov V et al (2000) The influence of cytotoxicity of macromolecules and of VEGF gene modulated vascular permeability on the enhanced permeability and retention effect in resistant solid tumors. Pharm Res 17(5):505–514. doi:10.1023/a:1007500412442
Mislick KA, Baldeschwieler JD (1996) Evidence for the role of proteoglycans in cation-mediated gene transfer. Proc Natl Acad Sci 93(22):12349–12354
Mitra AK, Zillhardt M, Hua Y et al (2012) MicroRNAs reprogram normal fibroblasts into cancer-associated fibroblasts in ovarian cancer. Cancer Discovery 2(12):1100–1108. doi:10.1158/2159-8290.cd-12-0206
Mitragotri S, Lahann J (2009) Physical approaches to biomaterial design. Nat Mater 8:15–23
Mizrahy S, Goldsmith M, Leviatan-Ben-Arye S et al (2014) Tumor targeting profiling of hyaluronan-coated lipid based-nanoparticles. Nanoscale 6(7):3742–3752. doi:10.1039/c3nr06102g
Mok S, Koya RC, Tsui C et al (2014) Inhibition of CSF-1 receptor improves the antitumor efficacy of adoptive cell transfer immunotherapy. Cancer Res 74(1):153–161. doi:10.1158/0008-5472.can-13-1816
Mora E, Guglielmotti A, Biondi G et al (2012) Bindarit: an anti-inflammatory small molecule that modulates the NFkB pathway. Cell Cycle 11(1):159–169. doi:10.4161/cc.11.1.18559

Mukaida N, Sasaki SI, Baba T (2014) Chemokines in cancer development and progression and their potential as targeting molecules for cancer treatment. Mediat Inflamm 2014, 170381. doi:10.1155/2014/170381

Murakami M, Ernsting MJ, Undzys E et al (2013) Docetaxel conjugate nanoparticles that target a-smooth muscle actin-expressing stromal cells suppress breast cancer metastasis. Cancer Res 73(15):4862–4871. doi:10.1158/0008-5472.can-13-0062

Muro S, Koval M, Muzykantov V (2004) Endothelial endocytic pathways gates for vascular drug delivery. Curr Vasc Pharmacol 2(3):281–299

Muro S, Garnacho C, Champion J et al (2008a) Control of endothelial targeting and intracellular delivery of therapeutic enzymes by modulating the size and shape of ICAM-1-targeted carriers. Mol Ther 16(8):1450–1458

Muro S, Garnacho C, Champion JA et al (2008b) Control of endothelial targeting and intracellular delivery of therapeutic enzymes by modulating the size and shape of ICAM-1-targeted carriers. Mol Ther 16(8):1450–1458, 10.1038/mt.2008.127 mt2008127 [pii]

Na Y-R, Yoon Y-N, Son D-I et al (2013) Cyclooxygenase-2 inhibition blocks M2 macrophage differentiation and suppresses metastasis in murine breast cancer model. PLoS One 8(5), e63451. doi:10.1371/journal.pone.0063451

Nakanishi Y, Nakatsuji M, Seno H et al (2011) COX-2 inhibition alters the phenotype of tumor-associated macrophages from M2 to M1 in ApcMin/+ mouse polyps. Carcinogenesis 32(9):1333–1339. doi:10.1093/carcin/bgr128

Napp J, Behnke T, Fischer L et al (2011) Targeted luminescent near-infrared polymer-nanoprobes for in vivo imaging of tumor hypoxia. Anal Chem 83(23):9039–9046. doi:10.1021/ac201870b

Ng SL, Such GK, Johnston APR et al (2011) Controlled release of DNA from poly(vinylpyrrolidone) capsules using cleavable linkers. Biomaterials 32(26):6277–6284, doi: http://dx.doi.org/10.1016/j.biomaterials.2011.05.011

Nie S, Xing Y, Kim G et al (2007) Nanotechnology: applications in cancer. Annu Rev Biomed Eng 9:257–288

Nugent LJ, Jain RK (1984) Extravascular diffusion in normal and neoplastic tissues. Cancer Res 44(1):238–244

Oh P, Borgström P, Witkiewicz H et al (2007) Live dynamic imaging of caveolae pumping targeted antibody rapidly and specifically across endothelium in the lung. Nat Biotechnol 25(3):327–337. doi:10.1038/nbt1292

Oldenborg PA, Zheleznyak A, Fang YF et al (2000) Role of CD47 as a marker of self on red blood cells. Science 288(5473):2051–2054. doi:10.1126/science.288.5473.2051

Olive KP, Jacobetz MA, Davidson CJ et al (2009) Inhibition of hedgehog signaling enhances delivery of chemotherapy in a mouse model of pancreatic cancer. Science 324(5933):1457–1461. doi:10.1126/science.1171362

Owens DE, Peppas NA (2006) Opsonization, biodistribution, and pharmacokinetics of polymeric nanoparticles. Int J Pharm 307(1):93–102, doi: http://dx.doi.org/10.1016/j.ijpharm.2005.10.010

Pack D, Putnam D, Langer R (2000) Design of imidazole-containing endosomolytic biopolymers for gene delivery. Biotechnol Bioeng 67(2):217–223

Palamà IE, Coluccia AML, Torre AD et al (2010) Multilayered polyelectrolyte capsules and coated colloids: cytotoxicity and uptake by cancer cells. Sci Adv Mater 2(2):138–150. doi:10.1166/sam.2010.1082

Panni RZ, Linehan DC, DeNardo DG (2013) Targeting tumor-infiltrating macrophages to combat cancer. Immunotherapy 5(10):1075–1087. doi:10.2217/imt.13.102

Panowksi S, Bhakta S, Raab H et al (2014) Site-specific antibody drug conjugates for cancer therapy. MAbs 6(1):34–45. doi:10.4161/mabs.27022

Pasqualini R, Ruoslahti E (1996) Organ targeting In vivo using phage display peptide libraries. Nature 380(6572):364–366. doi:10.1038/380364a0

Pasqualini R, Koivunen E, Ruoslahti E (1995) A peptide isolated from phage display libraries is a structural and functional mimic of an RGD-binding site on integrins. J Cell Biol 130(5):1189–1196

Patel T, Zhou J, Piepmeier JM et al (2012) Polymeric nanoparticles for drug delivery to the central nervous system. Adv Drug Deliv Rev 64(7):701–705, doi: http://dx.doi.org/10.1016/j.addr.2011.12.006

Peer D, Karp JM, Hong S et al (2007) Nanocarriers as an emerging platform for cancer therapy. Nat Nanotechnol 2(12):751–760, 10.1038/nnano.2007.387 nnano.2007.387 [pii]

Peggs KS, Quezada SA, Allison JP (2009) Cancer immunotherapy: co-stimulatory agonists and co-inhibitory antagonists. Clin Experimen Immunol 157(1):9–19. doi:10.1111/j.1365-2249.2009.03912.x

Perrault S, Walkey C, Jennings T et al (2009) Mediating tumor targeting efficiency of nanoparticles through design. Nano Lett 9(5):1909–1915

Pietras K, Pahler J, Bergers G et al (2008) Functions of paracrine PDGF signaling in the proangiogenic tumor stroma revealed by pharmacological targeting. PLoS Med 5(1), e19. doi:10.1371/journal.pmed.0050019

Pilon-Thomas S, Kodumudi KN, El-Kenawi AE et al (2016) Neutralization of tumor acidity improves antitumor responses to immunotherapy. Cancer Res 76(6):1381–1390. doi:10.1158/0008-5472.can-15-1743

Platt VM, Szoka FC (2008) Anticancer therapeutics: targeting macromolecules and nanocarriers to hyaluronan or CD44, a hyaluronan receptor. Mol Pharm 5(4):474–486. doi:10.1021/mp800024g

Pluen A, Boucher Y, Ramanujan S et al (2001) Role of tumor–host interactions in interstitial diffusion of macromolecules: cranial vs. subcutaneous tumors. Proc Natl Acad Sci 98(8):4628–4633. doi:10.1073/pnas.081626898

Poon Z, Chang D, Zhao X et al (2011) Layer-by-layer nanoparticles with a pH-sheddable layer for in vivo targeting of tumor hypoxia. ACS Nano 5(6):4284–4292. doi:10.1021/nn200876f

Pridgen EM, Langer R, Farokhzad OC (2007) Biodegradable, polymeric nanoparticle delivery systems for cancer therapy. Nanomedicine 2(5):669–680. doi:10.2217/17435889.2.5.669

Provenzano PP, Cuevas C, Chang AE et al (2012) Enzymatic targeting of the stroma ablates physical barriers to treatment of pancreatic ductal adenocarcinoma. Cancer Cell 21(3):418–429. doi:10.1016/j.ccr.2012.01.007

Qian SW, Li X, Zhang YY et al (2010) Characterization of adipocyte differentiation from human mesenchymal stem cells in bone marrow. BMC Dev Biol 10:47. doi:10.1186/1471-213x-10-47

Qiu Y, Liu Y, Wang L et al (2010) Surface chemistry and aspect ratio mediated cellular uptake of Au nanorods. Biomaterials 31(30):7606–7619

Quail DF, Joyce JA (2013) Microenvironmental regulation of tumor progression and metastasis. Nat Med 19(11):1423–1437. doi:10.1038/nm.3394

Quante M, Tu SP, Tomita H et al (2011) Bone marrow-derived myofibroblasts contribute to the mesenchymal stem cell niche and promote tumor growth. Cancer Cell 19(2):257–272. doi:10.1016/j.ccr.2011.01.020

Quante M, Varga J, Wang TC et al (2013) The gastrointestinal tumor microenvironment. Gastroenterology 145(1):63–78. doi:10.1053/j.gastro.2013.03.052

Räsänen K, Vaheri A (2010) Activation of fibroblasts in cancer stroma. Exp Cell Res 316(17):2713–2722. doi:10.1016/j.yexcr.2010.04.032

Rees P, Brown MR, Summers H et al (2011) A transfer function approach to measuring cell inheritance. BMC Syst Biol 5(1):31

Rejman J, Oberle V, Zuhorn I et al (2004a) Size-dependent internalization of particles via the pathways of clathrin- and caveolae-mediated endocytosis. Biochem J 377:159–169

Rejman J, Oberle V, Zuhorn I et al (2004b) Size-dependent internalization of particles via the pathways of clathrin and caveolae mediated endocytosis. J Biochem 377:159–169

Rejman J, Oberle V, Zuhorn IS et al (2004c) Size-dependent internalization of particles via the pathways of clathrin- and caveolae-mediated endocytosis. Biochem J 377(Pt 1):159–169, 10.1042/BJ20031253.BJ20031253 [pii]

Ries CH, Cannarile MA, Hoves S et al (2014) Targeting tumor-associated macrophages with anti-CSF-1R antibody reveals a strategy for cancer therapy. Cancer Cell 25(6):846–859. doi:10.1016/j.ccr.2014.05.016

Rodriguez PL, Harada T, Christian DA et al (2013) Minimal “Self” peptides that inhibit phagocytic clearance and enhance delivery of nanoparticles. Science 339(6122):971–975. doi:10.1126/science.1229568

Rolny C, Mazzone M, Tugues S et al (2011) HRG inhibits tumor growth and metastasis by inducing macrophage polarization and vessel normalization through downregulation of PlGF. Cancer Cell 19(1):31–44. doi:10.1016/j.ccr.2010.11.009

Rosenberg SA, Packard BS, Aebersold PM et al (1988) Use of tumor-infiltrating lymphocyts and interleukin-2 in the immunotherapy of patients with metastatic melanoma. A preliminary report. N Engl J Med 319(25):1676–1680

Rosenholm JM, Sahlgren C, Linden M (2010) Towards multifunctional, targeted drug delivery systems using mesoporous silica nanoparticles–opportunities & challenges. Nanoscale 2(10):1870–1883. doi:10.1039/c0nr00156b

Ruiter D, Bogenrieder T, Elder D et al (2002) Melanoma–stroma interactions: structural and functional aspects. Lancet Oncol 3(1):35–43, doi: http://dx.doi.org/10.1016/S1470-2045(01)00620-9

Sahay G, Alakhova DY, Kabanov AV (2010) Endocytosis of nanomedicines. J Control Release 145(3):182–195. doi:10.1016/j.jconrel.2010.01.036

Sanchez-Lopez E, Flashner-Abramson E, Shalapour S et al (2015) Targeting colorectal cancer via its microenvironment by inhibiting IGF-1 receptor-insulin receptor substrate and STAT3 signaling. Oncogene. doi:10.1038/onc.2015.326

Saraogi GK, Sharma B, Joshi B et al (2011) Mannosylated gelatin nanoparticles bearing isoniazid for effective management of tuberculosis. J Drug Target 19(3):219–227. doi:10.3109/1061186x.2010.492522

Satchi-Fainaro R, Puder M, Davies JW et al (2004) Targeting angiogenesis with a conjugate of HPMA copolymer and TNP-470. Nat Med 10(3):255–261. doi:10.1038/nm1002

Schipper M, Nakayama-Ratchford N, Davis C et al (2008) A pilot toxicology study of single-walled carbon nanotubes in a small sample of mice. Nat Nanotechnol 3:216–221

Schmid S (1997) Clathrin-coated vesicle formation and protein sorting an integrated process. Annu Rev Biochem 66:511–548

See V, Free P, Cesbron Y et al (2009) Cathepsin L digestion of nanobioconjugates upon endocytosis. ACS Nano 3(9):2461–2468. doi:10.1021/nn9006994

Serda R, Gu J, Bhavane R et al (2009) The association of silicon microparticles with endothelial cells in drug delivery to the vasculature. Biomaterials 30(13):2440–2448

Serda RE, Godin B, Blanco E et al (2010) Multi-stage delivery nano-particle systems for therapeutic applications. Biochim Biophys Acta. doi: S0304-4165(10)00127-3 [pii] 10.1016/j.bbagen.2010.05.004

Serda RE, Godin B, Blanco E et al (2011) Multi-stage delivery nano-particle systems for therapeutic applications. Biochim Biophys Acta Gen Subj 1810(3):317–329. doi:10.1016/j.bbagen.2010.05.004

Shi Q, Abbruzzese JL, Huang S et al (1999) Constitutive and inducible interleukin 8 expression by hypoxia and acidosis renders human pancreatic cancer cells more tumorigenic and metastatic. Clin Cancer Res 5(11):3711–3721

Shimoni O, Yan Y, Wang Y et al (2012) Shape-dependent cellular processing of polyelectrolyte capsules. ACS Nano 7(1):522–530. doi:10.1021/nn3046117

Simpson-Abelson MR, Purohit VS, Pang WM et al (2009) IL-12 delivered intratumorally by multilamellar liposomes reactivates memory T cells in human tumor microenvironments. Clin Immunol 132(1):71–82. doi:10.1016/j.clim.2009.03.516

Slowing I, Trewyn BG, Lin VS (2006) Effect of surface functionalization of MCM-41-type mesoporous silica nanoparticles on the endocytosis by human cancer cells. J Am Chem Soc 128(46):14792–14793. doi:10.1021/ja0645943

Sokolov K, Follen M, Aaron J et al (2003) Real-time vital optical imaging of precancer using anti-epidermal growth factor receptor antibodies conjugated to gold nanoparticles. Cancer Res 63(9):1999–2004
Sonawane N, Szoka F, Verkman A (2003) Chloride accumulation and swelling in endosomes enhances DNA transfer by polyamine-DNA polyplexes. J Biol Chem 278(45):44826–44831
Sounni NE, Noel A (2013) Targeting the tumor microenvironment for cancer herapy. Clin Chem 59(1):85–93. doi:10.1373/clinchem.2012.185363
Sparano JA, Bernardo P, Stephenson P et al (2004) Randomized phase III trial of marimastat versus placebo in patients with metastatic breast cancer who have responding or stable disease after first-line chemotherapy: eastern cooperative oncology group trial E2196. J Clin Oncol 22(23):4683–4690. doi:10.1200/jco.2004.08.054
Sprinzl MF, Reisinger F, Puschnik A et al (2013) Sorafenib perpetuates cellular anticancer effector functions by modulating the crosstalk between macrophages and natural killer cells. Hepatology 57(6):2358–2368. doi:10.1002/hep.26328
Srinivasan S, Sawyer PN (1970) Symposium on friction and static electricity in solid/fluid and solid/solid systemsRole of surface charge of the blood vessel wall, blood cells, and prosthetic materials in intravascular thrombosis. J Colloid Interface Sci 32(3):456–463, doi: http://dx.doi.org/10.1016/0021-9797(70)90131-1
Srinivasan S, Alexander JF, Driessen WH et al (2013) Bacteriophage associated silicon particles: design and characterization of a novel theranostic vector with improved payload carrying potential. J Mater Chem B 1(39):5218–5229. doi:10.1039/c3tb20595a
Stephan MT, Moon JJ, Um SH et al (2010) Therapeutic cell engineering using surface-conjugated synthetic nanoparticles. Nat Med 16(9):1035–1041. doi:10.1038/nm.2198
Sukhorukov GB, Rogach AL, Garstka M et al (2007) Multifunctionalized polymer microcapsules: novel tools for biological and pharmacological applications. Small 3(6):944–955. doi:10.1002/smll.200600622
Sun X, Rossin R, Turner J et al (2005) An assessment of the effects of shell cross-linked nanoparticle size, core composition, and surface PEGylation on in vivo biodistribution. Biomacromolecules 6(5):2541–2554
Sun T-M, Du J-Z, Yao Y-D et al (2011) Simultaneous delivery of siRNA and paclitaxel via a "Two-in-One" micelleplex promotes synergistic tumor suppression. ACS Nano 5(2):1483–1494. doi:10.1021/nn103349h
Takei K, Haucke V (2001) Clathrin-mediated endocytosis: membrane factors pull the trigger. Trends Cell Biol 11(9):385–391
Tanei T, Leonard F, Liu X et al (2016) Redirecting transport of nanoparticle albumin-bound paclitaxel to macrophages enhances therapeutic efficacy against liver metastases. Cancer Res 76(2):429–439. doi:10.1158/0008-5472.can-15-1576
Tangutoori S, Spring BQ, Mai Z et al (2016) Simultaneous delivery of cytotoxic and biologic therapeutics using nanophotoactivatable liposomes enhances treatment efficacy in a mouse model of pancreatic cancer. Nanomed Nanotechnol Biol Med 12(1):223–234. doi:10.1016/j.nano.2015.08.007
Tannock IF, Rotin D (1989) Acid pH in tumors and its potential for therapeutic exploitation. Cancer Res 49(16):4373–4384
Tao L, Hu W, Liu Y et al (2011) Shape-specific polymeric nanomedicine: emerging opportunities and challenges. Experimen Biol Med 236(1):20–29. doi:10.1258/ebm.2010.010243
Thiviyanathan V, Somasunderam AD, Gorenstein DG (2007) Combinatorial selection and delivery of thioaptamers. Biochem Soc Trans 35:50–52
Thorek DL, Tsourkas A (2008) Size, charge and concentration dependent uptake of iron oxide particles by non-phagocytic cells. Biomaterials 29(26):3583–3590. doi:10.1016/j.biomaterials.2008.05.015
Thurn KT, Thomas S, Moore A et al (2011) Rational therapeutic combinations with histone deacetylase inhibitors for the treatment of cancer. Future Oncol 7(2):263–283. doi:10.2217/fon.11.2

Tlsty TD, Coussens LM (2006) Tumor stroma and regulation of cancer development. Annu Rev Pathol 1:119–150

Togo S, Polanska UM, Horimoto Y et al (2013) Carcinoma-associated fibroblasts are a promising therapeutic target. Cancers 5(1):149–169. doi:10.3390/cancers5010149

Tommelein J, Verset L, Boterberg T et al (2015) Cancer-associated fibroblasts connect metastasis-promoting communication in colorectal cancer. Front Oncol 5:63. doi:10.3389/fonc.2015.00063

Torchilin VP (2005) Recent advances with liposomes as pharmaceutical carriers. Nat Rev Drug Discov 4(2):145–160, 10.1038/nrd1632 nrd1632 [pii]

Torchilin VP (2007) Micellar nanocarriers: pharmaceutical perspectives. Pharm Res 24(1):1–16. doi:10.1007/s11095-006-9132-0

Torchilin V (2011) Tumor delivery of macromolecular drugs based on the EPR effect. Adv Drug Deliv Rev 63(3):131–135. doi:10.1016/j.addr.2010.03.011

Tsagozis P, Eriksson F, Pisa P (2008) Zoledronic acid modulates antitumoral responses of prostate cancer-tumor associated macrophages. Cancer Immunol Immunother 57(10):1451–1459. doi:10.1007/s00262-008-0482-9

Tuerk C, Gold L (1990) Systematic evolution of ligands by exponential enrichment – Rna ligands to bacteriophage-T4 DNA-Polymerase. Science 249(4968):505–510

U'Ren L, Kedl R, Dow S (2006) Vaccination with liposome-DNA complexes elicits enhanced antitumor immunity. Cancer Gene Ther 13(11):1033–1044

van Broekhoven CL, Parish CR, Demangel C et al (2004) Targeting dendritic cells with antigen-containing liposomes: a highly effective procedure for induction of antitumor immunity and for tumor immunotherapy. Cancer Res 64(12):4357–4365. doi:10.1158/0008-5472.can-04-0138

van Zijl F, Krupitza G, Mikulits W (2011) Initial steps of metastasis: cell invasion and endothelial transmigration. Mutat Res 728(1–2):23–34. doi:10.1016/j.mrrev.2011.05.002

Venugopal J, Ramakrishna S (2005) Applications of polymer nanofibers in biomedicine and biotechnology. Appl Biochem Biotechnol 125(3):147–158, doi: ABAB:125:3:147 [pii]

Vergaro V, Scarlino F, Bellomo C et al (2011) Drug-loaded polyelectrolyte microcapsules for sustained targeting of cancer cells. Adv Drug Deliv Rev 63(9):847–864, doi: http://dx.doi.org/10.1016/j.addr.2011.05.007

Walter K, Omura N, Hong S-M et al (2010) Overexpression of smoothened activates the sonic hedgehog signaling pathway in pancreatic cancer associated fibroblasts. Clin Cancer Res 16(6):1781–1789. doi:10.1158/1078-0432.ccr-09-1913

Wang K, He Q, Yan X et al (2007) Encapsulated photosensitive drugs by biodegradable microcapsules to incapacitate cancer cells. J Mater Chem 17(38):4018–4021. doi:10.1039/b708477c

Wang B, Zhang L, Bae SC et al (2008) Nanoparticle-induced surface reconstruction of phospholipid membranes. Proc Natl Acad Sci 105(47):18171–18175. doi:10.1073/pnas.0807296105

Wang X-L, Xu R, Wu X et al (2009) Targeted systemic delivery of a therapeutic siRNA with a multifunctional carrier controls tumor proliferation in mice. Mol Pharm 6(3):738–746. doi:10.1021/mp800192d

Wang J, Tian S, Petros RA et al (2010) The complex role of multivalency in nanoparticles targeting the transferrin receptor for cancer therapies. J Am Chem Soc 132(32):11306–11313. doi:10.1021/ja1043177

Wang F, Yu L, Monopoli MP et al (2013) The biomolecular corona is retained during nanoparticle uptake and protects the cells from the damage induced by cationic nanoparticles until degraded in the lysosomes. Nanomed Nanotechnol Biol Med 9(8):1159–1168, doi: http://dx.doi.org/10.1016/j.nano.2013.04.010

Wang C, Xu L, Liang C et al (2014) Immunological responses triggered by photothermal therapy with carbon nanotubes in combination with anti-CTLA-4 therapy to inhibit cancer metastasis. Adv Mater 26(48):8154–8162. doi:10.1002/adma.201402996

Wight TN, Kinsella MG, Evanko SP et al (2014) Versican and the regulation of cell phenotype in disease. Biochim Biophys Acta 1840(8):2441–2451. doi:10.1016/j.bbagen.2013.12.028

Win K, Feng S-S (2005) Effects of particle size and surface coating on cellular uptake of polymeric nanoparticles for oral delivery of anticancer drugs. Biomaterials 26:2713–2722

Wong S, Pelet J, Putnam D (2007) Polymer systems for gene delivery – past, present, and future. Prog Polym Sci 32:799–837
Woods A, Couchman J (1994) Syndecan 4 heparan sulfate proteoglycan is a selectively enriched and widespread focal adhesion component. Mol Biol Cell 5:183–192
Xiang S, Scholzen A, Minigo G et al (2006) Pathogen recognition and development of particulate vaccines: does size matter? Methods Enzymol 40(1):1–9
Xu L, Fukumura D, Jain RK (2002) Acidic extracellular pH induces Vascular Endothelial Growth Factor (VEGF) in human glioblastoma cells via ERK1/2 MAPK signaling pathway: mechanism of low pH-induced VEGF. J Biol Chem 277(13):11368–11374. doi:10.1074/jbc.M108347200
Yadav AK, Mishra P, Jain S et al (2008) Preparation and characterization of HA-PEG-PCL intelligent core corona nanoparticles for delivery of doxorubicin. J Drug Target 16(6):464–478. doi:10.1080/10611860802095494
Yan X, Scherphof GL, Kamps JAAM (2005) Liposome opsonization. J Liposome Res 15(1–2):109–139. doi:10.1081/lpr-64971
Yan Y, Johnston APR, Dodds SJ et al (2010) Uptake and intracellular fate of disulfide-bonded polymer hydrogel capsules for doxorubicin delivery to colorectal cancer cells. ACS Nano 4(5):2928–2936. doi:10.1021/nn100173h
Yang X-Z, Du J-Z, Dou S et al (2012) Sheddable ternary nanoparticles for tumor acidity-targeted siRNA delivery. ACS Nano 6(1):771–781. doi:10.1021/nn204240b
Yokoi K, Godin B, Oborn CJ et al (2013) Porous silicon nanocarriers for dual targeting tumor associated endothelial cells and macrophages in stroma of orthotopic human pancreatic cancers. Cancer Lett 334(2):319–327. doi:10.1016/j.canlet.2012.09.001
Yoo J-W, Mitragotri S (2010a) Polymer particles that switch shape in response to a stimulus. Proc Natl Acad Sci 107(25):11205–11210. doi:10.1073/pnas.1000346107
Yoo J, Mitragotri S (2010b) Polymer particles that switch shape in response to a stimulus. Proc Natl Acad Sci U S A 107(25):11205–11210
Yuan H, Li J, Bao G et al (2010) Variable nanoparticle-cell adhesion strength regulates cellular uptake. Phys Rev Lett 105(13):138101
Zamboni W (2008) Concept and clinical evaluation of carrier-mediated anticancer agents. Oncologist 13:248–260
Zauner W, Farrow N, Haines A (2001) In vitro uptake of polystyrene microspheres: effect of particle size, cell line, and cell density. J Control Release 71(1):39–51
Zeisberger SM, Odermatt B, Marty C et al (2006) Clodronate-liposome-mediated depletion of tumour-associated macrophages: a new and highly effective antiangiogenic therapy approach. Br J Cancer 95(3):272–281. doi:10.1038/sj.bjc.6603240
Zhang J, Liu J (2013) Tumor stroma as targets for cancer therapy. Pharmacol Ther 137(2):200–215, doi: http://dx.doi.org/10.1016/j.pharmthera.2012.10.003
Zhang D, Neumann O, Wang H et al (2009) Gold nanoparticles can induce the formation of protein-based aggregates at physiological pH. Nano Lett 9(2):666–671
Zhang QW, Liu L, Gong CY et al (2012) Prognostic significance of tumor-associated macrophages in solid tumor: a meta-analysis of the literature. PLoS One 7(12), e50946. doi:10.1371/journal.pone.0050946
Zhou X, Huang L (1994) DNA transfection mediated by cationic liposomes containing lipopolylysine characterization and mechanism of action. Biochim Biophys Acta Biomembr 1189(2):195–203
Zhu X-D, Zhang J-B, Zhuang P-Y et al (2008) High expression of macrophage colony-stimulating factor in peritumoral liver tissue is associated with poor survival after curative resection of hepatocellular carcinoma. J Clin Oncol 26(16):2707–2716. doi:10.1200/jco.2007.15.6521

Part III
How to Extend the Circulation Time of Nanovehicles

Chapter 5
A New Approach to Decrease the RES Uptake of Nanodrugs by Pre-administration with Intralipid® Resulting in a Reduction of Toxic Side Effects

Li Liu and Chien Ho

Abstract The number of different nanoparticle drug-delivery systems currently being developed is increasing rapidly. A major limitation for the translation of nanodrugs to clinical applications is their rapid uptake by the cells of the reticuloendothelial system (RES), in particular liver and spleen. This can reduce the efficacy at the target tissue and increase the nano-toxicity to these off-target organs. Unfortunately, the accumulation of nanodrugs in tumors represents a small fraction of total injected dose (1–10 %). The majority (40–80 %) of injected nanodrugs end up in the RES. Many studies have been conducted to try to decrease the RES uptake and to increase the tumor targeting of nanodrugs by modifying nanoparticle characteristics, such as size, shape, charge, surface property, and composition. Our new strategy is to temporarily blunt the uptake of nanodrugs by the RES using Intralipid® 20 %, which has been used as a safe nutrition source for four decades. We have found that, in rodents, pre-treatment with Intralipid® 20 % (using a clinical route and dosage, 2 g/kg) can change the pharmacokinetic/pharmacodynamic profile of nanoparticles. A single dose of Intralipid® can reduce RES uptake by ~50 % and increase blood half-life by ~3-fold of nano- and micron-sized iron-oxide particles. Recently, we have applied this finding to the delivery of a platinum (Pt)-containing anti-cancer nanodrug. Pre-administration with Intralipid® can increase the bioavailability of the Pt-nanodrug and reduce the toxicity in liver, spleen, and kidney. Our method is a general one applicable to any approved and in-development nanodrugs without additional modification of the nano-carriers.

Keywords Reticuloendothelial system • Intralipid® • Magnetic resonance imaging • Cell tracking • Superparamagnetic iron-oxide nanoparticle • Immune cell labeling • Anti-cancer nanodrug • Platinum-containing nanodrug • Off-target accumulation • Toxic side effects • Hepatotoxicity • Nephrotoxicity • Enlarged spleen • Blood half-life • Bioavailability • Pharmacokinetic/pharmacodynamics profile

---

L. Liu • C. Ho (✉)
Department of Biological Sciences, Carnegie Mellon University,
4400 Fifth Avenue, Pittsburgh, PA 15213, USA
e-mail: chienho@andrew.cmu.edu

A. Prokop, V. Weissig (eds.), *Intracellular Delivery III*, Fundamental Biomedical Technologies 8, DOI 10.1007/978-3-319-43525-1_5

Abbreviations

ABC	Accelerated blood clearance
ALT	Alanine aminotransferase
BN	Brown Norway
DACHPt	Dichloro (1 2-diaminocyclohexane) platinum (II)
DACHPt/HANP	DACHPt-incorporated nanoparticles coated with hyaluronic acid polymer
EPR	Enhanced permeability and retention
FDA	Food and Drug Administration
FITC	Fluorescein isothiocyanate
HA	Hyaluronic acid polymer
H & E	Hematoxylin/eosin
ICP-MS	Inductively coupled plasma-mass spectrometry
MPIO	Micron-sized paramagnetic iron-oxide
MRI	Magnetic resonance imaging
NP	Nanoparticles
PBS	Phosphate-buffered-saline
PD	Pharmacodynamic
PEG	Polyethylene glycol
PK	Pharmacokinetic
POD	Post operation day
Ppm	Part per million
Pt	Platinum
RES	Reticuloendothelial system
SD	Sprague Dawley
SPIO	Superparamagnetic iron-oxide
TUNEL	Terminal deoxynucleotidyl transferase dUTP nick end labeling
USPIO	Ultra-small superparamagnetic iron-oxide

5.1 Introduction

Nanotechnology-based imaging agents and therapeutics have been investigated for more than 30 years. There are more than 200 nanodrugs that have been either approved or are under clinical investigation (Min et al. 2015). Their applications range from imaging (Bashir et al. 2015; Shapiro 2015), cancer (Wicki et al. 2015), inflammation (Getts et al. 2014), autoimmune diseases (Gharagozloo et al. 2015), regenerative medicine (Gao et al. 2015), to fungal infections (Hamill 2013), vaccines (Gregory et al. 2013; Stegmann et al. 1987), anesthetics (Gambling et al. 2005; Viscusi et al. 2005), and macular degeneration (Agosta et al. 2012). A major challenge for developing and translating nanodrugs to clinical use is that the nanodugs, after intravenous administration, are rapidly cleared from the circulation by the reticuloendothelial system (RES), especially by the liver and spleen (Albanese et al. 2012; Guo and Huang 2011; Onoue et al. 2014). This rapid uptake hinders the effective delivery of the loaded therapeutics at the target sites due to insufficient bioavailability in the blood and can cause toxic side effects due to off-target tissue distribution.

RES uptake is important for the pharmacokinetic (PK), pharmacodynamic (PD), efficacy, and for the toxicity of a nanodrug system. Nanodrugs can offer several pharmacokinetic advantages, such as specific drug delivery, improved bioavailability, long duration of action, high metabolic stability, and high membrane permeability (Onoue et al. 2014). Therefore, by altering the biopharmaceutic and PK properties of new drug candidates, nanodrug systems could be a promising approach to obtain the desired drug properties at the desired location. In the formulation of nanoparticles, it is crucial to minimize RES uptake and to prolong the circulation in the blood of the nanoparticles so as to increase their chances of reaching the desired target. This can be partially achieved by modifying nanoparticle characteristics and surface properties, such as size, shape, charge, composition, targeting moiety, and, the most commonly used surface modification with polyethylene glycol (PEG) (Albanese et al. 2012; Arvizo et al. 2011; Guo and Huang 2011; Kozlowska et al. 2009; Liu et al. 2014; Neuberger et al. 2005). Current results are that even for the best designed nanoparticles, a big portion (40–80 %) of the injected nanoparticles end up in the liver and spleen (Albanese et al. 2012). Moreover, from a toxicological perspective, each additional feature could exhibit an extra risk of toxicity, and the high complexity of nanocarriers can be counter-productive (Xue et al. 2014). The RES uptake and resulting nanotoxicity are still big challenges for development and clinical translation of nanodrugs, e.g., siRNA-based nanodrugs (Guo and Huang 2011; Onoue et al. 2014; Xue et al. 2014).

The toxicity in the RES has also been reported or warned for Food and Drug Administration (FDA) approved nanodrugs, e.g., Abraxane® (http://www.accessdata. fda.gov/drugsatfda_docs/label/2014/021660s040lbl.pdf), Marqibo® (www.marqibo.com), Doxil®/Caelyx® (Gabizon et al. 2002), and Onivyde® (http://www.accessdata.fda.gov/drugsatfda_docs/label/2015/207793lbl.pdf). It is difficult to modify these approved drugs to reduce RES toxicity, because each new modification calls for thorough toxicity, pharmacology, biomechanics, and efficacy studies before translating to a clinical setting.

Our novel strategy is to target the RES to temporarily blunt particle clearance. In particular, we set out to find FDA approved agents to achieve this aim. Intralipid® was approved by the FDA in 1972 as a source of parenteral nutrition for patients. Intralipid® is the brand name of the first safe fat emulsion for human use manufactured by Fresenius Kabi (Uppsala, Sweden). Intralipid® 20.0 % is composed of 20 % soybean oil, 1.2 % egg-yolk phospholipids, and 2.25 % glycerol (http://www.accessdata.fda.gov/drugsatfda_docs/label/2007/017643s072,018449s039lbl.pdf). The major fatty acid constituents are linoleic acid (44–62 %), oleic acid (19–30 %), palmitic acid (7–14 %), linolenic acid (4–11 %), and stealic acide (1.4–5.5 %). Kupffer cells in the liver play an important role in the uptake and metabolism of Intralipid® (Vilaro and Llobera 1988). Intralipid® infusion has been reported to inhibit RES function by possibly inhibiting peritoneal clearance and impairing the phagocytic activity of Kupffer cells (Nugent 1984). Our hypothesis is that the clearance of nanodrugs by the RES can be reduced by using agents, such as Intralipid®, which are also cleared by Kupffer cells and inhibit their phagocytic activities, prior to injection of the nanoparticles. In this chapter, we summarize our recent findings on using Intralipid® to decrease the RES uptake, improve the bioavailability, and reduce the toxic side effects of nano- or micron-sized imaging agents and chemotherapeutics.

5.2 Current Strategies and Limitations for Reducing the RES Clearance of Nanoparticles

One major accomplishment in this field is to make "stealth" nanoparticles, i.e., nanoparticles escaping the RES, through coating the surface with hydrophilic polymers/surfactants, and/or formulation with biodegradable copolymers with hydrophilic segments, such as PEG, poloxamer, poloxamine, polyethylene oxide, dextrane, and polysorbate 80 (Tween 80) (Guo and Huang 2011; Salmaso and Caliceti 2013). PEG is the most commonly used non-ionic hydrophilic polymer to make "stealth" nanoparticles in order to reduce their RES uptake and increase their blood circulation. The first approved PEGylated product, Doxil/Caelyx, has already been in the clinic for ~20 years (Lasic and Needham 1995; Woodle 1995). Clinical experience and laboratory research have shown not only the benefits of this polymer, but also drawbacks (Knop et al. 2010). The unfavorable effects of PEG polymers can be divided into the following categories: immunological response, e.g., hypersensitivity reactions, induced by the polymer itself or by its side products formed during synthesis (Chanan-Khan et al. 2003; Szebeni 2005); unexpected changes in pharmacokinetic behavior, i.e., the accelerated blood clearance (ABC) phenomenon (Abu Lila et al. 2013; Dams et al. 2000; Koide et al. 2008); toxic side products and an antagonism arising from degradation under stress or upon exposure to oxygen (Knop et al. 2010). Thus, although PEG is a very popular polymer to make "stealth" nanoparticles, scientists are still searching for new alternatives to reduce the RES uptake of nanodrugs.

Some other important findings are that smaller-sized particles are removed from the bloodstream via RES clearance slower than larger-sized particles (Neuberger et al. 2005); neutral and zwitterionic nanoparticles exhibit longer circulation times than negatively and positively charged nanoparticles (Arvizo et al. 2011; Liu et al. 2014). In addition, surface modification of nanodrugs, with specific proteins, antibodies, and other biomolecules, can be used to design drugs that act selectively on particular tissues. This approach has been employed to provide reduced liver and spleen uptake, improved therapeutic potential, and reduced side effects of some anticancer drugs (Kozlowska et al. 2009).

The above modifications are effective in reducing RES clearance of the nanoparticles; however, the majority of the injected nanomaterials still ends up in the liver and spleen, resulting in low delivery efficiency to the target site (Albanese et al. 2012). Moreover, each modification adds an extra risk of side effects and complications. The high complexity of nanocarriers might generate more drawbacks (Xue et al. 2014). Our Intralipid® approach is a simple and safe one, applicable to any approved and in-development nanodrugs to improve their bioavailability and to decrease their toxic side effects, with no need of new modification(s) of the nanodrugs and/or the nanocarriers.

5.3 Using Intralipid® to Reduce the RES Uptake of Nanoparticles for Imaging

Non-invasive *in-vivo* magnetic resonance imaging (MRI) of monocytes/macrophages labeled with MRI contrast agents may lead to a better understanding of the pathogenesis of many diseases, including graft rejection (Wu et al. 2006, 2009, 2011; Ye et al. 2008), atherosclerotic plaques (Tang et al. 2009; Tu et al. 2011), tumors (Daldrup-Link et al. 2011; Shih et al. 2011), abdominal aortic aneurysm (Miyama et al. 2012), renal ischemia (Jo et al. 2003), Alzheimer's disease (Beckmann et al. 2011), etc. MRI combined with superparamagnetic iron-oxide (SPIO) contrast agents is an effective cell-tracking method. Our laboratory has been able to monitor the infiltration of macrophages labeled with ultra-small superparamagnetic iron-oxide (USPIO, with particle size ~30 nm in diameter) and micron-sized paramagnetic iron-oxide (MPIO, ~0.9 μm in diameter) particles at rejecting kidneys, hearts, and lungs in our rat models for organ transplantation to study the immune responses during transplant rejection (Wu et al. 2006, 2009; Yang et al. 2001; Ye et al. 2002, 2008).

MRI contrast agents are taken up by the RES, in particular by liver Kupffer cells, which often contribute to a major loss of the agents in circulation (Chouly et al. 1996; Minchin and Martin 2010; Okon et al. 1994; Neuberger et al. 2005). Hepatic Kupffer cells are responsible for ~90 % of the total phagocytic capacity that is involved in maintaining homeostasis in circulation (van Rooijen and van Kesteren-Hendrikx 2002). Strategies that reduce liver uptake and prolong the residence time in the circulation of these MRI contrast agents can improve the *in-vivo* labeling efficiency of immune cells and lower the required effective dose.

When Kupffer cell activity needs to be reduced, e.g., in order to enhance the efficiency of gene delivery, clodronate liposomes have been applied to deplete Kupffer cells in the liver (Schughart et al. 1999). However, the depletion effects of clodronate liposomes should be transient, but the re-appearance of Kupffer cells could take up to one week (van Rooijen and van Kesteren-Hendrikx 2002). We have developed a safer way to temporarily inhibit Kupffer cell activity by using Intralipid® (Liu et al. 2013). In this study, USPIO and MPIO were used as our test cases.

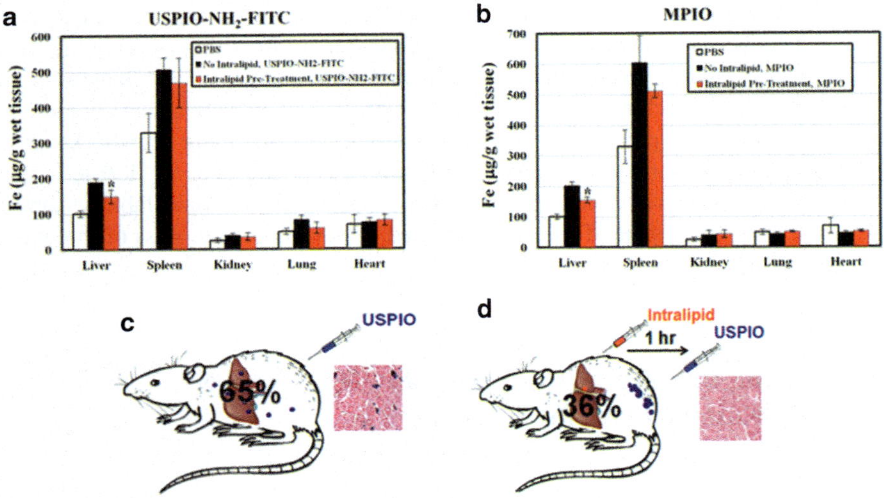

Fig. 5.1 Changes in the iron levels, as measured by inductively coupled plasma-mass spectrometry (ICP-MS), in different tissues at 48 h following administration of iron-oxide particles with and without Intralipid® pre-treatment: (**a**) USPIO-NH_2-FITC particles; and (**b**) MPIO particles. *$p<0.01$ compared with iron-oxide particles administered, but without Intralipid® pre-treatment; (**c**) A pictorial summary of the iron distribution in the liver from the data of USPIO-NH_2-FITC. Side figures: Perls' Prussian blue iron staining of liver tissues [Modified from Figs. 5.1 and 5.2 of (Liu et al. 2013); with permission]

5.3.1 *Treatment Protocol*

Male Brown Norway (BN; RT1^n) rats, with body weights between 250 and 280 g, were used in this study. Intralipid® 20 % was administered by intravenous injection (clinical route) at a dose of 2 g/kg (clinical dose). Phosphate-buffered-saline (PBS) was administered to control animals. USPIO-NH_2-fluorescein isothiocyanate (FITC) particles were injected at a dose of 4.5 mg Fe/kg body weight. MPIO particles were injected at a dose of 6 mg Fe/kg body weight.

5.3.2 *Intralipid® Changes the Biodistribution and Reduces the RES Uptake of Iron-Oxide Particles*

48-h post injection of iron-oxide particles, various tissues (liver, spleen, kidney, lung, and heart) were collected for iron-level determination and histological analysis. As shown in Fig. 5.1, greater fractions of nano- or micron-sized iron-oxide particles are found in the liver and spleen, particularly in the liver, than in the heart, kidney, or lung (Fig. 5.1a and b). Figure 5.1c shows a summary of liver accumulation of the nano-sized USPIO particle, based on the calculation of iron concentrations shown in Fig. 5.1a. With no Intralipid® pre-treatment, ~65 % of injected

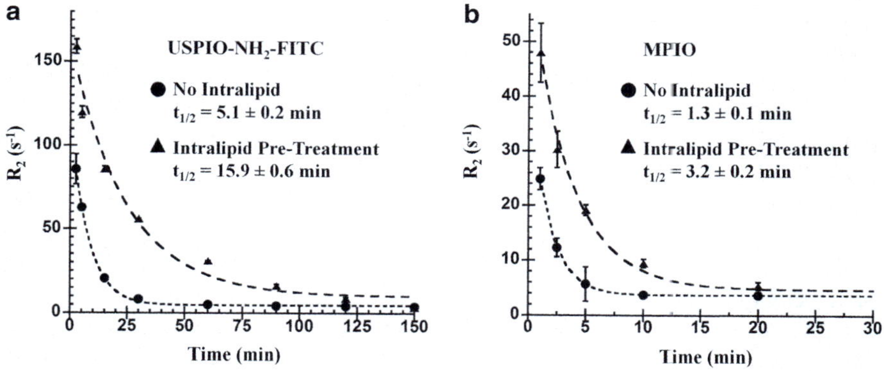

Fig. 5.2 Changes in blood clearance of iron-oxide particles upon Intralipid® pre-treatment: (**a**) USPIO-NH_2-FITC particles; and (**b**) MPIO particles [Taken from Fig. 5.3 of (Liu et al. 2013); with permission]

USPIO nanoparticles ends up in the liver. Upon Intralipid® pre-treatment, the liver only takes up ~36 % of injected USPIO nanoparticles. Thus, the liver uptake of USPIO particles significantly decreases by 45.1 ± 6.5 %. Representative light microscopic images of Perls' Prussian blue iron staining of liver tissues are shown in Fig. 5.1c, which confirm the results from the iron concentration measurement. Similarly, we have found that Intralipid® pre-treatment can reduce the liver uptake of MPIO particles by 49.2 ± 5.9 %.

5.3.3 *Intralipid® Increases the Blood Half-life ($t_{1/2}$) of Iron-Oxide Particles*

Intralipid® pre-treatment results in a 3- and 2.5-fold increase in the blood half-life of USPIO and MPIO particles, respectively, as determined by the changes of the transverse relaxation rate (R_2) values in whole blood following particle injection (Fig. 5.2a and b). As a consequence, the blood bioavailability of USPIO and MPIO particles increases ~3 and ~2 fold, respectively.

5.3.4 *Intralipid® Increases the Labeling of Monocytes in the Circulation and Improves the Sensitivity of Cellular MRI*

Escaping RES uptake and prolonging the *in-vivo* circulation of MRI contrast agents can lead to increased uptake in target tissues and cells, leading to increased sensitivity of MRI. Phagocytic immune cells (e.g., monocytes and macrophages) can be labeled *in vivo* by iron-oxide particles (Wu et al. 2006, 2009). We have found that pre-treatment with Intralipid® significantly enhances the labeling efficiency of blood monocytes/macrophages by nano- and micron-sized iron-oxide particles, as shown by flow cytometry (Fig. 5.3a and c). As summarized in Fig. 5.3b and d, Intralipid® can increase the monocyte labeling efficiency two to threefold for both USPIO and MPIO particles.

We have tested this Intralipid® methodology for *in-vivo* cell tracking by MRI using our rat heterotopic working heart and lung transplant model. Figure 5.4 shows representative *in-vivo* MR images of rejecting heart and lung through a T_2^*-weighted MRI technique. On post-operational day (POD) 6, when the allografts exhibit severe grade IV rejection, a large amount of monocytes/macrophages have infiltrated into the allograft heart and lung (Wu et al. 2009, 2013). In order to test the Intralipid® method, a half-dose of MPIO particles as in our previous study (Wu et al. 2006), was intravenously administered to the recipient on POD 3. As shown in Fig. 5.4a, with a low dose of the contrast agent, the allografts are not very hypointense and few macrophages can be detected by MRI. Upon Intralipid® pre-treatment (Fig. 5.4b), a significantly increased hypointensity can be detected in transplanted heart and lung, because Intralipid® treatment increases the efficiency of macrophage labeling by MPIO. These findings are encouraging because our approach can increase the efficiency of labeling of immune cells with nano- to micron-sized imaging agents, and therefore the dose of the contrast agents needed for *in-vivo* cell tracking and imaging experiments can be minimized.

5.4 Using Intralipid® to Reduce the RES Uptake of Anti-cancer Nanodrugs for Improved Drug Delivery

Cancer is a major public health problem worldwide and is the second leading cause of death in the United States. In 2016, 1,685,210 new cancer cases and 595,690 cancer deaths are projected to occur in the United States (Siegel et al. 2016). How to improve the delivery of anti-cancer drugs specifically to tumors and how to reduce the toxic side effects of these cytotoxic drugs in normal tissue are major challenges for cancer treatment. Nanomedicine holds great promise for generating new preventative, diagnostic, and therapeutic approaches to cancer in areas where improvements cannot be realized using existing technologies (http://nano.cancer.gov/). Anti-cancer nanodrugs show several advantages: nanodrugs tend to accumulate in solid tumors as a result of the enhanced permeability and retention (EPR) of macromolecules, thereby enhancing their anti-tumor or tumor-diagnosis activity; nanoformulation can help to overcome problems of solubility and chemical stability of conventional therapeutics, e.g., nucleic acids (siRNA and aptamers), thus improving the PK/PD profiles of the drugs (Williams et al. 2013; Xiang et al. 2015); and targeted nanomedicine therapeutics may decrease the resistance of tumors against anti-cancer drugs (Zhang and Li 2013).

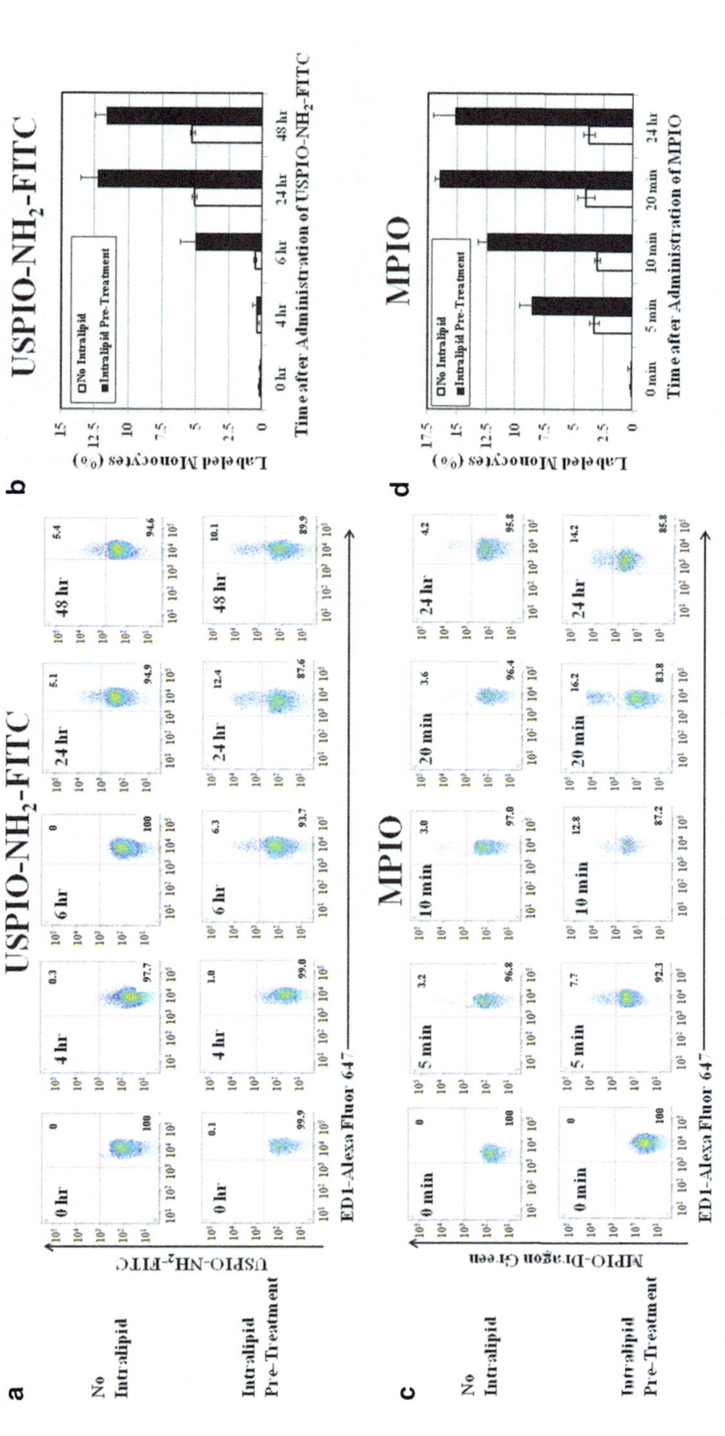

Fig. 5.3 Changes in blood monocyte labeling by iron-oxide particles upon Intralipid® pre-treatment. Representative flow cytometry dot plots showing the kinetics of blood monocyte labeled by (**a**) USPIO-NH_2-FITC particles and (**c**) MPIO particles upon Intralipid® pre-treatment. (**b**) and (**d**) are summary charts of the data shown in (**a**) and (**c**), respectively [Modified from Fig. 5.4 of (Liu et al. 2013); with permission]

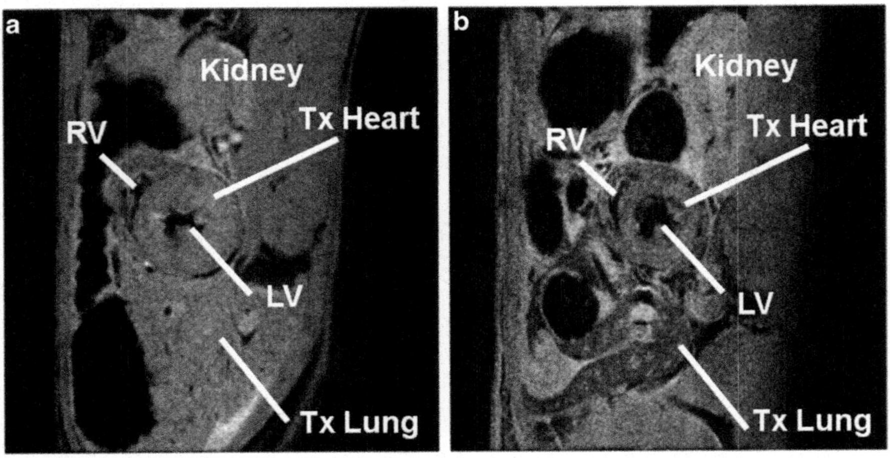

Fig. 5.4 More MPIO-labeled monocytes/macrophages can be detected at a rejecting rat heart and lung by *in-vivo* cellular MRI, upon Intralipid® pre-treatment. MPIO (3 mg Fe/kg) was intravenously administered into the transplant recipient rat on POD 3. These two MR images were acquired on POD 6. (**a**) No Intralipid® pre-treatment; (**b**) Intralipid® pre-treatment. *Tx* transplanted, *RV* right ventricle, *LV* left ventricle (Unpublished results)

Pt-based drugs, including cisplatin, carboplatin, and oxaliplatin (first, second, and third generation), have become the most potent as well as the most widely prescribed anti-cancer drugs (Wheate et al. 2010). Unfortunately, their continuous use is greatly limited by dose limiting toxicities, partial anti-tumor response in most patients, development of drug resistance, and tumor relapse (Kamaly et al. 2013; Ludwig et al. 2004; Pabla and Dong 2008; Stefanowicz et al. 2011; Ulusakarya et al. 2010; Wheate et al. 2010; Yao et al. 2007). Nanocarrier-based drug delivery may generate new therapeutic approaches for Pt-drugs. Studies have shown that the therapeutic performance of oxaliplatin can be improved by incorporating the central dichloro (1, 2-diaminocyclohexane) platinum (II) (DACHPt) motif into the core of these nanocarriers (Cabral et al. 2005; Murakami et al. 2011; Oberoi et al. 2012; Wu et al. 2014). Pt-based nanodrugs are providing encouraging preclinical and clinical results and may facilitate the development of the next generation of Pt chemotherapy (Cabral et al. 2005; Murakami et al. 2011; Oberoi et al. 2013).

A major limitation for both approved and in-development nanodrugs, e.g., Pt-based nanodrugs, is their rapid clearance by the cells of the RES, in particular liver and spleen. The accumulation of the cytotoxic drugs in the RES, will not only reduce their efficacy, but also increase their toxicity for these off-target organs. Strategies that decrease RES uptake and increase the bioavailability of nanomedicines can improve tumor targeting and decrease the side effects. With the rationale and hypothesis described above, we have extended our Intralipid® approach to delivery of an anti-cancer nanodrug (Liu et al. 2015). In this study, an improved Pt anti-cancer nanodrug, DACHPt-incorporated nanoparticles (NP), coated with hyaluronic acid polymer (HA) (DACHPt/HANP), was used as our test case.

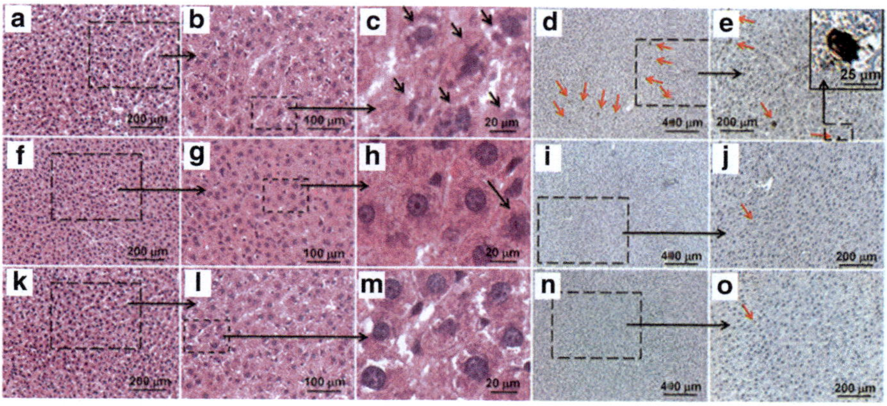

Fig. 5.5 Intralipid® reduces toxic side effects in liver caused by the anti-cancer nanodrug, DACHPt/HANP. Light microscopy images of **h & e** stained (**a**–**c**, **f**–**h**, and **k**–**m**) and TUNEL stained liver tissue (**d**–**e**, **i**–**j**, and **n**–**o**). (**a**–**e**) are from the liver tissues of DACHPt/HANP administered, but not Intralipid® treated, animals; (**f**–**j**) are from the liver tissues of Intralipid® pre-treated animals; (**k**–**o**) are from the liver tissues of naïve animals. (**c**, **h**, and **m**) are examples of enlarged views of (**b**, **g**, and **l**), which are enlarged from part of (**a**, **f**, and **k**). Similarly, (**e**, **j**, and **o**) are enlarged from (**d**, **i**, and **n**). Small *black arrows* on (**c** and **h**) indicate cell necrosis; *red arrows* on (**d** and **e**) indicate cell apoptosis [Taken from Fig. 5.1 of (Liu et al. 2015)]

5.4.1 Treatment Protocol

Male Sprague Dawley (SD) rats with an indwelling jugular vein catheter implanted were purchased from Harlan Laboratories (Indianapolis, IN). SD rats with body weights between 250 and 280 g were used. Intralipid® 20 % was administered to SD rats at the clinical dose (2 g/kg) using the clinical route (i.e., intravenously) one hour before intravenous injection of DACHPt/HANP (2 mg Pt/kg for PK/PD studies; 4 mg Pt/kg and 6 mg Pt/kg for toxicity studies). PBS was administered to control animals.

5.4.2 Intralipid® Reduces Toxic Side Effects of DACHPt/HANP in Liver, Spleen, and Kidney

Pathological Analysis and Terminal Deoxynucleotidyl Transferase dUTP Nick End Labeling (TUNEL) Assay for Apoptotic Cells in Liver, Spleen, and Kidney Light microscopic images of hematoxylin/eosin (H & E) stained and TUNEL stained liver, spleen, and kidney tissue sections are shown in Figs. 5.5, 5.6, and 5.7, respectively. The liver tissue sections from the Intralipid® pre-treated group are shown in Fig. 5.5f, g, h, i and j. Compared with the DACHPt/HANP administered, but no Intralipid® pre-treatment group (Fig. 5.5a, b, c, d, and e), the pathological changes

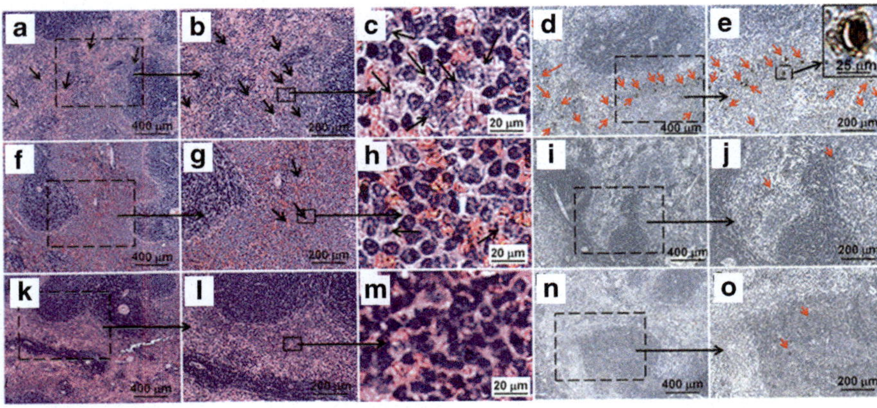

Fig. 5.6 Intralipid® reduces toxic side effects in spleen caused by DACHPt/HANP. Light microscopy images of **h & e** stained (**a**–**c**, **f**–**h**, and **k**–**m**) and TUNEL stained spleen tissue (**d**–**e**, **i**–**j**, and **n**–**o**). (**a**–**e**) are from the spleen tissues of DACHPt/HANP administered, but not Intralipid® treated, animals; (**f**–**j**) are from Intralipid® pre-treated animals; (**k**–**o**) are from naïve healthy animals. *Black arrows* on (**a**) indicate concurrent abnormal proliferation of mononuclear cells; *black arrows* on (**b**, **c**, **g**, and **h**) indicate cell necrosis; *red arrows* on (**d** and **e**) indicate cell apoptosis [Taken from Fig. 5.3 of (Liu et al. 2015)]

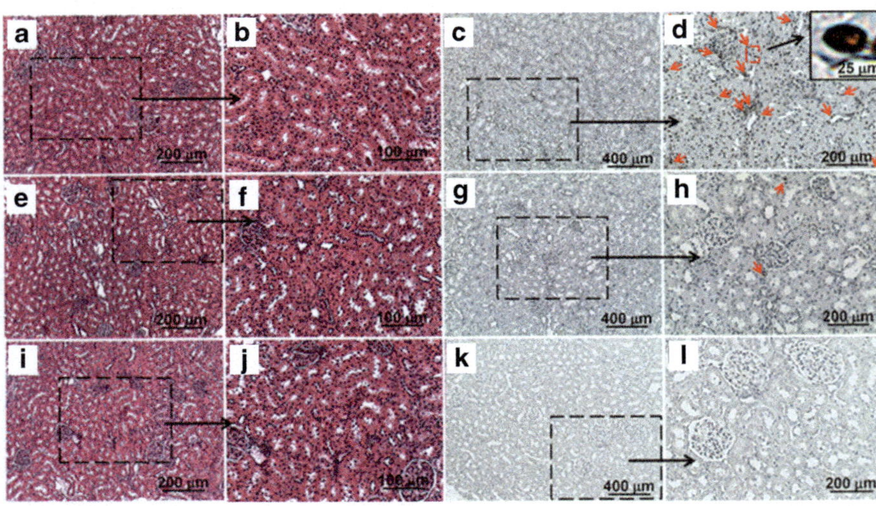

Fig. 5.7 Intralipid® reduces toxic side effects in kidney caused by DACHPt/HANP. Light microscopy images of **h & e** stained (**a**-**b**, **e**-**f**, and **i**-**j**) and TUNEL stained (**c**-**d**, **g**-**h**, and **k**-**l**) kidney tissue. (**a**-**d**) are from the kidney tissues of DACHPt/HANP administered, but not Intralipid® treated, animals; (**e**-**h**) are Intralipid® pre-treated animals; (**i**-**l**) are from naïve healthy animals. *Red arrows* on (**d** and **h**) indicate cell apoptosis [Taken from Fig. 5.4 of (Liu et al. 2015)]

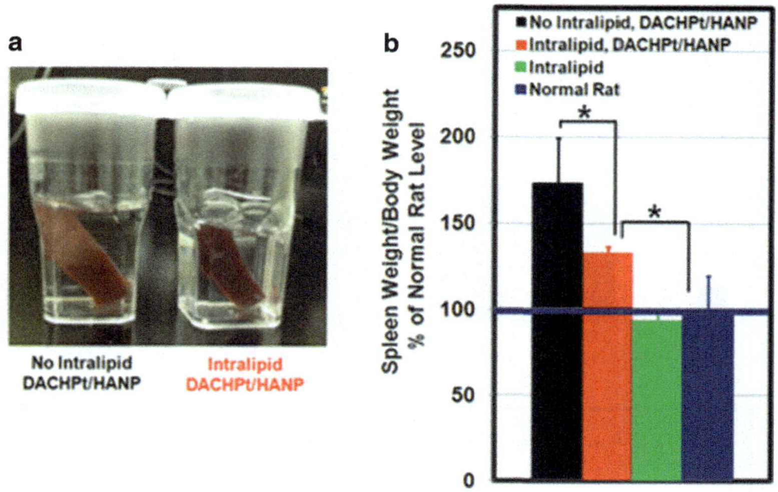

Fig. 5.8 Intralipid® pre-treatment can reduce spleen swelling significantly: (**a**) picture of the spleens from DACHPt/HANP treated, without or with Intralipid® treatment, SD rats; and (**b**) the changes in spleen weight/body weight ratio upon Intralipid® treatment. The ratio from a naïve SD rat is treated as 100 %. $^{*}p<0.05$ [Taken from Fig. 5.2 of (Liu et al. 2015)]

in the liver tissue with Intralipid® pre-treatment (Fig. 5.5f, g, h, i, and j) are characterized as showing fewer necrotic cells (black arrow in Fig. 5.5h) and apoptotic cells (red arrows in Fig. 5.5j), and are comparable to the liver tissues of naïve rats (Fig. 5.5k, l, m, n, and o).

As shown in Fig. 5.6, Intralipid® pre-treatment protects the spleen tissue from the damages caused by the anti-cancer nanodrug, DACHPt/HANP. With Intralipid® pre-treatment, the pathological changes in the spleen tissue (Fig. 5.6f, g, h, i, and j) are characterized as uniformly distributed mononuclear cells (Fig. 5.6f), fewer necrotic (Fig. 5.6g and h), and fewer apoptotic (Fig. 5.6i and j) cells, as compared with the no Intralipid® pre-treatment group (Fig. 5.6a, b, c, d, and e).

Interestingly, Intralipid® can protect not only the tissues in the RES, but also the tissue in kidney (Fig. 5.7). Nephrotoxicity is one of the most severe side effects of current Pt drugs (Ludwig et al. 2004; Negro et al. 2010; Pabla and Dong 2008; Stefanowicz et al. 2011; Wheate et al. 2010; Yao et al. 2007). The DACHPt/HANP nanodrug is designed to increase the concentration and prolong the half-life of DACHPt at tumor sites and to decrease the side effects like nephrotoxicity. Although our Intralipid® therapy was originally designed to decrease the RES uptake of the nanodrug, we also find that Intralipid® reduces cell apoptosis in kidney dramatically (Fig. 5.7d vs h, red arrows).

Intralipid® Reduces Spleen Swelling Spleen swelling and enlargement are observed from DACHPt/HANP-treated animals, when the animals are sacrificed 72-h post nanodrug administration (Fig. 5.8a left). Intralipid® pre-treatment reduces spleen swelling (Fig. 5.8a right). In Fig. 5.8b, the ratios of spleen weight/body weight are shown as the percentage of the normal level.

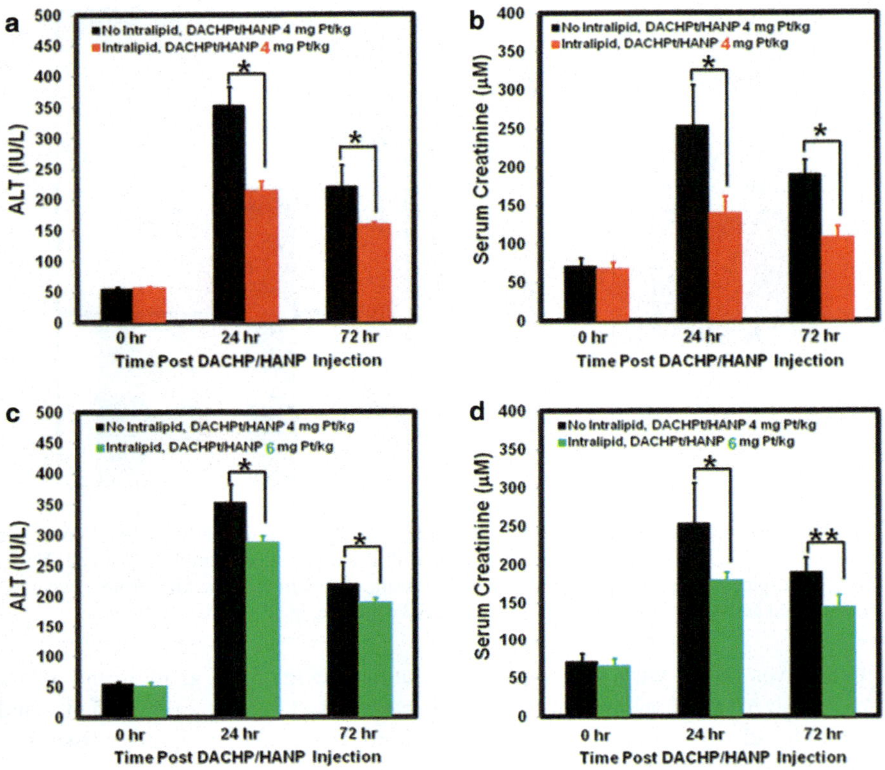

Fig. 5.9 Effects of Intralipid® pre-treatment on the serum ALT activities (**a**, **c**) and creatinine levels (**b**, **d**) in DACHPt/HANP treated rats. When the rats are treated with 4 mg Pt/kg of the nanodrug, the Intralipid® pre-treatment group shows significantly lower serum ALT activity and creatinine level (**a** and **b**). The group pre-treated with Intralipid® followed by a higher dosage (6 mg Pt/kg) of DACHPt/HANP, exhibits lower ALT activity and creatinine level than the group, which is treated with 4 mg Pt/kg of the nanodrug, but no Intralipid® (**c** and **d**). $^{*}p<0.001$; $^{**}p<0.05$ [Modified from Fig. 5.5 of (Liu et al. 2015)]

Serum Alanine Aminotransferase (ALT) Activity and Creatinine Colorimetric Assays to Assess Hepatotoxicity and Nephrotoxicity When the rats were treated with 4 mg Pt/kg of DACHPt/HANP, the Intralipid® pre-treatment group shows significantly lower serum ALT activity and creatinine level (Fig. 5.9a and b). These results are consistent with our findings in the liver and kidney histological studies as shown in Figs. 5.5 and 5.7.

In order to show the potency of Intralipid® protective effects, another group of rats were pre-treated with Intralipid® followed by a higher dosage, 6 mg Pt/kg, of DACHPt/HANP (Fig. 5.9c and d). It is found that both ALT activities and creatinine

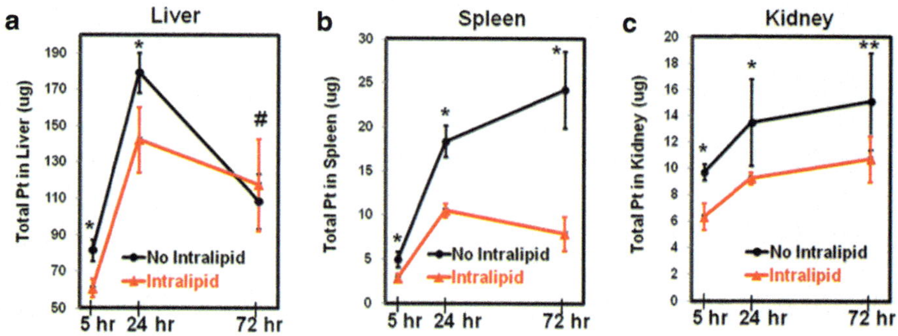

Fig. 5.10 Changes in Pt accumulation in liver (**a**), spleen (**b**), and kidney (**c**) with and without Intralipid® pre-treatment, at 5-, 24-, and 72-h post DACHPt/HANP administration. *P* values represent the differences between the amount of Pt in the tissue with and without Intralipid® pretreatment at the same time point. $^{*}p<0.001$; $^{**}p<0.01$; $^{\#}p>0.1$ [Modified from Fig. 5.6 of (Liu et al. 2015)]

levels from this group are significantly lower than the group treated with the lower dosage of the nanodrug (4 mg Pt/kg), but no Intralipid® pre-treatment. This result indicates that, using Intralipid®, the clinicians might be able to give the patients more anti-cancer nanodrugs to kill the tumors with less toxic side effects!

5.4.3 *Intralipid® Changes the Tissue Distribution and Blood Clearance of DACHPt/HANP*

As shown in Fig.5.10, one single administration of Intralipid® can significantly decrease Pt accumulation in liver, spleen, and kidney. Intralipid® can decrease liver accumulation of the Pt-containing drug by 24.9 % and 20.4 % at 5- and 24-h post injection, respectively (Fig. 5.10a). Intralipid® can significantly decrease spleen uptake of the Pt-containing drug by 40.1 %, 42.4, and 67.2 % at 5-, 24-, and 72-h post administration (Fig. 5.10b). Upon Intralipid® pre-treatment, Pt concentrations in the kidney decrease by 34.0, 31.2, and 28.7 % at 5-, 24-, and 72-h post DACHPt/HANP administration, respectively (Fig. 5.10c).

Changes in the Pt concentrations in circulation upon Intralipid® pre-treatment are shown in Fig. 5.11a. As a consequence of the decreased RES clearance, the bioavailability of the Pt-nanodrug increases by 18.7 % during the first 5 h and by 9.4 % during 24 h, respectively (Fig. 5.11b and c).

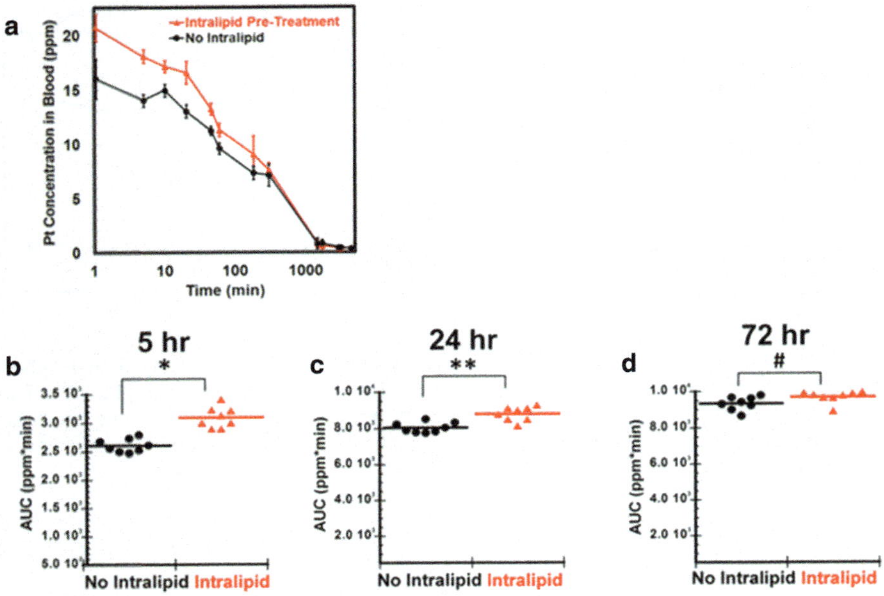

Fig. 5.11 Changes in the Pt concentrations in circulation and bioavailability upon Intralipid® pre-treatment during 72 h. (**a**) Changes of blood Pt concentration. X-axis represents the duration post DACHPt/HANP injection, in logarithmic scale (base: 10). (**b**, **c**, **d**) Changes in the bioavailability of DACHPt/HANP nanodrug [calculated by the area under the curve (AUC), by the trapezoidal rule, using KaleidaGraph 4.1 (Synergy Software, Reading, PA)]: (**b**) 5 h; (**c**) 24 h; (**d**) 72 h after the nanodrug administration. $^{*}p<0.0001$; $^{**}p<0.001$; $^{\#}p>0.05$ [Modified from Fig. 5.7 and Fig. S3 of (Liu et al. 2015)]

5.4.4 *A Pilot Study: Using Multi-doses of Intralipid® to Deliver Multi- and Over-dose of Nanodrugs*

The above experiments are designed as "proof of concept" studies showing that Intralipid® can change the PK/PD and toxicological profiles of an anti-cancer nanodrug. To increase and prolong the effectiveness of Intralipid®, the administration dosages, time courses, and frequencies of Intralipid® treatment need to be adjusted and optimized with the chemotherapy regimen. Multi-doses of Intralipid® are necessary.

When a pre-clinical animal study was conducted for another anti-cancer Pt-nanodrug, DACHPt/micelle, the dosing frequency was twice a week (Cabral et al. 2005; Murakami et al. 2011). In clinical practice, the clinicians desire to use high doses of anti-cancer nanodrug, but are limited by the toxic side effects. Thus, we have designed a multi- and over-dose experiment to treat rats with DACHPt/HANP (4 mg Pt/kg, intravenously) twice a week (on Day 1 and Day 4) (Fig. 5.12a). On Day 1 and Day 4, Intralipid® was intravenously administered (clinical route) one hour before the administration of the nanodrug at a dosage of 2 g/kg and one hour

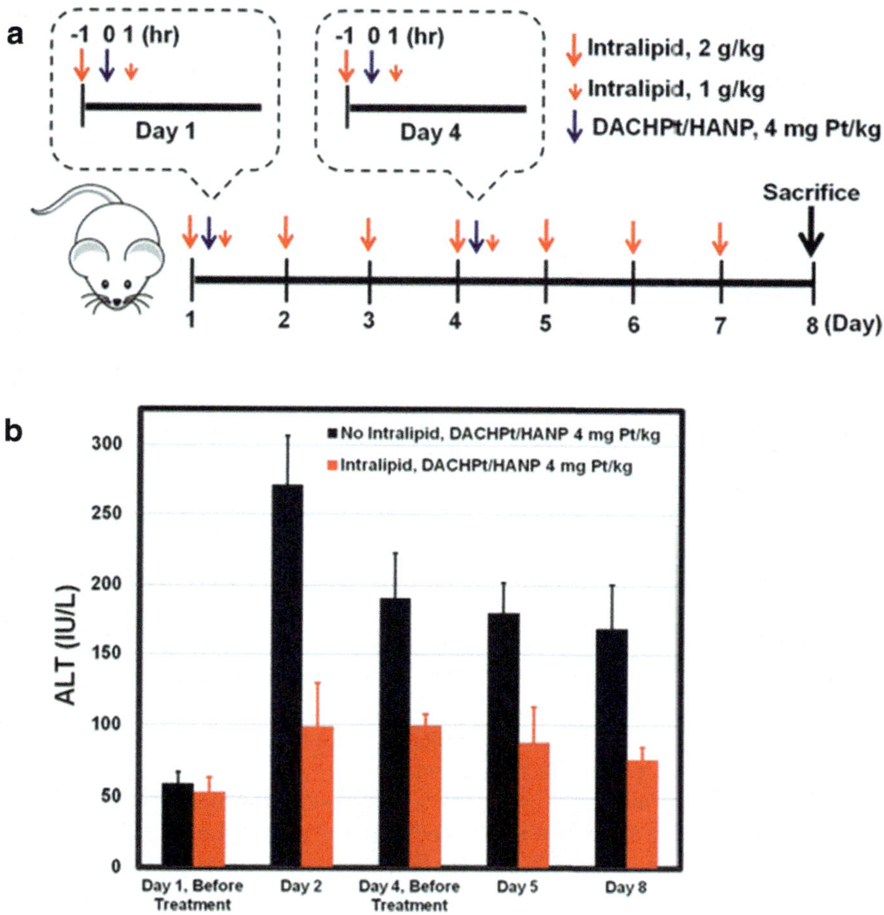

Fig. 5.12 Using multi-doses of Intralipid® to deliver multi- and over-doses of DACHPt/HANP. (**a**), treatment protocol, and (**b**), effects of Intralipid® treatments on the serum ALT activities (Unpublished results)

after the administration of the nanodrug at a dosage of 1 g/kg. On the other days, the rats were administered 2 g/kg of Intralipid® each day (Fig. 5.12a). PBS was administered to control animals. In this experiment, the dosage of Intralipid® is below the recommended maximum dosage which is 3 g/kg/day.

Figure 5.12b shows the changes of serum ALT activity upon Intralipid® treatment for this pilot study. The serum ALT activity is 58.9 ± 8.1 IU/L for naïve rats. Intralipid® treatment does not alter ALT activity (53.2 ± 10.5 IU/L) (Fig. 5.12b). With no Intralipid® protection, the serum ALT activities elevate to 271.2 ± 35.2 IU/L at 1-day post the first injection of the Pt-nanodrug. After 2 days of healing and recovery, the serum ALT activities decrease to 189.9 ± 32.1 IU/L. At 1-day and 4-day post the second injection of the Pt-nanodrug, the serum ALT activities are

179.3 ± 22.5 IU/L and 168.3 ± 32.2 IU/L, respectively. Intralipid® treatment significantly reduces the serum ALT activities (Fig. 5.12b). At the endpoint of the experiment (Day 8), the serum ALT activities are 168.3 ± 32.2 IU/L vs 76.2 ± 8.5 IU/L (no-Intralipid® vs Intralipid® treatment). These results indicate that Intralipid® can reduce the hepatocellular damage from the Pt-nanodrug in this multi- and over-dose experiment.

5.4.5 Using Intralipid® to Deliver Liposome-Based Nanodrugs

The liposome was the earliest nanoparticle delivery system in nanomedicine. Currently, at least 13 lipid nanoparticle-based therapeutics, including Doxil®, Marqibo®, and Onivyde®, have been approved for clinical use (Min et al. 2015). The "lipid"-coating exhibit several advantages as drug delivery vehicles. First, they can be used to deliver both hydrophobic and hydrophilic molecules, as the core of liposomes is hydrophilic, whereas there is a hydrophobic domain between the liposome's lipid layers. Second, the liposomal delivery can change the pharmacokinetics and biodistribution of the delivered therapeutic agent, thus improving therapeutic efficacy as well as reducing toxicity. For example, liposome-based anti-cancer nanodrug delivery system tends to specifically accumulate at the tumor site due to the EPR effect of macromolecules. Third, liposomes can protect the therapeutic agents from the *in-vivo* environment, thus improving the stability of the therapeutic. Fourth, liposomes can be engineered to escape lysosomes and deliver the therapeutic into the cytosol, e.g., gene therapy agent.

Intralipid® in our methodology shows a new function, which is to temporarily reduce the phagocytic activity of Kupffer cells, thus reducing the RES accumulation and the associated toxic side effects. We believe that our Intralipid® methodology can be applied for these liposome nanodrugs. However, we need to be aware of two problems. First, "lipid overload" should be avoided. Total amount of triglyceride (from both Intralipid® and liposome-based nanodrug) should be considered. Plasma triglyceride level needs to be monitored. Second, the drug-drug interaction is a critical issue. When there is a high concentration of more than one drug in the blood stream, some unexpected events could take place. For a specific nanodrug, Intralipid®-nanodrug interaction test should be conducted before setting up the chemotherapy regimen for clinical studies.

5.5 Conclusion

RES uptake is a big challenge for developing and translating nanodrug delivery systems, thus determining the PK/PD profile, efficacy, and toxicity of the nanodrug. Currently, "stealth" nanoparticles strategy is a major accomplishment to minimize the RES uptake. Our studies show that Intralipid® can be used to reduce the RES uptake of nanoparticles for imaging and for anti-cancer drug delivery purposes, so that: (i) *in-vivo* cell-labeling efficiency can be increased and imaging sensitivity can be improved; and (ii) off-target accumulation and resulting toxicities of anti-cancer nanodrugs can be reduced. Our Intralipid® methodology could be a valuable addition to the "stealth" strategies, i.e., generally applicable to any approved or in-developing nanodrugs to reduce the RES uptake.

Acknowledgement We thank Dr. Qing Ye, Dr. Yijen Wu, Mr. Yehuda Creeger, and Dr. Daniel J. Bain for advice and assistance in the surgery, pathology, MRI, flow cytometry, and ICP-MS experiments. We thank Drs. Maggie Lu, Ya-Chin Lo, Yuan-Hung Hsu, Ming-Cheng Wei, Yu-Hsiang Chen, Shen-Chuan Lo, and Shian-Jy Wang for providing us the DACHPt/HANP nanodrugs and valuable discussions. We also thank Dr. E. Ann Pratt for helpful discussions to improve our manuscript. This work was supported by grants from the National Institutes of Health (P41EB-001977) and from the Disruptive Health Technology Institute, Carnegie Mellon University (AO16076-HIGHMARK-HO).

References

Abu Lila AS, Kiwada H, Ishida T (2013) The accelerated blood clearance (ABC) phenomenon: clinical challenge and approaches to manage. J Control Release 172(1):38–47

Agosta E, Lazzeri S, Orlandi P et al (2012) Pharmacogenetics of antiangiogenic and antineovascular therapies of age-related macular degeneration. Pharmacogenomics 13(9):1037–1053

Albanese A, Tang PS, Chan WC (2012) The effect of nanoparticle size, shape, and surface chemistry on biological systems. Annu Rev Biomed Eng 14:1–16

Arvizo RR, Miranda OR, Moyano DF et al (2011) Modulating pharmacokinetics, tumor uptake and biodistribution by engineered nanoparticles. PLoS One 6(9), e24374

Bashir MR, Bhatti L, Marin D, Nelson RC (2015) Emerging applications for ferumoxytol as a contrast agent in MRI. J Magn Reson Imaging 41(4):884–898

Beckmann N, Gerard C, Abramowski D et al (2011) Noninvasive magnetic resonance imaging detection of cerebral amyloid angiopathy-related microvascular alterations using superparamagnetic iron oxide particles in APP transgenic mouse models of Alzheimer's disease: application to passive Abeta immunotherapy. J Neurosci 31(3):1023–1031

Cabral H, Nishiyama N, Okazaki S et al (2005) Preparation and biological properties of dichloro(1,2-diaminocyclohexane)platinum(II) (DACHPt)-loaded polymeric micelles. J Control Release 101(1–3):223–232

Chanan-Khan A, Szebeni J, Savay S et al (2003) Complement activation following first exposure to pegylated liposomal doxorubicin (Doxil): possible role in hypersensitivity reactions. Ann Oncol 14(9):1430–1437

Chouly C, Pouliquen D, Lucet I et al (1996) Development of superparamagnetic nanoparticles for MRI: effect of particle size, charge and surface nature on biodistribution. J Microencapsul 13(3):245–255

Daldrup-Link HE, Golovko D, Ruffell B et al (2011) MRI of tumor-associated macrophages with clinically applicable iron oxide nanoparticles. Clin Cancer Res 17(17):5695–5704

Dams ET, Laverman P, Oyen WJ et al (2000) Accelerated blood clearance and altered biodistribution of repeated injections of sterically stabilized liposomes. J Pharmacol Exp Ther 292(3):1071–1079

Gabizon A, Tzemach D, Mak L et al (2002) Dose dependency of pharmacokinetics and therapeutic efficacy of pegylated liposomal doxorubicin (DOXIL) in murine models. J Drug Target 10(7):539–548

Gambling D, Hughes T, Martin G et al (2005) A comparison of Depodur, a novel, single-dose extended-release epidural morphine, with standard epidural morphine for pain relief after lower abdominal surgery. Anesth Analg 100(4):1065–1074

Gao Y, Lim J, Teoh SH, Xu C (2015) Emerging translational research on magnetic nanoparticles for regenerative medicine. Chem Soc Rev 44(17):6306–6329

Getts DR, Terry RL, Getts MT et al (2014) Therapeutic inflammatory monocyte modulation using immune-modifying microparticles. Sci Transl Med 6(219):219ra217

Gharagozloo M, Majewski S, Foldvari M (2015) Therapeutic applications of nanomedicine in autoimmune diseases: from immunosuppression to tolerance induction. Nanomedicine 11(4):1003–1018

Gregory AE, Titball R, Williamson D (2013) Vaccine delivery using nanoparticles. Front Cell Infect Microbiol 3:13

Guo ST, Huang L (2011) Nanoparticles escaping res and endosome: challenges for siRNA delivery for cancer therapy. J Nanomater. Article ID 742895, 12 pages, doi:10.1155/2011/742895

Hamill RJ (2013) Amphotericin B formulations: a comparative review of efficacy and toxicity. Drugs 73(9):919–934

Jo SK, Hu X, Kobayashi H et al (2003) Detection of inflammation following renal ischemia by magnetic resonance imaging. Kidney Int 64(1):43–51

Kamaly N, Fredman G, Subramanian M et al (2013) Development and in vivo efficacy of targeted polymeric inflammation-resolving nanoparticles. Proc Natl Acad Sci U S A 110(16):6506–6511

Knop K, Hoogenboom R, Fischer D, Schubert US (2010) Poly(ethylene glycol) in drug delivery: pros and cons as well as potential alternatives. Angew Chem Int Edit 49(36):6288–6308

Koide H, Asai T, Hatanaka K et al (2008) Particle size-dependent triggering of accelerated blood clearance phenomenon. Int J Pharm 362(1–2):197–200

Kozlowska D, Foran P, MacMahon P et al (2009) Molecular and magnetic resonance imaging: the value of immunoliposomes. Adv Drug Deliv Rev 61(15):1402–1411

Lasic DD, Needham D (1995) The "Stealth" liposome: a prototypical biomaterial. Chem Rev 95(8):2601–2628

Liu L, Hitchens TK, Ye Q et al (2013) Decreased reticuloendothelial system clearance and increased blood half-life and immune cell labeling for nano- and micron-sized superparamagnetic iron-oxide particles upon pre-treatment with Intralipid. Biochim Biophys Acta 1830(6):3447–3453

Liu X, Li H, Chen Y et al (2014) Mixed-charge nanoparticles for long circulation, low reticuloendothelial system clearance, and high tumor accumulation. Adv Healthcare Mater 3(9):1439–1447

Liu L, Ye Q, Lu M et al (2015) A new approach to reduce toxicities and to improve bioavailabilities of platinum-containing anti-cancer nanodrugs. Sci Rep 5:10881

Ludwig T, Riethmuller C, Gekle M et al (2004) Nephrotoxicity of platinum complexes is related to basolateral organic cation transport. Kidney Int 66(1):196–202

Min Y, Caster JM, Eblan MJ, Wang AZ (2015) Clinical translation of nanomedicine. Chem Rev 115(19):11147–11190

Minchin RF, Martin DJ (2010) Nanoparticles for molecular imaging–an overview. Endocrinology 151(2):474–481

Miyama N, Dua MM, Schultz GM et al (2012) Bioluminescence and magnetic resonance imaging of macrophage homing to experimental abdominal aortic aneurysms. Mol Imaging 11(2):126–134

Murakami M, Cabral H, Matsumoto Y et al (2011) Improving drug potency and efficacy by nanocarrier-mediated subcellular targeting. Sci Transl Med 3(64):64ra62

Negro A, Grasselli C, Galli P (2010) Oxaliplatin-induced proximal renal tubular acidosis. Intern Emerg Med 5(3):267–268

Neuberger T, Schöpf B, Hofmann H et al (2005) Superparamagnetic nanoparticles for biomedical applications: possibilities and limitations of a new drug delivery system. J Magn Magn Mater 293:483–496
Nugent KM (1984) Intralipid effects on reticuloendothelial function. J Leukoc Biol 36(2):123–132
Oberoi HS, Nukolova N, Zhao Y et al (2012) Preparation and in vivo evaluation of Dichloro(1,2-Diaminocyclohexane)platinum(II)-loaded core cross-linked polymer micelles. Chemother Res Pract 2012:905796
Oberoi HS, Nukolova NV, Kabanov AV, Bronich TK (2013) Nanocarriers for delivery of platinum anticancer drugs. Adv Drug Deliv Rev 65(13–14):1667–1685
Okon E, Pouliquen D, Okon P et al (1994) Biodegradation of magnetite dextran nanoparticles in the rat. A histologic and biophysical study. Lab Invest 71(6):895–903
Onoue S, Yamada S, Chan HK (2014) Nanodrugs: pharmacokinetics and safety. Int J Nanomedicine 9:1025–1037
Pabla N, Dong Z (2008) Cisplatin nephrotoxicity: mechanisms and renoprotective strategies. Kidney Int 73(9):994–1007
Salmaso S, Caliceti P (2013) Stealth properties to improve therapeutic efficacy of drug nanocarriers. J Drug Deliv 2013:374252
Schughart K, Bischoff R, Hadji DA et al (1999) Effect of liposome-encapsulated clodronate pretreatment on synthetic vector-mediated gene expression in mice. Gene Ther 6(3):448–453
Shapiro EM (2015) Biodegradable, polymer encapsulated, metal oxide particles for MRI-based cell tracking. Magn Reson Med 73(1):376–389
Shih YYI, Hsu YH, Duong TQ et al (2011) Longitudinal study of tumor-associated macrophages during tumor expansion using MRI. NMR Biomed 24(10):1353–1360
Siegel RL, Miller KD, Jemal A (2016) Cancer statistics, 2016. CA Cancer J Clin 66(1):7–30
Stefanowicz J, Owczuk R, Izycka-Swieszewska E et al (2011) Nephrotoxicity of platinum derivatives in children – a review of the literature. Wspolczesna Onkol 15(2) 74–79
Stegmann T, Morselt HW, Booy FP et al (1987) Functional reconstitution of influenza virus envelopes. EMBO J 6(9):2651–2659
Szebeni J (2005) Complement activation-related pseudoallergy: a new class of drug-induced acute immune toxicity. Toxicology 216(2–3):106–121
Tang TY, Muller KH, Graves MJ et al (2009) Iron oxide particles for atheroma imaging. Arterioscler Thromb Vasc Biol 29(7):1001–1008
Tu C, Ng TSC, Sohi HK et al (2011) Receptor-targeted iron oxide nanoparticles for molecular MR imaging of inflamed atherosclerotic plaques. Biomaterials 32(29):7209–7216
Ulusakarya A, Misra S, Haydar M et al (2010) Acute renal failure related to oxaliplatin-induced intravascular hemolysis. Med Oncol 27(4):1425–1426
van Rooijen N, van Kesteren-Hendrikx E (2002) Clodronate liposomes: perspectives in research and therapeutics. J Liposome Res 12(1–2):81–94
Vilaro S, Llobera M (1988) Uptake and metabolism of Intralipid by rat liver: an electron-microscopic study. J Nutr 118(8):932–940
Viscusi ER, Martin G, Hartrick CT et al (2005) Forty-eight hours of postoperative pain relief after total hip arthroplasty with a novel, extended-release epidural morphine formulation. Anesthesiology 102(5):1014–1022
Wheate NJ, Walker S, Craig GE et al (2010) The status of platinum anticancer drugs in the clinic and in clinical trials. Dalton Trans 39(35):8113–8127
Wicki A, Witzigmann D, Balasubramanian V et al (2015) Nanomedicine in cancer therapy: challenges, opportunities, and clinical applications. J Control Release 200:138–157
Williams HD, Trevaskis NL, Charman SA et al (2013) Strategies to address low drug solubility in discovery and development. Pharmacol Rev 65(1):315–499
Woodle MC (1995) Sterically stabilized liposome therapeutics. Adv Drug Deliv Rev 16(2–3):249–265

Wu YL, Ye Q, Foley LM et al (2006) In situ labeling of immune cells with iron oxide particles: an approach to detect organ rejection by cellular MRI. Proc Natl Acad Sci U S A 103(6):1852–1857

Wu YL, Ye Q, Sato K et al (2009) Noninvasive evaluation of cardiac allograft rejection by cellular and functional cardiac magnetic resonance. JACC Cardiovasc Imaging 2(6):731–741

Wu YL, Ye Q, Ho C (2011) Cellular and functional imaging of cardiac transplant rejection. Curr Cardiovasc Imaging Rep 4(1):50–62

Wu YL, Ye Q, Eytan DF et al (2013) Magnetic resonance imaging investigation of macrophages in acute cardiac allograft rejection after heart transplantation. Circ Cardiovasc Imaging 6(6):965–973

Wu H, Cabral H, Toh K et al (2014) Polymeric micelles loaded with platinum anticancer drugs target preangiogenic micrometastatic niches associated with inflammation. J Control Release 189:1–10

Xiang D, Shigdar S, Qiao G et al (2015) Nucleic acid aptamer-guided cancer therapeutics and diagnostics: the next generation of cancer medicine. Theranostics 5(1):23–42

Xue HY, Liu S, Wong HL (2014) Nanotoxicity: a key obstacle to clinical translation of siRNA-based nanomedicine. Nanomedicine 9(2):295–312

Yang D, Ye Q, Williams M et al (2001) USPIO-enhanced dynamic MRI: evaluation of normal and transplanted rat kidneys. Magn Reson Med 46(6):1152–1163

Yao X, Panichpisal K, Kurtzman N et al (2007) Cisplatin nephrotoxicity: a review. AmJ Med Sci 334(2):115–124

Ye Q, Yang D, Williams M et al (2002) In vivo detection of acute rat renal allograft rejection by MRI with USPIO particles. Kidney Int 61(3):1124–1135

Ye Q, Wu YL, Foley LM et al (2008) Longitudinal tracking of recipient macrophages in a rat chronic cardiac allograft rejection model with noninvasive magnetic resonance imaging using micrometer-sized paramagnetic iron oxide particles. Circulation 118(2):149–156

Zhang Q, Li F (2013) Combating P-glycoprotein-mediated multidrug resistance using therapeutic nanoparticles. Curr Pharm Des 19(37):6655–6666

Part IV
Differences Between In Vivo Status in Men and Mice

Chapter 6
Authentic Vascular and Stromal Structure in Animal Disease Model for Nanomedicine

Hiroshi Nishihara and Mitsunobu R. Kano

Abstract The elucidation of pharmacokinetics of nanodrugs is indispensable for the further development of nanomedicine and the importance of vasculature and stroma cells for the delivery of drugs via nanocarrier systems is currently attracting a great deal of attention. In particular, the major discrepancy between the human and the experimental model in terms of vasculature and stromal environment must be well characterized when moving to the upper scale at translational medicine. This chapter discusses the histological discrimination of stromal elements in human pathological specimens and animal models in view of nanodrug-associated pathophysiology, which is necessary to understand the disease-specific and/or patient-specific pharmacokinetics of nanodrugs, and will lead to the promising efficacy of personalized nanomedicine.

Keywords Vasculature • Stromal cell • Animal model • Pathophysiology

Abbreviations

αSMA	alpha smooth muscle actin
CAFs	cancer associated fibroblasts
EPR	enhanced permeability and retention
GBM	glioblastoma multiforme
PDAC	pancreatic ductal adenocarcinoma
PDGFRβ	platelet-derived growth factor receptor beta
TAMs	tumor-associated macrophages

H. Nishihara (✉)
Department of Translational Pathology, Hokkaido University Graduate School of Medicine, Sapporo 060-8638, Japan

Translational Research Laboratory, Hokkaido University Hospital, Sapporo, Japan
e-mail: hnishihara@med.hokudai.ac.jp; hnishihara@s5.dion.ne.jp

M.R. Kano
Department of Pharmaceutical Biomedicine, Graduate School of Medicine, Dentistry, and Pharmaceutical Sciences, Okayama University, Okayama 700-8530, Japan

A. Prokop, V. Weissig (eds.), *Intracellular Delivery III*, Fundamental Biomedical Technologies 8, DOI 10.1007/978-3-319-43525-1_6

6.1 Introduction

Most of the nanodrugs are delivered through blood vessels; therefore, the role of vasculatures has been well-characterized, including morphological abnormalities such as fenestration or defects of endothelial cells (Daruwalla et al. 2009; Maeda et al. 2003; Skinner et al. 1990) (ADDR). In addition, the role of vascular pericytes in the permeation of nanoparticles was recently focused upon (Kano et al. 2009). Besides the vasculature, the stromal components such as fibroblasts and macrophages could be possible players to modulate the pharmacokinetics, not only for small molecules but also for nanodrugs, because the stromal cells are usually located between the blood vessels and the pharmacological targets such as cancer cells.

The development of clinically applicable "nanodrugs" is largely dependent on the local pharmacokinetics of nanodrugs, which are mostly validated through the animal disease models. The enhanced permeability and retention (EPR) effect interpret the efficacy of nanodrugs, especially in the solid tumor model (Fang et al. 2011; Torchilin 2011). However, the detailed pathophysiology associated with nano-sized materials, especially in pathologic lesions, has not been well-described. Moreover, the histological discrepancy of the microenvironment, such as vasculature and stromal cells, in between human pathological specimens and animal models has yet to be investigated.

Here we discusses the histological discrimination of stromal elements in human pathological specimens and animal models in view of nanodrug-associated pathophysiology, which is necessary to understand the disease-specific and/or patient-specific pharmacokinetics of nanodrugs, and will lead to the promising efficacy of personalized nanomedicine.

6.2 Vasculature in Human Pathology

6.2.1 *Endothelial Cells for Nanomedicine*

For the successful delivery of nanodrugs to the disease lesion, the entire vascular system is required to avoid leakage through the vascular wall. The internal layer is composed of a single layer of spindle-shaped endothelial cells in the normal vascular system, which are visualized by immunohistochemistry against CD31, CD34 and factor VIII-related antigen in human tissue (Fig. 6.1a, and data not shown) (Cotran 1999). Junctional complexes by tight, adherens or gap junctions are formed between endothelial cells; these are necessary to maintain vessel wall homeostasis and permeability (Bazzoni and Dejana 2004). Claudin family proteins including Claudin-5 within tight junctions play a pivotal role to create the barrier which regulates electrical resistance between cells (Fig. 6.1b) (Van Itallie and Anderson 2004; Sawada et al. 2003). Adherens junctions have a key role in contact inhibition and also regulate permeability to soluble molecules. Connexin family proteins form gap junctions which create channels between adjacent cells (Sohl and Willecke 2004).

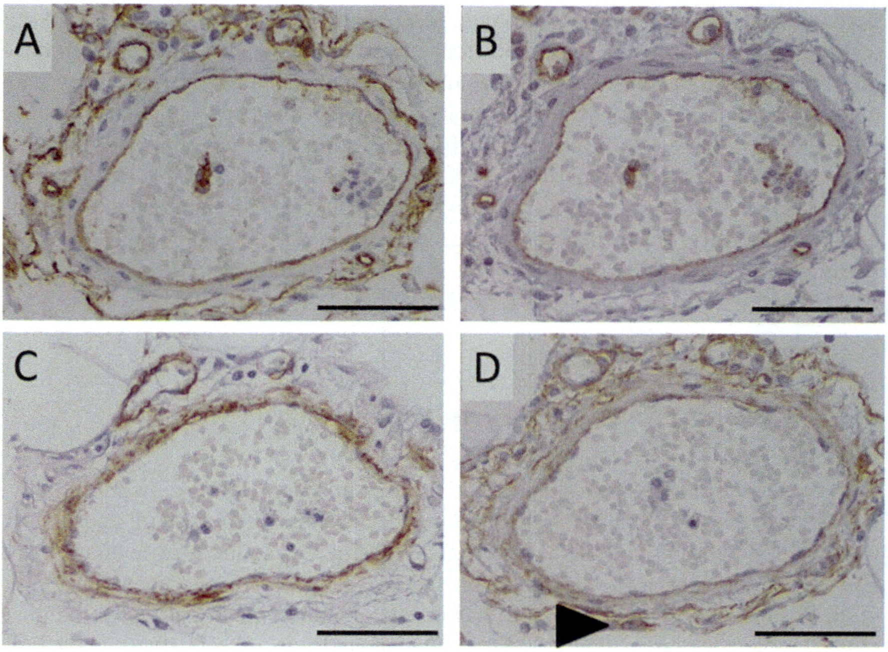

Fig. 6.1 Immunohistochemical appearance of vascular mural cells. The entire vascular system is lined internally by a single layer of spindle-shaped endothelial cells, which are identified immunohistochemically with antibodies against CD34 (**a**, duodenum). Claudin-5 is immunopositive in endothelial cells (**b**, duodenum), meaning the formation of tight junction barriers. Immunohistochemistry for αSMA visualizes vascular smooth muscle cells (**c**, duodenum). The pericytes are recognized by anti-PDGFRβ antibody (**d**, duodenum) which also reacts upon myofibroblasts adjacent to the vasculature (**d**, arrowhead). Antibodies employed here are as follows: anti-CD34 (Mouse Monoclonal clone: QBEnd10, DAKO), Claudin-5 (Rabbit Polyclonal, Abcam), anti-αSMA (Mouse Monoclonal clone:1A4, Thermo Scientific), anti-PDGFRβ (Rabbit Monoclonal clone:C82A3, CST)

Based on these molecular mechanisms, endothelial cells serve as a semipermeable membrane which controls the transfer of small and large molecules including nanoparticles through the walls of arterioles and capillaries and modulates vascular tone and blood flow which directly affect the accumulation of nanodrugs (Cotran 1999).

6.2.2 *Pericytes for Nanomedicine*

Pericytes are microvascular periendothelial mesenchymal cells including vascular smooth muscle cells (Van Itallie and Anderson 2004), and are embedded within the vascular basement membrane and form the mural component of the vascular wall (Bazzoni and Dejana 2004). Several molecules expressed in murine pericytes, such as NG2, PDGFRβ and RGS-5, have been reported, although these are not pericyte-specific (Van Itallie and Anderson 2004). In general human pathology, αSMA is the most reliable immunohistochemical marker for vascular smooth muscle cells (Fig. 6.1c), while PDGFRβ is another attractive candidate to identify pericytes in capillaries, although myofibroblasts adjacent to pericytes are also positive for PDGFRβ (Fig. 6.1d). In human specimens, desmin is negative in vascular mural cells except in large muscular arteries (Nishihara 2014). The principle roles of pericytes are (1) the fundamental for blood vessel formation and maintenance, (2) vascular contractility, (3) formation of the blood–brain barrier and (4) the supply of mesenchymal stem cells for tissue regeneration (Armulik et al. 2011), however, the role for nanomedicine is yet to be clarified.

6.2.3 Vascular Structure in Human Diseases

Cancer is one of the most popular target diseases for nanomedicine; in addition, we can obtain variable information from Cancer because the phenotype of signal transduction, cellular alternation and histological divergence is mostly accentuated. The EPR effect can explain the mechanism for drug delivery of nanomedicine (Fang et al. 2011). The tumor vasculature has the unique anatomical-pathophysiological nature; molecules larger than 40 kDa such as nanodrugs are selectively leaked out from tumor vessels, but not from normal vasculatures (Matsumura and Maeda 1986; Maeda 2001a, b; Fang et al. 2003; Maeda et al. 2009). In human pathology, however, the morphological appearance and the number of vasculatures vary according to the organ or histological subtypes of cancer or even the patient. More importantly, the structure of tumor vasculature appears to be different depending on each tissue. In glioblastoma of the brain, characteristic glomeruloid vessels consist of thick αSMA-positive pericytes, while the lining of endothelial cells seems to be discontinuous (Fig. 6.2e, f). On the contrary, the αSMA-positive pericytes were discontinuously observed in tumor vessels in ovarian adenocarcinoma (Fig. 6.2b, c) (Zhang et al. 2012). Recently, we proposed three types of structural variety of tumor bloods; type (1), thin, defective αSMA-positive pericyte coverage, type (2), thin, αSMA-positive pericyte coverage, and type (3), thick or glomeruloid, αSMA-positive pericyte coverage (Fig. 6.2g).

Inflammatory disease is the other luminous model to understand drug delivery system. In the inflammatory condition, endothelial dysfunction rapidly occurrs within minutes by variable chemical mediators such as histamine and serotonin that induce excessive vascular permeability (Cotran 1999). Endothelial activation will occur within several hours or even days by cytokines and bacterial products. During such reversible changes in the functional state of endothelial cells, the neovascularization of capillaries discontinuously covered with αSMA-positive pericytes is microscopically observed in inflammatory granulation tissue (Nishihara 2014). In addition, vascular remodeling is often observed in the later phase of specific inflammatory condition such as during the wound healing process or prolonged latent infection (Dulmovits and Herman 2012). However, a histopathological approach for

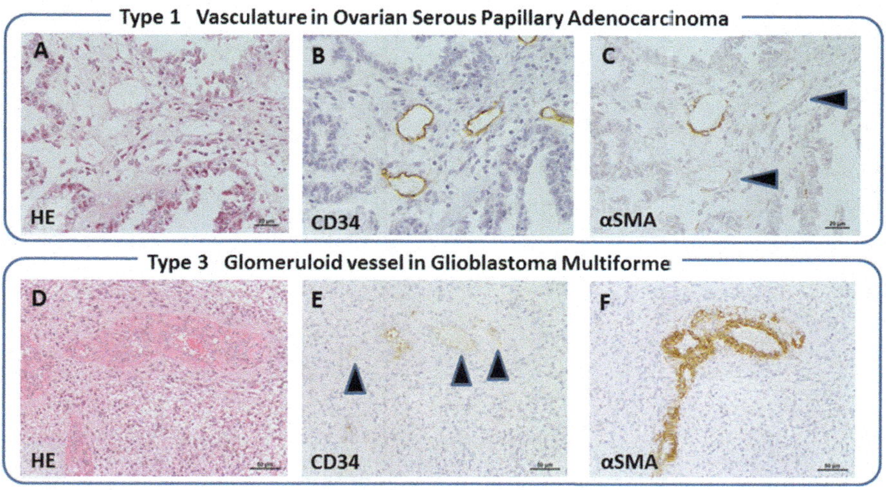

G Structural varieties of blood vessel (pericyte coverage)

Type	Thickness	Coverage	Representative cancer	Response for chemotherapy
1	thin	defective	ovarian, colon Ca	good
2	thin	continuous	almost all types of Ca	intermediate
3	thin	glomeruloid	PDAC, GBM	poor

Fig. 6.2 Structural variety of pericyte coverage. In ovarian serous papillary adenocarcinoma, tumor vasculatures consist of an entire single layer of endothelial cells (**b**, CD34) and discontinuous coverage of pericytes (**c**, αSMA), while the majority of mural cells in glomeruloid vessels in glioblastoma are αSMA-positive pericytes (**f**, αSMA) with scattered endothelial cells (**e**, CD34, arrowhead). Based on the coverage pattern of pericytes, three types of structural variety of tumor blood vessels are proposed (**g**). In the general oncological aspect, the cancers with lower pericyte coverage represent relatively higher chemosensitivity. **a**, **d**, Hematoxylin and Eosin stain. Scale Bars: A. B, C; 20 μm, D, E, F; 50 μm. *PDAC* pancreatic ductal adenocarcinoma, *GBM* glioblastoma multiforme, *Ca* carcinoma

vascular structure in inflammatory and infectious conditions is still being investigated to establish a new therapeutic strategy using nanodrugs against such diseases.

6.3 Stromal Structure in Human Pathological Conditions

6.3.1 Stromal Cells for Nanomedicine

Stromal tissue with the exception of vasculature consists of connective collagenous matrix, fibrotic cells including myofibroblasts and inflammatory cells. The second barrier to successful nanodrug delivery after leakage from vasculature would be such stromal tissue. The myofibroblasts were originally identified within experimental granulation tissue by electron micrographs (Gabbiani et al. 1971), and share spindle morphology common with those of fibroblasts and smooth muscle cells.

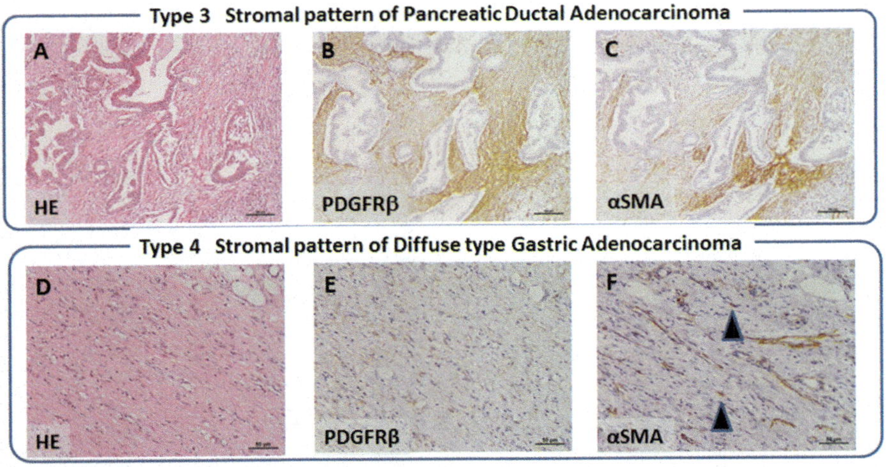

G Histological subtype of cancer stroma

Type	Desmoplasia	PDGFRβ	αSMA	Representative cancer	Response for chemotherapy
1	absent	-	-	early stage of colon, gastric Ca, ovarian Ca	good
2	dense fibrosis	+	-	majority of cancer	relatively poor
3	dense fibrosis	+	+	majority of cancer, PDAC	relatively poor
4	scar like	-	+/-	Diffuse type of gastric adenocarcinoma	poor

Fig. 6.3 Structural variety of cancer stroma. Desmoplastic cancer stroma including CAFs consists of spindle cells which are positive for αSMA (**c**, **f**) and/or PDGFRβ (**b**, **d**). Based on the expression pattern of the molecular maker in stromal cells, three types of structural variety of stromal cells are proposed (**g**). While the clinical manifestation for pattern of positivity of αSMA and/or PDGFRβ is still to be investigated, desmoplasia is one of the clinically relevant, unfavorable prognostic factors. The scar-like stroma with dense collagen bundles observed in diffuse type gastric adenocarcinoma is usually negative for both αSMA and PDGFRβ, while scattered αSMA-positive spindle cells, possible CAFs, are frequently identified (**f**, arrowhead). **a**, **d**, Hematoxylin and Eosin stain. Scale Bars: A. B, C; 100 μm, D, E, F; 50 μm. *PDAC* pancreatic ductal adenocarcinoma, *Ca* carcinoma

Myofibroblasts contain numerous bundles of cytoplasmic microfilaments (stress fibers) (Mills 2007), and the cells are connected by intermediate or adherens junctions (Hinz and Gabbiani 2003).

The stress fibers in myofibroblasts are immunohistochemically positive for αSMA (Fig. 6.3). Myofibroblasts can be observed in multiple normal human tissues (Mills 2007) such as mucosal epithelium (Boya et al. 1988; Guldner et al. 1972; Kaye et al. 1968; Sappino et al. 1989), lung (Kapanci et al. 1974), and bone marrow (Charbord et al. 1990). One of the biochemical functions of myofibroblasts is synthesis of several extracellular matrix components: collagen type I, III, IV and V (Zhang et al. 1994; Gabbiani et al. 1976; Rudolph et al. 1977; Barsky et al. 1982), glycoproteins (Gressner and Bachem 1990) and proteoglycans such as fibronectin (Berndt et al. 1994), which will affect the tissue permeability of nanoparticles (Nishihara 2014).

The inflammatory response occurs in stromal tissue by immunologic cells including neutrophils, macrophages, eosinophils, lymphocytes, and basophils

(Cotran 1999). Phagocytosis by macrophages is one of the fundamental protective responses against infection and foreign bodies including certain therapeutic agents; therefore the macrophages in atherosclerosis are now a novel target for nanomedicine (Psarros et al. 2012). In human pathological specimens, macrophages can be immunohistochemically identified with anti-CD68 antibody.

6.3.2 Stromal Structure in Human Diseases

Cancer-associated fibroblasts (CAFs), which are identified by the expression of actin and appear to originate from resident fibroblasts (Erez et al. 2010; Hinz et al. 2007), promote the proliferation and progression of cancer through the production of growth factors, angiogenetic factors and metalloproteinases, allowing invasion and metastasis of cancer cells (Li et al. 2007). In human pancreas adenocarcinoma which usually represents strong desmoplastic reaction of dense fibrotic tissue (Kalluri and Zeisberg 2006), stromal cells including CAFs were positive for αSMA and/or PDGFRβ. Interestingly, the amount of PDGFRβ-positive-, not αSMA-positive-, stromal cells was statistically correlated with poor prognosis of the patients with pancreatic ductal adenocarcinoma (Yuzawa et al. 2012), suggesting that molecular expression in stromal cells might be associated with the tumor progression and/or drug sensitivity in pancreatic cancer. Based on the clinicopathological evaluation, we recently proposed the following histological subtypes of cancer stroma (Fig. 6.3); (1) sparse desmoplasia, (2) dense desmoplasia with PDGFRβ expression, (3) dense desmoplasia with αSMA expression, and (4) scar-like desmoplasia composed of scattered αSMA-positive cells and collagen bundles (Nishihara 2014). Type (1) is sometimes observed in ovarian adenocarcinoma and early stages of gastric and colorectal adenocarcinoma, thus the association with favorable patients' prognosis, while (2) to (4) tend to be found in pancreatic ductal adenocarcinoma or the scirrhous type of gastric adenocarcinoma, meaning the poorer patients' outcome.

Tumor-associated macrophages (TAMs) are considered as stimulators of tumor progression and angiogenesis via production of multiple cytokines such as VEGF, HGF and IL-8 (Li et al. 2007). Two subtypes for activated macrophages have been advocated: the classically activated (M1) macrophages and the alternatively activated (M2) macrophages (Murray and Wynn 2011; Stein et al. 1992; Goerdt and Orfanos 1999). M1 macrophages express proinflammatory molecules such as IL-12, TNF-α, iNOS and MHCI/II, while M2 macrophages express anti-inflammatory molecules such as IL-10 and arginase (Mantovani et al. 2002). In human specimens, M1 and M2 are positive for anti-CD68 antibody, while M2 macrophages can be identified with anti-CD163 antibody. TAMs orchestrate various aspects of cancer promotion by affecting angiogenesis, immunomodulation, matrix deposition and remodeling (Hao et al. 2012), although the roles of M1 and M2 in tumors are still controversial (Ong et al. 2012; Dumont et al. 2008). In addition, phagocytosis must be considered for nanodrug efficacy, because the nanocarriers must be excluded from the reticulo-endothelial system, which destroys any foreign material though opsonization followed by phagocytosis (Danhier et al. 2010; Malam et al. 2009; Gullotti and Yeo 2009).

Considering the strategies for nanodrug-mediated treatment according to the type of stromal cells, dense and scar-like desmoplasia in cancer, which are also observed in inflammatory diseases, can induce stromal stiffness resulting in abrogation of nanodrug delivery, even though nanodrugs are selectively leaked out from vasculature by the EPR effect. In addition, further pharmacokinetic and clinicopathological aspects for nanodrugs in association with type and amount of TAMs in human pathological conditions including cancer and inflammatory disease are also required for the successful nanomedicine.

6.4 Structural Discrepancy of Vasculature and Stromal Structure Between Human and Experimental Models

The aim of using models in science is to extract an aspect of interest from the whole phenomenon, so that scientists can analyze or simulate the aspect of focus in a reproducible manner (Frigg 2012). Therefore, an appropriate model to achieve a certain goal would vary depending on the goal, or the viewpoint of a scientist. Note that a model only mimic a partial but most appropriate aspect of a phenomenon, and does not present the whole picture of the phenomenon. Just as an example, consider if you are asked to scientifically examine the reason for the cause of an aviation accident, what models you would use. The model to use may vary depending on a cause you may expect. If it were an engine trouble, you may need a model for the engine. If you may suspect a human error, what you may need might not even be rated to aircrafts, but can be neurological or psychological models.

What is the appropriate model, then, in terms of investigate the behavior of nanodrugs in the disease foci? To answer this question, let us first revisit the characteristics of nanomedicine. Nanomedicine is at least named after a differentiation by its size – "nano" – from other ordinary medicine. What is different is, however, not only the size, but its function to accumulate more in disease foci and less in other parts of the body. Nanomedicine has been reported as depositing mostly at cancer when administered systemically, by means of characteristic leakiness of tumor vessels: This is a theory named the EPR effect (Matsumura and Maeda 1986). If this is the case, what we should extract as a scientific model for investigation of the efficacy of nanomedicine is not only cancer cells, but also routes to reach the cells. The EPR effect consist of leakiness of tumor vasculature and immaturity of lymphatic vessels; not mentioning cancer cells. In other words, if one would like to investigate why a given nanomedicine is not efficacious on some cancer, the reasons could not only be attributed to cancer cells, as usually expected, but also the structure of the surrounding tissue including vessels, especially because it is nanomedicine (Kano 2014).

Most of the animal cancer models that are in use consist of fast growing cancer cells and leaky vessels, and omit other aspects of stromal structures described earlier in this section. Even if the cancer cells that are xenografted are derived from humans, or are intractable cancers such as pancreatic cancer, most of the models show similar characteristic (Weinberg 2013). This is understandable if we consider the original goal in using such animal models. In the development of an ordinary medicine, we have had an expectation that the medicine should reach the cancer cells without any trouble *en route*, and the efficacy of the medicine should be determined only by the sensitivity of the cells to the medicine. This might be also a basis to use a flat monolayer culture of cancer cells in developing an anti-tumor medicine. If we see the whole phenomenon from this viewpoint, what we need in the experimental model is the cancer cells and some kind of vessels.

Nanomedicine, however, is differentiated from the ordinary medicine by its size and pharmacokinetics as we discussed earlier. In other words, we focus on the challenges *en route*, not only the cancer cells. It is therefore natural that we need a different model to develop a nanomedicine, because the viewpoint is different. The challenge is then to realize such an appropriate model, by including components which should be in the experimental system as the model. We have been attempting for the inclusion of the structure of vasculature (Kano 2014; Kano et al. 2007) or the amount of fibrotic tissue (Hosoya et al. 2012; Sakai et al. 2016), and realize how it is difficult to establish new experimental systems which are easy to handle, reproducible, and credible. As an example, the pancreatic cancer model we have been using, BxPC3, needs almost 3 weeks after inoculation to grow before use, and the histological pattern varies considerably because most of its stromal components are derived from the animal (Fig. 6.4).

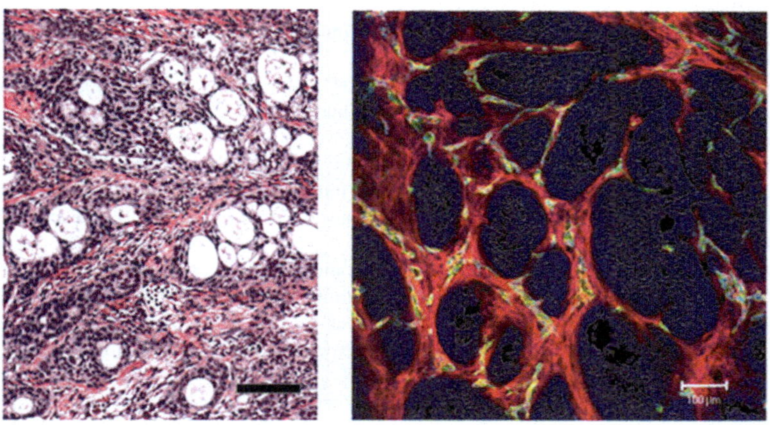

Fig. 6.4 Histology of PDAC xenograft in nude mouse. The histology of BxPC3 xenograft, used as a model of poorly differentiated pancreatic adenocarcinoma, is shown in H&E staining and immunohistochemistry. Examination revealed nests of tumor cells in gland-like structures, with areas rich in fibrotic components (filled by α-smooth muscle actin (SMA)-positive myofibroblasts, shown in *red*) between them. The tumor tissue also includes some PECAM-1-positive vessels (shown in *green*) in the interstitium, although almost no vasculature was observed inside the nests of tumor cells. The antibodies to PECAM-1 and VE-cadherin were from BD PharMingen (San Diego, CA), those to neuroglycan 2 and collagen IV were from Chemicon (Temecula, CA), and that to SMA was from Sigma–Aldrich (St. Louis, MO). (Scale bars, 100 μm.) (Kano et al. 2007)

6.5 Conclusions

Success in drug delivery of nanomedicine is highly prescribed by the permeability of vasculature and stromal barrier within the target lesion; therefore, it is extremely important to comprehend the vascular and stromal structure in the target lesion. However, the discrepancy between the human and the experimental model in terms of vasculature and stromal environment is an impediment for adequate preclinical evaluation of nanomedicine. The proper pathophysiological estimation of target disease and selection for an appropriate animal model will lead to the promising efficacy of personalized nanomedicine.

References

Armulik A, Genove G, Betsholtz C (2011) Pericytes: developmental, physiological, and pathological perspectives, problems, and promises. Dev Cell 21(2):193–215

Barsky SH, Rao CN, Grotendorst GR, Liotta LA (1982) Increased content of Type V Collagen in desmoplasia of human breast carcinoma. Am J Pathol 108(3):276–283

Bazzoni G, Dejana E (2004) Endothelial cell-to-cell junctions: molecular organization and role in vascular homeostasis. Physiol Rev 84(3):869–901

Berndt A, Kosmehl H, Katenkamp D, Tauchmann V (1994) Appearance of the myofibroblastic phenotype in Dupuytren's disease is associated with a fibronectin, laminin, collagen type IV and tenascin extracellular matrix. Pathobiology 62(2):55–58

Boya J, Carbonell AL, Martinez A (1988) Myofibroblasts in human palatal mucosa. Acta Anat (Basel) 131(2):161–165

Charbord P, Lerat H, Newton I, Tamayo E, Gown AM, Singer JW, Herve P (1990) The cytoskeleton of stromal cells from human bone marrow cultures resembles that of cultured smooth muscle cells. Exp Hematol 18(4):276–282

Cotran RS (1999) Pathologic basis of disease, 6th edn. Saunders, Philadelphia

Danhier F, Feron O, Preat V (2010) To exploit the tumor microenvironment: passive and active tumor targeting of nanocarriers for anti-cancer drug delivery. J Control Release 148(2):135–146

Daruwalla J, Greish K, Malcontenti-Wilson C, Muralidharan V, Iyer A, Maeda H, Christophi C (2009) Styrene maleic acid-pirarubicin disrupts tumor microcirculation and enhances the permeability of colorectal liver metastases. J Vasc Res 46(3):218–228

Dulmovits BM, Herman IM (2012) Microvascular remodeling and wound healing: a role for pericytes. Int J Biochem Cell Biol 44(11):1800–1812

Dumont P, Berton A, Nagy N, Sandras F, Tinton S, Demetter P, Mascart F, Allaoui A, Decaestecker C, Salmon I (2008) Expression of galectin-3 in the tumor immune response in colon cancer. Lab Invest 88(8):896–906

Erez N, Truitt M, Olson P, Arron ST, Hanahan D (2010) Cancer-associated fibroblasts are activated in incipient neoplasia to orchestrate tumor-promoting inflammation in an NF-kappaB-dependent manner. Cancer Cell 17(2):135–147

Fang J, Sawa T, Maeda H (2003) Factors and mechanism of "EPR" effect and the enhanced antitumor effects of macromolecular drugs including SMANCS. Adv Exp Med Biol 519:29–49

Fang J, Nakamura H, Maeda H (2011) The EPR effect: unique features of tumor blood vessels for drug delivery, factors involved, and limitations and augmentation of the effect. Adv Drug Deliv Rev 63(3):136–151

Frigg R, Hartmann S (2012) Models in science. In: Edward NZ (ed) The Stanford encyclopedia of philosophy, (Fall 2012 Edition). URL http://plato.stanford.edu/archives/fall2012/entries/models-science/

Gabbiani G, Ryan GB, Majne G (1971) Presence of modified fibroblasts in granulation tissue and their possible role in wound contraction. Experientia 27(5):549–550

Gabbiani G, Le Lous M, Bailey AJ, Bazin S, Delaunay A (1976) Collagen and myofibroblasts of granulation tissue. A chemical, ultrastructural and immunologic study. Virchows Arch B Cell Pathol 21(2):133–145

Goerdt S, Orfanos CE (1999) Other functions, other genes: alternative activation of antigen-presenting cells. Immunity 10(2):137–142

Gressner AM, Bachem MG (1990) Cellular sources of noncollagenous matrix proteins: role of fat-storing cells in fibrogenesis. Semin Liver Dis 10(1):30–46

Guldner FH, Wolff JR, Keyserlingk DG (1972) Fibroblasts as a part of the contractile system in duodenal villi of rat. Z Zellforsch Mikrosk Anat 135(3):349–360

Gullotti E, Yeo Y (2009) Extracellularly activated nanocarriers: a new paradigm of tumor targeted drug delivery. Mol Pharm 6(4):1041–1051

Hao NB, Lu MH, Fan YH, Cao YL, Zhang ZR, Yang SM (2012) Macrophages in tumor microenvironments and the progression of tumors. Clin Dev Immunol 2012:948098

Hinz B, Gabbiani G (2003) Cell-matrix and cell-cell contacts of myofibroblasts: role in connective tissue remodeling. Thromb Haemost 90(6):993–1002

Hinz B, Phan SH, Thannickal VJ, Galli A, Bochaton-Piallat ML, Gabbiani G (2007) The myofibroblast: one function, multiple origins. Am J Pathol 170(6):1807–1816

Hosoya H, Kadowaki K, Matsusaki M, Cabral H, Nishihara H, Ijichi H, Koike K, Kataoka K, Miyazono K, Akashi M, Kano MR (2012) Engineering fibrotic tissue in pancreatic cancer: a novel three-dimensional model to investigate nanoparticle delivery. Biochem Biophys Res Commun 419(1):32–37

Kalluri R, Zeisberg M (2006) Fibroblasts in cancer. Nat Rev Cancer 6(5):392–401

Kano MR (2014) Nanotechnology and tumor microcirculation. Adv Drug Deliv Rev 74:2–11

Kano MR, Bae Y, Iwata C, Morishita Y, Yashiro M, Oka M, Fujii T, Komuro A, Kiyono K, Kaminishi M, Hirakawa K, Ouchi Y, Nishiyama N, Kataoka K, Miyazono K (2007) Improvement of cancer-targeting therapy, using nanocarriers for intractable solid tumors by inhibition of TGF-beta signaling. Proc Natl Acad Sci U S A 104(9):3460–3465

Kano MR, Komuta Y, Iwata C, Oka M, Shirai YT, Morishita Y, Ouchi Y, Kataoka K, Miyazono K (2009) Comparison of the effects of the kinase inhibitors imatinib, sorafenib, and transforming growth factor-beta receptor inhibitor on extravasation of nanoparticles from neovasculature. Cancer Sci 100(1):173–180

Kapanci Y, Assimacopoulos A, Irle C, Zwahlen A, Gabbiani G (1974) "Contractile interstitial cells" in pulmonary alveolar septa: a possible regulator of ventilation-perfusion ratio? Ultrastructural, immunofluorescence, and in vitro studies. J Cell Biol 60(2):375–392

Kaye GI, Lane N, Pascal RR (1968) Colonic pericryptal fibroblast sheath: replication, migration, and cytodifferentiation of a mesenchymal cell system in adult tissue. II. Fine structural aspects of normal rabbit and human colon. Gastroenterology 54(5):852–865

Li H, Fan X, Houghton J (2007) Tumor microenvironment: the role of the tumor stroma in cancer. J Cell Biochem 101(4):805–815

Maeda H (2001a) The enhanced permeability and retention (EPR) effect in tumor vasculature: the key role of tumor-selective macromolecular drug targeting. Adv Enzym Regul 41:189–207

Maeda H (2001b) SMANCS and polymer-conjugated macromolecular drugs: advantages in cancer chemotherapy. Adv Drug Deliv Rev 46(1–3):169–185

Maeda H, Fang J, Inutsuka T, Kitamoto Y (2003) Vascular permeability enhancement in solid tumor: various factors, mechanisms involved and its implications. Int Immunopharmacol 3(3):319–328

Maeda H, Bharate GY, Daruwalla J (2009) Polymeric drugs for efficient tumor-targeted drug delivery based on EPR-effect. Eur J Pharm Biopharm 71(3):409–419

Malam Y, Loizidou M, Seifalian AM (2009) Liposomes and nanoparticles: nanosized vehicles for drug delivery in cancer. Trends Pharmacol Sci 30(11):592–599
Mantovani A, Sozzani S, Locati M, Allavena P, Sica A (2002) Macrophage polarization: tumor-associated macrophages as a paradigm for polarized M2 mononuclear phagocytes. Trends Immunol 23(11):549–555
Matsumura Y, Maeda H (1986) A new concept for macromolecular therapeutics in cancer chemotherapy: mechanism of tumoritropic accumulation of proteins and the antitumor agent smancs. Cancer Res 46(12 Pt 1):6387–6392
Mills SE (2007) Histology for pathologists, 3rd edn. Lippincott Williams & Wilkins, Philadelphia
Murray PJ, Wynn TA (2011) Obstacles and opportunities for understanding macrophage polarization. J Leukoc Biol 89(4):557–563
Nishihara H (2014) Human pathological basis of blood vessels and stromal tissue for nanotechnology. Adv Drug Deliv Rev 74:19–27
Ong SM, Tan YC, Beretta O, Jiang D, Yeap WH, Tai JJ, Wong WC, Yang H, Schwarz H, Lim KH, Koh PK, Ling KL, Wong SC (2012) Macrophages in human colorectal cancer are pro-inflammatory and prime T cells towards an anti-tumour type-1 inflammatory response. Eur J Immunol 42(1):89–100
Psarros C, Lee R, Margaritis M, Antoniades C (2012) Nanomedicine for the prevention, treatment and imaging of atherosclerosis. Maturitas 73(1):52–60
Rudolph R, Guber S, Suzuki M, Woodward M (1977) The life cycle of the myofibroblast. Surg Gynecol Obstet 145(3):389–394
Sakai S, Iwata C, Tanaka HY, Cabral H, Morishita Y, Miyazono K, Kano MR (2016) Increased fibrosis and impaired intratumoral accumulation of macromolecules in a murine model of pancreatic cancer co-administered with FGF-2. J Control Release 230:109–115
Sappino AP, Dietrich PY, Skalli O, Widgren S, Gabbiani G (1989) Colonic pericryptal fibroblasts. Differentiation pattern in embryogenesis and phenotypic modulation in epithelial proliferative lesions. Virchows Arch A Pathol Anat Histopathol 415(6):551–557
Sawada N, Murata M, Kikuchi K, Osanai M, Tobioka H, Kojima T, Chiba H (2003) Tight junctions and human diseases. Med Electron Microsc 36(3):147–156
Skinner SA, Tutton PJ, O'Brien PE (1990) Microvascular architecture of experimental colon tumors in the rat. Cancer Res 50(8):2411–2417
Sohl G, Willecke K (2004) Gap junctions and the connexin protein family. Cardiovasc Res 62(2):228–232
Stein M, Keshav S, Harris N, Gordon S (1992) Interleukin 4 potently enhances murine macrophage mannose receptor activity: a marker of alternative immunologic macrophage activation. J Exp Med 176(1):287–292
Torchilin V (2011) Tumor delivery of macromolecular drugs based on the EPR effect. Adv Drug Deliv Rev 63(3):131–135
Van Itallie CM, Anderson JM (2004) The molecular physiology of tight junction pores. Physiology (Bethesda) 19:331–338
Weinberg R (ed) (2013) Heterotypic interactions and the biology of angiogenesis, 2nd edn. Garland Science, New York
Yuzawa S, Kano MR, Einama T, Nishihara H (2012) PDGFRbeta expression in tumor stroma of pancreatic adenocarcinoma as a reliable prognostic marker. Med Oncol 29(4):2824–2830
Zhang K, Rekhter MD, Gordon D, Phan SH (1994) Myofibroblasts and their role in lung collagen gene expression during pulmonary fibrosis. A combined immunohistochemical and in situ hybridization study. Am J Pathol 145(1):114–125
Zhang L, Nishihara H, Kano MR (2012) Pericyte-coverage of human tumor vasculature and nanoparticle permeability. Biol Pharm Bull 35(5):761–766

Part V
Cell-Specific Targeting

Chapter 7
Ligand-targeted Particulate Nanomedicines Undergoing Clinical Evaluation: Current Status

Roy van der Meel*, Laurens J.C. Vehmeijer*, Robbert Jan Kok, Gert Storm, and Ethlinn V.B. van Gaal

Abstract Since the introduction of Doxil® on the market nearly 20 years ago, a number of nanomedicines have become part of treatment regimens in the clinic. With the exception of antibody-drug conjugates, these nanomedicines are all devoid of targeting ligands and rely solely on their physicochemical properties and the (patho)physiological processes in the body for their biodistribution and targeting capability. At the same time, many preclinical studies have reported on nanomedicines exposing targeting ligands, or ligand-targeted nanomedicines, yet none of these have been approved at this moment. In the present review, we provide a concise overview of 13 ligand-targeted particulate nanomedicines (ligand-targeted PNMs) that have progressed into clinical trials. The progress of each ligand-targeted PNM is discussed based on available (pre)clinical data. Main conclusions of these analyses are that (a) ligand-targeted PNMs have proven to be safe and efficacious in preclinical models; (b) the vast majority of ligand-targeted PNMs is generated for the treatment of cancer; (c) contribution of targeting ligands to the PNM efficacy is not unambiguously proven; and (d) targeting ligands do not cause localization of the PNM within the target tissue, but rather provide benefits in terms of target cell internalization and target tissue retention once the PNM has arrived at the target site.

---

*Author contributed equally with all other contributors.

Published from Advanced Drug Delivery Reviews 65: 1284–1298 (2013) in slightly modified form with kind permission of © Elsevier 2016.

R. van der Meel
Department of Pharmaceutics, Utrecht Institute for Pharmaceutical Sciences (UIPS), Utrecht University, Utrecht, The Netherlands

Department of Clinical Chemistry & Haematology, University Medical Center Utrecht, Utrecht, The Netherlands

Department of Biochemistry & Molecular Biology, University of British Columbia, Vancouver, BC, Canada

L.J.C. Vehmeijer • R.J. Kok
Department of Pharmaceutics, Utrecht Institute for Pharmaceutical Sciences (UIPS), Utrecht University, Utrecht, The Netherlands

A. Prokop, V. Weissig (eds.), *Intracellular Delivery III*, Fundamental Biomedical Technologies 8, DOI 10.1007/978-3-319-43525-1_7

Increased understanding of the *in vivo* fate and interactions of the ligand-targeted PNMs with proteins and cells in the human body is mandatory to rationally advance the clinical translation of ligand-targeted PNMs. Future perspectives for ligand-targeted PNM approaches include the delivery of drugs that are unable or inefficient in passing cellular membranes, treatment of drug resistant tumors, targeting of the tumor blood supply, the generation of targeted vaccines and nanomedicines that are able to cross the blood-brain barrier.

Keywords Nanomedicines • Targeting ligand • Clinical translation • Particulate nanocarrier • Liposome • Polymeric nanoparticle • Bacterial-derived minicell • Retrovector

Abbreviations

AD-PEG	Adamantane-conjugated polyethylene glycol
ADC	antibody-drug conjugate
APC	antigen-presenting cell
BBB	blood-brain barrier
CDP	cyclodextrin-containing polymer
CNS	central nervous system
CTL	cytotoxic T cells
DC	dendritic cell
DC-SIGN	dendritic cell-specific intracellular adhesion molecule 3-grabbing non-integrin
DLT	dose-limiting toxicity
DTXL	docetaxel
DOX	doxorubicin
EPR	enhanced permeability and retention
EGFR	epidermal growth factor receptor
GSH	glutathione

G. Storm (✉)
Department of Pharmaceutics, Utrecht Institute for Pharmaceutical Sciences (UIPS), Utrecht University, Utrecht, The Netherlands

Department of Targeted Therapeutics, University of Twente, Enschede, The Netherlands
e-mail: G.Storm@uu.nl

E.V.B. van Gaal
Department of Pharmaceutics, Utrecht Institute for Pharmaceutical Sciences (UIPS), Utrecht University, Utrecht, The Netherlands

Cristal Therapeutics, Maastricht, The Netherlands

OctoPlus, Leiden, The Netherlands

ILs	immunoliposomes
IFN-γ	interferon-gamma
LPS	lipopolysaccharide
RRM2	M2 subunit of ribonucleotide reductase
MDR	multi drug resistance
MPS	mononuclear phagocyte system
MTD	maximum tolerated dose
NGPE	N-glutaryl-phosphatidylethanolamine
L-OHP	oxaliplatin
PAC	paclitaxel
PNM	particulate nanomedicine
PLD	PEGylated liposomal doxorubicin
PK	pharmacokinetic
PLA	poly(d,l-lactide)
PEG	polyethylene glycol
PLGA	poly(lactic-co-glycolic acid)
PSMA	prostate-specific membrane antigen
ACUPA	S,S-2-[3-[5-amino-1-carboxypentyl]-ureido]-pentanedioic acid
scFv	single-chain antibody fragment
shRNA	short hairpin RNA
Tf	transferrin
Tf-AD-PEG	transferrin-conjugated adamantane-conjugated polyethylene glycol
TfR	transferrin receptor

7.1 Introduction

Nanomedicine is the science and application of nanotechnology for diagnosis, monitoring, prevention, treatment and understanding of disease to ultimately gain clinical benefit (European Medical Research Counsils 2004). The focus of the current review is on targeted nanomedicines developed to generate therapeutics that are more effective and/or less harmful to patients compared to conventional drugs. Interdisciplinary pioneering research over the last few decades that focused on colloidal systems, polymer chemistry and antibody technology, has led to the introduction of the term "nanomedicine"(Drexler et al. 1991), and has facilitated the rapid evolvement of the drug targeting and delivery field and subsequent clinical translation of targeted nanomedicines (Duncan and Gaspar 2011; Kamaly et al. 2012; Allen and Cullis 2013). The exploitation of nanocarriers for drug delivery has many potential advantages:

(1) improve unfavorable pharmacokinetics and tissue distribution of many drugs, (2) increase therapeutic efficacy by achieving higher accumulation of a drug in the target tissue, (3) reduce (dose-limiting) adverse effects by minimizing drug exposure to non-target tissues, (4) feasibility of combination therapy by targeted delivery

of multiple therapeutic agents in one nanomedicine, (5) ability to manipulate the nanocarrier surface with a range of molecules such as targeting moieties for increased target specificity or polymers to reduce interactions with plasma proteins and blood cells to improve circulation kinetics.

Targeted nanomedicines, either marketed or under development, are designed for the treatment of a broad range of indications such as infections (Huh and Kwon 2011), cardiovascular diseases (Lobatto et al. 2011), central nervous system diseases (Srikanth and Kessler 2012) and inflammatory diseases (Crielaard et al. 2012). The primary emphasis is however on the development of nanomedicines for the treatment of (mostly solid) malignancies (Peer et al. 2007; Davis et al. 2008; Jain and Stylianopoulos 2010). The discovery that macromolecules accumulate in solid tumors over time by virtue of the enhanced retention and permeability (EPR) effect (Box 7.1) (Matsumura and Maeda 1986), has greatly advanced the development and clinical translation of anti-cancer nanomedicines.

With the exception of antibody-drug conjugates (ADCs), all currently marketed nanomedicines are devoid of targeting ligands and their pharmacokinetic properties and biodistribution rely solely on physicochemical properties of the nanomedicine and subsequent interactions in the circulation and at tissue sites including the site of disease. Ligand-targeted particulate nanomedicines (ligand-targeted PNMs) (Box 7.1) are equipped with targeting ligands to increase the target specificity of ligand-lacking particulate nanomedicines (ligand-lacking PNMs). Additionally, ligand-targeted PNMs can be applied to target diseases where the EPR effect is not present.

Whereas several ligand-lacking PNMs have become part of treatment regimens in the clinic, only a small number of ligand-targeted PNMs have progressed into (early) clinical evaluation and none of them have thus far been approved (Svenson 2012). The aim of this review is to reveal and discuss the evidence for added delivery benefits of conjugating targeting ligands to PNMs currently undergoing clinical evaluation based on analysis of available (pre)clinical data (Table 7.1). The aim is not to provide a complete (historical) overview, discuss basic scientific issues regarding targeted drug delivery and/or perspectives of nanomedicines in general, which has already been discussed in several excellent reviews (Hoffman 2008; Petros and DeSimone 2010; Duncan and Gaspar 2011; Cheng et al. 2012; Kamaly et al. 2012; Svenson 2012; Etheridge et al. 2013). The ligand-targeted PNMs discussed in this review are defined by three components: *the particulate nanocarrier*, *targeting ligands* and *therapeutic agent*. This review therefore does not focus on other nanomedicines such as ligand-lacking PNMs (Svenson 2012), ADCs (Adair et al. 2012; Casi and Neri 2012; Sievers and Senter 2013) and stimuli-responsive nanomedicines (Ganta et al. 2008). The review consists of an objective presentation of available evidence for target localization, safety and efficacy of ligand-targeted PNMs. Based on these data, a scoring table was prepared which summarizes the main characteristics and research outcomes of the evaluated nanomedicines (Table 7.2).

Box 7.1 Definitions

Particulate nanomedicines

Particulate nanomedicines (PNMs) are drug-loaded submicrometer size delivery vehicles designed to improve the pharmacokinetic and biodistribution profiles of the encapsulated molecules. These molecules can be adsorbed, entrapped or dissolved in particulate nanocarriers via non-covalent interactions or via degradable or non-degradable covalent linkers (Petros and DeSimone 2010). Nanomedicines discussed in this review are defined as particulate nanocarriers developed to deliver therapeutic agents to sites of disease. Over the last few decades, many particulate nanocarriers have been developed for the delivery of therapeutics including liposomes, polymer-drug conjugates, micelles, polymeric nanoparticles, dendrimers and albumin nanoparticles. Currently, a few dozen first generation nanomedicines are routinely used in the clinic and it is estimated that approximately 250 nanomedicines are under (pre)clinical investigation (Svenson 2012; Etheridge et al. 2013).

PEGylation

Polyethylene glycol (PEG) is a hydrophilic polymer that has been widely used for the development of drug-polymer conjugates because it can improve protein solubility, stability and pharmacokinetic parameters (Abuchowski et al. 1977; Knop et al. 2010). In addition, coating the surface of PNMs with PEG provides 'stealth' properties by inhibiting blood protein adsorption. This effect inhibits subsequent clearance of PNMs from the circulation by the mononuclear phagocyte system (MPS). The discovery that PEGylation could greatly enhance the circulation time of nanocarriers such as polymeric nanoparticles (Gref et al. 1994) and liposomes (Blume and Cevc 1990; Klibanov et al. 1990) has greatly advanced the clinical translation of nanomedicines. Although the majority of PNMs clinically approved or under evaluation contains PEG, several issues regarding PEGylation remain such as decrease of drug release and cell uptake ('PEG dilemma') (Romberg et al. 2008), activation of the complement system (Moghimi et al. 2010) and accelerated blood clearance of consecutive administered doses (Dams et al. 2000; Laverman et al. 2001).

EPR effect

The enhanced permeability and retention (EPR) effect was first proposed by Matsumura and Maeda (Matsumura and Maeda 1986) and describes the phenomenon that macromolecules accumulate in tumors over time. Tumor vasculature is characterized by poorly developed leaky vasculature containing

(continued)

Box 7.1 (continued)
inter-endothelial gaps which allow for the extravasation of PNMs. In addition, tumors often fail to drain extravasated PNMs due to an impaired lymphatic system (Maeda et al. 2013). The EPR effect is exploited by most anti-cancer nanomedicines as it is expected to increase the therapeutic efficacy of chemotherapeutics due to the relative improvement in tumor accumulation of PNMs compared to small molecules.

Targeted drug delivery

In the field of nanomedicine, targeting refers to the design of therapeutic nanocarriers with the intention to increase accumulation at sites of disease in the body. This is fundamentally different from molecularly targeted drugs that are intended to specifically interact with a certain protein, but have not been designed to localize at specific sites in the body (Kamaly et al. 2012). Historically, the terms 'passive' and 'active' targeting were implemented to distinguish between nanomedicines without or equipped with targeting ligands, respectively. Passive targeting primarily refers to anti-cancer nanomedicines that accumulate in tumors due to a combination of the physicochemical properties of the PNMs and prolonged circulation half life, extravasation from the blood circulation and the pathophysiology of the tumor contributing to the EPR effect. Active targeting (also described as ligand-targeting or receptor-mediated targeting) involves the attachment of ligands to the surface of PNMs that bind to proteins overexpressed on diseased cells. Although in theory this can potentially improve PNM target specificity and improve therapeutic activity, it is believed that in the case of many pathologies, ligand-targeted nanomedicines are subjected to the same physiological localization as ligand-lacking nanomedicines and therefore have comparable biodistribution and accumulation profiles. However, targeting ligands may offer advantages at in terms of target cell uptake once arrived at the target site. It is increasingly recognized that the terms 'passive' and 'active' targeting are not correctly representing the real-life situation (Bae and Park 2011; Kwon et al. 2012; Lammers et al. 2012). We have therefore decided for the sake of clarity to use the terms 'ligand-lacking nanomedicines' (ligand-lacking PNMs) and 'ligand-targeted nanomedicines' (ligand-targeted PNMs). A third targeting strategy based on stimuli-responsive PNMs referred to as triggered drug release is currently receiving much attention but is beyond the scope of this manuscript (for review see (Ganta et al. 2008)).

Table 7.1 Overview of ligand-targeted nanomedicines undergoing clinical evaluation

Product name	Company	Approx. size (nm)	Payload	Ligand	Target	Clinical indication	Clinical phase
Lipid-based nanomedicines							
MBP-426	Mebiopharm	50–200	Oxaliplatin	Protein	Transferrin receptor	Metastatic gastric, gastro esophageal junction, esophageal adenocarcinoma	Phase II
SGT-53	SynerGene Therapeutics	90	p53 plasmid DNA	Antibody fragment (scFv)	Transferrin receptor	Solid tumors	Phase Ib
SGT-94	SynerGene Therapeutics	90	RB94 plasmid DNA	Antibody fragment (scFv)	Transferrin receptor	Solid tumors	Phase I
MM-302	Merrimack Pharmaceuticals	75–110	Doxorubicin	Antibody fragment (scFv)	ErbB2 (HER2)	Breast cancer	Phase I
Lipovaxin-MM	Lipotek		Melanoma antigens and IFNγ	Single domain antibody (dAb) fragment (VH)	DC-SIGN	Melanoma vaccine	Phase I
anti-EGFR ILs-DOX	University Hospital Basel	85	Doxorubicin	Antibody fragment (Fab')	EGFR	Solid tumors	Phase I
2B3-101	to-BBB Technologies		Doxorubicin	Protein	Gluthathione transporters	Solid tumors	Phase I/IIa

(continued)

Table 7.1 (continued)

MCC-465	Mitsubishi Pharma Corporation	140	Doxorubicin	Antibody fragment (F(ab)'2)	Not characterized	Advanced gastric cancer	Phase I (discontinued)
Polymer-based nanomedicines							
BIND-014	BIND Biosciences	100	Docetaxel	Small molecule	Prostate specific membrane antigen	Solid tumors	Phase II
CALAA-01	Calando Pharmaceuticals	50–70	RRM2 siRNA	Protein	Transferrin receptor	Solid tumors	Phase I
SEL-068	Selecta Biosciences	150–250	Nicotine antigen, T-helper cell peptide, TLR agonist	Small molecule	Antigen presenting cells	Smoking cessation vaccine	Phase I
Bacterially-derived minicell							
Erbitux®EDVsPAC	EnGeneIC	400	Paclitaxel	Antibody	EGFR	Solid tumors	Phase II
Retroviral vector							
Rexin-G	Epeius Biotechnologies	100	Cytocidal dominant negative cyclin-G1 DNA construct	Small molecule	Collagen	Sarcoma, osteosarcoma, pancreatic cancer	Phase II[a]

[a]Approved in the Republic of the Philippines under an expanded program as a first-line and adjuvant therapy for pancreatic and breast cancer, and as a second-line therapy for all chemotherapy-resistant solid malignancies

Table 7.2 Superior effect of ligand-targeted nanomedicines versus ligand-lacking nanomedicines in (pre)clinical studies

	Cellular studies					Animal studies				Clinical evaluation studies			
	Cell association[a]	Competition[b]	Target specificity[c]	Cell internalization[d]	Efficacy[e]	Target localization[f]	Target specificity[g]	Cell internalization[d]	Efficacy[h]	Target localization[i]	Target specificity	Cell internalization[d]	Efficacy
MBP-462 (Ishida et al. 2001; Suzuki et al. 2008)	+	+			+	+			+				‡
SGT-53 (Xu et al. 2001; Xu et al. 2002; Yu et al. 2004; Senzer et al. 2013)	+	+	+		+	+	+		+	+		+	
SGT-94 (Pirollo et al. 2008)	*	*	+		+	*	*		+				
MM-302 (Park et al. 1995; Kirpotin et al. 1997; Park et al. 2002; Kirpotin et al. 2006)	+	+	+	+	+	-	+	+	+				‡
Lipovaxin-MM (van Broekhoven et al. 2004)	+	+			+		+		+				
Anti-EGFR ILs-DOX (Mamot et al. 2003; Mamot et al. 2005, Mamot et al. 2012b)	+		+	+	+	-		+	+				‡

Table 7.2 (continued)

	Cellular studies					Animal studies				Clinical evaluation studies			
	Cell association[a]	Competition[b]	Target specificity[c]	Cell internalization[d]	Efficacy[e]	Target localization[f]	Target specificity[g]	Cell internalization[d]	Efficacy[h]	Target localization[i]	Target specificity	Cell internalization[d]	Efficacy
MCC-465 (Hosokawa et al. 2003; Hamaguchi et al. 2004; Hosokawa et al. 2004; Matsumura et al. 2004)		+	+	+	+		+		+				‡
BIND-014 (Farokhzad et al. 2004; Farokhzad et al. 2006; Cheng et al. 2007; Hrkach et al. 2012)	+	+	+		+	+	+		+				‡
CALAA-01 (Bellocq et al. 2003; Bartlett and Davis 2007; Davis et al. 2010)	+	+		+	+	-	+		+	+		+	‡
SEL-068													
2B3-101													

Erbitux®EDVsPAC (MacDiarmid et al. 2007; MacDiarmid et al. 2009)	+		+	+	+	+	+		+				
Rexin-G (Gordon et al. 2000; Gordon et al. 2001; Gordon et al. 2006)					+				+	+			‡

Scoring was based on studies reported in references between brackets. Positive scores (denoted by a + symbol) were granted if a superior effect mediated by the targeting ligand was observed. Negative scores (denoted by a – symbol) were granted when no superior effect mediated by the targeting ligand was observed. All scores (except clinical evaluation studies) were based on studies performed with the ligand-targeted nanomedicine (ligand-targeted PNM) and ligand-lacking nanomedicine (ligand-lacking PNM) as control. Open squares indicate that no ligand-lacking PNM was used as a control, the study was not performed or no data was accessible or available. Granted scores were based on the ligand-targeted platform which may differ from the formulation used for clinical evaluation. *These studies have not been reported for SGT-94 but have been performed with SGT-53 that exploits the same particulate nanocarrier. ‡ Since no ligand-targeted nanomedicines have yet been evaluated in large phase III studies, scoring represents reported signs of efficacy in phase I and II trials

[a]Association to target-expressing cells (radioactivity measurements, flow cytometry, fluorescence/confocal microscopy)

[b]Inhibition of cell association to target-expressing cells by addition of free targeting ligand or target substrate (radioactivity measurements, flow cytometry, fluorescence/confocal microscopy)

[c]No cell association to cells that do not express target (radioactivity measurements, flow cytometry, fluorescence/confocal microscopy)

[d]Target cell internalization of nanomedicine or therapeutic agent (confocal microscopy, reporter assays)

[e]Induction of tumor cell death, inhibition of tumor cell proliferation (cell viability/proliferation assays) or immune cell response (cytotoxic T lymphocyte assays)

[f]Nanomedicine accumulation in target tissue (radioactivity/fluorescence measurements, reporter assays)

[g]No accumulation or effect of nanomedicine in target-negative model or no accumulation or effect in target tissue of nanomedicine equipped with control targeting ligands (radioactivity/fluorescence measurements, reporter assays)

[h]Inhibition/regression of tumor growth, prolonged survival or induction of protective immune response in syngeneic/xenograft tumor models (caliper measurements, tumor foci count, imaging); [i] Accumulation in target tissue (reporter assays)

7.2 Ligand-Targeted Particulate Nanomedicines Under Clinical Evaluation

Up to date, 13 ligand-targeted particulate nanomedicines (PNMs) have progressed into clinical trials (Table 7.1). These systems include lipid- and polymer-based delivery vehicles, a retroviral vector and bacterially-derived minicells (Fig. 7.1).

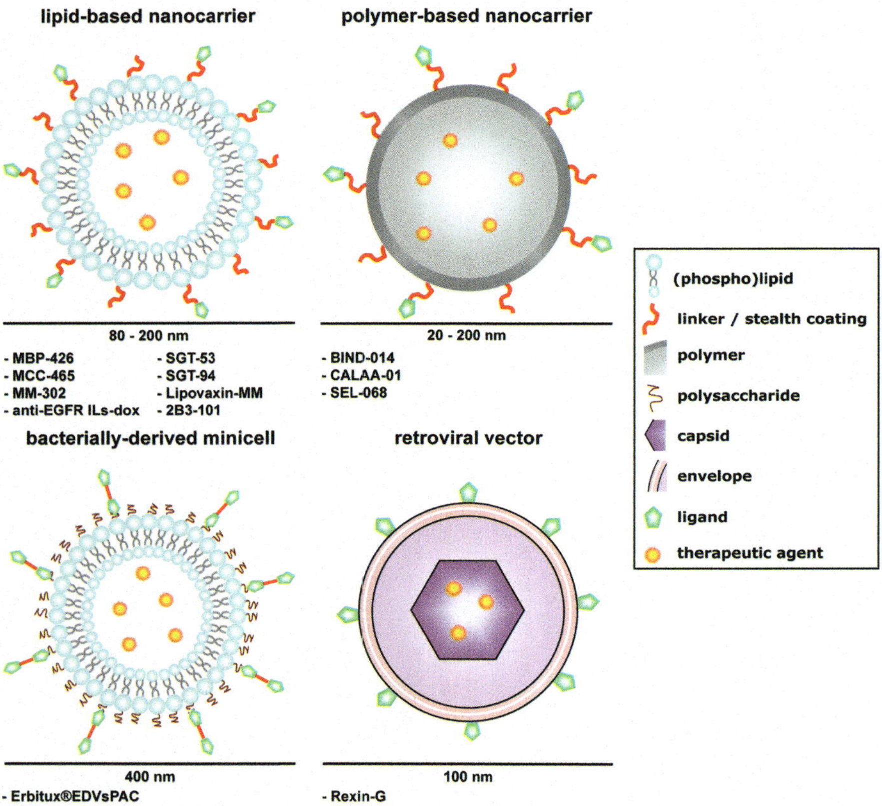

Fig. 7.1 Schematic overview of ligand-targeted nanomedicines undergoing clinical evaluation. The discussed ligand-targeted nanomedicines (ligand-targeted PNMs) are defined by three components: the particulate nanocarrier, targeting ligands and therapeutic agent. Utilized particulate delivery systems include lipid- and polymer based nanocarriers, bacterially-derived minicells and a retroviral vector. Targeting ligands conjugated to PNM include antibodies or antibody fragments, protein (transferrin) or small molecules. Therapeutically active cargo of the ligand-targeted PNMs includes chemotherapeutics, small interfering RNA, plasmid DNA or antigens and adjuvants

7.2.1 *Lipid-Based Nanomedicines*

Originally discovered by Bangham and colleagues (Bangham et al. 1965), liposomes were one of the first particulate nanocarriers utilized for the generation of nanomedicines. Liposomes are vesicular structures which consist of an aqueous core surrounded by a lipid bilayer. Doxorubicin (DOX) encapsulated in long-circulating PEGylated liposomes (Doxil®/Caelyx) was approved in 1995 and has been used in the clinic since then (Barenholz 2012). The lipid-based ligand-targeted PNMs discussed in this review feature either liposomes or formulations based on lipids such as lipoplexes.

7.2.1.1 MBP-426

MBP-426 (Mebiopharm) is a liposome loaded with oxaliplatin (L-OHP) currently undergoing phase Ib/II trials for treatment of second line gastric, gastroesophageal or esophageal adenocarcinomas in combination with leucovorin and fluorouracil (Mebiopharm Co. Ltd.). The liposome is conjugated to transferrin (Tf) for tumor targeting. Platinum binds irreversibly to plasma proteins and erythrocytes and encapsulation of L-OHP in nanocarriers can reduce these interactions thereby improving tumor accumulation and circulation time (Graham et al. 2000). Initial studies with empty PEGylated liposomes showed increased Tf-specific cell association and internalization in Tf-overexpressing murine colon carcinoma (Colon 26) cells of Tf-PEG-liposomes compared to ligand-lacking formulations (Ishida et al. 2001). Tf-PEG-liposomes loaded with L-OHP (EC_{50} 8 μg/mL L-OHP) were more cytotoxic compared to PEG-liposomes devoid of Tf (EC_{50} 18 μg/mL L-OHP) in Colon 26 cells. The cytotoxicity of L-OHP encapsulated in Tf-PEG liposomes could be inhibited by adding an excess of free Tf, indicating that that the cytotoxic effects were mediated by Tf-specific delivery (Suzuki et al. 2008). Studies in mice bearing Colon 26 tumors revealed similar plasma clearance values and biodistribution for ligand-targeted and ligand-lacking L-OHP-loaded PEG-liposomes indicating that the conjugation of Tf did not influence circulation times or uptake by the mononuclear phagocyte system (MPS). Although biodistribution for the ligand-lacking and ligand-targeted formulation was similar, L-OHP concentration in tumors 72 h after injection was ~2.5 times higher in mice treated with L-OHP encapsulated in Tf-PEG-liposomes when compared to PEG-liposomes (Suzuki et al. 2008). Tf-PEG-liposomes loaded with L-OHP significantly suppressed tumor growth compared to L-OHP encapsulated in ligand-lacking liposomes (Suzuki et al. 2008). Based on these results, Mebiopharm further developed this formulation for clinical evaluation. The original formulation was optimized and N-glutaryl-phosphatidylethanolamine (NGPE) is used to couple Tf. The use of NGPE causes the liposome to collapse in environments with low pH such as the endosome. In this way, MBP-426 releases L-OHP upon receptor mediated endocytosis and endosomal localization. In mice bearing human pancreas xenograft tumors, additive tumor growth inhibiting effects were observed when MBP-426 treatment was combined with either gemcitabine or erlotinib (Izbicka et al. 2007). Phase I studies in 39 patients with advanced solid or metastatic solid tumors revealed thrombocytopenia as dose limiting toxicity and a dose of 226 mg/m^2 was recommend for further studies (Mebiopharm Co. Ltd., Sankhala et al. 2009). Results of phase Ib trials in nine patients reported 170 mg/m^2 (versus free L-OHP 85 mg/m^2) as recommended dose for phase II studies and potential activity was observed in two L-OHP-resistant patients (Mebiopharm Co. Ltd., Senzer et al. 2009).

7.2.1.2 SGT-53 and SGT-94

SGT-53 (SynerGene Therapeutics) is a nanomedicine developed for the treatment of solid tumors. The formulation consists of cationic lipids that are complexed with plasmid DNA encoding wild-type p53 tumor suppressor protein. SGT-53 is targeted to the Tf receptor (TfR) on tumor cells via a single-chain antibody fragment (scFv) to achieve intracellular delivery of the plasmid DNA (Xu et al. 2001). Initial formulations contained Tf as targeting ligand (Xu et al. 1997) but the scFv has a smaller size than the Tf molecule and it allows large scale recombinant production and stricter quality control (Xu et al. 2002). TfRscFv-lipoplexes were shown to associate specifically with head and neck and prostate tumor cells (Xu et al. 2001). Using reporter assays and Western blotting it was demonstrated that transfection of tumor cells by TfRscFv-lipoplexes resulted in functional exogenous p53 expression *in vitro* and *in vivo* (Xu et al. 2001; Xu et al. 2002). Importantly, in a mouse tumor metastasis model treatment with TfRscFv-p53-lipoplexes combined with docetaxel (DTXL) resulted in a significant increase in survival compared to non-targeted p53-lipolexes combined with DTXL (Xu et al. 2001). Although these results were promising, rapid clearance of the TfRscFv-lipoplexes was observed. A sterically stabilized PEGylated lipoplex was designed to optimize circulation times *in vivo* (Yu et al. 2004). Although PEGylation of the lipoplexes resulted in reduced transfection efficiency *in vitro*, in a human xenograft prostate tumor model it was demonstrated that the targeted PEGylated lipoplexes induced approximately 7-fold more protein expression in tumors 96 h after treatment than non-PEGylated targeted lipoplexes, indicating the importance of lipoplex stability and circulation time (Yu et al. 2004). Recently reported results of a phase I trial with SGT-53 as a single agent in 11 patients with advanced solid tumors demonstrated no dose limiting toxicities and dose-dependent levels of the transgene were present in tumor biopsies of three patients. After 6 weeks of treatment, 7 of 11 patients had stable disease (Senzer et al. 2013). As SGT-53 is intended to be used in combination with standard radio/chemotherapy, it is now undergoing phase Ib trials to evaluate the safety of combinational therapy with DTXL and to establish a recommended dose for further studies (Synergene Therapeutics Inc.).

SGT-94 utilizes the same TfR-targeted platform as SGT-53 but its cargo consists of the gene that encodes the tumor suppressor protein RB94 (Pirollo et al. 2008). RB94 has broad anti-tumor activity and up to date no cytotoxicity with normal human cells or tumor cell resistance to RB94 has been observed (Xu et al. 1994; Xu et al. 1996; Zhang et al. 2003). *In vitro* cytotoxicity studies revealed that Tf-decorated RB94 lipoplexes increased chemosensitization of human bladder cancer cells 30-fold to gemcitabine and >55-fold to cisplatin compared to ligand-lacking formulations. Treatment of normal human endothelial cells did not result in significant sensitization which indicates that Tf mediated tumor cell specificity (Pirollo et al. 2008). RB94 protein expression was detected in tumors derived from mice injected with Tf-RB94-lipoplexes and TfRscFv-RB94-lipoplexes but not in mice injected with control formulations. Importantly, no detectable RB94 expression in the liver was observed as determined by WB, immunohistochemistry and DNA PCR (Pirollo et al. 2008). In efficacy studies with mice bearing human bladder carcinoma xenografts, treatment of mice with TfRscFv-RB94-lipoplexes combined with gemcitabine significantly inhibited tumor growth compared to ligand-lacking RB-94-lipoplexes and gemcitabine, and targeted formulations with a control vector combined with gemcitabine (Pirollo et al. 2008). SGT-94 has entered phase I trials to evaluate its safety and maximum tolerated dose (MTD) and to find evidence of RB94 expression in tumors after systemic administration (Synergene Therapeutics Inc.).

7.2.1.3 MM-302

MM-302 (Merrimack Pharmaceuticals) is a HER2-targeted nanomedicine that consists of PEGylated liposomes loaded with DOX and has progressed into phase I trials (Merrimack Pharmaceuticals Inc.). Tumor targeting is achieved by the attachment of HER2-targeted scFv antibody fragments to the surface of the liposomes (Park et al. 2001; Park et al. 2004, Nellis et al. 2005a, b). Since the first reports on HER2-targeted immunoliposomes (ILs) loaded with DOX emerged (Park et al. 1995; Kirpotin et al. 1997), many parameters of the formulation have been optimized such as liposomal composition, antibody construct and conjugation method (Park et al. 2001). These early studies have described the increase in HER2-positive ($HER2^+$) cell binding and internalization of anti-HER2 liposomes compared to control liposomes. Increased cell association could be reversed by addition of free anti-HER2 antibody fragments confirming HER2-mediated interactions of the ligand-targeted PNM. In addition, a HER2-negative cell line did not show detectable uptake of anti-HER2 liposomes (Park et al. 1995; Kirpotin et al. 1997). Importantly, biodistribution studies *in vivo* revealed that conjugation of anti-HER2 antibody fragments did not increase radiolabeled liposomal tumor accumulation of the nanomedicine compared to PEGylated liposomal DOX (PLD) (Kirpotin et al. 2006). At the same time, gold-labeled anti-HER2 liposomes localized intracellularly while ligand-lacking liposomes primarily distributed to the extracellular tumor stroma. In a HER2-negative xenograft model the intratumoral distribution of ligand-targeted and ligand-lacking liposomes was similar indicating that both formulations accumulate in the tumor but anti-HER2 ILs associated directly with tumor cells (Kirpotin et al. 2006). Pharmacokinetic (PK) studies in rats showed comparable circulation times for HER2-targeted ILs and control formulations indicating that the presence of an antibody fragment on the liposomes did not alter clearance rates or induced accelerated clearance after multiple doses (Park et al. 2001; Park et al. 2002). Anti-tumor efficacy of anti-HER2 ILs-DOX has been extensively evaluated in multiple studies in four different human HER2⁺ breast cancer xenograft models. Although liposomal formulations varied between studies with regards to PEGylation, antibody fragment and conjugation method, pooled results of all eight studies demonstrate that treatment with anti-HER2 ILs-DOX significantly inhibited tumor growth when compared to PLD. In one of the xenograft studies, anti-HER2 ILs-DOX demonstrated cure rates up to 50 % (Park et al. 2001; Park et al. 2002). Additionally, anti-HER2 ILs-DOX treatment also showed superior efficacy in a xenograft model when compared to combination treatment with either free DOX or PLD and trastuzumab. In a xenograft model expressing low levels of HER2, treatment with anti-HER2 ILs-DOX and PLD induced only modest anti-tumor effects, confirming anti-HER2 ILs-DOX *in vivo* selectivity and the requirement of a receptor density or activity threshold for effective drug delivery (Park et al. 2001; Park et al. 2002). These studies have resulted in an optimized formulation used for clinical evaluation that consists of the anti-HER2 scFv F5 conjugated to PEG-PE micelles which are incorporated into PLD (Park et al. 2001; Park et al. 2004, Nellis et al. 2005a, b). The last few years, updates were presented at conferences on the progress of MM-302 including cardiosafety, efficacy and PK studies in preclinical models (Wickham et al. 2010; Geretti et al. 2011; Klinz et al. 2011). Recently, preliminary data of the ongoing phase I trials were presented. So far, 34 patients with $HER2^+$ advanced breast cancer have enrolled of which 12 patients achieved stable disease and two patients have achieved partial response. MM-302 is tolerable in patients up to 40 mg/m^2 and plasma pharmacokinetics are similar to ligand-lacking PLD (Wickham and Futch 2012).

7.2.1.4 Anti-EGFR ILs-DOX

Generated by the same original developers as MM-302 (Hermes Biosciences), ILs loaded with DOX that target epidermal growth factor receptor (EGFR) overexpressing tumors via coupling of Fab' fragments of the anti-EGFR mAb cetuximab have also progressed into clinical trials (Noble et al. 2004). *In vitro* studies showed superior cell association and internalization of anti-EGFR ILs-DOX compared to ligand-lacking control formulations. For example, quantitative studies performed with pH-sensitive-loaded liposomes demonstrated ~30-fold more EGFR-positive cell internalization of anti-EGFR ILs compared to non-targeted PEGylated liposomes. In addition, cytotoxicity studies in EGFR-positive MDA-MB-468 cells showed that anti-EGFR ILs-DOX were 29-fold more effective than PLD (Mamot et al. 2003). Studies in rats showed similar pharmacokinetics of ligand-targeted and ligand-lacking liposomal DOX indicating that conjugation of antibody fragments did not alter liposomal stability or circulation time (Mamot et al. 2005). As observed with MM-302, biodistribution studies in mice showed no differences in tumor accumulation for EGFR-targeted liposomes and ligand-lacking formulations. However, quantitative flow cytometry analysis demonstrated that cellular accumulation of anti-EGFR liposomes was 6-fold higher when compared to ligand-lacking liposomes in tumor cells derived from mice (Mamot et al. 2005). In two EGFR-overexpressing tumor xenograft models, anti-EGFR ILs-DOX significantly inhibited tumor growth when compared to PLD (Mamot et al. 2005). Interestingly, in a drug resistant tumor xenograft model anti-EGFR ILs-DOX could significantly inhibit tumor growth when compared to PLD, suggesting that anti-EGFR ILs-DOX can overcome multidrug resistance (MDR) (Mamot et al. 2012a). In a recently finished phase I trial (University Hospital Basel Switzerland), 26 patients with EGFR-overexpressing advanced solid tumors were enrolled and treated with escalating doses of anti-EGFR ILs-DOX. One patient showed complete response, one partial response and ten patients had stable disease lasting 2–12 months. A recommended dose of 50 mg DOX per m^2 was recommended for phase II trials (Mamot et al. 2012b).

7.2.1.5 2B3-101

2B3-101 (to-BBB Technologies) is liposome loaded with DOX designed to cross the blood-brain barrier (BBB) for the treatment of glioma. The BBB is a physical, transport and metabolic barrier that poses challenges for drug delivery to the brain. Important criteria related to the BBB, nanocarrier and clinical translation have been proposed for the development of nanomedicines to treat central nervous system (CNS) diseases (Gaillard et al. 2012b). 2B3-101 makes use of glutathione (GSH) as a targeting ligand, with the aim to cross the BBB via glutathione transporters without disrupting the neuroprotective function of the BBB. In proof-of-concept studies in rats, it was demonstrated that increasing amounts of GSH conjugated to PEGylated liposomes loaded the antiviral drug ribavirin resulted in higher amounts of free ribavirin in the brain (Rip et al. 2010). In preclinical studies in rats, 2B3-101 showed similar PK values and toxicity profile as compared to PLD (Gaillard et al. 2012a). However, DOX retention in the brain of rats was significantly higher after repeated administrations of 2B3-101 compared to PLD (Gaillard et al. 2012a). In a human breast cancer xenograft model in mice, both 2B3-101 and PLD demonstrated significant anti-tumor efficacy. In mice bearing intracranial U87 xenograft tumors, treatment with 2B3-101 given at the MTD prolonged survival up to 60 % compared to controls (Gaillard et al. 2012a). 2B3-101 is currently undergoing phase I/IIa trials to determine the safety and PK of the ligand-targeted PNM as a single agent or in combination with trastuzumab (to-BBB technologies B.V.). Accessible and/or available (pre)clinical data of 2B3-101 has been mostly confined to conference abstracts and the developers' website.

7.2.1.6 MCC-465

MCC-465 (Mitsubishi Tanabe Pharma) is a DOX-loaded PEGylated liposome targeted to tumor cells via the conjugation of F(ab')2 of the human GAH antibody (Hosokawa et al. 2003). Although its target antigen has not been characterized, selective binding of GAH antibody was demonstrated as staining of viable tumor tissues and tissue sections stained positively while no staining was observed on non-cancerous tissues (Hamaguchi et al. 2004; Hosokawa et al. 2004). Confocal microscopy studies showed that fluorescently labeled GAH-conjugated ILs loaded with DOX internalized in human stomach cancer cells via GAH-mediated interactions, as the addition of free GAH in combination with GAH-ILs-DOX prevented cell uptake. Ligand-lacking control formulations were hardly internalized by the tumor cells (Hosokawa et al. 2003; Hosokawa et al. 2004). In a pulse-chase assay *in vitro*, GAH-ILs-DOX induced significantly stronger dose-dependent cytotoxicity in human gastric tumor cells compared to PLD. No significant cytotoxicity of GAH-ILs-DOX was observed in human endothelial cells. The anti-tumor efficacy of GAH-ILs-DOX in various human xenograft models in mice was significantly higher than ligand-lacking control PLD (Hosokawa et al. 2003; Hamaguchi et al. 2004; Hosokawa et al. 2004; Shimada et al. 2005). No significant anti-tumor efficacy was observed in xenograft studies with GAH-negative cell lines and it was suggested that GAH-ILs-DOX can overcome DOX resistance of tumor cells (Hosokawa et al. 2003; Hamaguchi et al. 2004). Results from a phase I study indicated that MCC-465 was well tolerated with an MTD of 45.5 mg/m^2 and a dose of 32.5 mg/m^2 in an equivalent amount of DOX was recommended for phase II studies. No anti-tumor effects were observed but stable disease was observed in 10 of 18 patients (Matsumura et al. 2004). Recent updates on MCC-465 are not available and it is uncertain whether development is discontinued.

7.2.1.7 Lipovaxin-MM

Lipovaxin-MM (Lipotek) is a lipid-based vaccine for immunotherapy of malignant melanoma. Lipovaxin-MM does not directly target melanoma cells, but instead its strategy is based on delivering melanoma antigens to dendritic cells (DC) which in turn activate tumor-specific CD8$^+$ cytotoxic T cells (CTL) (Altin and Parish 2006). The melanoma antigens in Lipovaxin-MM are derived from the membrane fraction of lysed MM200 melanoma cells. MM200 plasma derived membrane vesicles are isolated and subsequently fused with liposomes containing cytokines such as interferon-gamma (IFN-γ) or lipopolysaccharide (LPS) that provide a DC "danger" or maturation signal. The vaccine is targeted to DCs via engraftment of the domain antibody DMS5000 which is highly specific for DC-specific intracellular adhesion molecule 3-grabbing non-integrin (DC-SIGN) (Altin et al., Altin and Parish 2006). In proof of concept studies, cell association of DC-targeted vaccines *in vitro* was 4 to 8-fold higher than ligand-lacking control formulations. This effect could be reversed by pre-incubation of the cells with free targeting ligand demonstrating specific interactions of the ligand-targeted PNM. DC-targeting *in vivo* was demonstrated by determining the number of fluorescent-positive cells in a draining lymph node after injection with ligand-targeted and ligand-lacking formulations. DC-targeted vesicles induced 4-fold more fluorescent cells than ligand-lacking formulations (van Broekhoven et al. 2004). In addition, in a B16-OVA melanoma model immunization of mice with targeted vaccines induced strong CTL responses in splenic T cells, induced protective immunity against tumors and could inhibit tumor growth (van Broekhoven et al. 2004). According to the patent application, Lipovaxin-MM used for studies in non-human primates consists of 4 pre-mix components (MM200 membrane vesicles, lyophilized liposomes, IFN-gamma and DMS5000) that are formulated prior to administration (Altin et al.). Treatment of macaques with Lipovaxin-MM resulted in production of vaccine-specific antibodies but it is not certain if this effect is caused specifically by the antigens as a ligand-lacking control was not included in this study (Altin et al.). A phase I study in 12 melanoma patients to determine adverse events, immunogenicity and efficacy of Lipovaxin-MM was recently completed but results have not yet been made available (Lipotek Pty. Ltd.).

7.2.2 *Polymer-Based Nanomedicines*

The potential of polymers for drug delivery was demonstrated by pioneering work in the 1970s (Couvreur et al. 1979; Gros et al. 1981). Polymer-based particulate nanocarriers such as polymeric nanoparticles are produced by self-assembly or cross-linking of polymeric building blocks to obtain nanoparticles with favorable physicochemical characteristics.

7.2.2.1 BIND-014

BIND-014 (BIND Biosciences) is a polymeric nanoparticle developed for the treatment of solid tumors. BIND-014 is composed of poly(d,l-lactide) (PLA) and PEG block copolymers to form a hydrophobic core for the encapsulation of DTXL, and a hydrophilic surface for prolonged circulation (Hrkach et al. 2012). The ligand-targeted PNM is targeted to prostate-specific membrane antigen (PSMA) expressing cells using the small-molecule S,S-2-[3-[5-amino-1-carboxypentyl]-ureido]-pentanedioic acid (ACUPA) as targeting ligand (Maresca et al. 2009; Hrkach et al. 2012). PSMA is expressed by prostate tumor cells and additionally, by the neovasculature of other types of solid tumors but not on normal vasculature (Chang et al. 1999). BIND-014 was developed by a novel strategy in which a library was composed of more than 100 self-assembling nanoparticles to obtain a single ligand-targeted PNM with optimized physicochemical properties (Gu et al. 2008; Shi et al. 2011; Hrkach et al. 2012). Initial *in vitro* studies performed with the PSMA-targeting RNA aptamer A10 (Lupold et al. 2002) as targeting ligand demonstrated a 77-fold increase in cell association of PSMA-targeted formulations compared to ligand-lacking formulations. No cell association to PSMA-negative cells was observed for either of the formulations (Farokhzad et al. 2004). In mice bearing human PSMA-positive prostate xenograft tumors, targeted poly(lactic-co-glycolic acid) (PLGA)-based nanoparticles delivered 3.77-fold more chemotherapeutic agent to tumors compared to ligand-lacking control nanoparticles after 24 h (Cheng et al. 2007). PSMA-targeted nanoparticles loaded with DTXL were significantly more cytotoxic *in vitro* compared to control DTXL nanoparticles without targeting ligand. In xenograft studies, ligand-targeted PNM loaded with DTXL significantly inhibited tumor growth and increased survival compared to ligand-lacking DTXL nanoparticles (Farokhzad et al. 2006). In later preclinical studies, optimized BIND-014 treatment caused significant tumor growth inhibition in a mouse xenograft prostate tumor model compared to ligand-lacking controls. In contrast, no difference in anti-tumor effect was observed in PSMA-negative xenograft models (Hrkach et al. 2012). BIND-014 is currently undergoing a phase I clinical trial to determine the safety in patients with advanced or metastatic cancer (BIND Biosciences). Interim data in three patients demonstrated that DTXL plasma levels are two orders of magnitude higher when administered as BIND-014 compared to solvent-based DTXL. Preliminary signs of BIND-014 anti-tumor efficacy were observed in two patients (Hrkach et al. 2012). Full phase I results with BIND-014 in patients with advanced solid tumors were recently presented which included anti-tumor response in 9 out of 28 patients and a MTD of 60 mg/m^2 (Von Hoff et al. 2013). Phase II studies to evaluate the safety and efficacy of BIND-014 in patients with metastatic castration-resistant prostate cancer or as second-line therapy for patients with lung cancer have recently been initiated (BIND Biosciences, BIND Biosciences).

7.2.2.2 CALAA-01

CALAA-01 (Calando Pharmaceuticals) is a polymeric nanoparticle for siRNA-mediated treatment of solid tumors. This nanomedicine based on the RONDEL™ platform, which consists of four components that are mixed together and self-assemble into nanoparticles prior to administration: a linear cyclodextrin-containing polymer (CDP) backbone, adamantane-conjugated polyethylene glycol (AD-PEG), Tf-conjugated AD-PEG (Tf-PEG-AD) and siRNA (Davis 2009). CALAA-01 induces knockdown of the M2 subunit of ribonucleotide reductase (RRM2), which catalyzes the formation of deoxyribonucleotides from ribonucleotides for DNA synthesis (Cerqueira et al. 2005, Heidel et al. 2007a). Tf-nanoparticles were shown to associate with HeLa cells in a ligand density-dependent manner and cell uptake studies in the presence of free Tf demonstrated TfR-mediated cell internalization (Bartlett and Davis 2007). A multimodality imaging approach revealed no differences in tumor accumulation and tissue distribution between ligand-targeted PNM and ligand-lacking PNM siRNA formulations (Bartlett et al. 2007). However, using reporter assays it was shown that Tf-targeted nanoparticles did exhibit enhanced transfection efficiency in tumor bearing mice compared to ligand-lacking formulations (Bartlett et al. 2007). Increased inhibition of tumor growth in mice by Tf-siRNA-nanoparticles compared to ligand-lacking formulations was demonstrated in a mouse model of metastatic Ewing's sarcoma (Hu-Lieskovan et al. 2005). In addition, in mice bearing head and neck cancer xenografts, CALAA-01 treatment reduced RRM2 mRNA and protein levels resulting in significant inhibition of tumor growth compared to nanoparticles with control siRNA (Rahman et al. 2012). Multiple systemic doses of CALAA-01 in non-human primates were well-tolerated and no significant signs of toxicity were observed at siRNA doses up to 8 mg/kg (Heidel et al. 2007b). Phase I trials evaluating CALAA-01 are ongoing (Calando Pharmaceuticals) and early results in three patients with solid tumors showed dose-dependent intracellular localization in tumor cells but not in the adjacent epidermis. Decreased protein expression of RRM2 in the tumor was observed in at least one patient, suggesting evidence for RNAi in humans (Davis et al. 2010).

7.2.2.3 SEL-068

SEL-068 (Selecta Biosciences) is a nicotine vaccine developed for treatment of tobacco dependence (Goniewicz and Delijewski 2013). The self-assembling synthetic polymeric nanoparticle (Gu et al. 2008) contains encapsulated toll-like receptor agonist to reduce the production of inflammatory cytokines, encapsulated universal T-helper cell peptide to evoke T-cell responses and nicotine covalently conjugated to the surface of the nanoparticle as a B-cell antigen (Kishimoto et al. 2012; Pittet et al. 2012). Administration of SEL-068 in mice and cynomolgus monkeys induced high titers of anti-nicotine antibodies with high affinity (Kishimoto et al. 2012; Pittet et al. 2012). In this way, addictive effects of smoking are counteracted by largely preventing nicotine in the circulation to cross the blood brain barrier and bind to nicotine receptors. Although SEL-068 is currently undergoing phase I clinical trials to evaluate the safety in smokers and non-smokers (Selecta Biosciences Inc.), available and/or accessible data is largely limited to conference abstracts and the website of Selecta Biosciences.

7.2.3 Bacterially-Derived Minicells

A relatively new NC platform utilizes bacterially-derived minicells for drug delivery. These minicells are bacterial cells of approximately 400 nm, devoid of a nucleus and produced by mutants in which genes responsible for cell division have been inactivated (MacDiarmid and Brahmbhatt 2011).

7.2.3.1 Erbitux®EDVsPAC

Targeted minicells for the treatment of solid tumors are under development by EnGeneIc. A wide range of chemotherapeutic drugs can be incorporated in the minicells, including DOX, paclitaxel (PAC) and cisplatin (MacDiarmid et al. 2007).

Additionally, minicells can be loaded in a similar fashion with siRNA or with plasmid DNA encoding short hairpin RNA (shRNA) (MacDiarmid et al. 2009). Tumor targeting of minicells is achieved by bispecific antibodies which recognize both the O-polysaccharide component of the lipopolysaccharide present on the minicell surface and a cell surface receptor overexpressed on tumor cells such as EGFR (MacDiarmid et al. 2011). Cell specific association, uptake and toxicity of EGFR-targeted minicells loaded with DOX (^{EGFR}minicells$_{DOX}$) was demonstrated in EGFR-expressing MDA-MB-468 human breast cancer cells (MacDiarmid et al. 2007). In several human tumor xenograft models *in vivo* (breast, lung, ovarian, lung, leukemia), different minicell formulations including ^{EGFR}minicells$_{DOX}$, EGFR-targeted minicells loaded with PAC (^{EGFR}minicells$_{PAC}$) and HER2-targeted minicells loaded with DOX (^{HER2}minicells$_{DOX}$) demonstrated strong anti-tumor activity compared to ligand-lacking control formulations (MacDiarmid et al. 2007). For comparison, 100-fold higher doses of Doxil® (100 μg) were needed to achieve similar anti-tumor effects of ^{EGFR}minicells$_{DOX}$ (1 μg) in mice bearing breast cancer xenografts (MacDiarmid et al. 2007). The anti-cancer effect of DOX-loaded minicells was further demonstrated by tumor regression in two dogs with advanced T cell non-Hodgkin's lymphoma, and safety of minicells was demonstrated by multiple consecutive iv-injections in three healthy pigs (MacDiarmid et al. 2007). Most interestingly, drug-resistance of colon cancer cells could be reversed with sequential treatment of EGFR-targeted minicells loaded with shRNA specific for the MDR P glycoprotein MDR1 (^{EGFR}minicells$_{shMDR1}$) followed by targeted minicells loaded with chemotherapeutics. Furthermore, the sequential combination treatment effectively reversed MDR in colon, breast and uterine xenograft models *in vivo* (MacDiarmid et al. 2009). Intermediate results of a phase I safety and tolerability study were recently presented (Solomon et al. 2012). Multiple doses of intravenously administered ^{EGFR}minicells$_{PAC}$ were generally well tolerated in 28 patients with advanced solid tumors and a dose of 1×10^{10} minicells are recommended for phase II studies (Solomon et al. 2012). A phase I/II study with EGFR-targeted minicells loaded with DOX in patients with glioma is also planned (EnGeneIC Ltd. 2013).

7.2.4 Retroviral Vectors

The unraveling of the retroviral life cycle basic principles led to the introduction of replication-incompetent retroviruses in the 1980s (Mann et al. 1983). Non-replicating retroviral vectors are able to efficiently integrate their genetic payload in the DNA of the target cell, making them attractive nanocarriers for gene therapy.

7.2.4.1 Rexin-G

Rexin-G is murine leukemia virus-based nanomedicine for the treatment of osteosarcoma, soft tissue sarcoma and pancreatic cancer developed by Epeius Biotechnologies. The main issue with retroviral vectors has been the lack of tissue specificity (Hall et al. 2000). However, Rexin-G is the first retrovector targeted to tumors and associated neovasculature via a high-affinity collagen-binding motif derived from von Willebrand factor. Rexin-G elicits anti-tumor effects by interfering with cell cycle control with a mutant cyclin G1 gene (Gordon and Hall 2010b). In human tumor xenografts, Rexin-G markedly inhibited tumor growth and increased survival compared to ligand-lacking controls (Gordon et al. 2001). Results from early phase I/II clinical trials in the Philippines for the treatment of metastatic pancreatic cancer and other solid tumors showed that Rexin-G was well tolerated, did not induce organ damage and that there were signs of antitumor activity (Gordon et al. 2004, 2006). In phase I/II clinical trials in the U.S.A., for the treatment of advanced or metastatic pancreatic cancer, Rexin-G was well tolerated in phase I studies but there was no evidence of an anti-tumor response (Galanis et al. 2008). In phase II of these clinical trials which involved higher doses of Rexin-G, no dose-limiting toxicity was found. At none of the doses tested, organ-related toxicity, signs of an antibody response, off-target transfection or presence of replication-competent retrovirus were observed. A correlation between Rexin-G dosage and overall survival was established (Chawla et al. 2010). Similar results were found in phase I/II and phase II trials for the treatment of sarcoma and osteosarcoma (Chawla et al. 2009). Based on these results Rexin-G gained orphan drug status for treatment of soft tissue sarcoma, osteosarcoma and pancreatic cancer in the U.S.A. (Gordon et al. 2006). Of note, during clinical trials Rexin-G treatment was associated with improvement of physiological conditions (liver function, ascites, blood chemistry, wound healing) presumably due to the targeting of exposed collagen by Rexin-G (Gordon and Hall 2009).

7.3 Discussion

The scope of this review was to provide an overview of ligand-targeted PNMs undergoing clinical evaluation and to reveal the added delivery benefits of the conjugated targeting ligands. Although 13 ligand-targeted PNMs have progressed into clinical trials, the contribution of targeting ligands to therapeutic efficacy of PNMs in humans has not yet been unambiguously proven. Twelve ligand-targeted PNMs are currently under active evaluation while the development of MCC-465 appears to have been discontinued. Limited access to (pre)clinical data for SEL-068 and 2B3-101 prevents detailed discussion of these products.

With the exception of the anti-nicotine vaccine SEL-068, all of the described ligand-targeted PNMs have been developed for the treatment of solid malignant neoplasms. As cancer remains the leading cause of death in the world today, the medical need to design more effective and safer anti-cancer drugs is evident. The anti-tumor effect of ligand-lacking PNMs, largely mediated by the EPR effect, may be further enhanced by the addition of targeting ligands to increase target cell specificity and internalization (reviewed elsewhere (Peer et al. 2007; Davis et al. 2008; Jain and Stylianopoulos 2010)).

The encapsulated *therapeutic agent* in seven anti-cancer nanomedicines is an established chemotherapeutic compound such as doxorubicin (MM-302, anti-EGFR ILS-DOX, MCC-465, 2B3-101), oxaliplatin (MBP-426), docetaxel (BIND-014) or paclitaxel (Erbitux®EDVs$_{PAC}$). These compounds have been previously approved by the FDA either as free drug or formulated as ligand-lacking PNMs, thus lowering the development risk and reducing regulatory issues for the new ligand-targeted formulations in development. Four ligand-targeted PNMs contain plasmid DNA or siRNA (SGT-53, SGT-94, CALAA-01, Rexin-G). These molecules are unable to pass cell membranes and are dependent on ligand-induced receptor-mediated internalization for therapeutic activity. The two vaccine formulations targeted to antigen-presenting cells (APC) contain antigen and adjuvants to stimulate the immune system to produce cytotoxic T-cells (Lipovaxin-MM) or neutralizing antibodies (SEL-068).

Of the 13 discussed ligand-targeted PNMs, the exploited *particulate nanocarrier* of 8 formulations are lipid-based, 3 are based on polymeric NPs, 1 on a retroviral vector and 1 on a bacterial vector. The application of established lipid-based particulate nanocarriers is likely due to the clinical experience gained with these systems as ligand-lacking PNMs, and to reduce development risks and regulatory issues associated with novel nanocarrier systems. For example, MM-302 and anti-EGFR ILs-DOX consist of a similar formulation as Doxil® but targeting ligands are introduced by post-insertion of micelles bearing targeting ligands for tumor targeting (Nellis et al. 2005a, b, Mamot et al. 2012b). The rapid development of ligand-targeted PNMs based on polymers is noteworthy. Such systems are characterized by the production of self-assembling polymeric NP and high-throughput strategies giving advantages in terms of large-scale manufacturing and batch-to-batch variation. CALAA-01 is formulated prior to systemic administration by self-assembly of the (ligand-modfied) polymeric components and siRNA (Davis 2009). BIND-014 and SEL-068, based on the Accurins™ technology, were developed by the design of pre-functionalized triblock co-polymers to create a library of self-assembling targeted polymeric nanoparticles allowing efficient tailoring of physicochemical characteristics (Gu et al. 2008; Shi et al. 2011). Genetic engineering has led to the development of the replication incompetent retroviral vector Rexin-G, which is generated in human producer cells to generate a targeted biocompatible ligand-targeted PNM with a size of approximately 100 nm (Gordon and Hall 2009). Interestingly, bacterially-derived minicells employed for the generation of Erbitux®EDVs$_{PAC}$ are characterized by a larger size (400 nm) compared to other PNMs (MacDiarmid and Brahmbhatt 2011). It has been shown that the cut-off size of permeable tumor vasculature in the majority of tumors varies between 380 and 780 nm (Yuan et al. 1995; Hobbs et al. 1998). However, the size of synthetic nanocarriers is generally designed to remain below 200 nm to avoid rapid uptake by the MPS and to enhance tumor penetration.

With regard to physicochemical properties of PNMs, the effect of parameters such as total entrapped drug and free drug content, release kinetics and surface characteristics should be acknowledged. Many nanomedicines aim at increasing the MTD due to an improved safety profile. If the overall dose (entrapped + free drug) is significantly increased, the free fraction may become dose-limiting, especially in case of highly potent drugs. Therefore, free drug content and release profile represent important parameters to take into account. Disclosure of release kinetics under physiologically relevant conditions is therefore encouraged.

Regarding the *targeting ligand* utilized for the discussed ligand-targeted PNMs, 4 nanomedicines target the transferrin receptor (TfR). Specific tumor markers such as EGFR, HER2 and PSMA are targeted by four nanomedicines (the exact target receptor of MCC-465 is not known). In contrast, a tumor stromal target is exploited by one ligand-targeted PNM. Both vaccine formulations target APC and one formulation is designed for crossing the BBB via GSH transporters The TfR is a well-established target for cancer treatment by virtue of its overexpression on a range of tumors (Daniels et al. 2012). Attachment of transferrin to PNMs for targeting is exploited by MBP-426 and CALAA-01. The lipoplex formulations SGT-53 and SGT-94 also target the TfR but make use of antibody fragments instead of transferrin (Xu et al. 2002). Antibody fragments are smaller than transferrin and recombinant expression allows efficient large scale production and high quality control reducing batch-to-batch variation. When compared to full monoclonal antibodies, the use of antibody fragments for targeting is preferred because they lack the Fc part of the antibody, preventing rapid recognition by cells of the immune system and subsequent clearance of the ligand-targeted PNM. In the case of Erbitux®EDVs$_{PAC}$ minicells, the Fc region is present, but complement-mediated toxicity is inhibited as protein A/G blocks the Fc part of the conjugated monoclonal antibodies (MacDiarmid et al. 2011). However, an antibody response to the O-polysaccharide component of the bispecific antibody was observed in phase I trials (Solomon et al. 2012). Interestingly, while most ligand-targeted PNMs are directed to a single surface receptor overexpressed on tumor cells, Rexin-G is equipped with more promiscuous high-affinity collagen-binding motifs as targeting ligands, resulting in efficient drug delivery to tumor cells, stroma cells, neovasculature and sites of metastasis without apparent significant toxicity towards healthy tissues (Gordon and Hall 2010a). This indicates that proteins overexpressed on tumor cells can be used to discriminate between tumor and healthy cells, but it may be beneficial for robust anti-cancer effects to target the tumor stroma rather than solely tumor cells. The two vaccine products show that ligand-targeted PNMs can also be directed to antigen presenting cells for the generation of targeted vaccines. It is likely that their prolonged circulation time allows the nanomedicines to reach target sites and activate cells of the immune system. Besides general vaccine applications, the vaccine strategy can be applied to design effective anti-cancer nanomedicines that are not hampered by limitations of direct tumor cell targeting (Lammers et al. 2012).

It is generally believed that ligand-lacking and ligand-targeted PNMs have comparable PK parameters, biodistribution and tumor targeting profiles. However, surface characteristics play an important role in interaction with blood components and cell membranes. Modification of PNMs with targeting ligands may therefore alter PK and biodistribution profiles. In addition, in case of ligand-targeted PNMs specific effects of a carefully chosen ligand can be outweighed by aspecific interactions due to charge interactions and adsorbance of proteins to the nanomedicine shell.

Studies comparing pharmacokinetics and biodistribution of ligand-lacking versus ligand-modified PNMs are scarce and more importantly: impossible in the clinical setting.

Localization studies in animal models which compared the ligand-targeted formulation to the corresponding ligand-lacking one have only been reported for 7 ligand-targeted PNMs (Table 7.2). Of those 7, 4 reported increased target localization compared to ligand-lacking PNM *in vivo*, while the other three studies demonstrated comparable target localization values for ligand-lacking and ligand-targeted PNMs. For MM-302, anti-EGFR ILs-DOX and CALAA-01, it was demonstrated that in murine xenograft models overall tumor accumulation was similar for ligand-lacking and ligand-targeted PNMs. However, in studies performed with MBP-462, SGT-53, BIND-014 and Erbitux®EDVs$_{PAC}$, a higher degree of tumor localization of the ligand-targeted PNM relative to the ligand-lacking PNM was observed. In the case of 5 ligand-targeted PNMs, literature has reported on improved *in vitro* cellular internalization versus ligand-lacking PNMs. The publications on MM-302 and anti-EGFR ILs-DOX reported results for *in vivo* cell internalization versus ligand-lacking PNMs, and both showed improvement over ligand-lacking PNMs. Intermediate results from a phase I trial with CALAA-01 reported on target cell internalization in tumor biopsies of three patients (Davis et al. 2010). Dose dependent presence of the transgene was reported in biopsies from metastatic lesions of three patients treated with SGT-53, while no transgene presence was detected in skin biopsies (Senzer et al. 2013). In light of these results it is possible that, while ligand-targeted and ligand-lacking formulations are both dependent on extravasation from the circulation into the tumor, ligand-targeted PNMs are retained longer in the tumor than their ligand-lacking counterparts due to increased cellular internalization or other targeting-ligand mediated interactions within the target. However, the number of ligand-targeted PNMs tested for *in vivo* target cell internalization is too limited to provide conclusive evidence. Therefore, in tumors where the EPR effect is present, the use of targeting ligands may only be useful to increase cellular internalization or in cases where the targeting ligand itself has intrinsic anti-tumor effects.

Regarding efficacy, all ligand-targeted PNMs included in this overview have shown increased efficacy *in vitro* and *in vivo* compared to their ligand-lacking counterparts (with the exception of SEL-068). Although *in vitro* and *in vivo* model systems do not provide definite proof of efficacy in humans and no data beyond phase I and II trials have been reported as of yet, these results are encouraging for the concept of ligand-targeted PNMs. In some cases signs of efficacy in phase I and/or II trials were observed, but these studies did not include ligand-lacking PNM controls.

How improved efficacy is related to the presence of a targeting ligand cannot be resolved because the comparison of PK and distribution of ligand-targeted versus ligand-lacking PNMs is often not included. As mentioned previously, such comparisons are scarce and clinical trials are not designed to compare ligand-lacking and ligand-targeted PNMs.

In the majority of the cases described in this review, there is insufficient literature that has reported on cellular and animal studies in which the ligand-targeted PNMs have been compared to ligand-lacking PNMs regarding the parameters in Table 7.2. The only exception is MM-302, which reported superior results to ligand-lacking PNMs in all of these parameters except target localization in animal studies. This indicates that the improved efficacy of MM-302 might be due to improved cellular internalization but since several other ligand-targeted PNMs reported improved target localization in animal studies, this does not necessarily hold true for all ligand-targeted PNMs. Since phase I clinical trials for most of the ligand-targeted PNMs reported in this overview are still ongoing, not all results have been published as of yet. Treatment with ligand-targeted PNMs seemed to be well tolerated in patients in the studies that have been published so far.

In the majority of the discussed ligand-targeted PNMs, the toxicity seems comparable to that of ligand-lacking PNMs in terms of MTD and dose-limiting toxicities (DLT) (Table 7.3). For example, the MTD of MM-302 (40 mg/m^2) (Wickham and Futch 2012), anti-EGFR ILs-DOX (50 mg/m^2) (Mamot et al. 2012b) and MCC-465 (45.5 mg/m^2) (Matsumura et al. 2004) is comparable to that of Doxil® (50 mg/m^2) indicating that conjugating targeting ligands to PNMs does not seem to alter the toxicity profile. However, results from a phase I trial with Erbitux®EDVs$_{PAC}$ report a different DLT compared to albumin-bound paclitaxel which may be related to the bacterially-derived particulate nanocarrier or the bispecific antibody (discussed above). It has to be noted that the MTD values for most of the discussed ligand-targeted are based on results obtained from smaller phase I trials and may change after larger phase II/III trials.

7.4 Future Directions

Ligand-targeted PNMs may prove beneficial in increasing drug exposure due to increased target cell uptake and target tissue retention compared to ligand-lacking PNMs. Additionally, there are several applications where the use of ligand-targeted PNMs may have advantages over ligand-lacking PNMs.

1. Ligand-targeted approaches are crucial for molecules that need to localize intracellularly for therapeutic activity but are not capable of crossing cellular membranes, such as nucleic acids. As a consequence, the development of systemically administered gene (regulating) therapy is evolving concurrently with the development of efficient ligand-targeted particulate nanocarriers (Pecot et al. 2011). The therapeutic potential of RNA interference is illustrated by CALAA-01, which decreased target protein expression in a patient's tumor in a phase I trial (Davis et al. 2010). The feasibility of therapeutic DNA is demonstrated by Rexin-G, which has shown promising anti-tumor activity in patients including inhibition of metastatic lesions, angiogenesis and intractable or resistant tumors (Gordon and Hall 2009).
2. One common mechanism underlying MDR of tumors is the overexpression of drug-efflux pumps, which actively expel anti-cancer drugs. Ligand-targeted PNMs may be able to circumvent MDR by virtue of another cellular fate after

Table 7.3 Dose-limiting toxicities (DLT) and maximum tolerated doses (MTD) of free drug or formulated as ligand-lacking and ligand-targeted particulate nanomedicines

Drug	Dose limiting toxicities (DLT)	Maximum tolerated dose (MTD)
Doxorubicin[a]	Cardiomyopathy, myelosuppression	40–60 mg/m^2 every 3–4 weeks
Doxil® (Uziely et al. 1995; Gabizon 2001)[b]	Palmar plantar erythrodysaesthesia, mucositis	50 mg/m^2 every 4 weeks
MM-302 (Wickham and Futch 2012)[c]	Not reported	40 mg/m^2 every 4 weeks
Anti-EGFR ILs DOX (Mamot et al. 2012b)[d]	Myelosuppression	50 mg/m^2 every 4 weeks
2B3-101	Phase I ongoing	Phase I ongoing
MCC-465 (Matsumura et al. 2004)[e]	Myelosuppression, apetite loss[e]	45.5 mg/m^2 every 3 weeks
Oxaliplatin (single agent) (Raymond et al. 1998)[f]	Neurotoxicity	200 mg/m^2 every 3–4 weeks
MBP-426 (single agent) (Sankhala et al. 2009)[g]	Myelosuppression	226 mg/m^2 every 3 weeks
Oxaliplatin in combination with 5-FU/LV (Maindrault-Goebel et al. 2001)[h]	Neurotoxicity, myelosuppression	130 mg/m^2 every 2 weeks
MBP-426 in combination with 5-FU/LV (Senzer et al. 2009)[i]	Back pain, nausea-vomiting, myelosuppression	170 mg/m^2 every 3 weeks
Docetaxel (Taxotere®)[j]	Myelosuppression	75 mg/m^2 every 3 weeks
BIND-014 (Von Hoff et al. 2013)[k]	Myelosuppression, fatigue	60 mg/m^2 every 3 weeks
Paclitaxel (Taxol®)[l]	Myelosuppression	135–175 mg/m^2 every 3 weeks
Nab-paclitaxel(Abraxane®) (Ibrahim et al. 2002)[m]	Sensory neuropathy, stomatitis, superficial keratopathy	300 mg/m^2 every 3 weeks
Erbitux®EDVsPAC (Solomon et al. 2012)[n]	Reactive arthritis, hypotension, fever, elevated levels liver enzymes	1×10^{10} minicells/dose

Data obtained from studies reported in references between brackets
[a]Most commonly used dosage
[b]The MTD of Doxil® was established in phase I studies at 60 mg/m^2 every 4 weeks but Phase II trials redefined the recommended dose at 50 mg/m^2 every 4 weeks. Higher doses can be tolerated when the dosing interval is sufficiently prolonged. The MTD of a single injection Doxil® is 70 mg/m^2
[c]Preliminary results phase I trial
[d]DLT established in two patients receiving 60 mg/m^2 every 4 weeks
[e]Recommended dose for phase II trials 32.5 mg/m^2 every 3 weeks
[f]Recommended dose for phase II trials 130 mg/m^2
[g]DLT established in 2 patients receiving 400 mg/m^2 and 1 patient receiving 226 mg/m^2
[h]Most commonly used dosage is 85 mg/m^2 every 2 weeks (Oxaliplatin FDA)
[i]DLT established in patients receiving 226 mg/m^2 and 170 mg/m^2
[j]Most commonly used dosage
[k]Dose limiting toxicities in 2 patients receiving 75 mg/m^2 every 3 weeks
[l]Most commonly used dosage [m] DLT established in patients receiving 375 mg/m^2
[n]DLT established in patients receiving 10^9, $1{,}5 \times 10^{10}$, 2×10^{10} and 5×10^{10} minicells/dose. Corresponding paclitaxel concentration not reported

receptor-mediated endocytosis rather than passive diffusion over cell membranes of free drug released by ligand-lacking PNMs (Gao et al. 2012). For example, anti-EGFR ILs-DOX showed significantly enhanced antitumor activity in a MDR breast cancer xenograft tumor model compared to free DOX and PLD (Mamot et al. 2012a).

3. Ligand-targeted approaches can also be exploited to generate nanomedicines that exploit two therapeutic strategies simultaneously in order to achieve additive or synergistic anti-tumor effects. For example, DOX-loaded polymeric micelles decorated with intrinsically active anti-EGFR nanobodies significantly reduced tumor growth and prolonged survival of tumor-bearing mice when compared to DOX-loaded micelles without attached targeting ligands (Talelli et al. 2013).
4. An alternative approach to the targeted delivery of anti-cancer drugs to tumor cells is targeting of the tumor blood supply. The endothelial cells of the tumor vasculature are readily accessible to targeted nanomedicines circulating in the bloodstream and more genetically stable than tumor cells limiting the occurrence of drug resistance phenomena. Delivery of DOX by ligand-targeted PNMs targeting $\alpha\nu\beta3$ integrins overexpressed on tumor neovasculature reduced tumor growth of DOX-insensitive tumors while PLD did not (Schiffelers et al. 2003). In line with these results, it was shown that DOX-loaded ligand-targeted PNMs targeting $\alpha\nu\beta3$ integrins suppressed metastasis (Murphy et al. 2008).
5. Besides the development of ligand-targeted PNMs for cancer treatment, ligand-targeted approaches can be exploited for the generation of effective vaccines as demonstrated by the clinical evaluation of Lipovaxin-MM and SEL-068.
6. The development of effective nanomedicines for the treatment of CNS remains challenging due to the presence of the BBB. In addition to the physical barrier, metabolic barriers and drug-efflux transporters results in a restriction of drugs that are able to cross the BBB in adequate amounts to reach therapeutic activity (Wong et al. 2012). Ligand-targeted approaches may be more effective compared to unencapsulated drugs or ligand-lacking nanomedicines as they can improve drug delivery to the CNS via receptor-mediated transcytosis (Pinzon-Daza et al. 2013), exemplified by the clinical evaluation of 2B3-101.

To determine the feasibility of clinically relevant ligand-targeted PNMs, further preclinical studies focused on relation between physicochemical properties (nanocarrier type, size and surface characteristics) in combination with targeting ligand properties (type and size) and biodistribution, safety and efficacy are encouraged. Current knowledge of nanotechnology, tumor biology and interactions of nanomedicines in the human body is (too) limited. To advance the applicability of ligand-targeted PNMs, lessons learnt from their bench-to-bedside translation have revealed key issues that need to be addressed including *in vitro/in vivo* characterization of PNM physicochemical properties (Cho et al. 2013), choice of appropriate animal models (Lammers et al. 2012) and the influence of receptor expression levels on ligand-targeted PNM efficacy (Hendriks et al. 2013). The applicability of ligand-targeted PNMs is ultimately determined by the balance between clinical benefits *versus* safety and cost-effectiveness of the production process (Cheng et al. 2012).

The efficacy and safety of ligand-targeted PNMs has been shown in animals, but the evidence for the added delivery value of target ligand-coupling to nanomedicines in humans remains to be established. Progress of the ligand-targeted PNMs described in this review through clinical trials will reveal in the upcoming years if ligand-targeted PNMs will represent safe and efficacious drugs in the future.

References

Abuchowski A, McCoy JR, Palczuk NC, van Es T, Davis FF (1977) Effect of covalent attachment of polyethylene glycol on immunogenicity and circulating life of bovine liver catalase. J Biol Chem 252(11):3582–3586

Adair JR, Howard PW, Hartley JA, Williams DG, Chester KA (2012) Antibody-drug conjugates – a perfect synergy. Expert Opin Biol Ther 12(9):1191–1206

Allen TM, Cullis PR (2013) Liposomal drug delivery systems: from concept to clinical applications. Adv Drug Deliv Rev 65(1):36–48

Altin JG, Parish CR (2006) Liposomal vaccines – targeting the delivery of antigen. Methods 40(1):39–52

Altin J, Atmosukarto I, De Wildt RM, Parish C, Price J. Composition for targeting dendritic cells. US 2011/0212168 A1. November 2009 (filing date)

Bae YH, Park K (2011) Targeted drug delivery to tumors: myths, reality and possibility. J Control Release 153(3):198–205

Bangham AD, Standish MM, Watkins JC (1965) Diffusion of univalent ions across the lamellae of swollen phospholipids. J Mol Biol 13(1):238–252

Barenholz Y (2012) Doxil® – the first FDA-approved nano-drug: lessons learned. J Control Release 160(2):117–134

Bartlett DW, Davis ME (2007) Physicochemical and biological characterization of targeted, nucleic acid-containing nanoparticles. Bioconjug Chem 18(2):456–468

Bartlett DW, Su H, Hildebrandt IJ, Weber WA, Davis ME (2007) Impact of tumor-specific targeting on the biodistribution and efficacy of siRNA nanoparticles measured by multimodality in vivo imaging. Proc Natl Acad Sci U S A 104(39):15549–15554

Bellocq NC, Pun SH, Jensen GS, Davis ME (2003) Transferrin-containing, cyclodextrin polymer-based particles for tumor-targeted gene delivery. Bioconjug Chem 14(6):1122–1132

BIND Biosciences. A phase 1 open label, safety, pharmacokinetic and pharmacodynamic dose escalation study of BIND-014 (Docetaxel Nanoparticles for Injectable Suspension), given by intravenous infusion to patients with advanced or metastatic cancer. In: ClinicalTrials.gov [Internet], Bethesda (MD): National Library of Medicine (US). Retrieved July 2013. Available from http://www.clinicaltrials.gov/ct2/show/NCT01300533

BIND Biosciences. A phase 2 study to determine the safety and efficacy of BIND-014 (Docetaxel Nanoparticles for Injectable Suspension) as second-line therapy to patients with non-small cell lung cancer. In: ClinicalTrials.gov [Internet], Bethesda (MD), National Library of Medicine (US). Retrieved July 2013. Available from http://www.clinicaltrials.gov/ct2/show/NCT01792479

BIND Biosciences. A phase 2 study to determine the safety and efficacy of BIND-014 (Docetaxel Nanoparticles for Injectable Suspension), administered to patients with metastatic castration-resistant prostate cancer. In: ClinicalTrials.gov [Internet], Bethesda (MD), National Library of Medicine (US). Retrieved July 2013. Available from http://www.clinicaltrials.gov/ct2/show/NCT01812746.

Blume G, Cevc G (1990) Liposomes for the sustained drug release in vivo. Biochim Biophys Acta 1029(1):91–97

Calando Pharmaceuticals. A phase I, dose-escalating study of the safety of intravenous CALAA-01 in adults with solid tumors refractory to standard-of-care therapies. In: ClinicalTrials.gov [Internet], Bethesda (MD), National Library of Medicine (US). Retrieved July 2013. Available from http://www.clinicaltrials.gov/ct2/show/NCT00689065

Casi G, Neri D (2012) Antibody-drug conjugates: basic concepts, examples and future perspectives. J Control Release 161(2):422–428

Cerqueira NM, Pereira S, Fernandes PA, Ramos MJ (2005) Overview of ribonucleotide reductase inhibitors: an appealing target in anti-tumour therapy. Curr Med Chem 12(11):1283–1294

Chang SS, O'Keefe DS, Bacich DJ, Reuter VE, Heston WD, Gaudin PB (1999) Prostate-specific membrane antigen is produced in tumor-associated neovasculature. Clin Cancer Res 5(10):2674–2681

Chawla SP, Chua VS, Fernandez L, Quon D, Saralou A, Blackwelder WC, Hall FL, Gordon EM (2009) Phase I/II and phase II studies of targeted gene delivery in vivo: intravenous Rexin-G for chemotherapy-resistant sarcoma and osteosarcoma. Mol Ther 17(9):1651–1657

Chawla SP, Chua VS, Fernandez L, Quon D, Blackwelder WC, Gordon EM, Hall FL (2010) Advanced phase I/II studies of targeted gene delivery in vivo: intravenous Rexin-G for gemcitabine-resistant metastatic pancreatic cancer. Mol Ther 18(2):435–441

Cheng J, Teply BA, Sherifi I, Sung J, Luther G, Gu FX, Levy-Nissenbaum E, Radovic-Moreno AF, Langer R, Farokhzad OC (2007) Formulation of functionalized PLGA-PEG nanoparticles for in vivo targeted drug delivery. Biomaterials 28(5):869–876

Cheng Z, Al Zaki A, Hui JZ, Muzykantov VR, Tsourkas A (2012) Multifunctional nanoparticles: cost versus benefit of adding targeting and imaging capabilities. Science 338(6109):903–910

Cho EJ, Holback H, Liu KC, Abouelmagd SA, Park J, Yeo Y (2013) Nanoparticle characterization: state of the art, challenges, and emerging technologies. Mol Pharm 10(6):2093–2110

Couvreur P, Kante B, Roland M, Speiser P (1979) Adsorption of antineoplastic drugs to polyalkylcyanoacrylate nanoparticles and their release in calf serum. J Pharm Sci 68(12):1521–1524

Crielaard BJ, Lammers T, Schiffelers RM, Storm G (2012) Drug targeting systems for inflammatory disease: one for all, all for one. J Control Release 161(2):225–234

Dams ET, Laverman P, Oyen WJ, Storm G, Scherphof GL, van Der Meer JW, Corstens FH, Boerman OC (2000) Accelerated blood clearance and altered biodistribution of repeated injections of sterically stabilized liposomes. J Pharmacol Exp Ther 292(3):1071–1079

Daniels TR, Bernabeu E, Rodriguez JA, Patel S, Kozman M, Chiappetta DA, Holler E, Ljubimova JY, Helguera G, Penichet ML (2012) The transferrin receptor and the targeted delivery of therapeutic agents against cancer. Biochim Biophys Acta 1820(3):291–317

Davis ME (2009) The first targeted delivery of siRNA in humans via a self-assembling, cyclodextrin polymer-based nanoparticle: from concept to clinic. Mol Pharm 6(3):659–668

Davis ME, Chen ZG, Shin DM (2008) Nanoparticle therapeutics: an emerging treatment modality for cancer. Nat Rev Drug Discov 7(9):771–782

Davis ME, Zuckerman JE, Choi CH, Seligson D, Tolcher A, Alabi CA, Yen Y, Heidel JD, Ribas A (2010) Evidence of RNAi in humans from systemically administered siRNA via targeted nanoparticles. Nature 464(7291):1067–1070

Drexler KE, Peterson C, Pergamit G (1991) Unbounding the future: the nanotechnology revolution. William Morrow/Quill Books, New York

Duncan R, Gaspar R (2011) Nanomedicine(s) under the microscope. Mol Pharm 8(6):2101–2141

EnGeneIC Ltd. A phase 1/2 study to determine the effect of doxorubicin loaded EnGeneIC delivery vehicles on progression free survival in patients with Recurrent Glioblastoma Multiforme (GBM). In: ClinicalTrials.gov [Internet], Bethesda (MD), National Library of Medicine (US). Retrieved August 2013. Available from https://www.anzctr.org.au/Trial/Registration/TrialReview.aspx?id=363821

Etheridge ML, Campbell SA, Erdman AG, Haynes CL, Wolf SM, McCullough J (2013) The big picture on nanomedicine: the state of investigational and approved nanomedicine products. Nanomedicine 9(1):1–14

European Medical Research Counsils (2004) European science foundation forward look on Nanomedicine 2005
Farokhzad OC, Jon S, Khademhosseini A, Tran TN, Lavan DA, Langer R (2004) Nanoparticle-aptamer bioconjugates: a new approach for targeting prostate cancer cells. Cancer Res 64(21):7668–7672
Farokhzad OC, Cheng J, Teply BA, Sherifi I, Jon S, Kantoff PW, Richie JP, Langer R (2006) Targeted nanoparticle-aptamer bioconjugates for cancer chemotherapy in vivo. Proc Natl Acad Sci U S A 103(16):6315–6320
Gabizon AA (2001) Pegylated liposomal doxorubicin: metamorphosis of an old drug into a new form of chemotherapy. Cancer Investig 19(4):424–436
Gaillard PJ, Gladdines W, Appeldoorn CCM, Rip J, Boogerd WJ, Beijnen JH, Brandsma D, van Tellingen O (2012a) Development of glutathione pegylated liposomal doxorubicin (2B3-101) for the treatment of brain cancer [abstract]. In: Proceedings of the 103rd annual meeting of the American Association for Cancer Research, Chicago, Illinois. AACR; 2012. Abstract nr 5687
Gaillard PJ, Visser CC, Appeldoorn CC, Rip J (2012b) Targeted blood-to-brain drug delivery –10 key development criteria. Curr Pharm Biotechnol 13(12):2328–2339
Galanis E, Carlson SK, Foster NR, Lowe V, Quevedo F, McWilliams RR, Grothey A, Jatoi A, Alberts SR, Rubin J (2008) Phase I trial of a pathotropic retroviral vector expressing a cytocidal cyclin G1 construct (Rexin-G) in patients with advanced pancreatic cancer. Mol Ther 16(5):979–984
Ganta S, Devalapally H, Shahiwala A, Amiji M (2008) A review of stimuli-responsive nanocarriers for drug and gene delivery. J Control Release 126(3):187–204
Gao Z, Zhang L, Sun Y (2012) Nanotechnology applied to overcome tumor drug resistance. J Control Release 162(1):45–55
Geretti E, Reynolds J, Hendriks B, Eckelhofer I, Espelin C, Gaddy D, Klinz S, Lee H, Leonard S, Olivier K, Agresta S, Niyikiza C, Nielsen U, Wickham T (2011) HER2-targeted liposomal doxorubicin MM-302 has a favorable cardiosafety profile in preclinical models [abstract]. Molecular Targets and Cancer Therapeutics, San Fransisco, California: AACR-NCI-EORTC; 2011. Abstract nr C90
Goniewicz ML, Delijewski M (2013) Nicotine vaccines to treat tobacco dependence. Hum Vaccin Immunother 9(1):13–25
Gordon EM, Hall FL (2009) The 'timely' development of Rexin-G: first targeted injectable gene vector (review). Int J Oncol 35(2):229–238
Gordon EM, Hall FL (2010a) Noteworthy clinical case studies in cancer gene therapy: tumor-targeted Rexin-G advances as an efficacious anti-cancer agent. Int J Oncol 36(6):1341–1353
Gordon EM, Hall FL (2010b) Rexin-G, a targeted genetic medicine for cancer. Expert Opin Biol Ther 10(5):819–832
Gordon EM, Liu PX, Chen ZH, Liu L, Whitley MD, Gee C, Groshen S, Hinton DR, Beart RW, Hall FL (2000) Inhibition of metastatic tumor growth in nude mice by portal vein infusions of matrix-targeted retroviral vectors bearing a cytocidal cyclin G1 construct. Cancer Res 60(13):3343–3347
Gordon EM, Chen ZH, Liu L, Whitley M, Wei D, Groshen S, Hinton DR, Anderson WF, Beart RW Jr, Hall FL (2001) Systemic administration of a matrix-targeted retroviral vector is efficacious for cancer gene therapy in mice. Hum Gene Ther 12(2):193–204
Gordon EM, Cornelio GH, Lorenzo CC 3rd, Levy JP, Reed RA, Liu L, Hall FL (2004) First clinical experience using a 'pathotropic' injectable retroviral vector (Rexin-G) as intervention for stage IV pancreatic cancer. Int J Oncol 24(1):177–185
Gordon EM, Lopez FF, Cornelio GH, Lorenzo CC 3rd, Levy JP, Reed RA, Liu L, Bruckner HW, Hall FL (2006) Pathotropic nanoparticles for cancer gene therapy Rexin-G IV: three-year clinical experience. Int J Oncol 29(5):1053–1064
Graham MA, Lockwood GF, Greenslade D, Brienza S, Bayssas M, Gamelin E (2000) Clinical pharmacokinetics of oxaliplatin: a critical review. Clin Cancer Res 6(4):1205–1218

Gref R, Minamitake Y, Peracchia MT, Trubetskoy V, Torchilin V, Langer R (1994) Biodegradable long-circulating polymeric nanospheres. Science 263(5153):1600–1603

Gros L, Ringsdorf H, Schupp H (1981) Polymeric antitumor agents on a molecular and on a cellular level? Angew Chem Int Ed Engl 20(4):305–325

Gu F, Zhang L, Teply BA, Mann N, Wang A, Radovic-Moreno AF, Langer R, Farokhzad OC (2008) Precise engineering of targeted nanoparticles by using self-assembled biointegrated block copolymers. Proc Natl Acad Sci U S A 105(7):2586–2591

Hall FL, Liu L, Zhu NL, Stapfer M, Anderson WF, Beart RW, Gordon EM (2000) Molecular engineering of matrix-targeted retroviral vectors incorporating a surveillance function inherent in von Willebrand factor. Hum Gene Ther 11(7):983–993

Hamaguchi T, Matsumura Y, Nakanishi Y, Muro K, Yamada Y, Shimada Y, Shirao K, Niki H, Hosokawa S, Tagawa T, Kakizoe T (2004) Antitumor effect of MCC-465, pegylated liposomal doxorubicin tagged with newly developed monoclonal antibody GAH, in colorectal cancer xenografts. Cancer Sci 95(7):608–613

Heidel JD, Liu JY, Yen Y, Zhou B, Heale BS, Rossi JJ, Bartlett DW, Davis ME (2007a) Potent siRNA inhibitors of ribonucleotide reductase subunit RRM2 reduce cell proliferation in vitro and in vivo. Clin Cancer Res 13(7):2207–2215

Heidel JD, Yu Z, Liu JY, Rele SM, Liang Y, Zeidan RK, Kornbrust DJ, Davis ME (2007b) Administration in non-human primates of escalating intravenous doses of targeted nanoparticles containing ribonucleotide reductase subunit M2 siRNA. Proc Natl Acad Sci U S A 104(14):5715–5721

Hendriks BS, Klinz SG, Reynolds JG, Espelin CW, Gaddy DF, Wickham TJ (2013) Impact of tumor HER2/ERBB2 expression level on HER2-targeted liposomal doxorubicin-mediated drug delivery: multiple low-affinity interactions lead to a threshold effect. Mol Cancer Ther 12:1816

Hobbs SK, Monsky WL, Yuan F, Roberts WG, Griffith L, Torchilin VP, Jain RK (1998) Regulation of transport pathways in tumor vessels: role of tumor type and microenvironment. Proc Natl Acad Sci U S A 95(8):4607–4612

Hoffman AS (2008) The origins and evolution of "controlled" drug delivery systems. J Control Release 132(3):153–163

Hosokawa S, Tagawa T, Niki H, Hirakawa Y, Nohga K, Nagaike K (2003) Efficacy of immunoliposomes on cancer models in a cell-surface-antigen-density-dependent manner. Br J Cancer 89(8):1545–1551

Hosokawa S, Tagawa T, Niki H, Hirakawa Y, Ito N, Nohga K, Nagaike K (2004) Establishment and evaluation of cancer-specific human monoclonal antibody GAH for targeting chemotherapy using immunoliposomes. Hybrid Hybridomics 23(2):109–120

Hrkach J, Von Hoff D, Mukkaram Ali M, Andrianova E, Auer J, Campbell T, De Witt D, Figa M, Figueiredo M, Horhota A, Low S, McDonnell K, Peeke E, Retnarajan B, Sabnis A, Schnipper E, Song JJ, Song YH, Summa J, Tompsett D, Troiano G, Van Geen Hoven T, Wright J, LoRusso P, Kantoff PW, Bander NH, Sweeney C, Farokhzad OC, Langer R, Zale S (2012) Preclinical development and clinical translation of a PSMA-targeted docetaxel nanoparticle with a differentiated pharmacological profile. Sci Transl Med 4(128):128ra139

Huh AJ, Kwon YJ (2011) 'Nanoantibiotics': a new paradigm for treating infectious diseases using nanomaterials in the antibiotics resistant era. J Control Release 156(2):128–145

Hu-Lieskovan S, Heidel JD, Bartlett DW, Davis ME, Triche TJ (2005) Sequence-specific knockdown of EWS-FLI1 by targeted, nonviral delivery of small interfering RNA inhibits tumor growth in a murine model of metastatic Ewing's sarcoma. Cancer Res 65(19):8984–8992

Ibrahim NK, Desai N, Legha S, Soon-Shiong P, Theriault RL, Rivera E, Esmaeli B, Ring SE, Bedikian A, Hortobagyi GN, Ellerhorst JA (2002) Phase I and pharmacokinetic study of ABI-007, a Cremophor-free, protein-stabilized, nanoparticle formulation of paclitaxel. Clin Cancer Res 8(5):1038–1044

Ishida O, Maruyama K, Tanahashi H, Iwatsuru M, Sasaki K, Eriguchi M, Yanagie H (2001) Liposomes bearing polyethyleneglycol-coupled transferrin with intracellular targeting property to the solid tumors in vivo. Pharm Res 18(7):1042–1048

Izbicka E, Diaz A, Campos D, Wick M, Takimoto C, Nagasawa M, Hashimoto C (2007) Evaluation of antitumor activity and biomarkers for MBP-426 in human pancreatic cancer xenograft models in vivo [abstract]. In: Proceedings of the 98th annual meeting of the American Association for Cancer Research, Los Angeles, California (CA): AACR; 2007. Abstract nr 5586

Jain RK, Stylianopoulos T (2010) Delivering nanomedicine to solid tumors. Nat Rev Clin Oncol 7(11):653–664

Kamaly N, Xiao Z, Valencia PM, Radovic-Moreno AF, Farokhzad OC (2012) Targeted polymeric therapeutic nanoparticles: design, development and clinical translation. Chem Soc Rev 41(7):2971–3010

Kirpotin D, Park JW, Hong K, Zalipsky S, Li WL, Carter P, Benz CC, Papahadjopoulos D (1997) Sterically stabilized anti-HER2 immunoliposomes: design and targeting to human breast cancer cells in vitro. Biochemistry 36(1):66–75

Kirpotin DB, Drummond DC, Shao Y, Shalaby MR, Hong K, Nielsen UB, Marks JD, Benz CC, Park JW (2006) Antibody targeting of long-circulating lipidic nanoparticles does not increase tumor localization but does increase internalization in animal models. Cancer Res 66(13):6732–6740

Kishimoto K, Altreuter D, Johnston L Keller P, Pittet L (2012) SEL-068 A fully synthetic nanoparticle vaccine for smoking cessation and relapse prevention [abstract]. In: Society for research on nicotine and tobacco 18th annual meeting, Houston, Texas: SRNT; 2012. Abstract nr POS1-1

Klibanov AL, Maruyama K, Torchilin VP, Huang L (1990) Amphipathic polyethyleneglycols effectively prolong the circulation time of liposomes. FEBS Lett 268(1):235–237

Klinz S, Eckelhofer I, Gaddy D, Geretti E, Hendriks B, Lee H, Leonard S, Reynolds J, Wickham T (2011) MM-302, a HER2-targeted liposomal doxorubicin, shows binding/uptake and efficacy in HER2 2+ cells and xenograft models [abstract]. In: Proceedings of the 102nd annual meeting of the American Association for Cancer Research, Orlando, Florida: AACR; 2011. Abstract nr 3637

Knop K, Hoogenboom R, Fischer D, Schubert US (2010) Poly(ethylene glycol) in drug delivery: pros and cons as well as potential alternatives. Angew Chem Int Ed Engl 49(36):6288–6308

Kwon IK, Lee SC, Han B, Park K (2012) Analysis on the current status of targeted drug delivery to tumors. J Control Release 164(2):108–114

Lammers T, Kiessling F, Hennink WE, Storm G (2012) Drug targeting to tumors: principles, pitfalls and (pre-) clinical progress. J Control Release 161(2):175–187

Laverman P, Carstens MG, Boerman OC, Dams ET, Oyen WJ, van Rooijen N, Corstens FH, Storm G (2001) Factors affecting the accelerated blood clearance of polyethylene glycol-liposomes upon repeated injection. J Pharmacol Exp Ther 298(2):607–612

Lipotek Pty. Ltd. A phase I open-label study of the safety and immunogenicity of escalating doses of Lipovaxin-MM, a novel melanoma immunotherapeutic, in patients with metastatic melanoma. In: ClinicalTrials.gov [Internet], Bethesda (MD), National Library of Medicine (US). Retrieved July 2013. Available from http://www.clinicaltrials.gov/ct2/show/NCT01052142

Lobatto ME, Fuster V, Fayad ZA, Mulder WJ (2011) Perspectives and opportunities for nanomedicine in the management of atherosclerosis. Nat Rev Drug Discov 10(11):835–852

Lupold SE, Hicke BJ, Lin Y, Coffey DS (2002) Identification and characterization of nuclease-stabilized RNA molecules that bind human prostate cancer cells via the prostate-specific membrane antigen. Cancer Res 62(14):4029–4033

MacDiarmid JA, Brahmbhatt H (2011) Minicells: versatile vectors for targeted drug or si/shRNA cancer therapy. Curr Opin Biotechnol 22(6):909–916

MacDiarmid JA, Mugridge NB, Weiss JC, Phillips L, Burn AL, Paulin RP, Haasdyk JE, Dickson KA, Brahmbhatt VN, Pattison ST, James AC, Al Bakri G, Straw RC, Stillman B, Graham RM, Brahmbhatt H (2007) Bacterially derived 400 nm particles for encapsulation and cancer cell targeting of chemotherapeutics. Cancer Cell 11(5):431–445

MacDiarmid JA, Amaro-Mugridge NB, Madrid-Weiss J, Sedliarou I, Wetzel S, Kochar K, Brahmbhatt VN, Phillips L, Pattison ST, Petti C, Stillman B, Graham RM, Brahmbhatt H (2009) Sequential treatment of drug-resistant tumors with targeted minicells containing siRNA or a cytotoxic drug. Nat Biotechnol 27(7):643–651

Maeda H, Nakamura H, Fang J (2013) The EPR effect for macromolecular drug delivery to solid tumors: Improvement of tumor uptake, lowering of systemic toxicity, and distinct tumor imaging in vivo. Adv Drug Deliv Rev 65(1):71–79

Maindrault-Goebel F, de Gramont A, Louvet C, Andre T, Carola E, Mabro M, Artru P, Gilles V, Lotz JP, Izrael V, Krulik M (2001) High-dose intensity oxaliplatin added to the simplified bimonthly leucovorin and 5-fluorouracil regimen as second-line therapy for metastatic colorectal cancer (FOLFOX 7). Eur J Cancer 37(8):1000–1005

Mamot C, Drummond DC, Greiser U, Hong K, Kirpotin DB, Marks JD, Park JW (2003) Epidermal growth factor receptor (EGFR)-targeted immunoliposomes mediate specific and efficient drug delivery to EGFR- and EGFRvIII-overexpressing tumor cells. Cancer Res 63(12):3154–3161

Mamot C, Drummond DC, Noble CO, Kallab V, Guo Z, Hong K, Kirpotin DB, Park JW (2005) Epidermal growth factor receptor-targeted immunoliposomes significantly enhance the efficacy of multiple anticancer drugs in vivo. Cancer Res 65(24):11631–11638

Mamot C, Ritschard R, Wicki A, Kung W, Schuller J, Herrmann R, Rochlitz C (2012a) Immunoliposomal delivery of doxorubicin can overcome multidrug resistance mechanisms in EGFR-overexpressing tumor cells. J Drug Target 20(5):422–432

Mamot C, Ritschard R, Wicki A, Stehle G, Dieterle T, Bubendorf L, Hilker C, Deuster S, Herrmann R, Rochlitz C (2012b) Tolerability, safety, pharmacokinetics, and efficacy of doxorubicin-loaded anti-EGFR immunoliposomes in advanced solid tumours: a phase 1 dose-escalation study. Lancet Oncol 13(12):1234–1241

Mann R, Mulligan RC, Baltimore D (1983) Construction of a retrovirus packaging mutant and its use to produce helper-free defective retrovirus. Cell 33(1):153–159

Maresca KP, Hillier SM, Femia FJ, Keith D, Barone C, Joyal JL, Zimmerman CN, Kozikowski AP, Barrett JA, Eckelman WC, Babich JW (2009) A series of halogenated heterodimeric inhibitors of prostate specific membrane antigen (PSMA) as radiolabeled probes for targeting prostate cancer. J Med Chem 52(2):347–357

Matsumura Y, Maeda H (1986) A new concept for macromolecular therapeutics in cancer chemotherapy: mechanism of tumoritropic accumulation of proteins and the antitumor agent smancs. Cancer Res 46(12 Pt 1):6387–6392

Matsumura Y, Gotoh M, Muro K, Yamada Y, Shirao K, Shimada Y, Okuwa M, Matsumoto S, Miyata Y, Ohkura H, Chin K, Baba S, Yamao T, Kannami A, Takamatsu Y, Ito K, Takahashi K (2004) Phase I and pharmacokinetic study of MCC-465, a doxorubicin (DXR) encapsulated in PEG immunoliposome, in patients with metastatic stomach cancer. Ann Oncol 15(3):517–525

Mebiopharm Co. Ltd. A phase I, open label study of MBP-426 given by intravenous infusion in patients with advanced or metastatic solid tumors. In: ClinicalTrials.gov [Internet], Bethesda (MD), National Library of Medicine (US). Retrieved July 2013. Available from http://www.clinicaltrials.gov/ct2/show/NCT00355888

Mebiopharm Co. Ltd. A phase Ib/II study of MBP-426 in patients with second line gastric, gastro esophageal, or esophageal adenocarcinoma. In: ClinicalTrials.gov [Internet], Bethesda (MD): National Library of Medicine (US). Retrieved July 2013. Available from http://www.clinicaltrials.gov/ct2/show/NCT00964080

Merrimack Pharmaceuticals Inc. A phase 1, multi-center, open-label, dose-escalation, safety, and pharmacokinetic clinical study of intravenously administered MM-302 in patients with advanced breast cancer. In: ClinicalTrials.gov [Internet], Bethesda (MD), National Library of Medicine (US). Retrieved July 2013. Available from http://www.clinicaltrials.gov/ct2/show/NCT01304797

Moghimi SM, Andersen AJ, Hashemi SH, Lettiero B, Ahmadvand D, Hunter AC, Andresen TL, Hamad I, Szebeni J (2010) Complement activation cascade triggered by PEG-PL engineered nanomedicines and carbon nanotubes: the challenges ahead. J Control Release 146(2):175–181

Murphy EA, Majeti BK, Barnes LA, Makale M, Weis SM, Lutu-Fuga K, Wrasidlo W, Cheresh DA (2008) Nanoparticle-mediated drug delivery to tumor vasculature suppresses metastasis. Proc Natl Acad Sci U S A 105(27):9343–9348

Nellis DF, Ekstrom DL, Kirpotin DB, Zhu J, Andersson R, Broadt TL, Ouellette TF, Perkins SC, Roach JM, Drummond DC, Hong K, Marks JD, Park JW, Giardina SL (2005a) Preclinical manufacture of an anti-HER2 scFv-PEG-DSPE, liposome-inserting conjugate. 1. Gram-scale production and purification. Biotechnol Prog 21(1):205–220

Nellis DF, Giardina SL, Janini GM, Shenoy SR, Marks JD, Tsai R, Drummond DC, Hong K, Park JW, Ouellette TF, Perkins SC, Kirpotin DB (2005b) Preclinical manufacture of anti-HER2 liposome-inserting, scFv-PEG-lipid conjugate. 2. Conjugate micelle identity, purity, stability, and potency analysis. Biotechnol Prog 21(1):221–232

Noble CO, Kirpotin DB, Hayes ME, Mamot C, Hong K, Park JW, Benz CC, Marks JD, Drummond DC (2004) Development of ligand-targeted liposomes for cancer therapy. Expert Opin Ther Targets 8(4):335–353

Park JW, Hong K, Carter P, Asgari H, Guo LY, Keller GA, Wirth C, Shalaby R, Kotts C, Wood WI et al (1995) Development of anti-p185HER2 immunoliposomes for cancer therapy. Proc Natl Acad Sci U S A 92(5):1327–1331

Park JW, Kirpotin DB, Hong K, Shalaby R, Shao Y, Nielsen UB, Marks JD, Papahadjopoulos D, Benz CC (2001) Tumor targeting using anti-her2 immunoliposomes. J Control Release 74(1–3):95–113

Park JW, Hong K, Kirpotin DB, Colbern G, Shalaby R, Baselga J, Shao Y, Nielsen UB, Marks JD, Moore D, Papahadjopoulos D, Benz CC (2002) Anti-HER2 immunoliposomes: enhanced efficacy attributable to targeted delivery. Clin Cancer Res 8(4):1172–1181

Park JW, Benz CC, Martin FJ (2004) Future directions of liposome- and immunoliposome-based cancer therapeutics. Semin Oncol 31(6 Suppl 13):196–205

Pecot CV, Calin GA, Coleman RL, Lopez-Berestein G, Sood AK (2011) RNA interference in the clinic: challenges and future directions. Nat Rev Cancer 11(1):59–67

Peer D, Karp JM, Hong S, Farokhzad OC, Margalit R, Langer R (2007) Nanocarriers as an emerging platform for cancer therapy. Nat Nanotechnol 2(12):751–760

Petros RA, DeSimone JM (2010) Strategies in the design of nanoparticles for therapeutic applications. Nat Rev Drug Discov 9(8):615–627

Pinzon-Daza ML, Campia I, Kopecka J, Garzon R, Ghigo D, Rigant C (2013) Nanoparticle- and liposome-carried drugs: new strategies for active targeting and drug delivery across blood–brain barrier. Curr Drug Metab 14(6):625–640

Pirollo KF, Rait A, Zhou Q, Zhang XQ, Zhou J, Kim CS, Benedict WF, Chang EH (2008) Tumor-targeting nanocomplex delivery of novel tumor suppressor RB94 chemosensitizes bladder carcinoma cells in vitro and in vivo. Clin Cancer Res 14(7):2190–2198

Pittet L, Altreuter D, Ilyinskii P, Fraser C, Gao Y, Baldwin S, Keegan M, Johnston L, Kishimoto T (2012) Development and preclinical evaluation of SEL-068, a novel targeted Synthetic Vaccine Particle (tSVP™) for smoking cessation and relapse prevention that generates high titers of antibodies against nicotine [abstract]. In: Immunology 2012 – 99th meeting of The American Association of Immunologists, Boston, Massachusetts: AAI; 2012. Abstract nr 75.11

Rahman MA, Amin AR, Wang X, Zuckerman JE, Choi CH, Zhou B, Wang D, Nannapaneni S, Koenig L, Chen Z, Chen ZG, Yen Y, Davis ME, Shin DM (2012) Systemic delivery of siRNA nanoparticles targeting RRM2 suppresses head and neck tumor growth. J Control Release 159(3):384–392

Raymond E, Chaney SG, Taamma A, Cvitkovic E (1998) Oxaliplatin: a review of preclinical and clinical studies. Ann Oncol 9(10):1053–1071

Rip J, Appeldoorn C, Manca F, Dorland R, Van Kregten J, Gaillard P (2010) Receptor-mediated delivery of drugs across the blood-brain barrier [abstract]. Pharmacology and toxicology of the blood-brain barrier: state of the art, needs for future research and expected benefits for the EU, Brussel, Belgium: COST; 2010. Abstract nr 5464

Romberg B, Hennink WE, Storm G (2008) Sheddable coatings for long-circulating nanoparticles. Pharm Res 25(1):55–71

Sankhala KK, Mita AC, Adinin R, Wood L, Beeram M, Bullock S, Yamagata N, Matsuno K, Fujisawa T, Phan A (2009) A phase I pharmacokinetic (PK) study of MBP-426, a novel liposome encapsulated oxaliplatin [abstract]. ASCO Annual Meeting, Orlando, Florida: ASCO; 2009. Abstract nr 2535

Schiffelers RM, Koning GA, ten Hagen TL, Fens MH, Schraa AJ, Janssen AP, Kok RJ, Molema G, Storm G (2003) Anti-tumor efficacy of tumor vasculature-targeted liposomal doxorubicin. J Control Release 91(1–2):115–122

Selecta Biosciences Inc. A phase 1 randomized, double-blind, placebo-controlled study to assess the safety and pharmacodynamics of increasing subcutaneous doses of SEL-068 in healthy non-smoker and smoker adults. In: ClinicalTrials.gov [Internet], Bethesda (MD): National Library of Medicine (US). Retrieved July 2013. Available from http://www.clinicaltrials.gov/ct2/show/NCT01478893

Senzer NN, Matsuno K, Yamagata N, Fujisawa T, Wasserman E, Sutherland W, Sharma S, Phan A (2009) MBP-426, a novel liposome-encapsulated oxaliplatin, in combination with 5-FU/leucovorin (LV): Phase I results of a Phase I/II study in gastro-esophageal adenocarcinoma, with pharmacokinetics [abstract]. Molecular Targets and Cancer Therapeutics, Boston, Massachusetts: AACR-NCI-EORTC; 2009. Abstract nr C36

Senzer N, Nemunaitis J, Nemunaitis D, Bedell C, Edelman G, Barve M, Nunan R, Pirollo KF, Rait A, Chang EH (2013) Phase I study of a systemically delivered p53 nanoparticle in advanced solid tumors. Mol Ther 21(5):1096–1103

Shi J, Xiao Z, Kamaly N, Farokhzad OC (2011) Self-assembled targeted nanoparticles: evolution of technologies and bench to bedside translation. Acc Chem Res 44(10):1123–1134

Shimada K, Kasamatsu T, Hasegawa T, Kanai Y, Niki H, Tagawa T, Matsumura Y (2005) The cytotoxicity and intracellular distributions of MCC-465, pegylated liposomal doxorubicin tagged with newly developed monoclonal antibody GAH, in ovarian cancer cell lines [abstract]. In: Proceedings of the 96th Annual Meeting of the American Association for Cancer Research, Anaheim, California: AACR; 2005. Abstract nr 1427

Sievers EL, Senter PD (2013) Antibody-drug conjugates in cancer therapy. Annu Rev Med 64:15–29

Solomon B, Desai J, Scott A, Rosenthal M, McArthur G, Rischin D, Segelov E, MacDiarmid J, Brambhatt H (2012) First-in-man, multicenter, phase I trial evaluating the safety of first-in-class therapeutic, EGFR-targeted, paclitaxel-packaged minicells [abstract]. In: 24th EORTC – NCI – AACR symposium on molecular targets and cancer therapeutics, Dublin, Ireland: EORTC – NCI – AACR; 2012. Abstract nr 585

Srikanth M, Kessler JA (2012) Nanotechnology-novel therapeutics for CNS disorders. Nat Rev Neurol 8(6):307–318

Suzuki R, Takizawa T, Kuwata Y, Mutoh M, Ishiguro N, Utoguchi N, Shinohara A, Eriguchi M, Yanagie H, Maruyama K (2008) Effective anti-tumor activity of oxaliplatin encapsulated in transferrin-PEG-liposome. Int J Pharm 346(1–2):143–150

Svenson S (2012) Clinical translation of nanomedicines. Curr Opinion Solid State Mater Sci 16(6):287–294

Synergene Therapeutics Inc. A phase I open-label safety and pharmacokinetic study of escalating doses of SGT-53 for infusion in subjects with advanced solid tumors. In: ClinicalTrials.gov [Internet], Bethesda (MD), National Library of Medicine (US). Retrieved July 2013. Available from http://www.clinicaltrials.gov/ct2/show/NCT00470613

Synergene Therapeutics Inc. A phase I study of systemic gene therapy with SGT-94 in patients with solid tumors. In: ClinicalTrials.gov [Internet], Bethesda (MD), National Library of Medicine (US). Retrieved July 2013. Available from http://www.clinicaltrials.gov/ct2/show/NCT01517464

Talelli M, Oliveira S, Rijcken CJ, Pieters EH, Etrych T, Ulbrich K, van Nostrum RC, Storm G, Hennink WE, Lammers T (2013) Intrinsically active nanobody-modified polymeric micelles for tumor-targeted combination therapy. Biomaterials 34(4):1255–1260

to-BBB technologies B.V. An open-label, phase I/IIa, dose escalating study of 2B3-101 in patients with solid tumors and brain metastases or recurrent malignant glioma. In: ClinicalTrials.gov [Internet], Bethesda (MD): National Library of Medicine (US). Retrieved August 2013. Available from http://clinicaltrials.gov/ct2/show/NCT01386580

University Hospital Basel Switzerland. A phase I study of doxorubicin-loaded anti-EGFR immunoliposomes in patients with advanced solid tumors. In: ClinicalTrials.gov [Internet], Bethesda (MD): National Library of Medicine (US). Retrieved July 2013. Available from http://www.clinicaltrials.gov/ct2/show/NCT01702129/

Uziely B, Jeffers S, Isacson R, Kutsch K, Wei-Tsao D, Yehoshua Z, Libson E, Muggia FM, Gabizon A (1995) Liposomal doxorubicin: antitumor activity and unique toxicities during two complementary phase I studies. J Clin Oncol 13(7):1777–1785

van Broekhoven CL, Parish CR, Demangel C, Britton WJ, Altin JG (2004) Targeting dendritic cells with antigen-containing liposomes: a highly effective procedure for induction of antitumor immunity and for tumor immunotherapy. Cancer Res 64(12):4357–4365

Von Hoff DD, Mita M, Eisenberg P, LoRusso P, Weiss G, Sachdev J, Mita A, Low S, Hrkach J, Summa J, Berk G, Ramanathan R (2013) A phase I study of BIND-014, a PSMA-targeted nanoparticle containing docetaxel, in patients with refractory solid tumors [abstract]. In: Proceedings of the 104th annual meeting of the American Association for Cancer Research, Washington, DC: AACR; 2013. Abstract nr LB-203

Wickham T, Futch K (2012) A phase I study of MM-302, a HER2-targeted liposomal doxorubicin, in patients with advanced, HER2- positive breast cancer [abstract]. In: 35th annual CTRC-AACR San Antonio breast cancer symposium, San Antonio, Texas: CTRC-AACR; 2012. Abstract nr P5-18-09

Wickham T, Reynolds J, Drummond D, Kirpotin D, Lahdenranta J, Leonard S, Geretti E, Lee H, Klinz S, Hendriks B, Olivier K, Eckelhofer I, Park J, Benz C, Moyo V, Niyikiza C, Nielsen U (2010) Preclinical safety and activity of MM-302, a HER2-targeted liposomal doxorubicin designed to have an improved safety and efficacy profile over approved anthracyclines [abstract]. In: 33rd annual CTRC-AACR San Antonio breast cancer symposium, San Antonio, Texas: CTRC-AACR; 2010. Abstract nr P3-14-09

Wong HL, Wu XY, Bendayan R (2012) Nanotechnological advances for the delivery of CNS therapeutics. Adv Drug Deliv Rev 64(7):686–700

Xu HJ, Xu K, Zhou Y, Li J, Benedict WF, Hu SX (1994) Enhanced tumor cell growth suppression by an N-terminal truncated retinoblastoma protein. Proc Natl Acad Sci U S A 91(21):9837–9841

Xu HJ, Zhou Y, Seigne J, Perng GS, Mixon M, Zhang C, Li J, Benedict WF, Hu SX (1996) Enhanced tumor suppressor gene therapy via replication-deficient adenovirus vectors expressing an N-terminal truncated retinoblastoma protein. Cancer Res 56(10):2245–2249

Xu L, Pirollo KF, Chang EH (1997) Transferrin-liposome-mediated p53 sensitization of squamous cell carcinoma of the head and neck to radiation in vitro. Hum Gene Ther 8(4):467–475

Xu L, Tang WH, Huang CC, Alexander W, Xiang LM, Pirollo KF, Rait A, Chang EH (2001) Systemic p53 gene therapy of cancer with immunolipoplexes targeted by anti-transferrin receptor scFv. Mol Med 7(10):723–734

Xu L, Huang CC, Huang W, Tang WH, Rait A, Yin YZ, Cruz I, Xiang LM, Pirollo KF, Chang EH (2002) Systemic tumor-targeted gene delivery by anti-transferrin receptor scFv-immunoliposomes. Mol Cancer Ther 1(5):337–346

Yu W, Pirollo KF, Rait A, Yu B, Xiang LM, Huang WQ, Zhou Q, Ertem G, Chang EH (2004) A sterically stabilized immunolipoplex for systemic administration of a therapeutic gene. Gene Ther 11(19):1434–1440

Yuan F, Dellian M, Fukumura D, Leunig M, Berk DA, Torchilin VP, Jain RK (1995) Vascular permeability in a human tumor xenograft: molecular size dependence and cutoff size. Cancer Res 55(17):3752–3756

Zhang X, Multani AS, Zhou JH, Shay JW, McConkey D, Dong L, Kim CS, Rosser CJ, Pathak S, Benedict WF (2003) Adenoviral-mediated retinoblastoma 94 produces rapid telomere erosion, chromosomal crisis, and caspase-dependent apoptosis in bladder cancer and immortalized human urothelial cells but not in normal urothelial cells. Cancer Res 63(4):760–765

Chapter 8
Anti-angiogenic Therapy by Targeting the Tumor Vasculature with Liposomes

Yu Sakurai and Hideyoshi Harashima

Abstract In recent years, anti-angiogenic therapy has attracted considerable interest as a new class of anti-cancer medicine because it is more efficacious with a lower toxicity in comparison to traditional cyto-toxic anti-cancer drugs. the enhanced permeability and retention (EPR) effect is now a well-established strategy for targeting cancer cells. On the other hand, there is no generally accepted strategy for targeting tumor endothelial cells, which appear to be responsible for a pathological angiogenesis in cancerous tissues. In this article, we review the various tumor endothelial targeting nano drug delivery systems (DDSs) that involve the use of cationic lipids, peptides, antibodies sugar chains. In addition, the mechanisms responsible for how endothelial cells affect a tumor microenvironment for nanoparticle penetration in tumor tissues are reviewed.

Keywords Liposomes • Tumor endothelial cells • Anti-angiogenic therapy

Abbreviations

5-FU	5-fluorouracil
AMO	anti-micro RNA oligonucleotide
bFGF	basic gibroblast factor
BPD-MA	benzoporphyrin derivative mono-acid ring A
CDC	cell division cycle homologue
CPP	cell penetrating peptie
DCP-TEPA	dicetyl phosphate-tetraethylenepentamine
DDAB	dimethyldioctadecyl ammonium bromide
DDS	drug delivery system
DOTAP	1,2-dioleoyl-3-trimethylammonium propane
DOX	doxorubicin
EPR	enhanced permeability and retention

Y. Sakurai • H. Harashima (✉)
Faculty of Pharmaceutical Sciences, Hokkaido University,
Kita-12, Nishi-6, Kita-ku, Sapporo 060-0812, Japan
e-mail: harasima@pharm.hokudai.ac.jp

A. Prokop, V. Weissig (eds.), *Intracellular Delivery III*, Fundamental Biomedical Technologies 8, DOI 10.1007/978-3-319-43525-1_8

FDA	Food and Drug Administration
HIF	hypoxia inducible factor
HUVEC	Human umbilical vein endothelial cells
IFP	interstitial fluid pressure
IL	interleukin
miRNA	micro RNA
MKI	multi-kinase inhibitor
MRI	magnetic resonance imaging
MT1-MMP	membrane type-1 matrixmetalloproteinase
mTOR	mammarian target of rapamycin
NCT	neutron capture therapy
NRP-1	neuropilin-1
ODN	oligodeoxynucleotide
PDGFR	platelet-derived growth factor receptor
pDNA	plasmid DNA
PDT	photodynamic therapy
PEG	polyethylene glycol
PKN3	phosphonositide-3-kinase
PLK1	polo-like kinase
PTX	paclitaxel
ROS	reactive oxygen species
SELEX	Systematic Evolution of Ligands by Exponential enrichment
siRNA	small interfering RNA
SLX	sialyl lewis^X
TNF-α	tumor necrosis factor-α
VCAM-1	vacular cell adhesion molecule-1
VEGF	vascular endothelial cells growth factor
VEGFR	vascular endothelial cells growth factor receptor
WHO	World Health Organization

8.1 Introduction

Cancer is now one of most frequent causes of death worldwide and is estimated to account for 13 % of all deaths worldwide by the world health organization (WHO). Despite the remarkable progress in cancer biology, developing curative treatments for cancer hasvenot been realized thus far, possibly due to the low selectivity and efficiency of anti-cancer therapeutics to cancer cells. Moreover, most of the currently used anti-cancer therapeutics involve the use of cytotoxic agents that target proliferative cells, not only cancer cells, but also normal fast growing cells such as immune cells and intestinal epithelial cells, which results in severe adverse events. Therefore, the exclusive targeting of only cancer cells would be a double-edged sword.

Angiogenesis is accompanied by tumor growth. This hypothesis was led to the fact that a thyroid that was inoculated with melanoma cells did not grow to a size of over 1–2 mm^3 in diameter (Folkman et al. 1963; Zetter 2008). Since Dr. Folkman proposed the hypothesis in 1971 in the New England Journal of Medicine (Folkman 1971), great efforts have been made to discover angiogenic cascades and their inhibitors. The first identified angiogenic factor was basic fibroblast growth factor (bFGF), based on the observation that human vascular endothelial cells were more proliferative in the presence of bFGF (Gospodarowicz et al. 1978). In 1991, two different groups identified the human vascular endothelial cells growth factor (VEGF, previously called the vascular permeability factor; VPF) (Keck et al. 1989; Leung et al. 1989), which is currently known to be one of the major angiogenic factors in tumor tissue. As of this writing, a number of angiogenic factors and the mechanism responsible for the development of angiogenesis is now known (Carmeliet and Jain 2011). In addition, anti-angiogenic therapy has many advantages: (1) As tumor endothelial cells occupy a small population of the total cells in tumor tissue compared to cancer cells, less therapeutic molecules could be required. (2) Endothelial cells would not be expected to vary in different organs, which would allow one type of anti-angiogenesis drug to be used in treating many types of cancer. (3) Unlike cancer cells, tumor endothelial cells are genetically stable. Therefore, tumor endothelial cells would not acquire drug resistance against an anti-angiogenic therapy (Dey et al. 2015). In addition, since abnormal angiogenesis occurs especially strongly in tumor tissue, anti-angiogenic therapy could be a promising therapy that would not be accompanied by harmful effects in off-target organs instead of traditional proliferative cell-target chemotherapy. Prompted by these proposed advantages, an anti-VEGF humanized murine antibody (Avastin®) was developed, and approved by the Food and Drug Administration (FDA) in 2004. In addition, a multi-kinase inhibitor (MKI), a mammalian target of the rapamycin (mTOR), VEGF receptor (VEGFR) 2 antibody are now being clinically used as an anti-angiogenic medicine. Thus, anti-angiogenic therapy is currently a rapidly growing field in the field of cancer chemotherapy.

During clinical usage, anti-angiogenic therapy faces many unexpected difficulties. Recent studies revealed that endothelial cells are tremendously diverse and can vary with the source of the organs, and consequently they respond to anti-angiogenic therapy differently (Auerbach 1991; Zetter 1990). In addition, it has been shown that tumor endothelial cells are genetically unstable, as evidenced by transcriptome analyses (St Croix et al. 2000), and tumor endothelial cells are aneuploid and have abnormal multiple centrosomes (Hida et al. 2004). Even worse, unexpected adverse effects including hypertension, thrombosis, bowel perforation, can occur during an anti-angiogenic treatment. Accordingly, an innovative strategy for evading the above issues would be required for developing an anti-angiogenic medicine.

Doxorubicin (DOX) -loaded liposomes (namely Doxil®, Caelyx®) were developed to circumvent the side effects of traditional chemotherapy against cancer cells, and was first approved as a liposomal anti-cancer medicine. Doxil® is delivered to tumor tissue based on the enhanced permeability and retention (EPR) effect, which means that macromolecules with a prolonged circulation time passively accumulate in tumor tissue due to the immature blood and lymph vasculature of such tissues (Maeda et al. 2000). DOX frequently causes hair loss, myelosuppression and cardiotoxicity because of its non-specific distribution after systemic injection. Since liposomal formulations are not prone to accumulate in those tissues, Doxil provides an enhanced therapeutic effect and reduced toxicity. Since the discovery of the EPR effect by Dr. Maeda and Dr. Matsumura (Matsumura and Maeda 1986), a number of other liposomal drug delivery systems (DDSs) have been developed. The tumor endothelium has not, however, been a target of DDSs because there is no established strategy for targeting tumor endothelial cells. As endothelial cells are more susceptible to chemotherapeutic agents, such as 5-fluorouracil (5-FU) and DOX, an anti-angiogenic strategy would have the advantage of targeting tumor endothelial cells, not cancer cells, in cancer chemotherapy (Kalra and Campbell 2006). This chapter reviews a number of studies on tumor endothelial cells targeting DDSs and subcellular targeting liposomal DDSs.

8.2 Liposomes for Anti-angiogenic Therapy

8.2.1 Cationic Liposomes

Cationic liposomes have been widely used since the latter part of the twentieth century due to their high internalization characteristics and their spontaneous interaction with negatively charged nucleic acids-based therapeutics. Accordingly, cationic liposomes are popular for the delivery of therapeutics to various tissues or cells, such as cancer cells and hepatocytes. In 1998, it was reported that cationic liposomes were likely to accumulate in angiogenic endothelial cells (Thurston et al. 1998). In this report, Thurston G et al. indicated that systemically injected liposomes containing a cationic lipid, such as 1,2-dioleoyl-3-trimethylammonium propane (DOTAP) ot dimethyldioctadecyl ammonium bromide (DDAB), into tumor-baring mice or mice with chronic airway inflammation caused by a *Mycoplasm pulmonis*-infection were highly co-localized in angiogenic endothelial cells. Although the cationic liposomes were presumably located in fenestrae of tumor endothelial cells, the exact mechanism responsible for this heavy accumulation in tumor endothelium remained unknown. As these discoveries provoked much scientific interest, a number of cationic liposomes-mediated DDSs to tumor endothelial cells have been reported, as shown in Table 8.1. When paclitaxel (PTX)-loaded DOTAP/dioleoyl-*sn*-glycero-3-phosphocholine liposomes were injected into A375 (human melanoma)-bearing mice, the liposomal PTX exhibited a better therapeutic effect than free PTX (Kunstfeld et al. 2003). Another group, as evidenced by intravital microscopy, reported that liposomal PTX with a positively charged lipid DOTAP was co-localized in tumor endothelia cells in a melanoma model (Schmitt-Sody et al. 2003). Actually, the velocity of blood flow in the tumor microcirculation, including the velocity of the bloodstream, was decreased by anti-angiogenic therapeutics in which an anti-cancer drug was delivered by means of a cationic liposome (Strieth et al. 2004), which suggests that the inhibition of the abnormal angiogenic growth of tumor endothelial cells could lead to "vascular normalization".

Is delivering cationic liposome-mediated to tumor endothelial cells more effective than EPR effect-mediated delivery to cancer cells? Wu J et al. reported that a

Table 8.1 Cationic liposomes for targeting tumor endothelial cells

Lipid	Cancer type	Therapeutics	Effect	Ref
DOTAP/DOPC	Human melanoma	Paclitaxel	Increased anti-tumor effect	Kunstfeld et al. (2003)
DOTAP/DOPC	Hamster melanoma	Paclitaxel	Increased anti-tumor effect and prolonged metastasis-free survival	Schmitt-Sody et al. (2003)
DOTAP/DOPC	Hamster melanoma	Paclitaxel	Increased anti-tumor effect and decreased blood velocity and micro vessel density	Strieth et al. (2004)
DDAB/EPC/chol	Human squamous cell carcinoma	Doxorubicin	Prolonged survival time	Wu et al. (2007)
DC-6-14/HSPC/chol	Murine lung cancer	Oxaliplatin	Increased accumulation and anti-tumor effect	Abu Lila et al. (2009)
Hexadecanoyl PEI	Murine melanoma	BPD-MA (PDT)	Increased apoptosis of tumor endothelial cells and anti-tumor effect	Takeuchi et al. (2003)
AtuFect/DPhyPE	Lung metastasis	Anti-PKN3 siRNA	Increased anti-tumor effect and prolonged survival	Aleku et al. (2008a, b), and Santel et al. (2006, 2010)

cationic liposome with DOX encapsulated was superior to polyethylene glycol (PEG)-modified neutral liposomes encapsulating DOX in therapeutic efficacy, although cationic liposomes were subjected to rapid clearance from the circulation compared to PEGylated (that is, PEG-modified) neutral liposomes (Wu et al. 2007). Likewise, Abu et al. reported that oxaliplatin-loaded cationic liposomes containing O,O′-ditetradecanoyl-N-(α-trimethyl ammonio acetyl) diethanolamine chloride (DC-6-14) accumulated to a greater extent in tumor tissue and exhibited a better anti-tumor effect than neutral liposomes (Abu Lila et al. 2009).

Photodynamic therapy (PDT) is a promising cancer treatment, in which photosensitizers are delivered to the interior of a target cell, and the photosensitizers are then activated by a tissue-penetrating laser kills cancer cells via the production of reactive oxygen species (ROS) produced by the photosensitizers. To deliver photosensitizers to tumor endothelial cells, Takeuchi et al. developed original cationic liposomes, which were modified with octadecanol-conjugated polyethyleneimine (PCLs) (Takeuchi et al. 2003). When PCLs encapsulating the benzoporphyrin derivative mono-acid ring A (BPD-MA) were injected into melanoma-bearing mice and then exposed to laser energy (68 nm, 150 J/cm^2), tumor growth was substantially inhibited and the levels of CD31-positive tumor endothelial cells were significantly diminished. In addition to anti-cancer chemotherapy, PDT also holds great promise for being an effective cancer treatment.

Nucleic acids, such as anti-sense oligodeoxynucleotide (ODN), plasmid DNA (pDNA) and small interfering RNA (siRNA) are one of the more promising class of molecules for use in anti-angiogenic therapy. Nucleic acids include negatively charged phosphodiester bonds, which explains whycationic liposomes are popularly used for the delivery of nucleic acids to tumor endothelial cells. AtuPLEX (also called Atu027) is an example of the successful delivery of siRNA to tumor endothelial cells, and a Phase I clinical trial for patients with advanced solid tumors was initiated in 2014 (Schultheis et al. 2014). AtuPLEX is a lipoplex composed of siRNA and an original cationic lipid, AtuFECT01 (β-L-arginyl-2,3-L-diaminopropionic acid-N-palmitoyl-N-oleyl-amide trihydrochloride) and neutral lipid 1,2-diphytanoyl-sn-glycero-3-phosphoethanolamine (DPhyPE). AtuPLEX was reported to inhibit, not only tumor endothelial cells in a human prostate cancer xenograft model (Santel et al. 2006), but also normal endothelial cells in the lung, heart and liver (Aleku et al. 2008a; Gutbier et al. 2010; Santel et al. 2006). AtuPLEX encapsulating siRNA against protein kinase N3 (*PKN3*), which is a downstream effector gene of the phosphoinositide-3-kinase cascade, inhibited tumor growth in a human prostate ectopic model and a human prostate orthotropic model via an anti-angiogenic effect (Aleku et al. 2008b). In this report, they also showed that AtuPLEX could induce gene silencing in the lung endothelium of a non-human primate, a Cynomolgus monkey. In another article, AtuPLEX was reported to substantially inhibit tumor growth in an experimental metastatic model (intravenous injection of Lewis lung carcinoma cells) and a spontaneous metastatic model (local injection of human breast cancer cells to mammary fat pads) (Santel et al. 2010). AtuPLEX is currently under a clinical trial with gemcitabine for the treatment of patients with pancreatic cancer. AtuPLEX would be expected to be a prime candidate for anti-angiogenic therapy by siRNA.

DOTAP-based PTX-loaded cationic liposomes (EndoTAG-1) were indeed announced to be in the process of being evaluated in Phase 1/2 clinical trials for the treatment of advanced head and neck cancer and a Phase 2 clinical trial for HER2-negative breast cancer patients (Awada et al. 2014; Strieth et al. 2014). Cationic liposome-mediated anti-angiogenic therapy would be a promising strategy due to the simplicity of their preparation and the targeting method.

8.2.2 Peptides

To deliver DDSs and their cargo to objective cells, the modification of peptide ligands with a high affinity to a protein which is, preferably, specifically and highly expressed in target cells, is a popular strategy in cancer cell targeting. For cancer cell targeting, ligand conjugation is not necessarily required because liposomes with a prolonged circulation time tend to accumulate in tumor tissue via the EPR effect. For targeting tumor endothelial cells, such strategies are often utilized as shown in Table 8.2.

Table 8.2 Peptide ligands for targeting tumor endothelial cells

Peptides	Cancer type	Therapeutics	Effect	Ref
RGD Motif				
c(RGDf(ε-S-acetylthioacetyl)K	Murine colorectal cancer	Doxorubicin	Inhibiting tumor growth	Schiffelers et al. (2003)
c(RGDf(ε-S-acetylthioacetyl)K	HUVEC	^{10}B (NCT)	*in vitro* cell killing effect	Koning et al. (2004)
GARYCRGDCFDG	Murine colorectal cancer	Doxorubicin	Inhibiting tumor growth	Holig et al. (2004)
GARYCRGDCFDG	Human lung cancer	Paclitaxel	Disrupting tumor vasculature	Meng et al. (2011)
Cyclic RGDyK	Human glioblastoma Human breast cancer	Doxorubicin	Increased accumulation	Kim et al. (2014)
RGDK lipopeptide	Murine melanoma	Plasmid DNA coding p53	Inhibiting tumor growth	Pramanik et al. (2008)
RGDK lipopeptide	Murine melanoma	Curcumin	Inhibiting tumor growth	Mondal et al. (2013)
RGDK lipopeptide	Murine melanoma	Doxorubicin/curcumin	Inhibiting tumor growth	Barui et al. (2014)
RGDK lipopeptide	Murine melanoma	Anti-CDC20 siRNA	Inhibiting tumor growth via anti-angiogenic effect	Majumder et al. (2014)
Cyclic RGDfK	HUVEC/matrigel	AMO against miR296	Inhibiting angiogenesis	Liu et al. (2011)
Cyclic RGDfK	Human renal cell carcinoma	Anti-VEGFR2 siRNA	Inhibiting tumor growth	Sakurai et al. (2014b, 2016)

(continued)

Table 8.2 (continued)

Peptides	Cancer type	Therapeutics	Effect	Ref
Cyclic RGDfK	Human renal cell carcinoma	Doxorubicin	Inhibiting tumor growth via anti-angiogenic effect	Kibria et al. (2013)
NGR Motif				
Cyclic KNGRE	Human fibrosarcoma	Doxorubicin	*in vitro* cell killing effect	Negussie et al. (2010)
GNGRGGVRSSSRTPSDKYC	Human neuroblastoma	Doxorubicin	Inhibiting tumor growth	Pastorino et al. (2003)
GGCNGRC	Human fibrosarcoma Human breast cancer	Paclitaxel		Luo et al. (2013)
Others				
Anginex	HUVEC	Gd-DTPA	Increased *in vitro* binding	Brandwijk et al. (2007)
(galectin-1)	HUVEC	Anti-VEGFR2 siRNA	Decreased protein expression	Yousefi et al. (2014)
APRPG (VEGFR1)				
GPLPLR (MT1-MMP)	Murine colorectal cancer	DPP-CNDAC (PDT)	Increased accumulation (PET) and therapeutic effect	Kondo et al. (2004)
$RwP^{N\text{-}Me}PwLM$ (G-protein coupled receptor)	HUVEC	ODN against Bcl-2	Increased *in vitro* binding, but not delivery efficacy	Santos et al. (2010)

KRLYCKNGGFFLRIHPDGR	Murine melanoma	Doxorubicin	Increased and prolonged survival rate	Chen et al. (2010)
VDGVREKSDPHIKLQLQA				
EERGVVSIKGVCANRYLA				
MKEDGRLLASKCVTDEC				
FFFERLESNNYNTY				
(FGFR)				
KDEPQRRSARLSAKPAPPKPEPKPKKAPAKK (nucleolin)	Human prostate cancer	Anti-PLK1 siRNA	Increased *in vitro* cell killing effect	Gomes-da-Silva et al. (2013)
Apelin 14, sequence is not clarified (APj)	Murine melanoma, murine colorectal cancer	Florescence	Accumulation in tumor vasculature	Kawahara et al. (2013)
CRPPR (NRP-1)		Doxorubicin	Inhibiting tumor growth	Paoli et al. (2014)
ARKRLDRNC (IL4R)	Human lung cancer	Doxorubicin	Accumulation in tumor vasculature	Chi et al. (2015)
YSAYPDSVPMMSK (EphA)	Human breast cancer	Doxorubicin	Inhibiting primary tumor growth and metastasis	Guo et al. (2015)
CGNSNPKSC (GX1)	Human gastric cancer	Adenovirus	Expression of marker gene, EGFP	Xiong et al. (2015)

8.2.2.1 RGD Motif

RGD (Arg-Gly-Asp) peptides are one of most popular ligands for targeting tumor endothelial cells. The RGD motif is recognized by some types of integrins, which are cell adhesion receptors (Barczyk et al. 2010). The RGD motif binds particularly well to $\alpha_v\beta_3$ integrin heterodimers, which are highly expressed in angiogenic tumor endothelial cells and some types of cancer cells (Arosio and Casagrande 2015). Gurrath et al. reported that a three dimensional conformation of the RGD-containing peptide is important for the strong inhibition of integrins and their receptors, such as vitronectin and laminin (Gurrath et al. 1992). They identified cyclized Arg-Gly-Asp-*D*-Phe-Val (cyclic RGDfV) as a superior ligand for integrins because the conformation of the molecule is fixed by cyclization, resulting in an accurate match of the three-dimensional structure. To modify liposomes with a cyclic RGD peptide, they altered valine, lysine or cysteine residues. These altered sequences were covalently bound to the head of a PEG lipid via a carboxyl group or a maleimide group, respectively. Thus, cyclized forms appear to be more favorable structures in most cases.

Schiffelers et al. reported that DOX-loaded cyclized RGDf(ε-S-acetylthioacetyl) K liposomes are localized in tumor endothelial cells after the systemic injection of the liposome into a C26 murine colon carcinoma model, and that the RGD liposome inhibited tumor growth more efficiently than the PEGylated liposome (Schiffelers et al. 2003). Using the same sequence, Koning et al. attempted neutron capture therapy (NCT) (Koning et al. 2004). NCT is a type of radiotherapy, in which thermal neutrons are captured by a boron 10 (^{10}B) compound, with the release of alpha particles. The released alpha particles have such a high energy that adjacent cells are damaged. After internalization of the RGD-modified ^{10}B liposomal formulations and subsequent radiation, 75.4 % of the human umbilical vein endothelial cell (HUVEC) were found to be dead in comparison to 8.3 % in the presence of the free ^{10}B compound. Höliq et al. newly identified the GARYCRGDCFDG sequence though a phage display method against HUVEC and melanoma cells (Holig et al. 2004). They also confirmed that this sequence was equal to the original cyclic RGD in its affinity for targeting cells. Likewise, when GARYCRGDCFDG-modified liposomal paclitaxel was injected into a human lung cancer model, CD34-positive tumor endothelial cells were diminished (Meng et al. 2011). Kim et al. combined cyclic RGD-mediated targeting with thermosensitive liposomes (Kim et al. 2014). The authors modified DOX-loaded liposomes with a thermosensitive peptide, an elastin-like molecule, which shrinks in size above a certain temperature. In that report, they demonstrated that the release of doxorubicin is facilitated in an *in vitro* study and an increased accumulation in tumor tissue. In addition to these chemotherapy and radiotherapy procedures, many groups have reported on the delivery of nucleic acids. Pramanik et al. reported that liposomes containing a lipopeptide, which consists of RGDK tetrapeptides conjugated with two alkyl chains, efficiently delivered pDNA to $\alpha_5\beta_1$ integrin positive endothelial cells (Pramanik et al. 2008). In that study, they reported the delivery of a pDNA coding anti-cancer gene, p53, and thus succeeded in demonstrating an anti-tumor effect. In another report, the same group selected curcumin as a cargo (Mondal et al. 2013), which is known to have anti-angiogenic effects via several pathways (Aggarwal et al. 2006; Shishodia et al. 2005). As a result, curcumin-loaded RGDK-modified liposomes significantly inhibited tumor growth. They also revealed that the co-encapsulation of curcumin and doxorubicin into RGDK liposomes synergistically inhibited tumor growth (Barui et al. 2014). Madjumder et al. also studied the effect of the down-regulation of a cell division cycle homologue (CDC) 20, which is a highly conserved protein that is involved in cell cycle regulation. (Majumder et al. 2014). To suppress the expression of CDC20, they used siRNA. When liposomal formulations of anti-CDC20 siRNA were injected into B16F10-bearing mice, a robust inhibition in tumor growth was observed. They then clarified that the anti-CDC20 siRNA induced apoptosis of endothelial cells by the TUNEL staining of tumor sections. Liu et al. focused on anti-micro RNA (miRNA) oligonucleotides (AMOs). MiRNA molecules had a partially complementary sequence against several endogenous mRNAs, and thereby down-regulated or up-regulated some of the signal cascades. A number of miRNA preparations are known to be involved in cancer progression (Calin and Croce 2006). These investigators attempted the delivery of AMO against miRNA-296 (miR-296). MiR-296 plays a key role in angiogenesis via regulating the expression of the hepatocyte growth factor-regulated tyrosine kinase substrate (HGS), which degrades the VEGF receptor 2 (VEGFR2) and the platelet-derived growth factor receptor β (PDGFRβ) (Wurdinger et al. 2008). After the injection of the formulated AMOs, the CD31-positive cells in a HUVEC/matrigel plug decreased, accompanied by a decreased HGS expression. We also succeeded in delivering siRNA to tumor endothelial cells using the RGD peptide and our original lipid, YSK05. YSK05 is a cationic lipid, which induces membrane fusion between a liposome membrane and the endosome membrane responds to acidification after internalization, with the siRNA consequently being effectively released to the cytosol (Sato et al. 2012). With an acid dissociation constant (p*Ka*) of 6.5, the YSK05-containing liposome is neutral under physiological conditions, which allows YSK05-containing liposomes to evade immunosurveillance. Thus, siRNA-loaded YSK-liposomes can inhibit the expression of a target gene in the liver (Hayashi et al. 2014; Sato et al. 2015) and in cancer (Sakurai et al. 2013, 2014a). For targeting tumor endothelial cells, we modified the YSK-liposome cyclic RGD peptide (RGD-MEND). When the RGD-MEND is systemically injected into human renal cell carcinoma-bearing mice at a dose of 3.0 mg/kg, CD31 expression was significantly inhibited (Sakurai et al. 2014b, 2016). The continuous injection of the RGD-MEND encapsulating siRNA against VEGFR2 resulted in suppressed tumor growth and a decrease in the density of micro vessels (Sakurai et al. 2014b). Moreover, we succeeded in maximizing the siRNA delivery efficacy of the RGD-MEND by optimizing RGD modification methodology (Hada et al. 2015).

Some groups have investigated methods for improving the binding of cyclic RGD to the surface of liposomes to target αvβ3 integrin, which is dependent on multimeric interaction. In tumor tissue imaging via cyclic RGD, a radioisotope-cyclic RGD tetramer was more strongly bound to $\alpha_v\beta_3$ integrin than the radioisotope-cyclic RGD monomer and dimer (Shi et al. 2011; Wu et al. 2005), based on near-infrared fluorescence imaging (Wu et al. 2006). We revealed that the dissociation constant (K_D) large-sized liposome (>300 nm) was approximately 10-times higher than that of a normal-sized one (=100 nm) (Kibria et al. 2013). Ligand ability is highly dependent on the character of the liposome being used, and therefore formulations need to be optimized in order to achieve the maximum therapeutic effect.

8.2.2.2 The NGR Motif

The NGR motif was first identified by an *in vivo* phage display (Arap et al. 1998). In this report, they discovered that a CNGRCVSGCAGRC-expressing phage could accumulate in the tumor vasculature, and that binding was inhibited in the presence of a CNGRC peptide truncated from the whole sequence. Moreover, when DOX was conjugated with the CNGRC peptide, the survival rate of tumor bearing mice was increased. Pasqualini et al. identified animonopeptidase N (CD13) as a receptor of CNGRC (Pasqualini et al. 2000). Negussie et al. synthesized the cyclic KNGRE peptide in order to link the NGR motif to a PEG lipid (Negussie et al. 2010). The NGR-motif conjugate facilitated the internalization of conjugate-modified liposomes in a CD13-posirive cell line. Although there are a number of reports on NGR peptide-mediated delivery, most failed to determine which is the actual target, cancer cells or tumor endothelial cells. This might be because aminopeptidase N is also expressed, not only in tumor endothelial cells, but in cancer cells as well. Pastorino et al. used the long NGR-containing peptide GNGRGGVRSSSRTPSDKYC as a ligand to tumor endothelial cells (Pastorino et al. 2003). They showed that systemically injected NGR-modified DOX-loaded liposomes were delivered to tumor endothelial cells in a human orthotropic neuroblastoma model, and TUNEL staining clarified that the injection induced the apoptosis of endothelial cells, not cancer cells. Another group suggested that the PTX-loaded CNGRC-modified liposome likewise inhibited tumor progression via the disruption of tumor endothelial cells (Luo et al. 2013).

8.2.2.3 Others

Anginex

Griffioen et al. screened peptide sequences that have an inhibitory effect on angiogenesis (angiostatic) based on the fact that known anti-angiogenic proteins, including endostatin, platelet factor-4, TNF and the bactericidal-permeability increasing protein commonly have a β-sheet structure (Griffioen et al. 2001). As a result of the screening, they identified ANIKLSVQMKLFKRHLKWKIIVKLNDGRELSLD as an endothelial-specifically cytotoxic peptide. A target protein of Anginex is Galectin-1, a glycoprotein that engages in cancer cell proliferation and angiogenesis via dimerization (Thijssen et al. 2015; Thijssen et al. 2006). As a reslt of a comparison of Anginex-modified liposomes with RGD-modified liposomes in delivering a magnetic resonance imaging (MRI) contrast agent, chelateing Gd^{3+}, meculumine gadopentate (Gd-DTPA) to HUVEC, Anginex-modifed liposomes could be used to distinctly visualize HUVEC pellets (Brandwijk et al. 2007). They also succeeded in delivering anti-VEGFR2 siRNA to HUVEC, whose efficacy was superior to the commercially available reagent, Lipofectamine2000® (Yousefi et al. 2014). Additionally, they showed that Anginex peptides themselves inhibited tumor growth (van der Schaft et al. 2002).

APRPG Peptide

The APRPG peptide was identified by Oku et al. as a consensus sequence in a phage display library against tumor xenografts (Oku et al. 2002). They first synthesized a APRPG-PEG-distearoyl-*sn*-gryceroethanolamine (DSPE) conjugate, and this conjugate permitted the liposome to accumulate in tumor tissue after systemic injection (Maeda et al. 2004a). An APRPG-liposome formulated with DOX significantly suppressed tumor growth in a murine melanoma model (Maeda et al. 2004b), a DOX-resistant murine leukemia model (Shimizu et al. 2005), a human orthotropic pancreatic model (hypovascular) (Yonezawa et al. 2007). Maeda et al. revealed that an APRPG-modified liposome encapsulating DOX accumulated in tumor endothelial cells (Maeda et al. 2006). They also carried out anti-angiogenic therapy successfully using PDT therapeutics BPD-MA (Ichikawa et al. 2005), cytotoxic nucleotide CNDAC-palmitoyl derivative (DPP-CNDAC) (Asai et al. 2002, Asai et al. 2008) and VEGFR2 phosphorylation inhibitor SU5416 (Katanasaka et al. 2008b) and SU1498 (Katanasaka et al. 2008a). Moreover, they attempted to deliver nucleic acids using their original cationic lipid, dicetyl phosphate-tetraethylenepentamine (DCP-TEPA). Asai et al. examined a complex of cholesterol-conjugated siRNA and a DCP-TEPA based liposome (TEPA-PCL) with an APRPG peptide and found that it could be delivered to tumor tissue after administration (Asai et al. 2011). They reported an siRNA against mTOR in tumor tissue (Koide et al. 2011) and a metastatic site (Koide et al. 2015). In addition to siRNA, the delivery of miRNA was also studied. When a murine colorectal cancer model was treated with miR-499 encapsulated in TEPA-PCL, tumor growth was substantially suppressed (Ando et al. 2014).

Kondo et al. delivered the PDT therapeutic DPP-CNDAC to tumor endothelial cells using substrate sequences of membrane type-1 matrixmetalloproteinase (MT1-MMP)-modified liposomes (Kondo et al. 2004) MT1-MMP is involved in the activation of MMP-2, which plays a key role in invasion and angiogenesis (Itoh and Seiki 2006). They conjugated stearic acid with a consensus substrate for MT1-MMP, GPLPLR peptides. Positron emission tomography (PET) imaging revealed that the accumulation of GPLPLR-liposomes was increased in tumor tissue. When murine colorectal cancer-bearing mice were treated with DPP-CNDAC-loaded GPLPLR-liposomes, a substantial delay in tumor growth was observed.

Santos et al. modified pH-sensitive liposomes with the G-protein coupled receptor antagonist hexapeptide, RwPN-MePwLM (lower cases mean a D-form), and encapsulated ODNs against Bcl-2 (Santos et al. 2010). They reported that peptide modification resulted in cellular binding, but not efficient silencing. Chen et al. studied a truncated 86 mer peptide against the human FGF receptor (FGFR) (Chen et al. 2010). They succeeded in inhibiting tumor growth by injecting DOX-loaded liposomes, but they did not demonstrate which cells are the fundamental target, i.e., cancer cells or tumor endothelial cells. Gomes-daSilva et al. reported that the F3 peptide targeting nucleolin, which is expressed at high levels in cancer cells and tumor endothelial cells, enhanced the efficiency of delivery of anti polo-like kinase 1 (PLK1) (Gomes-da-Silva et al. 2013). Kawahara et al. investigated apelin-modified liposomes, which were produced from endothelial cells and controlled the arrangement of the vasculature (Baffert et al. 2006). Short form apelin (13 mer)-modified liposomes were taken up by NIH3T3 cells that had been transfected with an apelin receptor (APJ). They also revealed that apelin-modified liposomes were delivered to the tumor vasculature of murine melanoma and murine colorectal cancer tissue (Kawahara et al. 2013). Paoli et al. reported that a CRPPR peptide with a high affinity against neropilin-1 (NRP-1) (Paoli et al. 2014), which was identified by phage display screening by Teesalu et al. (Teesalu et al. 2009). NRP-1 is a key mediator for tumor angiogenesis, and is expressed in tumor endothelial cells at high levels (Chaudhary et al. 2014). CRPPR-modified DOX-loaded liposomes succeeded in drastically inhibiting breast cancer growth. Chi et al. reported on the interleukin (IL)-4 receptor (IL4R) –targeted delivery (Chi et al. 2015). IL4R is known to be expressed in some types of cancer. In their study, they showed that IL4R is expressed in the tumor vasculature in TNF-α activated HUVEC and clinical patient specimens, and that IL4R-targeted liposomes were superior to DOX-loaded conventional liposomes. Guo et al. developed an ephrin type-A receptor (EphA2)-mediated DDS for targeting both cancer cells and tumor endothelial cells (Guo et al. 2015). EphA2 is a 130 kDa, transmembrane glycoprotein that is overexpressed in the angiogenic vasculature (Brantley-Sieders et al. 2005). They modified DOX-loaded liposomes with YSAYPDSVPMMSK, which was found to function as a ligand for EphA2 by a phage display, and found a substantial inhibition of breast cancer growth. Xiong et al. developed a unique DDSs. They encapsulated the adenovirus vector into anionic liposomes modified with CGNSNPKSC (GX-1), which was discovered as a consensus sequence binding to the human gastric tumor vasculature (Xiong et al. 2015). They demonstrated that adenovirus-loaded anionic GX-1-liposomes were efficiently taken up by HUVEC. Thus, a number of peptide-mediated DDSs for anti-angiogenic therapy have now been developed.

8.2.3 Nucleic Acids Aptamer

Nucleic acid aptamers are DNA or RNA, that can bind to a target receptor through assembling into a characteristic secondary structure. Aptamers have several advantages, in that they are less toxic and immunogenic compared to antibodies or peptides (Keefe et al. 2010). In most cases, Aptamers are identified by the Systematic

Table 8.3 Aptamers for targeting tumor endothelial cells

Name	Target	Cancer type	Effect	Ref
ESTA	E-selectin	Human breast cancer	Not yet	Mann et al. (2010, 2011)
AraHH001	Troponin T	Human renal cell carcinoma	Not yet	Ara et al. (2012, 2014a, b)

Evolution of Ligands by Exponential enrichment (SELEX) method, in which sequences with a high affinity for target proteins or cells are discovered from combinatory libraries by repeated *in vitro* selection procedures. Two different groups independently reported on this methodology in the early 1990s (Ellington and Szostak 1990; Tuerk and Gold 1990). Aptamers for anti-angiogenic therapy have already been approved as pegaptanib (Macugen® from Pfizer) in 2004 by the FDA, which is used against the patients with age-related macular degeneration. Thus, aptamers are expected to be a new class of therapeutics themselves or supportive devices for DDSs as shown in Table 8.3. Man et al. identified DNA aptamers against E-selectin (ESTA), in which a phosphodiester bond was altered with a phosphorothioate in order to increase binding affinity and protect the sequence from the action of nucleases (Mann et al. 2010). They proved that ESTA was taken up by human microvascular endothelial cells with tetracycline-inducible E-selectin in the presence of doxycycline. They then coupled ESTA with PEGylated liposomes and performed *in vivo* experiments with a human breast cancer model (Mann et al. 2011). When fluorescently labeled ESTA-liposomes were systemically injected into MDA-MB-231 bearing mice, ESTA-liposomes were co-localized in E-selectin positive tumor endothelial cells. We also discovered DNA aptamers with a high affinity for tumor endothelial cells by the cell-based SELEX method, in which we used isolated tumor endothelial cells and skin endothelial cells as a positive and negative control, respectively (Ara et al. 2012). Hida et al. isolated endothelial cells from tumor tissue and established a method for their culture (Hida et al. 2004), and kindly provided us with two cell lines. Of these identified sequences, AraHH001 had a strong binding affinity (K_D 43.8 nM ± 13.7 nM) and inhibited tube formation in tumor endothelial cells. We then explored a target receptor of AraHH001 using a peptide mass fingerprinting assay. As a result of matrix assisted laser desorption ionization – time of flight – mass spectrometry, we identified troponin T (Ara et al. 2014a), which is expressed in cardiomyocytes at physiological conditions as a target protein (Giannoni et al. 2009). This finding suggested that troponin T could be a specific marker for the presence of cancerous tissue in the body, and that such methodology could paradoxically contribute to the identification of a new marker. To assess the ability of AraHH001 as a ligand for tumor endothelial cells, we coupled AraHH001 with a PEG-lipid conjugate via a thioester bond. When liposomes are modified with AraHH001 aptamers, the liposomes tended to accumulated in the tumor endothelium, not in the extracellular space or in cancer cells (Ara et al. 2014b).

Table 8.4 Antibody conjugated liposomal carriers for angiogenic endothelial cells

Antibody	Cancer type	Therapeutics	Effect	Ref
$\alpha_V\beta_3$ integrin				
LM609	Rabbit cancer	MRI contrast agent	MRI vialization	Sipkins et al. (1998)
VEGFR2				
3G2	NIH-3 T3 transfectant	Fluorescence	*in vitro* binding	Benzinger et al. (2000)
DC101	Human colorectal cancer and human breast cancer	Doxorubicin	Inhibiting tumor growth via anti-angiogenic effct	Wicki et al. (2012)
scFv	Human melanoma and leukemia	Fluorescence	*in vitro* internalization	Rubio Demirovic et al. (2005)
VCAM1				
MK271	Humanb colorectal cancer	Fluorescence	Accumulation in tumor tissue	Gosk et al. (2008)
E-selectin				
H18/7	HUVEC	Doxorubicin	*in vitro* cell killing effect	Spragg et al. (1997)
E-selectin antibody	Ehrlich ascites	Fluorescence	Fuorescent imaging	Hirai et al. (2010a)
SAINT-O-Some	Activated HUVEC	siRNA	*in vitro* silencing effect	Asgeirsdottir et al. (2010), Kowalski et al. (2013, 2014); Leus et al. (2014a, b), and van der Gun et al. (2007)

8.2.4 Antibodies

Antibody-conjugated liposomes (immunoliposomes) are also used for tumor endothelial cell targeting. To our knowledge, a monoclonal antibody against $\alpha_V\beta_3$ integrin (clone LM609) conjugated liposome is the first report of an *in vivo* antibody targeting (Sipkins et al. 1998). In this report, the authors encapsulated Gd^{3+} with a chelator lipid, and visualized tumor tissue by MRI. In this section, we introduce examples of endothelial targeting through antibody conjugation, as summarized in Table 8.4.

8.2.4.1 VEGFR2

VEGFR2 plays a dominant role in tumor pathological angiogenesis (Brekken and Thorpe 2001), and is highly expressed in angiogenic tumor endothelial cells. Benzinger et al. conjugated an antibody against VEGFR2 (clone 3G2) with liposomes (3G2-IL), and performed *in vitro* studies with CHO cells that stably express VEGFR2. The uptake of 3G2-IL is increased in VEGFR2-expressing CHO cells, but not in normal CHO cells (Benzinger et al. 2000). This increase was inhibited by the addition of free VEGF, indicating that the uptake of 3G2-IL is mediated by VEGFR2. The anti-angiogenic ability of VEGFR2 antibody conjugated liposomes were assessed in a xenograft model and a spontaneous model. Wicki et al. concluded that a DOX-loaded antiVEGFR2 antibody (clone DC101) conjugated-liposome can inhibit tumor growth, and that the inhibition was accompanied by a decrease in the density of micro vessels and the proliferation of tumor endothelial cells in RipTag2 mice (a spontaneous pancreatic cancer model), PyMT mice (a spontaneous breast cancer model) and HT-29-bearing mice (a human colorectal cancer xenograft) (Wicki et al. 2012). Another group identified single chain Fv (scFv) against VEGFR2 (Rubio Demirovic et al. 2005).

8.2.4.2 VCAM-1

The vascular cell adhesion molecule-1 (VCAM-1) is an immunoglobulin-like type I membrane protein glycoprotein, which is observed in inflammatory sites, tumor tissue and pro-metastatic niches (Schlesinger and Bendas 2015). Gosk et al. conjugated the VCAM-1 antibody (clone MK271) with liposomes via a cyanur group (Gosk et al. 2008). When MK-271-conjugated liposomes were systemically injected into human colorectal cancer Colo677-bearing mice, the liposomes accumulated in the tumor CD31 positive endothelium, not in normal tissue (lung and liver).

8.2.4.3 E-Selectin

Inflammatory cytokines, such as tumor necrosis factor-α (TNF-α), interleukin-1α (IL-1α), or -1β, induce E-selectin expression. In normal tissue, E-selectin plays a role in leukocyte adhesion and on endothelial cells in an inflammatory cite via interactions between E-selectin and its ligand, such as Siaryl Lewis X (SLX) (Bird et al. 1997). On the other hand, E-selectins, which are located on the surface of tumor endothelial cells, are involved in metastasis in target organs, such as colorectal and breast cancer (Eichbaum et al. 2004; Ye et al. 1995). Thus, the specific targeting of E-selectin by an antibody can be used for selective delivery to tumor endothelial cells. It was reported that an anti-E-selectin murine antibody (H18/7)-conjugated liposome can deliver DOX to IL-1β activated HUVEC, and induced cell death (Spragg et al. 1997). In the article, they compared four types of liposomes (classical, PEGylated, cationic, pH-sensitive) with respect to binding ability. Of these, the classical liposome had the most enhancing effect by antibody conjugation. On the other hand, cationic liposomes were taken up, even in non-activated HUVEC. Considering *in vivo* experiments, anti-E-selectin mouse whole antibody conjugation to a liposome encapsulating fluorescence dye, Cy3 or Cy5.5, was used to visualize tumor tissue by *in vivo* imaging (Hirai et al. 2010a).

An antibody-mediated liposomal delivery has also been developed, although its target is inflammatory endothelial cells by cytokines, not tumor endothelial cells. A cationic amphilphlic lipid, 1-methyl-4-(*cis*-9-dioleyl)methyl-pyridinium chloride (SAINT) was designed by van der Gun BT et al. as a protein carrier (van der Gun et al. 2007). They revealed that a combination of an SAINT-containing siRNA lipoplex (SAINT-O-Somes) and an antibody against E-selectin inhibited gene expression, even in HUVEC (Asgeirsdottir et al. 2010), which remains difficult for siRNA transfection by liposomal transfection reagents in *in vitro* circumstances (Martin and Murray 2000). They argued that SAINT-O-Some conjugated with VCAM-1 also induced gene silencing in TNF-α activated HUVEC and human aortic endothelial cells (HAEC) (Kowalski et al. 2013). When they systemically injected TNF-α activated mice with SAINT-O-Somes, the SAINT-O-Somes accumulated in inflamed endothelial cells (Kowalski et al. 2014). Endothelial gene expression was measured by isolating endothelium from tissue sections using laser microdissection. As a result, endothelial cell specific gene vascular endothelial cell cadherin Ve-cadherin and RelA expression was suppressed by the SAINT-O-Somes injection. They then developed PEGylated SAINT-O-Somes and attempted to use them in an *in vivo* circumstance. PEGylated SAINT-O-Somes suppressed VE-cadherin production in TNF-α activated HUVEC, although its silencing effect was slightly weaker than the non-PEGylated preparation (Leus et al. 2014b). However, when TNF-α activated mice were intravenously injected with PEGylated SAINT-O-Somes, no silencing effect was observed in lungs or liver endothelial cells in spite of a long circulation time and a high accumulation of PEGylated SAINT-O-Somes (Leus et al. 2014a). They argued that PEGylation inhibited intracellular trafficking through membrane fusion between the SAINT-O-Somes and endosomal membranes. Although, as of this writing, they did not apply SAINT-O-Somes to the delivery of siRNA to tumor endothelial cells, we hope that their carrier successfully induced effective gene silencing in tumor endothelial cells, which are present in inflammatory conditions, for anti-angiogenic therapy.

8.2.5 Sialyl Lewis^X

Sialyl Lewis^X (SLX) is a tetrasaccharide, which consists of N-acetylneuraminic acid, two galactoses, N-acetylglucosamine. SLX was first discovered as a human gastric cancer antigen, which recognizes E-selectin on the surface of endothelial cells, specifically under inflammatory conditions (Kannagi 2007).

According to these properties of SLX, it was reported that SLX lipid derivatives on the surface of liposomes can efficiently inhibit the binding of SLX or SLX-expressing cells to E-selectin and E-selectin expressing cells (Bruehl et al. 2001; Zeisig et al. 2004). Minematsu et al. revealed that a colloidal gold-loaded SLX-liposome accumulated in tumor endothelial cells, but not in cancer cells after systemic injection (Minematsu et al. 2011). There are some reports concerning anti-tumor effects of SLX-liposomes. SLX-liposomes encapsulating the glycoglycerolipid monogalactosyl diacylglycerol, which selectively diminished DNA replication exclusively in cancer cells, is an example (Mizushina et al. 2012). The authors showed that the systemic injection of glycoglycerolipid monogalactosyl diacylglycerol -loaded SLX-liposomes inhibited tumor growth in a human colorectal cancer model without a decrease in body weight. Another group reported that the injection of cisplatin-loaded SLX-liposomes led to a substantial inhibition in tumor growth, and that cisplatin accumulation in tumor tissue was significantly increased by SLX modification (Hirai et al. 2010b). Regarding the intracellular trafficking of SLX-liposomes, it was reported that TNF-α activated HUVEC, which is the main population of SLX-liposomes is rapidly taken up via a clathrin-independent pathway (Alekseeva et al. 2015). In addition, after the internalization of the SLX-liposomes. they accumulate in the Golgi apparatus, and then move to the endoplasmic reticulum. In the near future, SLX-liposomes promise to function as the best types of anti-angiogenic therapy.

8.2.6 *Dual-Targeting*

Dual-targeting is a general term for describing nanocarriers that are modified with two ligands that can interact with different receptors, as summarized in Table 8.5. Takara et al. proposed a new concept of dual targeting, as shown in Fig. 8.1 (Sakurai et al. 2015; Takara et al. 2010). In their study, they modified liposomes with both

Table 8.5 Dual ligand systems for targeting tumor endothelial cells

Ligand 1	Target 1			
Ligand 2	Target 2	Cancer type	Effect	Ref
Tetraarginine (R4)	Electrostatic	Human renal cell carcinoma	Tumor growth inhibition	Takara et al. (2010, 2012)
CYGGRGNG	CD13			
Octaarginine (R8)	Electrostatic	Human renal cell carcinoma	*in vitro* uptake	Kibria et al. (2011b)
Cyclic RGD	$\alpha_V\beta_3$ integrin			
R8	Electrostatic	Human renal cell carcinoma	*in vitro* uptake	Kibria et al. (2011a)
KYND	Tumor endothelial marker 8			
Cyclic RGD	$\alpha_V\beta_3$ integrin	Murine melanoma	*in vivo* targeting (MRI contrast agent and fluorescence)	Kluza et al. (2012)
Anginex	Galectin-1			
APRPG	VEGFR1	Colorectal cancer	Tumor growth inhibition	Murase et al. (2010)
CNGRG	CD13			
APRPG	VEGFR1	Colorectal cancer	*in vitro* binding	Sugiyama et al. (2013)
GRGDS	$\alpha_V\beta_3$ integrin			

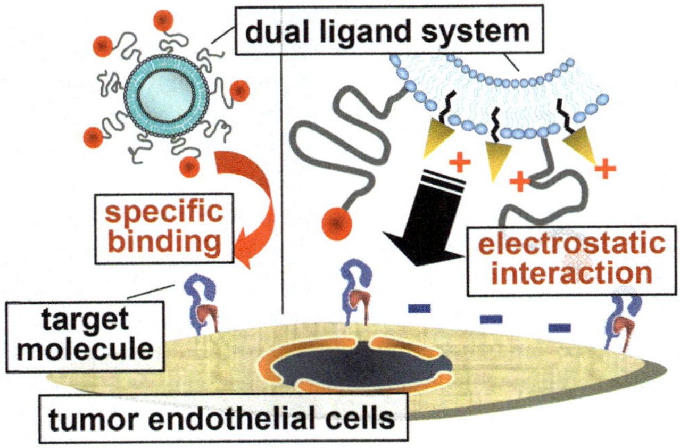

Fig. 8.1 Schematic diagram of dual-targeting liposomes by Hatakeyama et al. The dual-ligand system is equipped with oligoarginine and a specific ligand. Liposomes initially approach the target cells via association between a specific ligand and the target receptor, and oligoarginine on the surface of the liposome then forces the liposome to be internalized through electrostatic interactions (Reproduced from Sakurai et al. 2015 with permission from The Royal Society of Chemistry)

ligand peptides for targeting and oligoarginine peptides (RX; X = 2 ~ 8), which are known to be cell penetrating peptides (CPPs) (Raucher and Ryu 2015). Their dual-ligand liposomes first approached a target cell via an interaction between ligand peptides and target receptors on the cells, after which they become strongly internalized into target cells through electrostatic interactions between oligoarginine and cellular membranes. The uptake of NGR/R4 dually modified liposomes was observed only in CD13-positive tumor endothelial cells, not in CD13-negative skin endothelial cells. They then indicated that size controlled (>300 nm) dual-ligand liposomes exhibited a better anti-angiogenic effect in renal cell carcinoma models (Takara et al. 2012). They also attempted to combine octaarginine peptides (R8) with cyclic RGD peptides (Kibria et al. 2011b) or KYND peptides (Kibria et al. 2011a). The KYND peptides bind to tumor endothelial marker 8, which is specifically expressed in tumor endothelial cells (Bradley et al. 2001).

Kluza et al. attempted to combine Anginex peptides and cyclic RGD peptides for use as an MRI contrast agent, Gd-DTPA (Kluza et al. 2012). Anginex/cyclic RGD dually modified liposomes delivered larger amounts of Gd-DTPA in comparison with single modified liposomes. Murase et al. reported that DOX-loaded liposomes modified with both APRPG and CNGRC peptides accumulated in tumor tissue to a greater extent and exhibited a better therapeutic effect in a murine colorectal cancer model (Murase et al. 2010). They then explored the issue of whether dual ligands truly contributed to the synergistic binding to the target cells. Sugiyama et al. reported that dual-targeting liposomes (APRPG and GRGDC peptides) were more strongly associated with immobilized target protein than single-targeting liposomes by surface plasmon resonance (Sugiyama et al. 2013).

8.3 Impact of Tumor Endothelial Cells on Tumor Microenvironments

Anti-angiogenic therapy can contribute to cancer therapy via the direct suppression of neo vasculature formation, but also via an indirect improvement of chemotherapeutics concomitant with anti-angiogenic drugs. Wildiers et al. suggested that anti-angiogenic therapy facilitates the delivery of cytotoxic agents to tumor tissue (Wildiers et al. 2003). They measured the concentration of CPT-11 in the presence and absence of anti-VEGF therapy using an antibody. The CPT-11 concentration in the co-treatment group was significantly higher than the CPT-11 single treatment group, and the co-treatment decreased tumor growth in a HT29 colon cancer model in a synergistic manner. Such an enhancing effect was also observed in clinical trials. In Phase Ib trials, a combination therapy comprised of 3 mg/kg of intravenously injected anti-VEGF antibody and doxorubicin 50 mg/m^2 or carboplatin 175 mg/m^2 or fluorouracil 500 mg/m^2 with 20 mg/m^2 of leucovorin exhibited a synergistic therapeutic effect on patients with advanced cancer (Margolin et al. 2001). Tong et al. reported that VEGF signal inhibition by anti-VEGFR2 antibody (DC101) decreased interstitial fluid pressure (IFP), which was an inverted pressure from tumor tissue to the tumor vasculature and substantially restricted the EPR-mediated delivery of nanoparticles (Khawar et al. 2015). As a result, tumor blood vessels were rendered mature by the DC101 treatment and the intratumoral penetration of systemically injected albumin increased. Thus, tumor endothelial cells-targeting DDSs altered the tumor microenvironment in EPR-mediated cancer cell-targeting drugs. Okamoto et al. reported that the delivery of mircroRNA-499 by APRPG-liposomes increased tumor blood flow, resulting in a significant facilitation of the DOX therapeutic (Okamoto et al. 2016). These facts suggest that anti-angiogenic therapy synergistically potentiates therapeutic effects with traditional cytotoxic agents. Similarly, Abu et al. indicated that oxaliplatin-loaded cationic liposomes significantly improved the therapeutic effect of subsequent chemotherapy (Abu Lila et al. 2010). On the other hand, it has been suggested that the excessive inhibition of angiogenesis could lead to oxygen shortage (hypoxia), and consequently facilitate cancer metastasis (Ebos and Kerbel 2011; Loges et al. 2009). Moreover, that hypoxic environment affected cellular organelles, including a decrease in the number of mitochondria via hypoxic inducible factor (HIF)-1α (Keunen et al. 2011). Accordingly, a nanocarrier that can control organelle function by delivering therapeutics needs to be developed for a complete understanding of anti-angiogenic therapy and the consequent changes.

Acknowledgements We appreciate Dr. Milton S. Feather for this helpful advice in writing the English manuscript. No potential conflicts of interest were disclosed.

References

Abu Lila AS, Kizuki S, Doi Y et al (2009) Oxaliplatin encapsulated in PEG-coated cationic liposomes induces significant tumor growth suppression via a dual-targeting approach in a murine solid tumor model. J Control Release 137:8–14
Abu Lila AS, Doi Y, Nakamura K et al (2010) Sequential administration with oxaliplatin-containing PEG-coated cationic liposomes promotes a significant delivery of subsequent dose into murine solid tumor. J Control Release 142:167–173
Aggarwal S, Ichikawa H, Takada Y et al (2006) Curcumin (diferuloylmethane) down-regulates expression of cell proliferation and antiapoptotic and metastatic gene products through suppression of IkappaBalpha kinase and Akt activation. Mol Pharmacol 69:195–206
Alekseeva A, Kapkaeva M, Shcheglovitova O et al (2015) Interactions of antitumour Sialyl Lewis X liposomes with vascular endothelial cells. Biochim Biophys Acta 1848:1099–1110
Aleku M, Fisch G, Mopert K et al (2008a) Intracellular localization of lipoplexed siRNA in vascular endothelial cells of different mouse tissues. Microvasc Res 76:31–41
Aleku M, Schulz P, Keil O et al (2008b) Atu027, a liposomal small interfering RNA formulation targeting protein kinase N3, inhibits cancer progression. Cancer Res 68:9788–9798
Ando H, Asai T, Koide H et al (2014) Advanced cancer therapy by integrative antitumor actions via systemic administration of miR-499. J Control Release 181:32–39
Ara MN, Hyodo M, Ohga N et al (2012) Development of a novel DNA aptamer ligand targeting to primary cultured tumor endothelial cells by a cell-based SELEX method. PLoS One 7:e50174
Ara MN, Hyodo M, Ohga N et al (2014a) Identification and expression of troponin T, a new marker on the surface of cultured tumor endothelial cells by aptamer ligand. Cancer Med 3:825–834
Ara MN, Matsuda T, Hyodo M et al (2014b) An aptamer ligand based liposomal nanocarrier system that targets tumor endothelial cells. Biomaterials 35:7110–7120
Arap W, Pasqualini R, Ruoslahti E (1998) Cancer treatment by targeted drug delivery to tumor vasculature in a mouse model. Science 279:377–380
Arosio D, Casagrande C (2015) Advancement in integrin facilitated drug delivery. Adv Drug Deliv Rev 97:111–143
Asai T, Shimizu K, Kondo M et al (2002) Anti-neovascular therapy by liposomal DPP-CNDAC targeted to angiogenic vessels. FEBS Lett 520:167–170
Asai T, Miyazawa S, Maeda N et al (2008) Antineovascular therapy with angiogenic vessel-targeted polyethyleneglycol-shielded liposomal DPP-CNDAC. Cancer Sci 99:1029–1033
Asai T, Matsushita S, Kenjo E et al (2011) Dicetyl phosphate-tetraethylenepentamine-based liposomes for systemic siRNA delivery. Bioconjug Chem 22:429–435
Asgeirsdottir SA, Talman EG, de Graaf IA et al (2010) Targeted transfection increases siRNA uptake and gene silencing of primary endothelial cells in vitro–a quantitative study. J Control Release 141:241–251
Auerbach R (1991) Vascular endothelial cell differentiation: organ-specificity and selective affinities as the basis for developing anti-cancer strategies. Int J Radiat Biol 60:1–10
Awada A, Bondarenko IN, Bonneterre J et al (2014) A randomized controlled phase II trial of a novel composition of paclitaxel embedded into neutral and cationic lipids targeting tumor endothelial cells in advanced triple-negative breast cancer (TNBC). Ann Oncol 25:824–831
Baffert F, Le T, Sennino B et al (2006) Cellular changes in normal blood capillaries undergoing regression after inhibition of VEGF signaling. Am J Physiol Heart Circ Physiol 290:H547–H559
Barczyk M, Carracedo S, Gullberg D (2010) Integrins. Cell Tissue Res 339:269–280
Barui S, Saha S, Mondal G et al (2014) Simultaneous delivery of doxorubicin and curcumin encapsulated in liposomes of pegylated RGDK-lipopeptide to tumor vasculature. Biomaterials 35:1643–1656
Benzinger P, Martiny-Baron G, Reusch P et al (2000) Targeting of endothelial KDR receptors with 3G2 immunoliposomes in vitro. Biochim Biophys Acta 1466:71–78

Bird MI, Foster MR, Priest R et al (1997) Selectins: physiological and pathophysiological roles. Biochem Soc Trans 25:1199–1206

Bradley KA, Mogridge J, Mourez M et al (2001) Identification of the cellular receptor for anthrax toxin. Nature 414:225–229

Brandwijk RJ, Mulder WJ, Nicolay K et al (2007) Anginex-conjugated liposomes for targeting of angiogenic endothelial cells. Bioconjug Chem 18:785–790

Brantley-Sieders DM, Fang WB, Hicks DJ et al (2005) Impaired tumor microenvironment in EphA2-deficient mice inhibits tumor angiogenesis and metastatic progression. FASEB J 19:1884–1886

Brekken RA, Thorpe PE (2001) Vascular endothelial growth factor and vascular targeting of solid tumors. Anticancer Res 21:4221–4229

Bruehl RE, Dasgupta F, Katsumoto TR et al (2001) Polymerized liposome assemblies: bifunctional macromolecular selectin inhibitors mimicking physiological selectin ligands. Biochemistry 40:5964–5974

Calin GA, Croce CM (2006) MicroRNA signatures in human cancers. Nat Rev Cancer 6:857–866

Carmeliet P, Jain RK (2011) Molecular mechanisms and clinical applications of angiogenesis. Nature 473:298–307

Chaudhary B, Khaled YS, Ammori BJ et al (2014) Neuropilin 1: function and therapeutic potential in cancer. Cancer Immunol Immunother 63:81–99

Chen X, Wang X, Wang Y et al (2010) Improved tumor-targeting drug delivery and therapeutic efficacy by cationic liposome modified with truncated bFGF peptide. J Control Release 145:17–25

Chi L, Na MH, Jung HK et al (2015) Enhanced delivery of liposomes to lung tumor through targeting interleukin-4 receptor on both tumor cells and tumor endothelial cells. J Control Release 209:327–336

Dey N, De P, Brian LJ (2015) Evading anti-angiogenic therapy: resistance to anti-angiogenic therapy in solid tumors. Am J Transl Res 7:1675–1698

Ebos JM, Kerbel RS (2011) Antiangiogenic therapy: impact on invasion, disease progression, and metastasis. Nat Rev Clin Oncol 8:210–221

Eichbaum MH, de Rossi TM, Kaul S et al (2004) Serum levels of soluble E-selectin are associated with the clinical course of metastatic disease in patients with liver metastases from breast cancer. Oncol Res 14:603–610

Ellington AD, Szostak JW (1990) In vitro selection of RNA molecules that bind specific ligands. Nature 346:818–822

Folkman J (1971) Tumor angiogenesis: therapeutic implications. N Engl J Med 285:1182–1186

Folkman J, Long DM Jr, Becker FF (1963) Growth and metastasis of tumor in organ culture. Cancer 16:453–467

Giannoni A, Giovannini S, Clerico A (2009) Measurement of circulating concentrations of cardiac troponin I and T in healthy subjects: a tool for monitoring myocardial tissue renewal? Clin Chem Lab Med 47:1167–1177

Gomes-da-Silva LC, Ramalho JS, Pedroso de Lima MC et al (2013) Impact of anti-PLK1 siRNA-containing F3-targeted liposomes on the viability of both cancer and endothelial cells. Eur J Pharm Biopharm 85:356–364

Gosk S, Moos T, Gottstein C et al (2008) VCAM-1 directed immunoliposomes selectively target tumor vasculature in vivo. Biochim Biophys Acta 1778:854–863

Gospodarowicz D, Brown KD, Birdwell CR et al (1978) Control of proliferation of human vascular endothelial cells. Characterization of the response of human umbilical vein endothelial cells to fibroblast growth factor, epidermal growth factor, and thrombin. J Cell Biol 77:774–788

Griffioen AW, van der Schaft DW, Barendsz-Janson AF et al (2001) Anginex, a designed peptide that inhibits angiogenesis. Biochem J 354:233–242

Guo Z, He B, Yuan L et al (2015) Dual targeting for metastatic breast cancer and tumor neovasculature by EphA2-mediated nanocarriers. Int J Pharm 493:380–389

Gurrath M, Muller G, Kessler H et al (1992) Conformation/activity studies of rationally designed potent anti-adhesive RGD peptides. Eur J Biochem 210:911–921

Gutbier B, Kube SM, Reppe K et al (2010) RNAi-mediated suppression of constitutive pulmonary gene expression by small interfering RNA in mice. Pulm Pharmacol Ther 23:334–344

Hada T, Sakurai Y, Harashima H (2015) Optimization of a siRNA carrier modified with a pH-sensitive cationic lipid and a cyclic RGD peptide for efficiently targeting tumor endothelial cells. Pharmaceutics 7:320–333

Hayashi Y, Suemitsu E, Kajimoto K et al (2014) Hepatic monoacylglycerol O-acyltransferase 1 as a promising therapeutic target for steatosis, obesity, and type 2 diabetes. Mol Ther Nucleic Acids 3, e154

Hida K, Hida Y, Amin DN et al (2004) Tumor-associated endothelial cells with cytogenetic abnormalities. Cancer Res 64:8249–8255

Hirai M, Hiramatsu Y, Iwashita S et al (2010a) E-selectin targeting to visualize tumors in vivo. Contrast Media Mol Imaging 5:70–77

Hirai M, Minematsu H, Hiramatsu Y et al (2010b) Novel and simple loading procedure of cisplatin into liposomes and targeting tumor endothelial cells. Int J Pharm 391:274–283

Holig P, Bach M, Volkel T et al (2004) Novel RGD lipopeptides for the targeting of liposomes to integrin-expressing endothelial and melanoma cells. Protein Eng Des Sel 17:433–441

Ichikawa K, Hikita T, Maeda N et al (2005) Antiangiogenic photodynamic therapy (PDT) by using long-circulating liposomes modified with peptide specific to angiogenic vessels. Biochim Biophys Acta 1669:69–74

Itoh Y, Seiki M (2006) MT1-MMP: a potent modifier of pericellular microenvironment. J Cell Physiol 206:1–8

Kalra AV, Campbell RB (2006) Development of 5-FU and doxorubicin-loaded cationic liposomes against human pancreatic cancer: implications for tumor vascular targeting. Pharm Res 23:2809–2817

Kannagi R (2007) Carbohydrate antigen sialyl Lewis a–its pathophysiological significance and induction mechanism in cancer progression. Chang Gung Med J 30:189–209

Katanasaka Y, Ida T, Asai T et al (2008a) Effective delivery of an angiogenesis inhibitor by neovessel-targeted liposomes. Int J Pharm 360:219–224

Katanasaka Y, Ida T, Asai T et al (2008b) Antiangiogenic cancer therapy using tumor vasculature-targeted liposomes encapsulating 3-(3,5-dimethyl-1H-pyrrol-2-ylmethylene)-1,3-dihydro-indol-2-one, SU5416. Cancer Lett 270:260–268

Kawahara H, Naito H, Takara K et al (2013) Tumor endothelial cell-specific drug delivery system using apelin-conjugated liposomes. PLoS One 8:e65499

Keck PJ, Hauser SD, Krivi G et al (1989) Vascular permeability factor, an endothelial cell mitogen related to PDGF. Science 246:1309–1312

Keefe AD, Pai S, Ellington A (2010) Aptamers as therapeutics. Nat Rev Drug Discov 9:537–550

Keunen O, Johansson M, Oudin A et al (2011) Anti-VEGF treatment reduces blood supply and increases tumor cell invasion in glioblastoma. Proc Natl Acad Sci U S A 108:3749–3754

Khawar IA, Kim JH, Kuh HJ (2015) Improving drug delivery to solid tumors: priming the tumor microenvironment. J Control Release 201:78–89

Kibria G, Hatakeyama H, Harashima H (2011a) A new peptide motif present in the protective antigen of anthrax toxin exerts its efficiency on the cellular uptake of liposomes and applications for a dual-ligand system. Int J Pharm 412:106–114

Kibria G, Hatakeyama H, Ohga N et al (2011b) Dual-ligand modification of PEGylated liposomes shows better cell selectivity and efficient gene delivery. J Control Release 153:141–148

Kibria G, Hatakeyama H, Ohga N et al (2013) The effect of liposomal size on the targeted delivery of doxorubicin to Integrin alphavbeta3-expressing tumor endothelial cells. Biomaterials 34:5617–5627

Kim MS, Lee DW, Park K et al (2014) Temperature-triggered tumor-specific delivery of anticancer agents by cRGD-conjugated thermosensitive liposomes. Colloids Surf B: Biointerfaces 116:17–25

Kluza E, Jacobs I, Hectors SJ et al (2012) Dual-targeting of alphavbeta3 and galectin-1 improves the specificity of paramagnetic/fluorescent liposomes to tumor endothelium in vivo. J Control Release 158:207–214
Koide H, Asai T, Furuya K et al (2011) Inhibition of Akt (ser473) phosphorylation and rapamycin-resistant cell growth by knockdown of mammalian target of rapamycin with small interfering RNA in vascular endothelial growth factor receptor-1-targeting vector. Biol Pharm Bull 34:602–608
Koide H, Asai T, Kato H et al (2015) Susceptibility of PTEN-positive metastatic tumors to small interfering RNA targeting the mammalian target of rapamycin. Nanomedicine 11:185–194
Kondo M, Asai T, Katanasaka Y et al (2004) Anti-neovascular therapy by liposomal drug targeted to membrane type-1 matrix metalloproteinase. Int J Cancer 108:301–306
Koning GA, Fretz MM, Woroniecka U et al (2004) Targeting liposomes to tumor endothelial cells for neutron capture therapy. Appl Radiat Isot 61:963–967
Kowalski PS, Lintermans LL, Morselt HW et al (2013) Anti-VCAM-1 and anti-E-selectin SAINT-O-Somes for selective delivery of siRNA into inflammation-activated primary endothelial cells. Mol Pharm 10:3033–3044
Kowalski PS, Zwiers PJ, Morselt HW et al (2014) Anti-VCAM-1 SAINT-O-Somes enable endothelial-specific delivery of siRNA and downregulation of inflammatory genes in activated endothelium in vivo. J Control Release 176:64–75
Kunstfeld R, Wickenhauser G, Michaelis U et al (2003) Paclitaxel encapsulated in cationic liposomes diminishes tumor angiogenesis and melanoma growth in a "humanized" SCID mouse model. J Invest Dermatol 120:476–482
Leung DW, Cachianes G, Kuang WJ et al (1989) Vascular endothelial growth factor is a secreted angiogenic mitogen. Science 246:1306–1309
Leus NG, Morselt HW, Zwiers PJ et al (2014a) VCAM-1 specific PEGylated SAINT-based lipoplexes deliver siRNA to activated endothelium in vivo but do not attenuate target gene expression. Int J Pharm 469:121–131
Leus NG, Talman EG, Ramana P et al (2014b) Effective siRNA delivery to inflamed primary vascular endothelial cells by anti-E-selectin and anti-VCAM-1 PEGylated SAINT-based lipoplexes. Int J Pharm 459:40–50
Liu XQ, Song WJ, Sun TM et al (2011) Targeted delivery of antisense inhibitor of miRNA for antiangiogenesis therapy using cRGD-functionalized nanoparticles. Mol Pharm 8:250–259
Loges S, Mazzone M, Hohensinner P et al (2009) Silencing or fueling metastasis with VEGF inhibitors: antiangiogenesis revisited. Cancer Cell 15:167–170
Luo LM, Huang Y, Zhao BX et al (2013) Anti-tumor and anti-angiogenic effect of metronomic cyclic NGR-modified liposomes containing paclitaxel. Biomaterials 34:1102–1114
Maeda H, Wu J, Sawa T et al (2000) Tumor vascular permeability and the EPR effect in macromolecular therapeutics: a review. J Control Release 65:271–284
Maeda N, Takeuchi Y, Takada M et al (2004a) Synthesis of angiogenesis-targeted peptide and hydrophobized polyethylene glycol conjugate. Bioorg Med Chem Lett 14:1015–1017
Maeda N, Takeuchi Y, Takada M et al (2004b) Anti-neovascular therapy by use of tumor neovasculature-targeted long-circulating liposome. J Control Release 100:41–52
Maeda N, Miyazawa S, Shimizu K et al (2006) Enhancement of anticancer activity in antineovascular therapy is based on the intratumoral distribution of the active targeting carrier for anticancer drugs. Biol Pharm Bull 29:1936–1940
Majumder P, Bhunia S, Bhattacharyya J et al (2014) Inhibiting tumor growth by targeting liposomally encapsulated CDC20siRNA to tumor vasculature: therapeutic RNA interference. J Control Release 180:100–108
Mann AP, Somasunderam A, Nieves-Alicea R et al (2010) Identification of thioaptamer ligand against E-selectin: potential application for inflamed vasculature targeting. PLoS One 5:e13050
Mann AP, Bhavane RC, Somasunderam A et al (2011) Thioaptamer conjugated liposomes for tumor vasculature targeting. Oncotarget 2:298–304

Margolin K, Gordon MS, Holmgren E et al (2001) Phase Ib trial of intravenous recombinant humanized monoclonal antibody to vascular endothelial growth factor in combination with chemotherapy in patients with advanced cancer: pharmacologic and long-term safety data. J Clin Oncol 19:851–856

Martin SG, Murray JC (2000) Gene-transfer systems for human endothelial cells. stewart.martin@nottingham.ac.uk. Adv Drug Deliv Rev 41:223–233

Matsumura Y, Maeda H (1986) A new concept for macromolecular therapeutics in cancer chemotherapy: mechanism of tumoritropic accumulation of proteins and the antitumor agent smancs. Cancer Res 46:6387–6392

Meng S, Su B, Li W et al (2011) Integrin-targeted paclitaxel nanoliposomes for tumor therapy. Med Oncol 28:1180–1187

Minematsu H, Otani T, Oohashi T et al (2011) Development of an active targeting liposome encapsulated with high-density colloidal gold for transmission electron microscopy. J Electron Microsc (Tokyo) 60:95–99

Mizushina Y, Hada T, Yoshida H (2012) In vivo antitumor effect of liposomes with sialyl Lewis X including monogalactosyl diacylglycerol, a replicative DNA polymerase inhibitor, from spinach. Oncol Rep 28:821–828

Mondal G, Barui S, Saha S et al (2013) Tumor growth inhibition through targeting liposomally bound curcumin to tumor vasculature. J Control Release 172:832–840

Murase Y, Asai T, Katanasaka Y et al (2010) A novel DDS strategy, "dual-targeting", and its application for antineovascular therapy. Cancer Lett 287:165–171

Negussie AH, Miller JL, Reddy G et al (2010) Synthesis and in vitro evaluation of cyclic NGR peptide targeted thermally sensitive liposome. J Control Release 143:265–273

Okamoto A, Asai T, Ryu S et al (2016) Enhanced efficacy of Doxorubicin by microRNA-499-mediated improvement of tumor blood flow. J Clin Med 5:10

Oku N, Asai T, Watanabe K et al (2002) Anti-neovascular therapy using novel peptides homing to angiogenic vessels. Oncogene 21:2662–2669

Paoli EE, Ingham ES, Zhang H et al (2014) Accumulation, internalization and therapeutic efficacy of neuropilin-1-targeted liposomes. J Control Release 178:108–117

Pasqualini R, Koivunen E, Kain R et al (2000) Aminopeptidase N is a receptor for tumor-homing peptides and a target for inhibiting angiogenesis. Cancer Res 60:722–727

Pastorino F, Brignole C, Marimpietri D et al (2003) Vascular damage and anti-angiogenic effects of tumor vessel-targeted liposomal chemotherapy. Cancer Res 63:7400–7409

Pramanik D, Majeti BK, Mondal G et al (2008) Lipopeptide with a RGDK tetrapeptide sequence can selectively target genes to proangiogenic alpha5beta1 integrin receptor and mouse tumor vasculature. J Med Chem 51:7298–7302

Raucher D, Ryu JS (2015) Cell-penetrating peptides: strategies for anticancer treatment. Trends Mol Med 21:560–570

Rubio Demirovic A, Marty C, Console S et al (2005) Targeting human cancer cells with VEGF receptor-2-directed liposomes. Oncol Rep 13:319–324

Sakurai Y, Hatakeyama H, Sato Y et al (2013) Gene silencing via RNAi and siRNA quantification in tumor tissue using MEND, a liposomal siRNA delivery system. Mol Ther 21:1195–1203

Sakurai Y, Hatakeyama H, Akita H et al (2014a) Improvement of doxorubicin efficacy using liposomal anti-polo-like kinase 1 siRNA in human renal cell carcinomas. Mol Pharm 11:2713–2719

Sakurai Y, Hatakeyama H, Sato Y et al (2014b) RNAi-mediated gene knockdown and anti-angiogenic therapy of RCCs using a cyclic RGD-modified liposomal-siRNA system. J Control Release 173:110–118

Sakurai Y, Kajimoto K, Harashima H (2015) Anti-angiogenic nanotherapy via active targeting systems to tumors and adipose tissue vasculature. Biomater Sci 3:1253–1265

Sakurai Y, Hada T, Harashima H (2016) Preparation of a cyclic RGD: modified liposomal SiRNA formulation for use in active targeting to tumor and tumor endothelial cells. Methods Mol Biol 1364:63–69

Santel A, Aleku M, Keil O et al (2006) A novel siRNA-lipoplex technology for RNA interference in the mouse vascular endothelium. Gene Ther 13:1222–1234
Santel A, Aleku M, Roder N et al (2010) Atu027 prevents pulmonary metastasis in experimental and spontaneous mouse metastasis models. Clin Cancer Res 16:5469–5480
Santos AO, da Silva LC, Bimbo LM et al (2010) Design of peptide-targeted liposomes containing nucleic acids. Biochim Biophys Acta 1798:433–441
Sato Y, Hatakeyama H, Sakurai Y et al (2012) A pH-sensitive cationic lipid facilitates the delivery of liposomal siRNA and gene silencing activity in vitro and in vivo. J Control Release 163:267–276
Sato S, Li K, Kameyama T et al (2015) The RNA sensor RIG-I dually functions as an innate sensor and direct antiviral factor for hepatitis B virus. Immunity 42:123–132
Schiffelers RM, Koning GA, ten Hagen TL et al (2003) Anti-tumor efficacy of tumor vasculature-targeted liposomal doxorubicin. J Control Release 91:115–122
Schlesinger M, Bendas G (2015) Vascular cell adhesion molecule-1 (VCAM-1)–an increasing insight into its role in tumorigenicity and metastasis. Int J Cancer 136:2504–2514
Schmitt-Sody M, Strieth S, Krasnici S et al (2003) Neovascular targeting therapy: paclitaxel encapsulated in cationic liposomes improves antitumoral efficacy. Clin Cancer Res 9:2335–2341
Schultheis B, Strumberg D, Santel A et al (2014) First-in-human phase I study of the liposomal RNA interference therapeutic Atu027 in patients with advanced solid tumors. J Clin Oncol 32:4141–4148
Shi J, Zhou Y, Chakraborty S et al (2011) Evaluation of in-labeled cyclic RGD peptides: effects of peptide and linker multiplicity on their tumor uptake, excretion kinetics and metabolic stability. Theranostics 1:322–340
Shimizu K, Asai T, Fuse C et al (2005) Applicability of anti-neovascular therapy to drug-resistant tumor: suppression of drug-resistant P388 tumor growth with neovessel-targeted liposomal adriamycin. Int J Pharm 296:133–141
Shishodia S, Amin HM, Lai R et al (2005) Curcumin (diferuloylmethane) inhibits constitutive NF-kappaB activation, induces G1/S arrest, suppresses proliferation, and induces apoptosis in mantle cell lymphoma. Biochem Pharmacol 70:700–713
Sipkins DA, Cheresh DA, Kazemi MR et al (1998) Detection of tumor angiogenesis in vivo by alphaVbeta3-targeted magnetic resonance imaging. Nat Med 4:623–626
Spragg DD, Alford DR, Greferath R et al (1997) Immunotargeting of liposomes to activated vascular endothelial cells: a strategy for site-selective delivery in the cardiovascular system. Proc Natl Acad Sci U S A 94:8795–8800
St Croix B, Rago C, Velculescu V et al (2000) Genes expressed in human tumor endothelium. Science 289:1197–1202
Strieth S, Eichhorn ME, Sauer B et al (2004) Neovascular targeting chemotherapy: encapsulation of paclitaxel in cationic liposomes impairs functional tumor microvasculature. Int J Cancer 110:117–124
Strieth S, Dunau C, Michaelis U et al (2014) Phase I/II clinical study on safety and antivascular effects of paclitaxel encapsulated in cationic liposomes for targeted therapy in advanced head and neck cancer. Head Neck 36:976–984
Sugiyama T, Asai T, Nedachi YM et al (2013) Enhanced active targeting via cooperative binding of ligands on liposomes to target receptors. PLoS One 8:e67550
Takara K, Hatakeyama H, Ohga N et al (2010) Design of a dual-ligand system using a specific ligand and cell penetrating peptide, resulting in a synergistic effect on selectivity and cellular uptake. Int J Pharm 396:143–148
Takara K, Hatakeyama H, Kibria G et al (2012) Size-controlled, dual-ligand modified liposomes that target the tumor vasculature show promise for use in drug-resistant cancer therapy. J Control Release 162:225–232
Takeuchi Y, Kurohane K, Ichikawa K et al (2003) Induction of intensive tumor suppression by antiangiogenic photodynamic therapy using polycation-modified liposomal photosensitizer. Cancer 97:2027–2034

Teesalu T, Sugahara KN, Kotamraju VR et al (2009) C-end rule peptides mediate neuropilin-1-dependent cell, vascular, and tissue penetration. Proc Natl Acad Sci U S A 106:16157–16162

Thijssen VL, Postel R, Brandwijk RJ et al (2006) Galectin-1 is essential in tumor angiogenesis and is a target for antiangiogenesis therapy. Proc Natl Acad Sci U S A 103:15975–15980

Thijssen VL, Heusschen R, Caers J et al (2015) Galectin expression in cancer diagnosis and prognosis: a systematic review. Biochim Biophys Acta 1855:235–247

Thurston G, McLean JW, Rizen M et al (1998) Cationic liposomes target angiogenic endothelial cells in tumors and chronic inflammation in mice. J Clin Invest 101:1401–1413

Tuerk C, Gold L (1990) Systematic evolution of ligands by exponential enrichment: RNA ligands to bacteriophage T4 DNA polymerase. Science 249:505–510

van der Gun BT, Monami A, Laarmann S et al (2007) Serum insensitive, intranuclear protein delivery by the multipurpose cationic lipid SAINT-2. J Control Release 123:228–238

van der Schaft DW, Dings RP, de Lussanet QG et al (2002) The designer anti-angiogenic peptide anginex targets tumor endothelial cells and inhibits tumor growth in animal models. FASEB J 16:1991–1993

Wicki A, Rochlitz C, Orleth A et al (2012) Targeting tumor-associated endothelial cells: anti-VEGFR2 immunoliposomes mediate tumor vessel disruption and inhibit tumor growth. Clin Cancer Res 18:454–464

Wildiers H, Guetens G, De Boeck G et al (2003) Effect of antivascular endothelial growth factor treatment on the intratumoral uptake of CPT-11. Br J Cancer 88:1979–1986

Wu Y, Zhang X, Xiong Z et al (2005) microPET imaging of glioma integrin {alpha}v{beta}3 expression using (64)Cu-labeled tetrameric RGD peptide. J Nucl Med 46:1707–1718

Wu Y, Cai W, Chen X (2006) Near-infrared fluorescence imaging of tumor integrin alpha v beta 3 expression with Cy7-labeled RGD multimers. Mol Imaging Biol 8:226–236

Wu J, Lee A, Lu Y et al (2007) Vascular targeting of doxorubicin using cationic liposomes. Int J Pharm 337:329–335

Wurdinger T, Tannous BA, Saydam O et al (2008) miR-296 regulates growth factor receptor overexpression in angiogenic endothelial cells. Cancer Cell 14:382–393

Xiong D, Liu Z, Bian T et al (2015) GX1-mediated anionic liposomes carrying adenoviral vectors for enhanced inhibition of gastric cancer vascular endothelial cells. Int J Pharm 496:699–708

Ye C, Kiriyama K, Mistuoka C et al (1995) Expression of E-selectin on endothelial cells of small veins in human colorectal cancer. Int J Cancer 61:455–460

Yonezawa S, Asai T, Oku N (2007) Effective tumor regression by anti-neovascular therapy in hypovascular orthotopic pancreatic tumor model. J Control Release 118:303–309

Yousefi A, Bourajjaj M, Babae N et al (2014) Anginex lipoplexes for delivery of anti-angiogenic siRNA. Int J Pharm 472:175–184

Zeisig R, Stahn R, Wenzel K et al (2004) Effect of sialyl Lewis X-glycoliposomes on the inhibition of E-selectin-mediated tumour cell adhesion in vitro. Biochim Biophys Acta 1660:31–40

Zetter BR (1990) The cellular basis of site-specific tumor metastasis. N Engl J Med 322:605–612

Zetter BR (2008) The scientific contributions of M. Judah Folkman to cancer research. Nat Rev Cancer 8:647–654

Chapter 9
Accessing Mitochondrial Targets Using NanoCargos

Ru Wen, Afoma C. Umeano, and Shanta Dhar

Abstract Mitochondria are membrane bound organelles that play essential roles for cell life, including energy production, apoptosis, redox balance, and regulation of calcium. Mitochondrial dysfunction is a hallmark for various diseases ranging from well-known diseases like cancer to rare genetic disorders like Barth's syndrome. Accordingly, mitochondria have been identified as key targets for therapeutic intervention. Mitochondria targeting strategies using nanocargos are rapidly growing tools for delivery of therapeutic and/or diagnostic payloads to mitochondria. In this chapter, we will highlight specific mitochondrial targets for nanotechnology-based delivery vehicles, NanoCargos, and discuss intracellular uptake mechanisms for NanoCargos, as well as technological methods for investigating mechanism for NanoCargo internalization into mitochondria.

Keywords NanoCargo • Mitochondria targeting • Mitochondrial uptake • Technological methods

Abbreviations

AFM	Atomic force microscopy
ANT	Adenosine nucleotide translocator
ATP	Adenosine triphosphate
CPP	Cell-penetrating peptide
CDs	Carbon dots
CPZ	Chlorpromazine
DOX	Doxorubicin
FCCP	Carbonylcyanide-p-(trifluoromethoxy)phenylhydrazone
FRET	Förster Resonance Energy Transfer
GQDs	Graphene quantum dots

R. Wen • A.C. Umeano • S. Dhar (✉)
NanoTherapeutics Research Laboratory, Department of Biochemistry and Molecular Biology, Sylvester Comprehensive Cancer Center, Miller School of Medicine, University of Miami, Miami, FL 33136, USA
e-mail: shantadhar@med.miami.edu

A. Prokop, V. Weissig (eds.), *Intracellular Delivery III*, Fundamental Biomedical Technologies 8, DOI 10.1007/978-3-319-43525-1_9

HDL	High density lipoprotein
HGC	Hydrophobic modified glycol chitosan
HK	Hexokinase
ICP-MS	Inductively coupled plasma mass spectrometry
IMM	Inner mitochondrial membrane
IMS	Intermembrane space
IVIS	In vivo image system
JC-1	5,5'6,6'-tetrachloro-1,1',3,3'-tetraethylbenzimidazolcarbocyanine iodide
LDH	Layered double hydroxide
MM	Mitochondrial matrix
mPTPC	Mitochondrial permeability transition pore complex
mtDNA	Mitochondrial DNA
MTS	Mitochondria targeting sequence
nDNA	Nuclear DNA
OMM	Outer mitochondrial membrane
OXPHOS	Oxidative phosphorylation
PEG	Polyethyleneglycol
QDs	Quantum dots
ROS	Reactive oxygen species
SEM	Scanning electron microscopy
SERS	Surface-enhanced Raman scattering
TEM	Transmission electron microscopy
TPP	Triphenyphosphonium cation
TMRM	Tetramethylrhodamine methyl ester
TMRE	Tetramethylrhodamine ethyl ester
VDAC	Voltage dependent anion channel

9.1 Introduction

Mitochondria were the first subcellular organelles to be isolated from a eukaryotic cell (Ernster and Schatz 1981). As membrane bound organelles, mitochondria exist in almost all living organisms with eukaryotic cells. In humans, each cell contains hundreds of mitochondria. However, the exact number varies regarding to cell type, tissue and species. For example, muscle fibers have more mitochondria than adipocytes. Mitochondria range from 0.5 to 1.0 μm in diameter but provide most of a cell's energy as adenosine triphosphate (ATP), a source of chemical energy, *via* oxidative phosphorylation (OXPHOS) (Ernster and Schatz 1981). Specially, mitochondria differentiate themselves from other organelles by accommodating DNA, so called "mitochondrial DNA" (mtDNA). Because each mitochondrion contains dozens of copies of mtDNA, all associated with unique mutations, each cell contains thousands of unique copies of mtDNA. Located in the mitochondrial matrix,

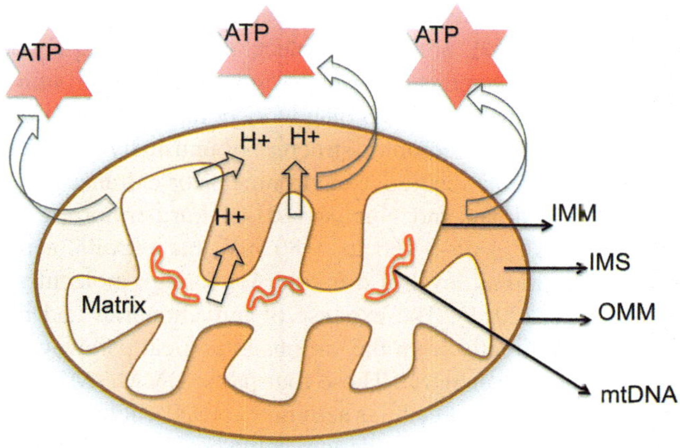

Fig. 9.1 The structure of the mitochondrion. *IMS* intermembrane space, *IMM* inner mitochondrial membrane, *OMM* outer mitochondrial membrane

mtDNA is made of 16,569 base pairs and 37 genes encoding 13 proteins, 2 rRNAs, and 22 tRNA associated with their own code for unique amino acids (Dimauro and Davidzon 2005). Mitochondria are heteroplasmic; mtDNA is much more likely to be mutated than nuclear DNA (nDNA) and variations in mtDNA occur among mitochondria in a cell, as well as, within a single mitochondrion. Unlike nDNA, mtDNA is exclusively maternally inherited, meaning offspring have no redundancy for protection against genetic mutation (DiMauro et al. 2001). Vulnerability of mtDNA is further conferred through its proximity to damaging agents. Because mtDNA lies in close proximity to reactive oxygen species (ROS) formed during various mitochondrial functions such as OXPHOS and mitochondria lack efficient DNA repair mechanisms, mtDNA undergoes frequent mutations, which may be passed down from mother to offspring (DiMauro et al. 2001). Thus, mitochondria play a major role in disease and disease progression. Mitochondria dysfunction has been associated with a vast array of diseases, ranging from rare hereditary diseases—like Barth's disease, myoclonic epilepsy with ragged-red fibers, and epilepsy—to common idiopathic diseases—like cancer, diabetes, and Alzheimer's diseases (Dimauro and Davidzon 2005; Federico et al. 2012; Pathak et al. 2015; Vasquez-Trincado et al. 2016; Wen et al. 2016). In addition to its unique genetic characteristics, mitochondria also possess unique membrane phospholipids, cardiolipin (CL), which contribute to a large array of mitochondrial processes, including, initiation of apoptosis, redox balance, and calcium homeostasis (Houtkooper and Vaz 2008; Santucci et al. 2014).

The structure of the mitochondrion is also complex: two lipid membranes, the outer mitochondrial membrane (OMM) and inner mitochondrial membrane (IMM), separated by an intermembrane space (IMS), which together enclose a mitochondrial matrix (MM) (Fig. 9.1). The OMM contains 8–10 % of the total proteins, mostly protein translocators, pore forming proteins, and mitochondrial fusion and fission proteins encoded by the nucleus and synthesized using cytosolic ribosomes (Walther and Rapaport 2009). Small low molecular weight compounds can diffuse across OMM and large high molecular weight compounds use protein translocators and pores (Endo and Yamano 2009; Stojanovski et al. 2012; Tatsuta et al. 2014). IMM is densely packed with a complex structure of unusually high protein to phospholipid ratio and provides a restrictive environment for entrance of chemical entities into the MM (Friedman and Nunnari 2014). Furthermore, a mitochondrial membrane potential ($\Delta\Psi_m$) of~−160 to −180 mV that prevails across the membranes imposes an additional level of difficulty for foreign molecules to cross into the matrix (Perry et al. 2011). The expanse of mitochondrial and mitochondrial-related diseases stems directly from the various biological pathways and functions that are dependent on mitochondria. These four parts: OMM, IMS, IMM, and MM participate in various important cellular activities carried out by mitochondria, such as, glucose metabolism, fatty-acid oxidation, heme biosynthesis, cell signaling, differentiation, senescence, homeostasis, and cell growth.

The vast scope of mitochondrial and mitochondrial-related diseases stems directly from the various biological pathways and functions that are dependent on mitochondria. Mitochondria are essential participants in all cellular and biologic function, including energy production, blood sugar modulation, and immune protection. Mitochondrial drug targets have been identified in the four major components towards addressing several diseases that feature mitochondrial damage or dysfunction, such as, cancer and aging, metabolic syndromes, cardiovascular disease, and neurological diseases (Pathak et al. 2015; Wen et al. 2016). Currently, techniques and therapeutics that reverse, modify and/or mitigate existing mitochondrial dysfunction to improve overall health are being developed (Marrache and Dhar 2012; Marrache and Dhar 2013; Marrache et al. 2013; Marrache et al. 2014; Pathak et al. 2014; Feldhaeusser et al. 2015; Marrache and Dhar 2015; Pathak and Dhar 2015; Pathak et al. 2015; Kalathil et al. 2016; Wen et al. 2016). Several protein targets have been identified for mitochondrial regulation in the context of disease and are listed in Table 9.1 (Milane et al. 2015). Here, we are expanding on the therapeutic potential of certain targets and how they relate to cellular and mitochondrial functions.

Although mitochondria undergo replication using various associated fusion/fission proteins, mitochondrial fusion/fission processes and associated proteins are used in various other functions, such as trafficking, mitophagy, etc. Mitochondria are degraded through autophagy or, more specifically "mitophagy." Mitophagy involves trafficking of healthy substituents away from damaged areas of mitochondria, or compartmentalizing damaged areas, and discarding these damages compartments for degradation, while keeping the functional section of mitochondria intact. Mitophagy ensures mitochondria quality control and clearance of ROS. Damaged portions of a mitochondrion inefficiently produce ATP and, thus, have significantly increased ROS production. Separation and destruction of damaged areas, protects remaining healthy mitochondrion and cell from damaging ROS and protects mtDNA from mutation. In a therapeutic context, mitophagy plays a critical role in important immunologic systems, especially in the presentation of antigens to antigen presenting

Table 9.1 Mitochondrial proteins as possible targets for various disease, including cancer, cardiovascular disease, and other mitochondria-related diseases

Mitochondrial targets for therapeutic intervention		
Outer mitochondrial membrane (OMM)	Inner mitochondrial membrane (IMM)	Mitochondrial Matrix (MM)
Fusion and fission proteins	Electron Transport Proteins (complexes I-V, etc.)	TCA cycle proteins
Hexokinase	Adenine nucleotide translocators (ANTs) or ADP/ ATP carriers	Fatty acid oxidation
K-voltage dependent anion channel (VDAC)	Mitochondrial potassium channels (K_{ATP} channels)	
Anti-apoptotic proteins (e.g. Bcl-2)	Uncoupling proteins	mtDNA
Peripheral benzodiazepine receptor		Cyclophilin D
Mitochondrial P-gp drug efflux pump		
Mitochondrial p53		
Mitochondrial migratory proteins		
Mitochondrial transporters, importers and pores		

cells via major histocompatibility complex II for Th1/2-mediated immune response. Fusion and fission proteins are also significant in MM maintenance and turn over, as well as in the delivery of proteins from the endoplasmic reticulum.

Targeting fusion and fission proteins involved in mitophagy could have profound therapeutic implications (Kroemer 2006; Maximo et al. 2009; Scatena 2012; Wallace 2012; Verschoor et al. 2013). For example, PTEN-induced kinase 1, a protein involved in mitophagy in OMM is able to recruit Parkin, a protein subunit of a ubiquitin ligase and the causative gene for autosomal recessive Parkinson's disease, for damaged mitochondria selection (Mizuno et al., Greenamyre and Hastings 2004; Chen and Dorn 2013; Matsuda et al. 2013; Han et al. 2015; Kazlauskaite and Muqit 2015; Pathak et al. 2015). Pink1 and Parkin may be targeted for pharmaceutical gain by engineering therapeutics that alter mitochondrial contacts to the endoplasmic reticulum involved in biosynthesis. Altering aberrant mitochondria of cancer cell via fusion pathways could be a possible strategy for efficient chemotherapy by driving cancer cells towards intrinsic apoptotic processes. On the other hand, increasing fission can sensitize cancer cells to extrinsic apoptotic pathways, such as those involved in immunologic derived cancer protection (e.g. cancer vaccines), and alter bioenergetics and host-response to improve cancer therapy (Milane et al. 2015).

Proteins involved in signal transduction, biosynthetic and OXPHOS have been associated with high-impact diseases, like diabetes, cardiovascular disease, mitochondrial dysfunction, and neurological disease (DiMauro et al. 2001; Roberts and Sindhu 2009; Milane et al. 2015; Vasquez-Trincado et al. 2016), and provide interesting and challenging targets for metabolic control of cells using NanoCargos (Pathak et al. 2014; Marrache and Dhar 2015). The first step of glycolysis—ATP mediated phosphorylation of glucose—is initiated by hexokinase (HK) catalyzation. Located in OMM, HK-II is related to ATP acquisition from potassium voltage dependent anion channels (VDAC)) and, interestingly, when HK-II binds to the OMM through VDAC, the interaction spatially inhibits pro-apoptotic proteins of Bcl-2 family and protects against cell death *via* mitochondrial apoptotic pathways. Recently, metabolic reprogramming of cancer cells using 3-bromopyruvate, a HKII inhibitor, was accomplished by mitochondria-targeted gold nanoparticles (Au-NPs) coated with triphenyphosphonium cation (TPP) as a mitochondrial targeting moiety (Marrache and Dhar 2015). Furthermore, these metabolic reprogramming Au-NPs displayed therapeutic efficacy in cancer cell killing (Marrache and Dhar 2015). Mitochondrial permeability transition pore complex (mPTPC) consists of VDAC, adenine nucleotide translocators (ANT) and cyclophilin-D (CyP-D) which expand between IMM and OMM (Mathupala et al. 2006). The mPTPC is formed under conditions associated with mitochondrial death and apoptosis and regulates permeabilization of both IMM and OMM (Vieira et al. 2000). Transporting pores, biosynthetic pathways, and channels are especially desirable targets in concerns to cancer because they destabilize cell bioenergetics and facilitate apoptosis. Mitochondrial P-gp drug efflux pump is an especially important target for cancer drug resistance. P-gp is responsible for pumping chemotherapeutics out of the mitochondria and cell. Any cellular stress, such as hypoxia, causes increased expression of P-gp. Furthermore, multiple-drug resistant cancer cells have shown high levels of P-gp in OMM (Solazzo et al. 2006).

There are various and important mitochondrial proteins. The challenge for research is in designing reliable methods and tools for targeting and mitigating aberrant mitochondrial functions. One solution is the use of nanotechnology-based delivery vehicles, "NanoCargos". NanoCargos provide a means of targeting mitochondrial membranes and delivering payloads (Yousif et al. 2009; Marrache et al. 2013; Sakhrani and Padh 2013; Pathak et al. 2015; Weinberg and Chandel 2015; Wen et al. 2016) (Fig. 9.2). Research in nanotechnology has been extremely promising towards answering the challenges of cancer, cardiovascular disease, diabetes, mitochondrial dysfunction related diseases, and neurological disease (Wen et al. 2016).

9.2 Mitochondrial Dysfunctions in Various Diseases

Mitochondria generate ROS as a byproduct of their role in energy production via OXPHOS (Cooper 2000). ROS production causes somatic mtDNA mutations which can produce an amplifying loop of increased respiration dysregulation resulting in accumulation of more ROS generated mutations. ROS production causes tissue aging by interrupting normal cellular metabolism, increasing cell death, and decreasing

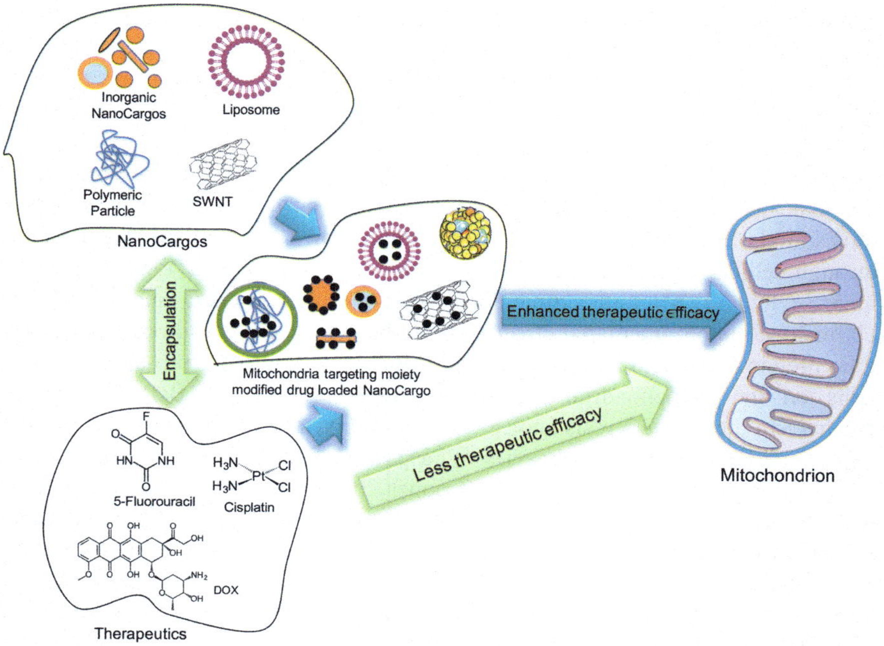

Fig. 9.2 A schematic diagram for NanoCargo and mitochondrial association

cellular capacity to replicate the genome (Roberts and Sindhu 2009; Federico et al. 2012; Milane et al. 2015). Our body's regenerative capacity—which underlies youthful features—is inherently dependent on our cells' replicative and functional capacity. Commercially, this is why most "age-defying" cosmetics and skincare products feature high levels of antioxidants such as tocopherols and ascorbic acid. ROS damage is not exclusive to aging. Currently, a number of NanoCargos incorporate antioxidant moieties in order to treat diseases like cancer and neurodegeneration (Yamada et al. 2015b).

In 2011, D. Hanahan and R. Weinberg expanded on the six hallmarks of cancer: (1) resistance against cell death, (2) sustained proliferation, (3) enhanced replicative endurance, (4) ability to escape from growth suppressors, (5) inducing metastasis, and (6) angiogenesis development by adding two additional hallmarks: reprogrammed energy metabolism and evasion of tumor destruction and two enabling features: genome instability and inflammation (Hanahan and Weinberg 2011). All these hallmarks are either directly or indirectly associated with mitochondria.

The dysfunctional mitochondria in cancer cells is attributed to mtDNA mutations, ROS production, and enhanced glycolysis for ATP generation (Armstrong 2006; Fadeel et al. 2008). Unlike normal mitochondrial genomes that are *heteroplasmic*, indicative of random ROS damage producing random mutations in a given mitochondrial genome, cancer cell mtDNA mutations are *homoplasmic*, meaning that all mitochondrial genomes are *identically* mutated. Several studies have highlighted associations between various cancers and specific mitochondrial mtDNA mutations. However, the mechanism of how mtDNA relates to a cancer cell's evolutionary gain has not yet been fully understood (Brandon et al. 2006; Chatterjee et al. 2006; van Gisbergen et al. 2015).

Dysfunction in mitochondria-mediated apoptosis inherently makes cells resistant to death. Generally, cancers cells are resistant to intrinsic and/or extrinsic apoptotic stimulators. The intrinsic pathway for programmed death is mediated primarily through cellular or mitochondrial damage (an "apoptotic signal") that triggers an increase in p53, a tumor suppressor protein (Wickramasekera and Das 2014). Ultimately, the p53-mediated effect is the reduction of Bcl-2 and other anti-apoptotic protein levels and an increase in Bid, Bax, Bim and other pro-apoptotic proteins on the mitochondrial membrane. Overexpression of Bcl-2 is another mechanism that confers resistance to apoptosis. Leakage of cytochrome c through mitochondrial pores triggers a series of caspase-mediated interactions that creates an apoptosome, a destroyer of cells (Cooper 2000). In comparison, the extrinsic pathway for apoptosis is relatively simple. A pro-apoptotic ligand binds to death receptors, which directly trigger caspase-mediated interactions that create apoptosomes (Cooper 2000). Many cancer nanotherapeutic methods focus on commonalities and interactions between intrinsic and extrinsic apoptotic pathways (Mkandawire et al. 2015; Zhang et al. 2015). Several drugs such as 5-fluorouracil, cisplatin were encapsulated into NanoCargos for altering cancer metabolism in order to destabilize cancers and induce apoptosis (Sakhrani and Padh 2013; Pathak et al. 2014; Rin Jean et al. 2014; Marrache and Dhar 2015; Milane et al. 2015; Pathak et al. 2015; Weinberg and Chandel 2015; Wen et al. 2016). Furthermore, mitochondria-targeting NanoCargos demonstrate significant therapeutic effect (Marrache et al. 2014; Pathak et al. 2014; Marrache and Dhar 2015).

Type 2 diabetes mellitus is associated with dysfunctional mitochondria which results in deregulated glucose pathway and excessive ROS accumulation in cells (e.g. pancreatic beta cells and hepatocytes) *via* OXPHOS (Ernster and Schatz 1981). Together, these traits produce systemic insulin resistance. Mitochondrial dysfunction is most overtly displayed in glucose-stimulated insulin secretion processes. For instance, low activity of NADH shuttles in beta cells of type 2 diabetic models results in impaired K_{ATP}-dependent and -independent GSIS glucose-sensing (Weksler-Zangen et al. 2008; Halperin et al. 2012). Such mechanisms provide insight into clinical manifestations of hyperglycemia and poor glucose control. Systemic ROS accumulation coincides with the inflammatory state associated with diabetes, characterized by macrophage recruitment in adipose and hepatic cells and increased inflammatory markers in the blood. Inflammatory markers in the liver, especially TNF-α, induce gluconeogenesis and further lead to elevated blood glucose levels (Flemming et al. 2015; Sharma 2015). NanoCargos have shown potential *via* encapsulation of therapeutics, such as plasmid DNA (Basarkar and Singh 2009), insulin (Damgé et al. 2010), and antioxidants (Ratnam et al. 2009). NanoCargo-based delivery of therapeutics to the mitochondria is a more precise targeting and promising strategy to improve uptake and therapeutic effect.

Neurological diseases have increasingly attracted great attention in the field. Several NanoCargo formulations have been made to target oxidative damage in neurological diseases (Wen et al. 2016). Liposomes, polymeric nanoparticles, solid nanoparticles and metal nanoparticles have been used as NanoCargos to deliver anticonvulsants and antioxidants. Furthermore, studies of mitochondria targeting NPs demonstrated significantly enhanced therapeutic effects such as mitochondria targeted ceria NanoCargos (Kwon et al. 2016) and aspirin containing NanoCargos (Kalathil et al. 2016). Intensive studies of mitochondria targeting NanoCargos are needed in order to potentiate clinical applications.

Apoptosis often occurs due to ROS mediated oxidation of proteins, membrane phospholipids and mtDNA, and eventual release of cytochrome c and apoptotic inducing factor (AIF). High density lipoprotein (HDL) mimicking NanoCargos were developed by our group for potential application in atherosclerosis. We developed mimics of HDL which can functionally perform reverse cholesterol transport (Marrache and Dhar 2013). These polymer-based HDL NanoCargos contained encapsulated contrast agents such as quantum dots (QDs) for detection of vulnerable plaques, apolipoprotein (apo)A-1 mimentic 4 F peptide, cholesteryl oleate for preventative treatment, and triphenylphosphonium cations as mitochondria targeting moieties (Marrache and Dhar 2013). HDL mimic NanoCargos were non-immunogenic and demonstrated promising therapeutic properties for combating atherosclerosis.

9.3 NanoCargo Intracellular Uptake Mechanisms for Targeting Mitochondria

Endocytosis is a common energy-dependent pathway used by cells to communicate with biological surroundings and for the uptake of ions, biomolecules, and/or nutrients. Endocytosis can be classified into: (i) clathrin mediated endocytosis, (ii) caveolae dependent endocytosis, (iii) phagocytosis, pinocytosis, (iv) macropinocytosis, (v) flotillin dependent pathway, (vi) circular dorsal ruffles, (vii) CLIC/GEEC-type, (viii) entosis pathway and so forth (Doherty and McMahon 2009).

Clathrin-mediated pathway is characterized by trafficking molecules from the plasma membrane to early endosomes using clathrin-coated vesicles (~100 nm in diameter). Adaptor proteins initiate clathrin-coated pits by promoting clathrin assembly, which transforms the plasma membrane into a deeply bended clathrin-coated pit for molecule recruitment (Ungewickell and Hinrichsen 2007). Caveolae are glycolipid raft invaginations of plasma membrane that are sensitive to cholesterol depletion and dynamin inhibition. Cholesterol depleting agents (e.g. filipin, nystatin, or methyl-cyclodextrin) and dynamin inhibitors can be used to investigate caveolae-mediated endocytosis (Nabi and Le 2003). Endocytosis inhibitors are commonly used for mechanistic uptake studies of NP internalization. NPs are internalized mainly through clathrin- and caveolae-mediated endocytosis (Qaddoumi et al. 2003; Harush-Frenkel et al. 2007; Wang et al. 2009; Voigt et al. 2014), which are receptor-mediated pathways such as mannose complement/Fcγ/scavenger receptor with proteins directly in contact with NanoCargos (Oh and Park 2014). Oh *et al.* reported that layered double hydroxide (LDH) NanoCargos were internalized through clathrin-mediated endocytosis (Oh et al. 2006). The colocalization of LDH NPs with dynamin, eps15, clathrin, and Tf further validated clathrin-mediated endocytosis of LDH NPs.

Macropinocytosis is a signal-dependent process, involving the formation of macropinosomes (>200 nm in diameter) and frequently occurs in antigen presenting cells (Andersen et al. 2014). Macropinosomes are larger than clathrin-coated vesicles and allow for non-specific internalization of large quantities of solute, membrane, nutrients, and antigens (Lim and Gleeson 2011). Iversen et al. reported a macropinocytosis-like mechanism for the uptake of ricinB QD NPs in HeLa cells (Iversen et al. 2012). RicinB-QD NPs uptake was inhibited in dynamin-negative HeLa cells in the presence of macropinocytosis inhibitor amiloride analog EIPA. The depletion of cholesterol by methyl-β-cyclodextrin and use of cytochalasin D (inhibitors of actin polymerization) reduced the uptake of RicinB-QD NPs. The colocalization of RicinB-QD NPs and dextran, a marker of fluid-phase uptake by macropinocytosis, confirmed a macropinocytosis-like mechanism for NP internalization. Fernando et al. reported that PFBT NPs were taken up through a macropinocytosis pathway, demonstrated by colocalization of NPs with Texas Red dextran and inhibition of uptake in cells treated with phosphoinositide 3-kinase inhibitors (Fernando et al. 2010).

Phagocytosis is an internalization process characterized by formation of particles-loaded invaginations, phagosomes, in the plasma membrane (Doherty and McMahon 2009). Fontana et al. reported phagocytic uptake of amoxicillin-loaded polyethylcyanoacrylate NPs (200–300 nm) using surface modifications (Fontana et al. 2001). Polyethyleneglycol (PEG) is an FDA-approved polymer. Coating NPs with PEG can yield a 50-70 % reduction in phagocytosis.

NPs may also be taken up through several different endocytotic pathways simultaneously. For example, Nam et al. reported that hydrophobic modified glycol chitosan (HGC) NPs were internalized through clathrin-mediated, caveolae-based and macropinocytosis pathways simultaneously (Nam et al. 2009). HGC NP uptake was decreased with chlorpromazine (CPZ), filipin III, and amiloride by 20, 40, and 30 %, respectively. Synergistic inhibitory effect was observed by CPZ and filipin III with 60 % reduction of HGC NPs internalization. Some HGC NPs were observed in late endosomes, lysosomes, and endoplasmic reticulum by TEM in colocalization studies. Dausend et al. reported modulation of uptake pathway *via* surface charge in polymeric NPs (Dausend et al. 2008). Polymeric NanoCargos with positive charge were internalized *via* macropinocytosis and partially clathrin-mediated pathways, microtubules, and cyclooxygenases. The uptake of negatively charged NPs, which may involve an unidentified dynamin-independent process, was inhibited to a lesser extent by dynamin than NPs with positive charge.

Direct penetration of the plasma membrane is another mechanism for NP uptake and is associated with NPs modified with cationic cell-penetrating peptide (CPP) or CPP-like molecules. It is a passive and energy independent process that involves the formation of a pore and/or inverted micelle, membrane thinning models, and/or carpet-like models (Herce et al. 2009; Madani et al. 2011). CPPs can be used for modification of NPs carriers for intracellular delivery of therapeutic agents (Lindgren et al. 2000; Deshayes et al. 2005; Wang and Melosh 2012). Verma *et al.* compared the plasma membrane penetration of NPs with similar composition (Verma et al. 2008). NanoCargos with highly ordered functional groups on their surfaces could penetrate plasma membrane with the bilayer intact, where as NanoCargos with accidental distribution of the same functional groups were recruited in endosomes.

Once NanoCargos enter cells, they are usually trapped into endosomes and trafficked through lysosomes into mitochondria *via* various mechanisms, including potential-mediated mechanisms, protein machinery pathways, and/or membrane fusion processes. A summary of NP internalization pathways is shown in Fig. 9.3. NPs modified with a cationic groups, like TPP (Marrache and Dhar 2012; Marrache and Dhar 2013; Pathak et al. 2014), and methyl-TPP (Smith et al. 2003), mainly undergo potential mediated pathways for mitochondrial uptake. Owing to the presence of $\Delta\Psi_m$, lipophilic cation decorated NanoCargos are able to have mitochondria association properties. Lipid soluble cations adsorb to the mitochondrial surface and take advantage of mitochondrial membrane potential for their association (Ross et al. 2005).

Our group has designed several mitochondria targeted NP systems exploiting this mechanism (Marrache and Dhar 2012; Marrache and Dhar 2013). Targeted NPs were able to undergo endosomal escape, and then associate with mitochondria *via* a potential mediated pathway (Marrache and Dhar 2012). QD-TPPs is reported to internalize in mitochondria *via* a $\Delta\Psi_m$ mediated process (Chakraborty and Jana 2015) upon entering the cell *via* lipid raft or caveolae-mediated endocytosis.

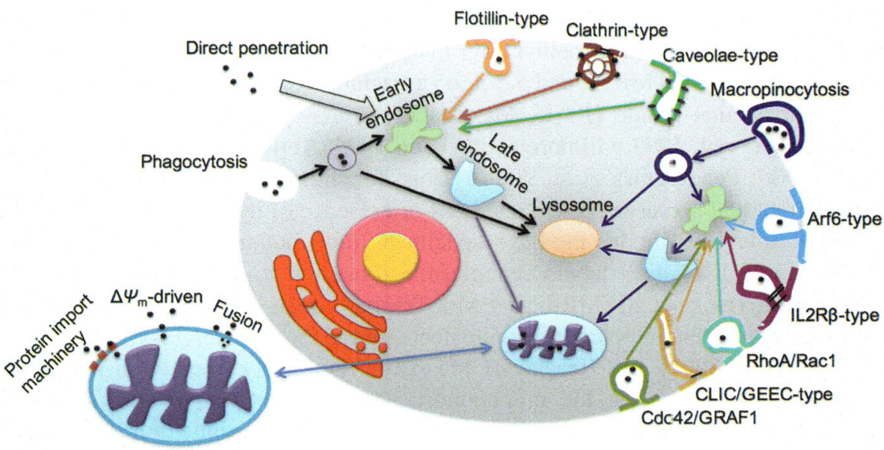

Fig. 9.3 Possible intracellular pathway of NP association with mitochondria

Protein import machinery was identified by the existence of mitochondria-penetrating peptides and/or mitochondria targeting sequences (MTSs), which consists of about 20–40 amino acids (Zhang et al. 2011). MTSs can be used for mitochondria-targeted NP design. Amyloid β-peptide is reported to import to mitochondria through translocation of OMM machinery (Petersen et al. 2008). Membrane fusion process of NanoCargo internalization in mitochondria can be achieved by a liposome based carrier MITO-Porter (Yamada et al. 2008). MITO-Porter based NanoCargos were brought into cells *via* macropinocytosis and underwent macropinosomal escape before binding to mitochondria thourgh electrostatic interactions. The membrane fusion was studied using sphingomyelin in MITO-Porter for delivery of macromolecules to the IMS, IMM, and MM.

9.4 Detection Techniques for Nanoparticle Internalization into Mitochondria

The most common methods for tracking NanoCargo internalization is achieved by labeling these nanomaterials and mitochondrial components with different labels. Figure 9.4 depicts techniques which are used for mitochondrial trafficking of NPs. A comparison of these techniques used in mitochondrial uptake studies for NanoCargos in terms of advantages and disadvantages is highlighted in Table 9.2.

9.4.1 Fluorescence Based Detection

Labels are commonly fluorophores that have measurable fluorescence in response to specific stimulation. Small organic molecules can be used in mitochondria labeling to visualize dynamic cellular or organismic behaviors. Organic fluorophores like tetramethylrhodamine methyl ester (TMRM), rhodamine 123, tetramethylrhodamine ethyl ester (TMRE), and 5,5',6,6'-tetrachloro-1,1',3,3'-tetraethylbenzimidazolcarbocyanine iodide (JC-1), MitoTracker®, are sensitive to $\Delta\Psi_m$ (Fig. 9.5) (Cossarizza et al. 1993; Gilmore and Wilson 1999; Keij et al. 2000; Nicholls and Ward 2000; Pendergrass et al. 2004). Thus, functional mitochondria with intact $\Delta\Psi_m$ attract and preserve these dyes. Loss or presence of dye may be used to indicate $\Delta\Psi_m$ integrity. Rhodamine 123 is specific for mitochondria labeling, but TMRE, TMRM, and JC-1 dyes labels endoplasmic reticulum to some extent (Chazotte 2009). MitoTracker CMTMRos and MitoTracker Green, are $\Delta\Psi_m$-insensitive and able to remain in mitochondria with decreased $\Delta\Psi_m$ (Pendergrass et al. 2004; Agnello et al. 2008). MitoTracker®CMTMRos chemically reacts with thiol moieties and can remain in the mitochondria after cell fixation (Chazotte 2009). MitoTracker Green is one of the most widely used labeling agents for mitochondria. MitoTracker Orange is reported to induce mitochondrial permeability transition and inhibit complex I, causing depolarization in liver mitochondria (Scorrano et al. 1999).

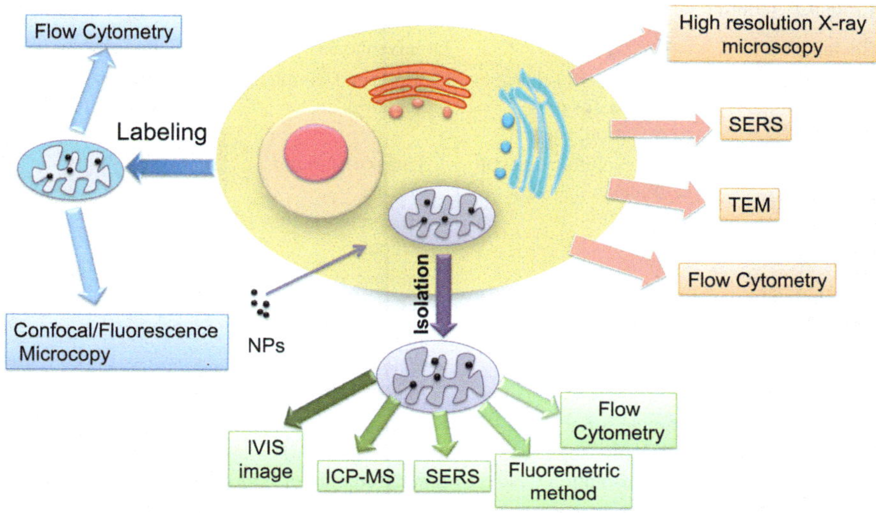

Fig. 9.4 Diagram of techniques used for intracellular trafficking of NPs into mitochondria. *ICP-MS* inductively coupled plasma mass spectrometry, *IVIS* in vivo image system, *SERS* surface-enhanced Raman scattering, *TEM* transmission electron microscopy

Table 9.2 A summary of advantages and disadvantages for techniques used in mitochondrial uptake studies for NanoCargos

Techniques	Advantages	Disadvantages	Comments
Confocal microscopy	Provides 3-D structures of sample	Limited resolution attributed to diffraction	Widely used in NP uptake studies
	Can monitor dynamic information in living organism	Can not provide detailed information at the subcellular level	Sample needs to be fluorescently stained
		Fluorophore may suffer from photobleaching and autofluorescence	
		Fixation of samples may cause artifacts	
Fluorescence microscopy	Provides the pattern/ location of NPs in a specific component in living/fixed cell	Suffers from photobleaching and autofluorescence	Sample needs to be fluorescently stained
		Fixation of samples may cause artifacts	Less widely used compared to confocal microscopy
		Can not provide detailed information at subcellular level	

(continued)

Table 9.2 (continued)

Techniques	Advantages	Disadvantages	Comments
TEM	Provides high accuracy of NPs status (e.g. aggregation, shape)	Complicated sample preparation.	Mostly used for metallic NPs
	High resolution and can distinguish cellular components	Time-consuming process.	
		Interpretation may be difficult.	
		Can not visualize the live cells.	
		Expensive	
		Only provides 2D images.	
SEM	Provide surface details of sample	Dehydration of samples may create artifacts	Limited application on NP uptake studies
	High resolution	Surface coating of sample is required	
ICP-MS	Superior detection limits of low concentration of elements	High cost	Widely used for *in vivo* tracing studies and need metallic component
		Can not quantify the elements in live samples.	
	Efficient and fast analysis of multiple elements	Expensive	
SERS	Non-destructive	Low sensitivity	A few studies has been made for cellular uptake
	Non-invasive	The fluorescence of samples may affect the results	
	No need of complicated sample preparation	Spectra interpretation may be difficulty	
	Can analyze samples both *in vivo* and *in vitro*		
	High spatial resolution.		
High resolution X-ray microscopy	High resolution	Unable to visualize relatively thick samples	A few studies was made for cellular uptake
	Can analyze small NP status in the cell samples	Complicated sample preparation and time-consuming process	
IVIS imaging	Non-invasive	Provides only 2D image	Widely used *in vivo*
	Easy operation	Same position is required for a group of samples	
	Non irradiative	Image saturation may be susceptible to inappropriate background setup	
	Enable both fluorescence and bioluminescence detection		

(continued)

Table 9.2 (continued)

Techniques	Advantages	Disadvantages	Comments
Fluorescence analyses	Relatively simple	Less sensitive and require high concentration of sample	Sample needs to be fluorescently stained
	Low cost		Widely used
FRET assay	Provides real-time monitoring of live cell dynamics	The properties of fluorescent probes are subjected to any physical changes	Has been used for vesicle-mitochondria fusion studies
	Good sample penetration	The process often causes protein fusion, mutation and/or alternation, which may affect the properties of molecules	
	Extremely high sensitivity		
	Can provide molecular distances		
Flow cytometry	Multiparameter data acquisition at the same time	Expensive	Widely used
	Relatively fast	User operation may be critical to obtain reliable data	
	Quantitative	Individual cell behavior may be difficult	
	Accurate and reproducible	Low cell throughput rate	
AFM	Provides 3D image	Limited area size for image	Some studies have been made for cellular uptake
	Simple sample preparation	Low scanning speed	
	High resolution	Subject to image artifacts and piezoelectric material	
		Expensive cantilevers	

The uptake of MitoTracker Red is controlled by $\Delta\Psi_m$, and its retention is associated with thiol groups of mitochondrial proteins after cell fixation, thus MitoTracker Red can be used for mitochondria-specific observation (Poot and Pierce 1999). Buckman *et al.* reported several MitoTracker dyes for mitochondria labeling in the central nervous system (Buckman et al. 2001). Mitochondria targeted GFP (MitoGFP) labeling is $\Delta\Psi_m$-independent and remains in the mitochondria. MitoGFP can easily be detected by confocal or fluorescence microscopy, and can be used for investigating mitochondrial dynamics in living cells (Hill et al. 2014).

NanoCargos can be fluorescently labeled by encapsulating a fluorescent probe. Yamada *et al.* reported the use of NBD and rhodamine for delivery vehicle internalization studies (Yamada and Harashima 2013). The liposome MITO-Porter was modified by MTS and labeled with NBD-DOPE and rhodamine-DOPE for Förster Resonance Energy Transfer (FRET) analysis. The modified MITO-Porter liposomes displayed efficient delivery of bioactive macromolecules to mitochondria.

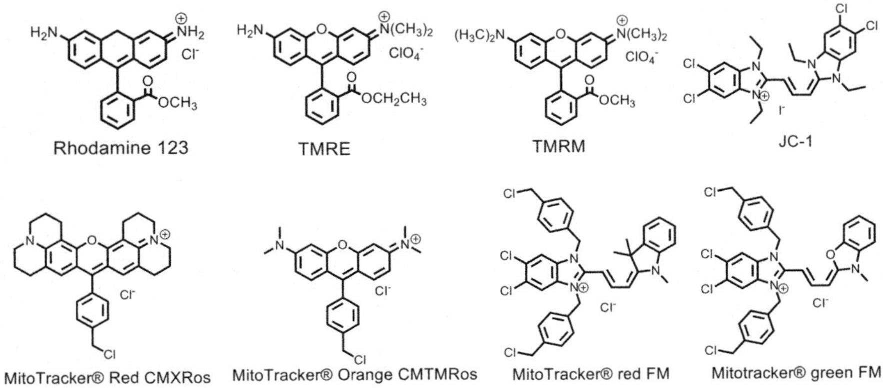

Fig. 9.5 Structure of $\Delta\Psi_m$-dependent fluorescent labelling reagents for mitochondria. TMRM-tetramethylrhodamine methyl ester, TMRE-tetramethylrhodamine ethyl ester, JC-1-5, 5', 6, 6'-tetrachloro-1, 1', 3, 3'-tetraethylbenzimidazolcarbocyanine iodide (JC-1)

Several therapeutics are fluorescent such as doxorubicin (DOX) and can be directly used for delivery vehicle tracking. Qu *et al.* reported delivery of DOX loaded silica NanoCargos to mitochondria for cancer therapy (Qu et al. 2015). The fluorescence of DOX and MitoTracker green was used for NanoCargo and mitochondria tracking, respectively. These NanoCargos were successfully delivered to the mitochondria by surface functionalization with TPP, evidenced by colocalization of DOX and MitoTracker green fluorescence using both fluorescence analysis and confocal microscopy.

Autofluorescence can come from biological molecules. Extracellular matrix may also contribute to autofluorescence in tissues (Monici 2005). Hence, fluorescent probes that avoid or minimize autofluorescence should be selected (Mosiman et al. 1997). Reduction of autofluorescence can also be accomplished by adding a reducing agent. $NaBH_4$ can decrease the interference of cellular autofluorescence upon cell fixation (Clancy and Cauller 1998). Schnell reported a solution composed of $CuSO_4$ (1–10 mM) and ammonium acetate (50 mM) buffer at pH 5 or Sudan Black B (1 %) in 70 % ethanol reduced autofluorescence from lipofuscin in tissue sections (Schnell et al. 1999). Photobleaching is attributed to unstable properties inherent in fluorophore and/or non-specific binding molecules. Prolong exposure to stimulus light can damage dye and reduce its capability for fluorescence (Bernas et al. 2005). Minimizing intensity and exposure during preparation and measurement is a key step towards solving this problem. By labeling NPs and mitochondria, NP intercellular trafficking can be investigated using instruments such as confocal microscopy, fluorescence microscopy, *in vivo* image system (IVIS), fluorescence analysis, FRET, and flow cytometry. Confocal microcopy is an optical imaging technique that reduces noise and enhances resolution compared to conventional fluorescence microscopy (White et al. 1987), and is most commonly used in NP intracellular uptake studies. Table 9.3 summarizes current mitochondria labeling strategies by

Table 9.3 Summary of mitochondria labeling strategies for fluorescence imaging

Label	Examples	Model	NanoCargo label	Comments	Reference
Mitotracker green	HDL mimicking NanoCargo	Live cell	Quantum dots (QDs)	QDs were well encapsulated into the NPs	Marrache and Dhar (2013)
MitoFluor Green	Stearyltriphenyl phosphonium (STPP)-modified NPs	Fixed cell	Rhodamine (Rh)	Rhodamine labeled phosphatidylethanolamine was encapsulated	Boddapati et al. (2008)
MitoTracker orange	Fe_2O_3-loaded NPs QD-TPP	Live cell	QDs	QD-TPP and Fe_2O_3-TPP NPs were toxic at high concentration	Chakraborty and Jana (2015)
MitoTracker Red	SWNT-PL-PEG-FITC	Live cell	Fluorescein isothiocyanate (FITC)	SWNTs located in mitochondria of tumor and normal cells	Zhou et al. (2010)
MitoFluor Red 589	MITO-Porter	Live cell	NBD, GFP	MITO-Porter incorporated with GFP or NBD–DOPE	Yamada et al. (2008)
MitoTracker Deep Red	Iridium (III) complexes	Live cell, Zebrafish model	Iridium (III)	Iridium complexes exhibited green to red emission under irradiation	Li et al. (2013)
MitoTracker Red, MitoTracker green	Fe_3O_4@mSiO_2-TPP/Carbon dots (CDs)	Fixed cell	CDs	Fe_3O_4@mSiO_2-TPP/CDs	Zhang et al. (2015)

confocal microscopy technique. Our group reported mitochondrial association of targeted PLGA-*b*-PEG-TPP/PLGA-*b*-PEG-QD NPs in HeLa cells *via* confocal microscopy (Marrache and Dhar 2012). Targeted NPs were able to access mitochondria after efficient endosomal escape. Non-targeted NPs remained in the lysosomes and cytoplasm. By using the same labeling method, we reported mitochondrial internalization of polymer-lipid hybrid NPs (Marrache and Dhar 2013). The internalization of targeted HDL NPs was confirmed by confocal microscopy in healthy RAW 264.7 cells. When the cells were treated with carbonylcyanide-p-(trifluoromethoxy)phenylhydrazone (FCCP) to induce depolarization of $\Delta\Psi_m$, targeted HDL NPs did not accumulate in mitochondria, as seen by IVIS analyses of mitochondrial and cytosolic fractions. Chakraborty *et al.* investigated internalization mechanisms for QD-TPP in HeLa cells using fluorescence microscopy (Chakraborty and Jana 2015). QD-TPP showed strong subcellular uptake with TPP functionalization, whereas, QD only demonstrated internalization on a cellular level. Colocalization of QD-TPP and mitochondria was confirmed by confocal microscopy using MitoTracker Orange. The cellular uptake mechanism was elucidated by flow cytometry studies using pathway-specific inhibitors to block the various endosytosis pathways. QD-TPP accessed cells mainly *via* caveolae or lipid raft mediated endocytosis in both CHO and HeLa cells. Qu *et al.* reported the colocalization of MSNP-PPh_3-DOX and mitochondria by confocal microscopy (Qu et al. 2015). The delivery of DOX to mitochondria was confirmed by fluorescence analysis of FITC and DOX in isolated mitochondria. Yamada *et al.* reported FRET analysis for investigating liposome NP fusion activity with mitochondria, in which two different organic dyes, NBD and rhodamine, were incorporated into NPs (Yamada et al. 2015a). The energy transfer in isolated mitochondria was evaluated by relative fluorescence intensity. The capability of nucleic acid delivery to MM was quantified by fluorescence analysis of Cy-5 RNA oligomer in mitochondrial subfractionation.

9.4.2 Mass Spectrometry Based Detection

Quantification of small amount of NPs in the mitochondria requires highly sensitive techniques like inductively coupled plasma mass spectrometry (ICP-MS). ICP-MS procures concentration of NPs by an elemental analysis, such as Cd concentration determination for Cd-Se QDs. Another strategy is to replace fluorescent probes with inorganic nanocrystals like quantum dots (QDs), carbon dots (CDs), iron oxide, and gold (Au), with unique optical, physical, and chemical properties. CdSe and CdTe are the most commonly used NP tracking tools in biological imaging applications (Resch-Genger et al. 2008). QDs may cause cytotoxicity to some degree since they consist of toxic metals (Smith and Nie 2009). Our group has demonstrated successful application of QDs for NPs tracking both *in vitro* and *in vivo* (Marrache and Dhar 2012, 2013; Marrache et al. 2014; Pathak and Dhar 2015; Pathak et al. 2015; Wen et al. 2016), as well as the use of Au-NPs (Marrache and Dhar 2015). Recently developed, CDs are carbon NPs with tunable fluorescence emissions and potential competitors to conventional QDs (Lim et al. 2015). Yang *et al.* reported successful *in vivo* studies for use of CDs as an imaging agent (Yang et al. 2009). Graphene quantum dots (GQDs) are being explored for NP labeling. Recently, Chong *et al.* calculated *in vivo* and *in vitro* safety profiles for GQDs (Chong et al. 2014).

9.4.3 Transmission Electron Microscopy (TEM) Based Methods

Transmission electron microscopy (TEM) is widely utilized for internalization studies of NPs as it is of extremely high resolution providing high accuracy of NPs status (e.g. shape, aggregation) and allows for distinguishing cellular organelles (Klein et al. 2015). Sample preparation is complicated compared to fluorescence imaging. Inorganic nanocrystals, such as QDs, iron oxide, Au, silver (Ag), are possible labeling agents for NPs. Our group reported the application of TEM for mitochondrial localization of AuNP based NanoCargos (Marrache and Dhar 2015). Mitochondria targeted Au-NPs were localized in MM and non targeted Au-NPs were found outside the mitochondria of PC3 cells. Karataş *et al.* depicted the interaction of Au NPs with mitochondrial membrane using TEM (Karataş et al. 2009).

9.4.4 Miscellaneous Methods

Karataş *et al.* also introduced surface-enhanced Raman scattering (SERS) techniques for further detailing interactions between Au-NPs and mitochondria (Karataş et al. 2009). Lung cancer cells and corresponding isolated mitochondria were treated with Au colloidal suspension. Atomic force microscopy (AFM) showed intact mitochondria interacting with Au NPs. Bressan *et al.* reported the absence of Ag NPs inside the mitochondria and, instead, the close association of Ag NPs to OMM in patient-derived human cells using TEM analysis (Bressan et al. 2013). A relative abundance of Ag NPs were aggregated in the perinuclear zone of cytoplasm and around mitochondria. Peckys et al. was able to trace Au NPs uptake in live cells by liquid scanning TEM (STEM) (Peckys and de Jonge 2011). Au NPs with round or oval shapes were bound to the 67 % of vesicles' membrane surface area and remained aggregated after 24 h. Important to note, STEM did not produce any significant damage to the cells. Chen et al. evaluated the distribution of NPs in cells by high resolution x-ray microscopy (Chen et al. 2011). High-resolution x-ray microscopy generates x-ray micro-images through phase and absorption contrast with two-dimensional (2-D) projection and 3-D tomographic reconstructions.

9.5 Conclusions

Mitochondria provide promising targets for new therapeutics aimed towards treating mitochondrial dysfunction associated diseases such as aging, cancer, heart disease, and neurodegeneration. Mitochondria targeted NanoCargos have shown to enhance therapeutic efficacy by directing biomolecules to mitochondria *in vitro* and *in vivo*. Such NanoCargos are able to enter cells through endocytosis and/or direct penetration, and internalize in mitochondria *via* $\Delta\Psi_m$-mediated mechanisms, protein machinery pathways, and/or membrane fusion processes. Endocytosis inhibitors and localization of NanoCargos to specific subcellular compartments like, mitochondria are widely investigated using many techniques, such as confocal microscopy, IVIS imaging, fluorescence analysis, flow cytometry, mass spectrometry, and electron microscopy, for identifying the mechanism of NanoCargo uptake. Real time imaging and tracking of NanoCargos in subfractions of mitochondria are still in need for intensive investigation. The comprehensive understanding of these uptake pathways will help to design efficient mitochondria-targeting drug delivery systems for manipulating conventional or pathological processes occurring in subcellular compartments and developing therapies for mitochondrial dysfunctions related diseases.

Acknowledgements We are thankful to the Department of Defense for a Prostate Cancer Idea award (W81XWH-12-1-0406); American Heart Association for a National Scientist Award (14SDG18690009); National Heart, Lung, and Blood Institute of National Institutes of Health (NIH) R56 high priority bridge award (Award Number. R56HL121392); National Institute of Neurological Disorders and Stroke of NIH R01NS093314 award, Georgia Research Alliance, and Sylvester Comprehensive Cancer Center for providing financial supports to conduct research in our lab in the area of nanomedicine.

References

Agnello M, Morici G, Rinaldi AM (2008) A method for measuring mitochondrial mass and activity. Cytotechnology 56(3):145–149

Andersen H, Parhamifar L, Moghimi SM (2014) Uptake and intracellular trafficking of nanocarriers. In: Prokop A, Iwasaki Y, Harada A (eds) Intracellular delivery II: fundamentals and applications, 1st edn. Springer, Dordrecht, pp 117–138

Armstrong JS (2006) Mitochondria: a target for cancer therapy. Br J Pharmacol 147(3):239–248

Basarkar A, Singh J (2009) Poly (lactide-co-glycolide)-polymethacrylate nanoparticles for intramuscular delivery of plasmid encoding interleukin-10 to prevent autoimmune diabetes in mice. Pharm Res 26(1):72–81

Bernas T, Robinson JP, Asem EK, Rajwa B (2005) Loss of image quality in photobleaching during microscopic imaging of fluorescent probes bound to chromatin. J Biomed Opt 10(6):064015-064015-9

Boddapati SV, D'Souza GGM, Erdogan S, Torchilin VP, Weissig V (2008) Organelle-targeted nanocarriers: specific delivery of liposomal ceramide to mitochondria enhances its cytotoxicity in vitro and in vivo. Nano Lett 8(8):2559–2563

Brandon M, Baldi P, Wallace DC (2006) Mitochondrial mutations in cancer. Oncogene 25(34):4647–4662

Bressan E, Ferroni L, Gardin C, Rigo C, Stocchero M, Vindigni V, Cairns W, Zavan B (2013) Silver nanoparticles and mitochondrial interaction. Int Dent J 2013:312747

Buckman JF, Hernández H, Kress GJ, Votyakova TV, Pal S, Reynolds IJ (2001) MitoTracker labeling in primary neuronal and astrocytic cultures: influence of mitochondrial membrane potential and oxidants. J Neurosci Methods 104(2):165–176

Chakraborty A, Jana NR (2015) Design and synthesis of triphenylphosphonium functionalized nanoparticle probe for mitochondria targeting and imaging. J Phys Chem C 119(5):2888–2895

Chatterjee A, Mambo E, Sidransky D (2006) Mitochondrial DNA mutations in human cancer. Oncogene 25(34):4663–4674

Chazotte B (2009) Labeling mitochondria with fluorescent dyes for imaging. Dig Cold Spring Harb Protoc. doi:10.1101/pdb.prot4948.

Chen Y, Dorn GW (2013) PINK1-phosphorylated mitofusin 2 is a Parkin receptor for culling damaged mitochondria. Science 340(6131):471–475

Chen H-H, Chien C-C, Petibois C, Wang C-L, Chu YS, Lai S-F, Hua T-E, Chen Y-Y, Cai X, Kempson IM (2011) Quantitative analysis of nanoparticle internalization in mammalian cells by high resolution X-ray microscopy. J Nanobiotechnol 9(14). doi:10.1186/1477-3155-9-14

Chong Y, Ma Y, Shen H, Tu X, Zhou X, Xu J, Dai J, Fan S, Zhang Z (2014) The in vitro and in vivo toxicity of graphene quantum dots. Biomaterials 35(19):5041–5048

Clancy B, Cauller L (1998) Reduction of background autofluorescence in brain sections following immersion in sodium borohydride. J Neurosci Methods 83(2):97–102

Cooper GM (2000) The cell – a molecular approach, 2nd edn. Sinauer Associates, Sunderland

Cossarizza A, Baccaranicontri M, Kalashnikova G, Franceschi C (1993) A New method for the cytofluorometric analysis of mitochondrial membrane potential using the J-aggregate forming lipophilic cation 5,5′,6,6′-tetrachloro-1,1′,3,3′-tetraethylbenzimidazolcarbocyanine iodide (JC-1). Biochem Biophys Res Commun 197(1):40–45

Damgé C, Socha M, Ubrich N, Maincent P (2010) Poly (ε-caprolactone)/eudragit nanoparticles for oral delivery of aspart-insulin in the treatment of diabetes. J Pharm Sci 99(2):879–889

Dausend J, Musyanovych A, Dass M, Walther P, Schrezenmeier H, Landfester K, Mailänder V (2008) Uptake mechanism of oppositely charged fluorescent nanoparticles in HeLa cells. Macromol Biosci 8(12):1135–1143

Deshayes S, Morris M, Divita G, Heitz F (2005) Cell-penetrating peptides: tools for intracellular delivery of therapeutics. Cell Mol Life Sci 62(16):1839–1849

Dimauro S, Davidzon G (2005) Mitochondrial DNA and disease. Ann Med 37(3):222–232

DiMauro S, Andreu AL, Musumeci O, Bonilla E (2001) Diseases of oxidative phosphorylation due to mtDNA mutations. Semin Neurol 21(3):251–260

Doherty GJ, McMahon HT (2009) Mechanisms of endocytosis. Annu Rev Biochem 78:857–902

Endo T, Yamano K (2009) Multiple pathways for mitochondrial protein traffic. Biol Chem 390(8):723–730

Ernster L, Schatz G (1981) Mitochondria: a historical review. J Cell Biol 91(3):227–255

Fadeel B, Ottosson A, Pervaiz S (2008) Big wheel keeps on turning: apoptosome regulation and its role in chemoresistance. Cell Death Differ 15(3):443–452

Federico A, Cardaioli E, Da Pozzo P, Formichi P, Gallus GN, Radi E (2012) Mitochondria, oxidative stress and neurodegeneration. J Neurol Sci 322(1–2):254–262

Feldhaeusser B, Platt SR, Marrache S, Kolishetti N, Pathak RK, Montgomery DJ, Reno LR, Howerth E, Dhar S (2015) Evaluation of nanoparticle delivered cisplatin in beagles. Nanoscale 7(33):13822–13830

Fernando LP, Kandel PK, Yu J, McNeill J, Ackroyd PC, Christensen KA (2010) Mechanism of cellular uptake of highly fluorescent conjugated polymer nanoparticles. Biomacromolecules 11(10):2675–2682

Flemming N, Gallo L, Forbes JM, Ward MS (2015) Tapping into mitochondria to find novel targets for diabetes complications. Curr Drug Targets 17(12):1341–1349
Fontana G, Licciardi M, Mansueto S, Schillaci D, Giammona G (2001) Amoxicillin-loaded polyethylcyanoacrylate nanoparticles: Influence of PEG coating on the particle size, drug release rate and phagocytic uptake. Biomaterials 22(21):2857–2865
Friedman JR, Nunnari J (2014) Mitochondrial form and function. Nature 505(7483):335–343
Gilmore K, Wilson M (1999) The use of chloromethyl-X-rosamine (Mitotracker red) to measure loss of mitochondrial membrane potential in apoptotic cells is incompatible with cell fixation. Cytometry 36(4):355–358
Greenamyre JT, Hastings TG (2004) Parkinson's–divergent causes, convergent mechanisms. Science 304(5674):1120–1122
Halperin F, Lopez X, Manning R, Kahn CR, Kulkarni RN, Goldfine AB (2012) Insulin augmentation of glucose-stimulated insulin secretion is impaired in insulin-resistant humans. Diabetes 61(2):301–309
Han JY, Kang MJ, Kim KH, Han PL, Kim HS, Ha JY, Son JH (2015) Nitric oxide induction of Parkin translocation in PTEN-induced putative kinase 1 (PINK1) deficiency: functional role of neuronal nitric oxide synthase during mitophagy. J Biol Chem 290(16):10325–10335
Hanahan D, Weinberg RA (2011) Hallmarks of cancer: the next generation. Cell 144(5):646–674
Harush-Frenkel O, Debotton N, Benita S, Altschuler Y (2007) Targeting of nanoparticles to the clathrin-mediated endocytic pathway. Biochem Biophys Res Commun 353(1):26–32
Herce H, Garcia A, Litt J, Kane R, Martin P, Enrique N, Rebolledo A, Milesi V (2009) Arginine-rich peptides destabilize the plasma membrane, consistent with a pore formation translocation mechanism of cell-penetrating peptides. Biophys J 97(7):1917–1925
Hill JH, Chen Z, Xu H (2014) Selective propagation of functional mitochondrial DNA during oogenesis restricts the transmission of a deleterious mitochondrial variant. Nat Genet 46(4):389–392
Houtkooper RH, Vaz FM (2008) Cardiolipin, the heart of mitochondrial metabolism. Cell Mol Life Sci 65(16):2493–2506
Iversen TG, Frerker N, Sandvig K (2012) Uptake of ricinB-quantum dot nanoparticles by a macropinocytosis-like mechanism. J Nanobiotechnology 10(1): 33
Kalathil AA, Kumar A, Banik B, Ruiter TA, Pathak RK, Dhar S (2016) New formulation of old aspirin for better delivery. Chem Commun 52(1):140–143
Karataş ÖF, Sezgin E, Aydın Ö, Çulha M (2009) Interaction of gold nanoparticles with mitochondria. Colloids Surf B 71(2):315–318
Kazlauskaite A, Muqit MM (2015) PINK1 and Parkin – mitochondrial interplay between phosphorylation and ubiquitylation in Parkinson's disease. FEBS J 282(2):215–223
Keij JF, Bell-Prince C, Steinkamp JA (2000) Staining of mitochondrial membranes with 10-nonyl acridine orange MitoFluor Green, and MitoTracker Green is affected by mitochondrial membrane potential altering drugs. Cytometry 39(3):203–210
Klein ND, Hurley KR, Feng ZV, Haynes CL (2015) Dark field transmission electron microscopy as a tool for identifying inorganic nanoparticles in biological matrices. Anal Chem 87(8):4356–4362
Kroemer G (2006) Mitochondria in cancer. Oncogene 25(34):4630–4632
Kwon HJ, Cha M-Y, Kim D, Kim DK, Soh M, Shin K, Hyeon T, Mook-Jung I (2016) Mitochondria-targeting ceria nanoparticles as antioxidants for Alzheimer's disease. ACS Nano 10(2):2860–2870
Li SP-Y, Lau CT-S, Louie M-W, Lam Y-W, Cheng SH, Lo KK-W (2013) Mitochondria-targeting cyclometalated iridium (III)–PEG complexes with tunable photodynamic activity. Biomaterials 34(30):7519–7532
Lim JP, Gleeson PA (2011) Macropinocytosis: an endocytic pathway for internalising large gulps. Immunol Cell Biol 89(8):836–843
Lim SY, Shen W, Gao Z (2015) Carbon quantum dots and their applications. Chem Soc Rev 44(1):362–381

Lindgren M, Hällbrink M, Prochiantz A, Langel Ü (2000) Cell-penetrating peptides. Trends Pharmacol Sci 21(3):99–103
Madani F, Lindberg S, Langel Ü, Futaki S, Gräslund A (2011) Mechanisms of cellular uptake of cell-penetrating peptides. Dig Biophys J. doi:10.1155/2011/414729
Marrache S, Dhar S (2012) Engineering of blended nanoparticle platform for delivery of mitochondria-acting therapeutics. Proc Natl Acad Sci 109(40):16288–16293
Marrache S, Dhar S (2013) Biodegradable synthetic high-density lipoprotein nanoparticles for atherosclerosis. Proc Natl Acad Sci 110(23):9445–9450
Marrache S, Dhar S (2015) The energy blocker inside the power house: mitochondria targeted delivery of 3-bromopyruvate. Chem Sci 6(3):1832–1845
Marrache S, Kumar Pathak R, Darley KL, Choi JH, Zaver D, Kolishetti N, Dhar S (2013) Nanocarriers for tracking and treating diseases. Curr Med Chem 20(28):3500–3514
Marrache S, Pathak RK, Dhar S (2014) Detouring of cisplatin to access mitochondrial genome for overcoming resistance. Proc Natl Acad Sci 111(29):10444–10449
Mathupala SP, Ko YH, Pedersen PL (2006) Hexokinase II: cancer's double-edged sword acting as both facilitator and gatekeeper of malignancy when bound to mitochondria. Oncogene 25(34):4777–4786
Matsuda S, Kitagishi Y, Kobayashi M (2013) Function and characteristics of PINK1 in mitochondria. Dig Oxid Med Cell Longev. doi:10.1155/2013/601587
Maximo V, Lima J, Soares P, Sobrinho-Simoes M (2009) Mitochondria and cancer. Virchows Arch 454(5):481–495
Milane L, Trivedi M, Singh A, Talekar M, Amiji M (2015) Mitochondrial biology, targets, and drug delivery. J Control Release 207:40–58
Mizuno Y, Hattori N, Mori H, Suzuki T, Tanaka K (2001) Parkin and Parkinson's disease. Curr Opin Neurol 14(4):477–482
Mkandawire MM, Lakatos M, Springer A, Clemens A, Appelhans D, Krause-Buchholz U, Pompe W, Rodel G, Mkandawire M (2015) Induction of apoptosis in human cancer cells by targeting mitochondria with gold nanoparticles. Nanoscale 7(24):10634–10640
Monici M (2005) Cell and tissue autofluorescence research and diagnostic applications. Biotechnol Annu Rev 11:227–256
Mosiman VL, Patterson BK, Canterero L, Goolsby CL (1997) Reducing cellular autofluorescence in flow cytometry: an in situ method. Cytometry 30(3):151–156
Nabi IR, Le PU (2003) Caveolae/raft-dependent endocytosis. J Cell Biol 161(4):673–677
Nam HY, Kwon SM, Chung H, Lee S-Y, Kwon S-H, Jeon H, Kim Y, Park JH, Kim J, Her S, Oh Y-K, Kwon IC, Kim K, Jeong SY (2009) Cellular uptake mechanism and intracellular fate of hydrophobically modified glycol chitosan nanoparticles. J Control Release 135(3):259–267
Nicholls DG, Ward MW (2000) Mitochondrial membrane potential and neuronal glutamate excitotoxicity: mortality and millivolts. Trends Neurosci 23(4):166–174
Oh N, Park J-H (2014) Endocytosis and exocytosis of nanoparticles in mammalian cells. D Int J Nanomed. doi:10.2147/IJN.S26592
Oh J-M, Choi S-J, Kim S-T, Choy J-H (2006) Cellular uptake mechanism of an inorganic nanovehicle and its drug conjugates: enhanced efficacy due to clathrin-mediated endocytosis. Bioconjug Chem 17(6):1411–1417
Pathak RK, Dhar S (2015) A nanoparticle cocktail: temporal release of predefined drug combinations. J Am Chem Soc 137(26):8324–8327
Pathak RK, Marrache S, Harn DA, Dhar S (2014) Mito-DCA: a mitochondria targeted molecular scaffold for efficacious delivery of metabolic modulator dichloroacetate. ACS Chem Biol 9(5):1178–1187
Pathak RK, Kolishetti N, Dhar S (2015) Targeted nanoparticles in mitochondrial medicine. WIREs Nanomed Nanobiotechnol 7(3):315–329
Peckys DB, de Jonge N (2011) Visualizing gold nanoparticle uptake in live cells with liquid scanning transmission electron microscopy. Nano Lett 11(4):1733–1738

Pendergrass W, Wolf N, Poot M (2004) Efficacy of MitoTracker Green™ and CMXRosamine to measure changes in mitochondrial membrane potentials in living cells and tissues. Cytometry A 61(2):162–169
Perry SW, Norman JP, Barbieri J, Brown EB, Gelbard HA (2011) Mitochondrial membrane potential probes and the proton gradient: a practical usage guide. Biotechniques 50(2):98–115
Petersen CAH, Alikhani N, Behbahani H, Wiehager B, Pavlov PF, Alafuzoff I, Leinonen V, Ito A, Winblad B, Glaser E (2008) The amyloid β-peptide is imported into mitochondria via the TOM import machinery and localized to mitochondrial cristae. Proc Natl Acad Sci 105(35):13145–13150
Polyak K, Li Y, Zhu H, Lengauer C, Willson JKV, Markowitz SD, Trush MA, Kinzler KW, Vogelstein B (1998) Somatic mutations of the mitochondrial genome in human colorectal tumours. Nat Genet 20(3):291–293
Poot M, Pierce RC (1999) Detection of apoptosis and changes in mitochondrial membrane potential with chloromethyl-X-rosamine. Cytometry 36(4):359–360
Qaddoumi MG, Gukasyan HJ, Davda J, Labhasetwar V, Kim K-J, Lee V (2003) Clathrin and caveolin-1 expression in primary pigmented rabbit conjunctival epithelial cells: role in PLGA nanoparticle endocytosis. Mol Vis 9:559–568
Qu Q, Ma X, Zhao Y (2015) Targeted delivery of doxorubicin to mitochondria using mesoporous silica nanoparticle nanocarriers. Nanoscale 7(40):16677–16686
Ratnam DV, Chandraiah G, Meena A, Ramarao P, Ravi Kumar M (2009) The co-encapsulated antioxidant nanoparticles of ellagic acid and coenzyme Q10 ameliorates hyperlipidemia in high fat diet fed rats. J Nanosci Nanotechnol 9(11):6741–6746
Resch-Genger U, Grabolle M, Cavaliere-Jaricot S, Nitschke R, Nann T (2008) Quantum dots versus organic dyes as fluorescent labels. Nat Methods 5(9):763–775
Rin Jean S, Tulumello DV, Wisnovsky SP, Lei EK, Pereira MP, Kelley SO (2014) Molecular vehicles for mitochondrial chemical biology and drug delivery. ACS Chem Biol 9(2):323–333
Roberts CK, Sindhu KK (2009) Oxidative stress and metabolic syndrome. Dig Life Sci. doi:10.1016/j.lfs.2009.02.026
Ross M, Kelso G, Blaikie F, James A, Cocheme H, Filipovska A, Da Ros T, Hurd T, Smith R, Murphy M (2005) Lipophilic triphenylphosphonium cations as tools in mitochondrial bioenergetics and free radical biology. Biochemistry 70(2):222–230
Sakhrani NM, Padh H (2013) Organelle targeting: third level of drug targeting. Drug Des Devel Ther 7:585–599
Santucci R, Sinibaldi F, Polticelli F, Fiorucci L (2014) Role of cardiolipin in mitochondrial diseases and apoptosis. Curr Med Chem 21(23):2702–2714
Scatena R (2012) Mitochondria and cancer: a growing role in apoptosis, cancer cell metabolism and dedifferentiation. Adv Exp Med Biol 942:287–308
Schnell SA, Staines WA, Wessendorf MW (1999) Reduction of lipofuscin-like autofluorescence in fluorescently labeled tissue. J Histochem Cytochem 47(6):719–730
Scorrano L, Petronilli V, Colonna R, Di Lisa F, Bernardi P (1999) Chloromethyltetramethylrosamine (Mitotracker OrangeTM) induces the mitochondrial permeability transition and inhibits respiratory complex I implication for the mechanism of cytochrome c release. J Biol Chem 274(35):24657–24663
Sharma K (2015) Mitochondrial hormesis and diabetic complications. Diabetes 64(3):663–672
Smith AM, Nie S (2009) Next-generation quantum dots. Nat Biotechnol 27(8):732–733
Smith RA, Porteous CM, Gane AM, Murphy MP (2003) Delivery of bioactive molecules to mitochondria in vivo. Proc Natl Acad Sci 100(9):5407–5412
Solazzo M, Fantappie O, Lasagna N, Sassoli C, Nosi D, Mazzanti R (2006) P-gp localization in mitochondria and its functional characterization in multiple drug-resistant cell lines. Exp Cell Res 312(20):4070–4078
Stojanovski D, Bohnert M, Pfanner N, van der Laan M (2012) Mechanisms of protein sorting in mitochondria. Dig Cold Spring Harb Protoc. doi:10.1101/cshperspect.a011320

Tatsuta T, Scharwey M, Langer T (2014) Mitochondrial lipid trafficking. Trends Cell Biol 24(1):44–52

Ungewickell EJ, Hinrichsen L (2007) Endocytosis: clathrin-mediated membrane budding. Curr Opin Cell Biol 19(4):417–425

van Gisbergen MW, Voets AM, Starmans MHW, de Coo IFM, Yadak R, Hoffmann RF, Boutros PC, Smeets HJM, Dubois L, Lambin P (2015) How do changes in the mtDNA and mitochondrial dysfunction influence cancer and cancer therapy? Challenges, opportunities and models. Mutat Res 764:16–30

Vasquez-Trincado C, Garcia-Carvajal I, Pennanen C, Parra V, Hill JA, Rothermel BA, Lavandero S (2016) Mitochondrial dynamics, mitophagy and cardiovascular disease. J Physiol 594(3):509–525

Verma A, Uzun O, Hu Y, Hu Y, Han H-S, Watson N, Chen S, Irvine DJ, Stellacci F (2008) Surface-structure-regulated cell-membrane penetration by monolayer-protected nanoparticles. Nat Mater 7(7):588–595

Verschoor ML, Ungard R, Harbottle A, Jakupciak JP, Parr RL, Singh G (2013) Mitochondria and cancer: past, present, and future. Biomed Res Int 2013:612369

Vieira HL, Haouzi D, El Hamel C, Jacotot E, Belzacq AS, Brenner C, Kroemer G (2000) Permeabilization of the mitochondrial inner membrane during apoptosis: impact of the adenine nucleotide translocator. Cell Death Differ 7(12):1146–1154

Voigt J, Christensen J, Shastri VP (2014) Differential uptake of nanoparticles by endothelial cells through polyelectrolytes with affinity for caveolae. Proc Natl Acad Sci 111(8):2942–2947

Wallace DC (2012) Mitochondria and cancer. Nat Rev Cancer 12(10):685–698

Walther DM, Rapaport D (2009) Biogenesis of mitochondrial outer membrane proteins. BBA-Mol Cell Res 1793(1):42–51

Wang A, Melosh N (2012) Direct penetration of cell-penetrating peptides across lipid bilayers. Biophys J 102(3):487a

Wang Z, Tiruppathi C, Minshall RD, Malik AB (2009) Size and dynamics of caveolae studied using nanoparticles in living endothelial cells. ACS Nano 3(12):4110–4116

Weinberg SE, Chandel NS (2015) Targeting mitochondria metabolism for cancer therapy. Nat Chem Biol 11(1):9–15

Weksler-Zangen S, Raz I, Lenzen S, Jorns A, Ehrenfeld S, Amir G, Oprescu A, Yagil Y, Yagil C, Zangen DH, Kaiser N (2008) Impaired glucose-stimulated insulin secretion is coupled with exocrine pancreatic lesions in the Cohen diabetic rat. Diabetes 57(2):279–287

Wen R, Banik B, Pathak RK, Kumar A, Kolishetti N, Dhar S (2016) Nanotechnology inspired tools for mitochondrial dysfunction related diseases. Adv Drug Deliv Rev 99(A): 52–69

White J, Amos W, Fordham M (1987) An evaluation of confocal versus conventional imaging of biological structures by fluorescence light microscopy. J Cell Biol 105(1):41–48

Wickramasekera NT, Das GM (2014) Tumor suppressor p53 and estrogen receptors in nuclear–mitochondrial communication. Mitochondrion 16:26–37

Yamada Y, Harashima H (2013) Enhancement in selective mitochondrial association by direct modification of a mitochondrial targeting signal peptide on a liposomal based nanocarrier. Mitochondrion 13(5):526–532

Yamada Y, Akita H, Kamiya H, Kogure K, Yamamoto T, Shinohara Y, Yamashita K, Kobayashi H, Kikuchi H, Harashima H (2008) MITO-Porter: A liposome-based carrier system for delivery of macromolecules into mitochondria via membrane fusion. BBA Biomembranes 1778(2):423–432

Yamada Y, Fukuda Y, Harashima H (2015a) An analysis of membrane fusion between mitochondrial double membranes and MITO-Porter, mitochondrial fusogenic vesicles. Mitochondrion 24:50–55

Yamada Y, Nakamura K, Abe J, Hyodo M, Haga S, Ozaki M, Harashima H (2015b) Mitochondrial delivery of Coenzyme Qvia systemic administration using a MITO-Porter prevents ischemia/reperfusion injury in the mouse liver. J Control Release 213:86–95

Yang S-T, Cao L, Luo PG, Lu F, Wang X, Wang H, Meziani MJ, Liu Y, Qi G, Sun Y-P (2009) Carbon dots for optical imaging in vivo. J Am Chem Soc 131(32):11308–11309

Yousif LF, Stewart KM, Kelley SO (2009) Targeting mitochondria with organelle-specific compounds: strategies and applications. ChemBioChem 10(12):1939–1950

Zhang E, Zhang C, Su Y, Cheng T, Shi C (2011) Newly developed strategies for multifunctional mitochondria-targeted agents in cancer therapy. Drug Discov Today 16(3–4):140–146

Zhang Y, Shen Y, Teng X, Yan M, Bi H, Morais PC (2015) Mitochondria-targeting nanoplatform with fluorescent carbon dots for long time imaging and magnetic field-enhanced cellular uptake. ACS Appl Mater Interfaces 7(19):10201–10212

Zhou F, Xing D, Wu B, Wu S, Ou Z, Chen WR (2010) New insights of transmembranal mechanism and subcellular localization of noncovalently modified single-walled carbon nanotubes. Nano Lett 10(5):1677–1681

Chapter 10
Redox-Responsive Nano-Delivery Systems for Cancer Therapy

Amit Singh, Thanh-Huyen Tran, and Mansoor M. Amiji

Abstract Stimuli-responsive nano-carrier systems have been pursued with great interest due to their advantages such as controlled drug release, improved pharmacokinetics and pharmacodynamics, and reduced side effects of the drugs. These nano-carriers have potential to accumulate effectively into the tumor due to "*enhanced permeability and retention*" (EPR) effect and advancements in surface science further allow active targeting strategies. Development of some such nano-vectors is driven by hallmark characteristics of tumor microenvironment such as hypoxia, acidic pH, and reducing biologically-relevant milieu This chapter highlights the features of self-regulated internally controlled redox-responsive nano-carrier systems, opportunities presented by them and the promise these "intelligent" delivery vectors offer in improving existing cancer therapeutic approaches.

Keywords Stimuli-responsive nanoparticles • Redox • Targeted delivery • Cancer

Abbreviations

ABC	ATP binding casette
cGLP	Current Good Laboratory Practices
cGMP	Current Good Manufacturing Practices
CNC	Cellulose nanocrystals
CPP-SA	1,3-bis(carboxyphenoxy) propane-sebacic acid
DOPE	Dioleoyl phosphatidylethanolamine
DTT	Dithiothreitol
ECM	Extracellular matrix

A. Singh • T.-H. Tran • M.M. Amiji (✉)
Department of Pharmaceutical Sciences, School of Pharmacy, Bouve College of Health Sciences, Northeastern University, 140 The Fenway Building, 360 Huntington Avenue, Boston, MA 02115, USA
e-mail: m.amiji@neu.edu

A. Prokop, V. Weissig (eds.), *Intracellular Delivery III*, Fundamental Biomedical Technologies 8, DOI 10.1007/978-3-319-43525-1_10

EGFR	Epidermal growth factor receptor
EPR	Enhanced permeability and retention
FDA	Food and Drug Administration
GSH	Glutathione
HIF	Hypoxia inducible factor
MMP	Matrix metalloproteinase
PDMAEMA	Poly(2-(dimethylamino) ethyl methacrylate)
PEG	Poly(ethylene glycol)
TMBQ	Trimethyl-locked benzoquinone
VCAM	Vascular cell adhesion molecule
VEGF	Vascular endothelial growth factor

10.1 Introduction

Drug delivery systems have come a long way since their advent and have shown tremendous promise in improving the pharmacokinetics and pharmacodynamics of drug with poor bioavailability while reducing the potential side effects. The previous generation of drug delivery systems would accumulate into the tumor due to "*enhanced permeability and retention*" (EPR) effect and strategies to selectively target further facilitate higher intracellular uptake with potential to improve efficacy profile. The new generations of molecular assemblies however have moved far ahead of the conventional drug delivery vectors and there is a growing interest in developing stimuli-responsive systems to control the release profile of the payload. Such stimuli-responsive delivery vehicles sequester the payload, protect it from direct interaction with the elements of the body and poly(ethylene) glycol (PEG) modification further assures increased drug half-life and residence time in the body. Additionally, the release of the encapsulated guests is minimal in the absence of the appropriate stimulus but in the presence of the stimulus, the drug is released. The motivation to developing such delivery vectors is guided by the tumor microenvironment, payload characteristics, desired rate of release and mechanism of release. To this end, stimuli-responsive systems offer great advantage in precise control over drug release profile, temporally as well as spatially by external or internal stimuli. Tumor microenvironment characteristically shows lower pH due to aerobic glycolysis and hypoxia induced reducing local milieu, which are major drivers of pH and redox responsive delivery systems respectively. Alternatively, the stimulus for drug release can be external where the nano-carrier responds on exposure to temperature, ultrasound, magnetic field, light or electric field (Fig. 10.1) (Iyer et al. 2013). Though the true clinical significance of the stimuli-responsive delivery systems is yet to be established, the indications from platform technologies show tremendous promise in overcoming some of the biological and pathophysiological challenges associated with cancer therapeutics.

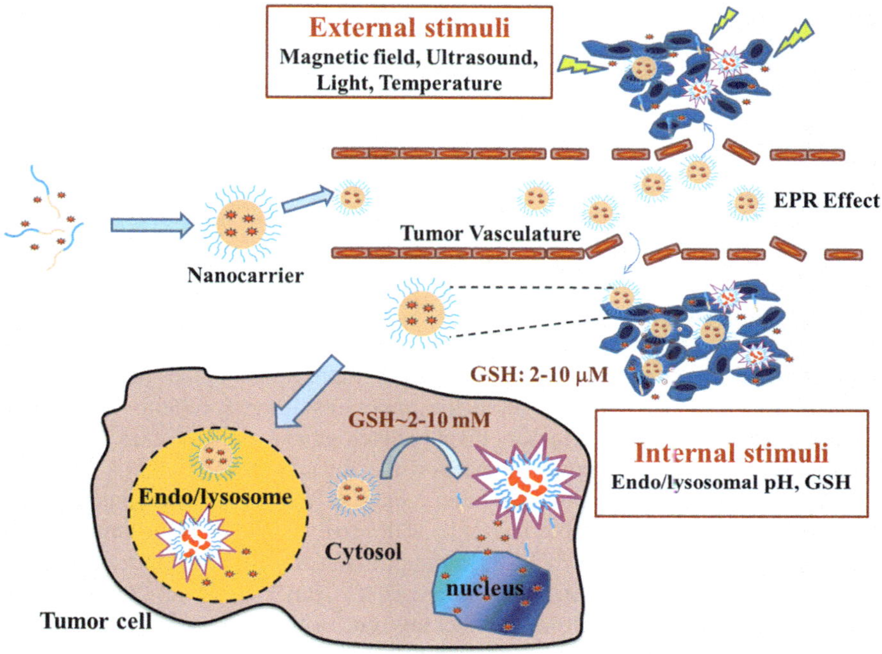

Fig. 10.1 Schematic representation of the fate of a stimuli-responsive nano-carrier in the systemic circulation

10.2 Stimuli-Responsive Systems and Cancer

Tumor microenvironment is a complex system that involves the interaction between malignant and neighbor stroma cells (fibroblast, endothelia, and inflammatory cells) embedded in a mesh of extracellular matrix (ECM) components that strongly influence tumor initiation, growth and invasion. The tumor microenvironment has several characteristics that distinguish it from the corresponding normal tissue including abnormal vasculature, low oxygen level or hypoxia, low extracellular pH, and high interstitial fluid pressure. These subtle difference in the microenvironment within the tumor facilitate an ideal internal stimulus for the self-regulated redox-responsive system to disseminate their payload (Iyer et al. 2013). A comprehensive general review on stimuli-responsive drug delivery systems has been published previously and is recommended for the reader (Ganta et al. 2008). This book chapter exclusively focuses on redox-responsive systems for cancer therapy with special emphasis on factors contributing to the difference in reduction potential within the tumors.

10.2.1 Hypoxia

Hypoxia refers to the condition in which tissues are deprived of oxygen or the partial pressure of oxygen falls below critical levels leading to hindered biological functioning of cells or organs. Within normal tissues, O_2 supply matches metabolic needs, but in tumors, demand far exceeds supply. Normal tissue pO_2 ranges between 10 and 80 mmHg depending on tissue type, whereas tumors contain regions where the pO_2 is <5 mmHg (Hockel and Vaupel 2001). Tumor hypoxia is generally attributed to chaotic and poorly organized blood vessels within the tumor tissues. The diffusion limit for O_2 is approximately 100–200 μm from a blood capillary (Carol Ward et al. 2013). Consequently, as tumor growth reaches 1–2 mm^3, abnormal tumor vasculature along with its leakiness and compression, reduces blood flow and limits delivery of oxygen throughout the tumor resulting in regions of hypoxia. Generally, 50–60 % of solid tumors display heterogeneous distribution of hypoxic regions within tumor. Because of inaccessible location, hypoxic cells have less chance to accumulate chemotherapeutic agents. In addition, hypoxia can trigger activation of hypoxia inducible factor-1α (HIF-1α) which has been implicated in regulating gene transcription of ATP Binding Cassette (ABC) transporters in cancer cells and plays a key role in tumor progression (Comerford et al. 2002). Hence, tumor hypoxia is resistant to chemotherapy due to the over-expression of ABC transporter, which pumps therapeutic agents out of cancer cells.

10.2.2 Low Extracellular pH

The tumors exhibit a lower extracellular pH than the normal tissue, which results from the over-production of lactic acid and hydrolysis of ATP by tumor cells. Tumor cells require abnormally high glycolysis to meet their fast metabolism and proliferation. Accordingly, a large amount of lactic acid in the tumor interstitium is generated from the glycolysis. In addition, carbon dioxide, the end product of cell respiration also contributes to the acidic extracellular pH in the tumor microenvironment. Hypoxia-induced expression of carbonic anhydrase IX also contributes to exacerbate the pH gradient between the intra- and extra-cellular tumor compartment (Romero-Garcia et al. 2011). The pH within the tumor is highly heterogeneous such that areas with optimal blood supply show pH closer to normal physiological pH of 7.4 while distant areas with poor or no vasculature show lower pH. The abnormal pH gradient continues from extracellular microenvironment to intracellular organelles, such as pH 5.5 in endosomes and pH < 5.5 in lysosomes that has been exploited to design pH sensitive delivery systems.

10.2.3 Redox-Responsive Systems

The basic principle of redox-responsive systems is to utilize the distinct difference in redox potential between tumor and normal tissues. Glutathione (GSH), an abundant reducing agent in living cells, has a higher interacellular concentration (~2–10 mM), compared to that in blood and extracellular matrix (~2–10 μM). Moreover, tumor tissues has about seven-fold higher GSH concentration than that of normal tissues, and is often elevated in multidrug resistance cancer cells (Meng et al. 2009). Additionally, nano-carriers internalized in the cells are initially sequestered in the endosomal/lysosomal compartment where presence of enzymes such as gamma interferon-inducible lysosomal thiol reductase (GILT) leads to cleavage of disulfide linkage (Arunachalam et al. 2000). These factors have been exploited to design redox-responsive nano-carriers containing disulfide bonds for releasing the therapeutic agents inside the tumor cell. Drugs that are not encapsulated into the vehicles could be bound onto the carriers via disulfide bond. The intracellular reduction is most likely mediated by disruption of disulfide bond that can be converted reversibly to thiols and undergo disulfide exchange in the presence of GSH and other thiols (Danhier et al. 2010). Besides broadly used disulfide linkages, diselenides are other functional groups used for redox-responsive drug release. Diselenide-containing polymer had the ability to inhibit tumor proliferation and triggered drug release even at low GSH concentration of 0.01 mg/mL.

10.2.4 Abnormal Tumor Vasculature

Abnormal tumor vasculature is the most defining characteristics of the tumor microenvironment. In normal tissues, the balance between proangiogenic and antiangiogenic molecules regulates the development of vascular system, which maintains the efficient and order network of blood vessels and lymphatic system. However, the balance of growth factors in the tumor is perturbed, which leads to the development of a disorganized vasculature and a complete lack of lymphatics. The architecture of the vasculature is also highly disorganized. The mean vascular density of tumor is generally lower than in normal tissues, and diffusion distances are greater. The vascular network often contains blind-ends and arteriolar-venous shunts. Finally, tumor vessels are generally long and highly chaotic with pericyte deficiency and aberrant basement membrane formation leading to an enhanced vascular permeability (Kumagai et al. 2002; Cairns et al. 2006). Nano-carriers exploit the high permeability of the tumor vasculature to extravasate and accumulate specifically into the tumor tissues, a phenomenon popularly termed as EPR effect.

10.2.5 Tumor Targeting Strategy

Tumor targeting strategies have been divided into two categories: passive targeting and active targeting. Passive targeting relies on the properties of the delivery system and the tumor vasculature to preferentially accumulate the drug in the tumor tissues and avoid the distribution to healthy tissues. Due to the leaky tumor vasculature and insufficient lymphatic drainage, nano-carriers can extravasate and be retained in the tumor. This phenomenon has been called EPR effect, first identified in the 1980s by Maeda et al (Matsumura and Maeda 1986). For passive targeting, nanoparticles have to circulate in blood for sufficient long periods. This goal can be achieved through optimal stealth properties, typically by incorporation of polymers like poly(ethylene glycol) (PEG) into the shell of the nano-carriers. Nano-carriers in the 10–100 nm size range and with a surface charge either slightly positive or slightly negative are often not cleared by renal clearance and reticuloendothelial system, thus accumulate at the tumors after prolonged circulation. However, successful implementation of EPR effect is dependent upon a number of factors including degree of capillary disorder, blood flow, and lymphatic drainage rate, making effective management difficult. Furthermore, the accumulation of drugs in tumor tissue does not always guarantee successful therapy if the drug does not reach the target site of the tumor cell such as the cell membrane, cytosol, or nucleus. In addition, high interstitial fluid pressure (IFP) in tumor could hinder the nano-carrier retention in tumor tissues. For effective cancer therapy, it is critical to precisely guide nano-carriers to a specific cell type or a specific non-cellular component in the tumor microenvironment, which could be achieved by actively targeted delivery systems.

For active targeting, nano-carriers generally need to first localize to the tumor mass; therefore, initial design considerations are similar for passive targeting. Active targeting then exploits tumor cell characteristics, such as over-expression of some specific antigens or receptors on their surfaces that are at low levels in normal tissue cells. Active drug targeting is achieved by chemical attachment of nano-carriers to a targeting ligand such as antibodies, peptides, nucleic acid aptamers, carbohydrates and small molecules that strongly interact with antigens (or receptors) displayed on the surface of tumor cells. Binding affinity of the targeting ligands with their receptors influences the tumor penetration of the nano-carriers. For targets in which cells are readily accessible, typically the tumor vasculature, high affinity binding is preferable. In the active targeting strategy, two cellular targets can be distinguished: the targeting of cancer cells, and the targeting of tumor endothelial cells. The aim of targeting cancer cells is to enhance cellular uptake of the nano-carriers. Thus, the active targeting of cancer cells is particularly attractive for the intracellular delivery of DNA, siRNA, and protein. The enhanced cellular uptake rather than an increased tumor accumulation is responsible for the anticancer efficacy of the actively targeted nano-carriers. The popular receptors for cancer cell targeting include transferrin, folate, glycoprotein, and epidermal growth factor receptor (EGFR). For targeting of tumor endothelial cells, the receptors of vascular endothelia growth factors (VEGF), the $\alpha_v\beta_3$ integrin, vascular cell adhesion molecule-1 (VCAM-1), and the matrix metalloproteinase (MMPs) are targeted. Among targeting delivery systems, $\alpha_v\beta_3$ integrin targeted nano-carriers could be considered as double targeting systems because $\alpha_v\beta_3$ integrin are over-expressed in both tumor cells and angiogenic endothelial cells (Desgrosellier and Cheresh 2010).

10.3 Illustrative Examples of Redox-Responsive Delivery Systems

10.3.1 Polymeric Systems

Disulfide linkages have been most extensively used for designing polymeric redox responsive delivery systems though selenium from the same group has also been explored especially due to lower bond energy of diselenide and carbon-selenide bond (Se–Se ~ 172 $kJmol^{-1}$, C–Se ~ 244 $kJmol^{-1}$). Dithiodipropionic acid, bis(2,2'-hydroxyethyl) disulfide, cystamine and their derivatives are most commonly used chemical entities for creating disulfide linkage to enable crosslinking of the monomers. Some recent reports however have explored beyond these linkages and novel redox-responsive functional groups have been developed such as platinum (IV)-coordinate polymers where derivatives of platinum-drugs such as cisplatin are introduced into polymer backbone and the resulting micelles formed could be used for encapsulating a secondary drug (Huo et al. 2014). Trimethyl-locked benzoquinone (TMBQ) and 4-N-amino-2,2,6,6-tetramethylpiperidin-1-oxyl-4-yl are two relatively novel functional groups that show sensitivity to redox environment but are yet to be explored for drug delivery application. Wang et al. developed an amphiphilic PEG-based ether-anhydride copolymer (m-PEG-SS-CPP-SA) system where the mPEG is connected with the 1,3-bis(carboxyphenoxy) propane-sebacic acid (CPP-SA) vis disulfide linkage. The copolymer self-assembles to form micellar nanoparticles of an average size of 69 nm in an aqueous environment to encapsulate the hydrophobic drug (curcumin) and releases the drug under the redox environment. The *in vivo* biodistribution confirmed enhanced bioavailability of the drug and subsequent anticancer activity was enhanced compared to free drug as well as drug loaded in non-redox responsive nanoparticles (Wang et al. 2014a).

Initial attempts used block copolymer scaffolds to design redox-responsive controlled drug delivery vehicles but recent endeavors have led to application of natural polymers such as gelatin or cotton for such applications. Hu *et al.*, have used the cotton derived cellulose nanocrystals (CNCs) to design a polycation derived redox-responsive system for a gene/drug combination therapy. The backbone of CNCs could be functionalized with poly(2-(dimethylamino) ethyl methacrylate) (PDMAEMA) using disulfide linkage, thereby imparting the nanocomposite with genes encapsulation capability and sensitivity to redox environment (Hu et al. 2015a). Similarly, thiolated gelatin could be crosslinked to form nanoparticles, which could be successfully loaded with nucleic acids to be delivered to a desired target using suitable ligand and PEG modification for long circulation. PEG modified thioloated gelatin nanoparticles loaded with gene encoding soluble form of vascular endothelial growth factor could effectively curb tumor growth and angiogenesis in human breast cancer orthotopic model (Iyer et al. 2013). More recent work from our group has demonstrated that these redox-responsive gelatin nanoparticles can be decorated with targeting ligands to selectively target a receptor on the cancer cell's surface to enhance efficacy and mitigate non-specific toxicity. The *in vitro* and *in vivo* studies of EGFR-targeted redox-responsive gelatin nanoparticles in subcutaneous as well as orthotopic pancreatic cancer mouse models demonstrate enhanced efficacy and safety (Xu et al. 2014; Singh et al. 2015). The detailed *in vitro* payload release studies confirm that these nanoparticles could efficiently delivery their payload when triggered through a redox environment simulated by using either glutathione or dithiothreitol and therefore cement the superiority of this approach over conventional nanoparticle based drug delivery for anticancer therapy.

In a recent report, Ma *et al.,* fabricated diselenide bond modified block copolymer to assemble micellar structure that demonstrate redox response even under mild conditions (Ma et al. 2010). More recently, efforts have been made to impart dual (or multiple) stimuli-responsive functionality within a single delivery system to benefit from other internal and external stimuli that persist in the tumor microenvironment. Redox-responsive linkages have been coupled with temperature sensitive and pH sensitive materials to impart additional control over payload release and have been successfully demonstrated to application in delivery of small molecules and biologics (Cheng et al. 2013). Dai *et al.,* for example, developed highly packed interlayer-crosslinked micelles (HP-IPM) with dual pH and redox sensitivity that were loaded with doxorubicin as an intrinsically fluorescent anticancer drug. Compact packing of the nanoparticles could provide a stable drug loading with minimum leakage and in vivo studies revealed that the nanoparticles could successfully accumulate in Bel-7402 hepatocellular xenograft to show improved tumor growth inhibition effect (Fig. 10.2) (Dai et al. 2011). A better understanding of the contribution of the molecular and physiological tumor heterogeneity towards therapeutic challenges and realization of the advantages of subtle internal physiological differences such as redox-potential towards improvement in design of delivery vectors has led to a surge in the scientific focus and literature publications (Wang et al. 2013; Han et al. 2014; Shao et al. 2014; Zhuang et al. 2014; Hu et al. 2015b; Xiao et al. 2015). A recent comprehensive review summarizing the design considerations and application of redox-responsive polymeric nanoparticles for drug delivery application has been published and is highly recommended for a detailed insight on the topic (Huo et al. 2014).

10.3.2 Lipid-Based Systems

Lipid-based delivery systems, especially liposomes have been studied for a long time for therapy applications and there is a tremendous interest in developing such stimuli-responsive systems. Interestingly, there have been fewer endeavors to

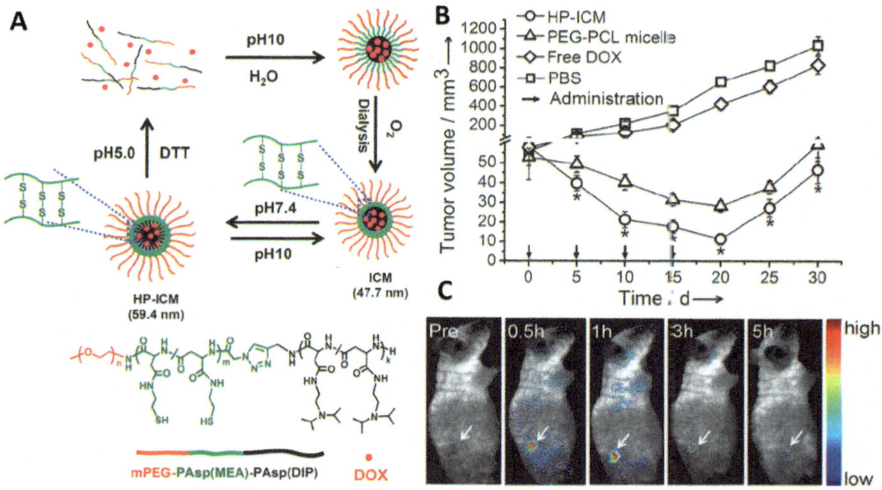

Fig. 10.2 (**a**) Formation and structural transitions of the dual-sensitive HP-ICM. Polymer composition based on ^1H NMR spectroscopy: n = 45, m = 15, k = 14. DOX = doxorubicin. (**b**) Tumor growth inhibition in nude mice bearing the Bel-7402 tumor after tail-vein injection of different formulations (**n** = 20; dose: 5 mg DOX per kg body weight per injection for DOX or DOX-loaded micelles). (**c**) *In vivo* DOX fluorescence images showing passive tumor accumulation of DOX-loaded HP-ICMs after tail-vein injection into nude mice bearing the Bel-7402 xenograft (dose: 5 mg DOX per kg body weight) (Reprinted with permission from Wiley-VCH (Dai et al. 2011))

realize such a system, probably due to the labile nature of the liposomes that are amenable to loss in structural integrity with smallest of change in interaction between the phospholipids. To this end, designing redox-responsive liposomes involves modifying constituent lipids with redox sensitive groups that compromises permeability/structural integrity of the nanoparticles in the suitable environment to facilitate drug release. Ong *et al.* modified dioleoyl phosphatidylethanolamine (DOPE) lipids with redox-sensitive quinone groups (1-Q3), which facilitate structural transition from lamellar liposomal (L_α) to hexagonal columnar (H_{II}) inverted micelle shape, thereby releasing the encapsulated content (Fig. 10.3). Though demonstrated on surrogate molecule calcein, the delivery system shows rapid payload release upon exposure to sodium hydrosulfite and thus have great potential as bioreductive delivery vehicle (Ong et al. 2008). Similarly, Zhang et al. prepared gallate derivative of palmitoyl oleoyl phosphoethanolamine (POPE), which was used with palmitoyl oleoyl phosphatidylcholine (POPC) to make large unilamellar vesicles that could be crosslinked by click chemistry to give a pH and redox dual-responsive delivery system. The vesicles show minimal drug release in the absence of a reducing agent (~20 % in 7 h) but exposure to dithiothreitol (DTT) resulted in rapid release of the drug (~90 % within 1 h) (Zhang and Zhao 2011). Redox-responsive lipidic systems are yet to be explored to their true potential but the initial results are certainly indicative of their beneficial application.

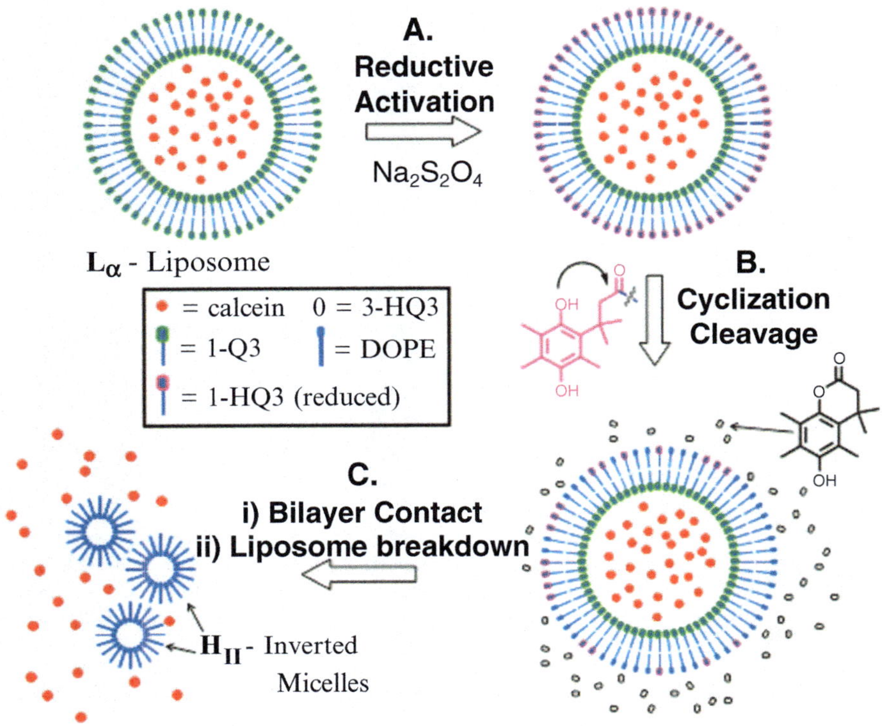

Fig. 10.3 Proposed mechanism for the conversion of liposomal 1-Q3 from lamellar (L_α) liposomes to inverted hexagonal (H_{II}) DOPE micelles. Reprinted with permission from ACS (Ong et al. 2008)

10.3.3 Hybrid Systems

A hybrid system is a relatively new concept of incorporating two or more components of different origin such as polymers, lipids or inorganic materials to design an integrated delivery vehicle. In any typical hybrid delivery vehicle, one (or more) component contributes as building block for structures to house and protect the therapeutic moiety while other components impart desired characteristics including stimuli-responsiveness. Such systems have inherent advantage since strengths of different constituting components can be simultaneously exploited to design more powerful and controlled delivery system. Wang *et al.,* developed a PEG-based polymeric lipid vesicle (PPLV) system with pH-sensitive hydrazone linkage and redox-responsive disulfide bond. *In vitro* results confirmed active uptake in acidic environment (pH ~5), rapid drug release profile triggered by a redox environment (10 mM glutathione) and therefore an efficient cytotoxic impact on the cells (Wang et al. 2014b). Mesoporous silica nanoparticles (MSNs) has been one of the most studied hybrid redox-responsive delivery system where the silica particles are used as nanoreservoirs of the drug with suitable gate keepers such as cadmium sulfide

(CdS) or iron oxide (Fe_3O_4) nanoparticles, cyclodextrin, poly(amido amine) dendrimer etc to control drug release via disulfide-based redox characteristics (Yang et al. 2012). As an example, Lai *et al.* capped MSNs with 2-(propyldisulfanyl) ethylamine groups, which functionalized the surface with mercaptoacetic acid coated CdS nanoparticles as gatekeepers of the pores (Fig. 10.4). Disulfide crosslinks ensure minimal release of the loaded cargo (<1 %) for up to 12 h in the absence of suitable stimulus but in the presence of DTT, the drug was rapidly release. Their study further revealed that the rate of drug release could be controlled precisely by tailoring the concentration of DTT (Lai et al. 2003).

More interestingly, redox-sensitive iron oxide capped MSN not only provides a controlled drug release characteristics to the delivery vehicle but also aids in magnetic responsiveness, which could be used to externally guide the delivery vehicle to the desired target site in the body. Giri *et al.*, successfully developed an iron oxide coated redox-responsive MSN system that could release its payload in response to a reducing environment and demonstrated excellent response to externally applied magnetic field (Giri et al. 2005). Such hybrid systems have performed exceptionally in the *in vitro* settings but there is a dearth of reports demonstrating the strengths of these vehicles *in vivo*. These initials reports however have shown encouraging results and careful design of these multifunctional systems would have potential in improving the outcome of the existing therapies against cancer (Fig. 10.5).

10.4 Conclusions and Future Perspective

Significant strides have been made towards developing and demonstrating the strengths of the redox responsive nanosystems in drug delivery. Their application in cancer therapeutics specially is very promising since these stimuli responsive systems exploit the inherent tumor microenvironment to trigger the release of the payload, assisting in site-specific delivery. Suitable surface modification can further provide "stealth properties" and target selectivity by using PEG and targeting ligands respectively. However, despite considerable advancement in design and significant pre-clinical success of the redox-responsive system, there is a dearth of studies at the clinical level. Material safety, clearance profile and biocompatibility remain major concern and even though most preclinical reports perform safety assessment, these studies are conducted in animal models, which may not be translatable to human subjects due to intra and inter-species physiological differences. Mass scale production for commercial application of these systems is another challenge that needs to be addressed. The US food and drug administration (FDA) has laid strict guidelines for good laboratory and manufacturing practices (cGLP/cGMP) that have to me met to ascertain batch-to-batch consistency of the formulation prior to their use of clinical trials in human subjects. Much of these critical issues are still to be addressed for redox-responsive delivery systems but the preclinical success does indicate considerable promise in application of these systems in addressing the challenges in cancer therapy.

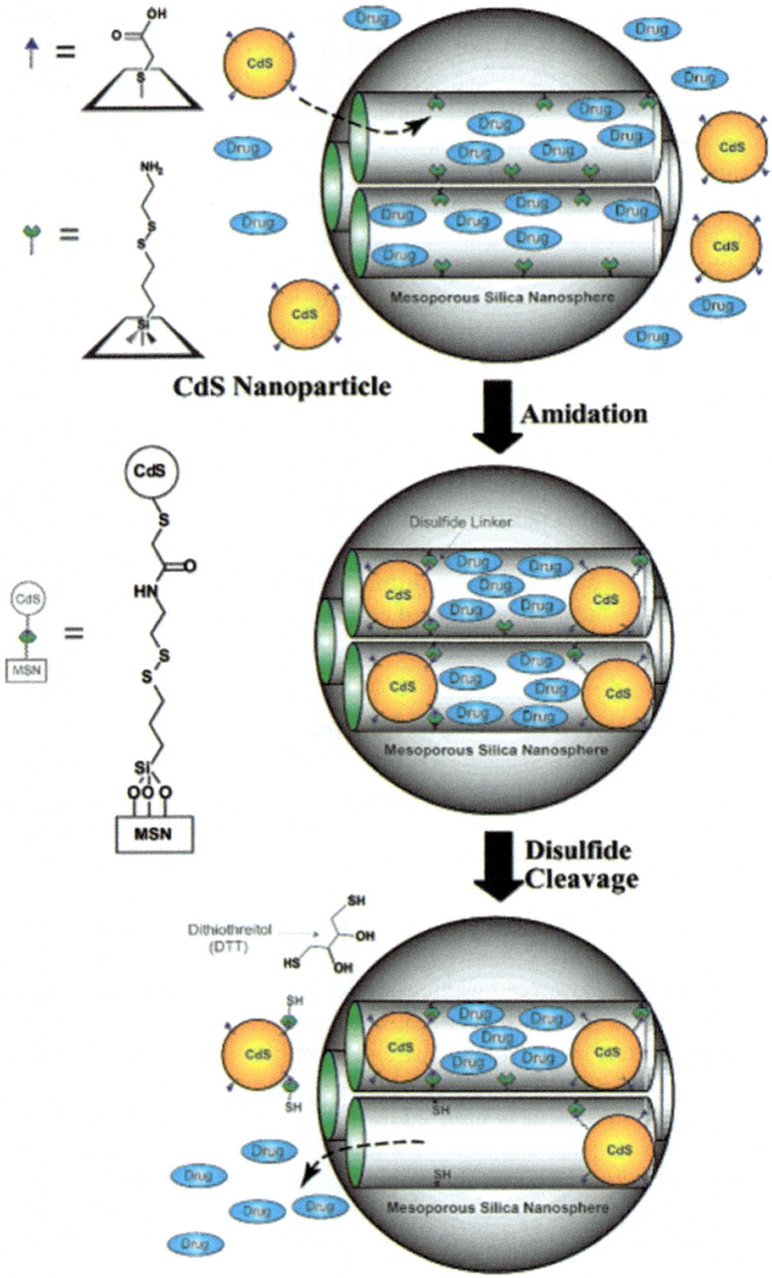

Fig. 10.4 Schematic representation of the CdS nanoparticle-capped MSN-based drug/neurotransmitter delivery system. The controlled-release mechanism of the system is based on chemical reduction of the disulfide linkage between the CdS caps and the MSN hosts (Reprinted with permission from ACS (Lai et al. 2003)

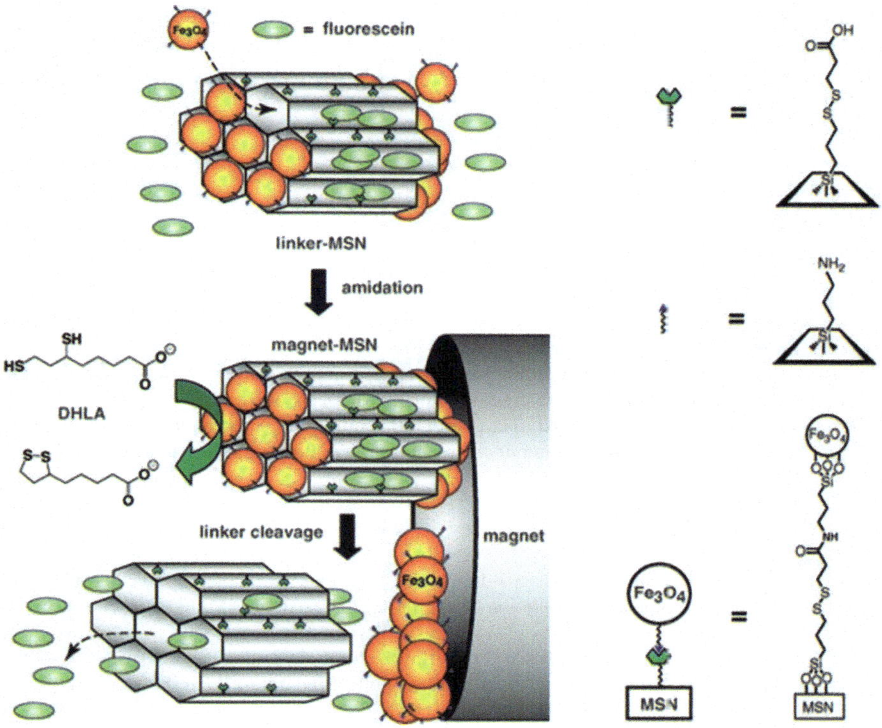

Fig. 10.5 Schematic of the stimuli-responsive delivery system (magnet-MSN) based on mesoporous silica nanorods capped with superparamagnetic iron oxide nanoparticles. The controlled-release mechanism of the system is based on reduction of the disulfide linkage between the Fe_3O_4 nanoparticle caps and the linker-MSN hosts by reducing agents such as DHLA (Giri et al. 2005)

10.5 Summary

- Tumor microenvironment has high concentration of glutathion (GSH) and elevated intracellular GSH which has been recognized as an ideal internal stimulus for rapid destabilization of nano-carriers inside cells to accomplish efficient intracellular drug release.
- Various redox-responsive nano-carriers have been designed based on chemical linkages which are stable in the circulation due to low GSH concentration but undergoes destabilization in the tumor microenvironment to trigger the payload release. Passive and active targeted delivery strategies can be used for tumor-specific accumulation upon systemic administration. Advances in material engineering can then allow for incorporation of additional stimuli-responsive properties.
- Despite several advances, numerous challenges such as absence of degradability or insufficient biocompatibility, the complexity of their architecture design and difficulties in the scaling-up of their synthesis persist, which limit the chances for clinical success of redox-responsive systems.

- In future, more efforts should be directed towards understanding the materials-biology interface to address limitation associated with biodistribution, long-term compatibility and stability and improved efficacy to achieve clinically relevant cancer therapy.

References

Arunachalam B, Phan UT, Geuze HJ, Cresswell P (2000) Enzymatic reduction of disulfide bonds in lysosomes: characterization of a gamma-interferon-inducible lysosomal thiol reductase (GILT). Proc Natl Acad Sci U S A 97(2):745–750

Cairns R, Papandreou I, Denko N (2006) Overcoming physiologic barriers to cancer treatment by molecularly targeting the tumor microenvironment. Mol Cancer Res 4(2):61–70

Carol Ward SPL, Mullen P, Harris AL, Harrison DJ, Claudiu IHK, Supuran T (2013) New strategies for targeting the hypoxic tumour microenvironment in breast cancer. Cancer Treat Rev 39:171–179

Cheng R, Meng F, Deng C, Klok HA, Zhong Z (2013) Dual and multi-stimuli responsive polymeric nanoparticles for programmed site-specific drug delivery. Biomaterials 34(14):3647–3657

Comerford KM, Wallace TJ, Karhausen J, Louis NA, Montalto MC, Colgan SP (2002) Hypoxia-inducible factor-1-dependent regulation of the multidrug resistance (MDR1) gene. Cancer Res 62(12):3387–3394

Dai J, Lin S, Cheng D, Zou S, Shuai X (2011) Interlayer-crosslinked micelle with partially hydrated core showing reduction and pH dual sensitivity for pinpointed intracellular drug release. Angew Chem Int Ed Engl 50(40):9404–9408

Danhier F, Feron O, Preat V (2010) To exploit the tumor microenvironment: Passive and active tumor targeting of nanocarriers for anti-cancer drug delivery. J Control Release 148(2):135–146

Desgrosellier JS, Cheresh DA (2010) Integrins in cancer: biological implications and therapeutic opportunities. Nat Rev Cancer 10(1):9–22

Ganta S, Devalapally H, Shahiwala A, Amiji M (2008) A review of stimuli-responsive nanocarriers for drug and gene delivery. J Control Release 126(3):187–204

Giri S, Trewyn BG, Stellmaker MP, Lin VS (2005) Stimuli-responsive controlled-release delivery system based on mesoporous silica nanorods capped with magnetic nanoparticles. Angew Chem Int Ed Engl 44(32):5038–5044

Han Y, Li J, Zan M, Luo S, Ge Z, Liu S (2014) Redox-responsive core cross-linked micelles based on cypate and cisplatin prodrugs-conjugated block copolymers for synergistic photothermal–chemotherapy of cancer. Polym Chem 5(11):3707–3718

Hockel M, Vaupel P (2001) Tumor hypoxia: definitions and current clinical, biologic, and molecular aspects. J Natl Cancer Inst 93(4):266–276

Hu H, Yuan W, Liu FS, Cheng G, Xu FJ, Ma J (2015a) Redox-responsive polycation-functionalized cotton cellulose nanocrystals for effective cancer treatment. ACS Appl Mater Interfaces 7(16):8942–8951

Hu K, Zhou H, Liu Y, Liu Z, Liu J, Tang J, Li J, Zhang J, Sheng W, Zhao Y, Wu Y, Chen C (2015b) Hyaluronic acid functional amphipathic and redox-responsive polymer particles for the co-delivery of doxorubicin and cyclopamine to eradicate breast cancer cells and cancer stem cells. Nanoscale 7(18):8607–8618

Huo M, Yuan J, Tao L, Wei Y (2014) Redox-responsive polymers for drug delivery: from molecular design to applications. Polym Chem 5(5):1519–1528

Iyer AK, Singh A, Ganta S, Amiji MM (2013) Role of integrated cancer nanomedicine in overcoming drug resistance. Adv Drug Deliv Rev 65(13–14):1784–1802

Kumagai Y, Toi M, Inoue H (2002) Dynamism of tumour vasculature in the early phase of cancer progression: outcomes from oesophageal cancer research. Lancet Oncol 3(10):604–610

Lai CY, Trewyn BG, Jeftinija DM, Jeftinija K, Xu S, Jeftinija S, Lin VS (2003) A mesoporous silica nanosphere-based carrier system with chemically removable CdS nanoparticle caps for stimuli-responsive controlled release of neurotransmitters and drug molecules. J Am Chem Soc 125(15):4451–4459

Ma N, Li Y, Xu H, Wang Z, Zhang X (2010) Dual redox responsive assemblies formed from diselenide block copolymers. J Am Chem Soc 132(2):442–443

Matsumura Y, Maeda H (1986) A new concept for macromolecular therapeutics in cancer chemotherapy: mechanism of tumoritropic accumulation of proteins and the antitumor agent smancs. Cancer Res 46(12 Pt 1):6387–6392

Meng F, Hennink WE, Zhong Z (2009) Reduction-sensitive polymers and bioconjugates for biomedical applications. Biomaterials 30(12):2180–2198

Ong W, Yang Y, Cruciano AC, McCarley RL (2008) Redox-triggered contents release from liposomes. J Am Chem Soc 130(44):14739–14744

Romero-Garcia S, Lopez-Gonzalez JS, Baez-Viveros JL, Aguilar-Cazares D, Prado-Garcia H (2011) Tumor cell metabolism: an integral view. Cancer Biol Ther 12(11):939–948

Shao Y, Shi C, Xu G, Guo D, Luo J (2014) Photo and redox dual responsive reversibly cross-linked nanocarrier for efficient tumor-targeted drug delivery. ACS Appl Mater Interfaces 6(13):10381–10392

Singh A, Xu J, Mattheolabakis G, Amiji M (2015) EGFR-targeted gelatin nanoparticles for systemic administration of gemcitabine in an orthotopic pancreatic cancer model. Nanomedicine 12(3):589–600

Wang H, Tang L, Tu C, Song Z, Yin Q, Yin L, Zhang Z, Cheng J (2013) Redox-responsive, core-cross-linked micelles capable of on-demand, concurrent drug release and structure disassembly. Biomacromolecules 14(10):3706–3712

Wang J, Yang G, Guo X, Tang Z, Zhong Z, Zhou S (2014a) Redox-responsive polyanhydride micelles for cancer therapy. Biomaterials 35(9):3080–3090

Wang S, Zhang S, Liu J, Liu Z, Su L, Wang H, Chang J (2014b) pH- and reduction-responsive polymeric lipid vesicles for enhanced tumor cellular internalization and triggered drug release. ACS Appl Mater Interfaces 6(13):10706–10713

Xiao D, Jia HZ, Ma N, Zhuo RX, Zhang XZ (2015) A redox-responsive mesoporous silica nanoparticle capped with amphiphilic peptides by self-assembly for cancer targeting drug delivery. Nanoscale 7(22):10071–10077

Xu J, Singh A, Amiji MM (2014) Redox-responsive targeted gelatin nanoparticles for delivery of combination wt-p53 expressing plasmid DNA and gemcitabine in the treatment of pancreatic cancer. BMC Cancer 14:75

Yang P, Gai S, Lin J (2012) Functionalized mesoporous silica materials for controlled drug delivery. Chem Soc Rev 41(9):3679–3698

Zhang S, Zhao Y (2011) Controlled release from cleavable polymerized liposomes upon redox and pH stimulation. Bioconjug Chem 22(4):523–528

Zhuang Y, Su Y, Peng Y, Wang D, Deng H, Xi X, Zhu X, Lu Y (2014) Facile fabrication of redox-responsive thiol-containing drug delivery system via RAFT polymerization. Biomacromolecules 15(4):1408–1418

Part VI
Improved Imaging

Chapter 11
Nano-emulsions for Drug Delivery and Biomedical Imaging

Nicolas Anton, François Hallouard, Mohamed F. Attia, and Thierry F. Vandamme

Abstract Over the last decade, nano-emulsion has gained a considerable interest in biomedical applications. The reason is simple, and lies in the combination of several advantages of this nano-carrier. Nano-emulsion consists of a dispersion of oil nano-droplets in a water phase, sizing from 20 to 200 nm. First advantage of nano-emulsions is their huge stability; Second is their very simple formulation; Third is their non-toxicity; And fourth is their very important loading capability of lipophilic or oil-soluble molecules in the oily core of the nano-droplets. On the other hand, if the formulation is easy, tailoring the nano-emulsions for a given application, optimizing the processes, functionalizing the droplets surface, and targeting organs or cancerous tumors remains a challenge. This chapter aims to draw an overview of the different aspects of nano-emulsions formulations and applications. A first part regards the different fabrication processes, followed by biomedical applications of nano-emulsions, *in vivo* fate, biodistribution, pharmacokinetics, targeting, applications as nanomedicines and drug delivery systems. Clinical applications of nano-emulsions are also discussed, as well as their applications as contrast agent for the main different imaging modalities, X-ray imaging, fluorescence imaging and magnetic resonance imaging.

N. Anton (✉) • T.F. Vandamme
CNRS UMR 7199, Laboratoire de Conception et Application de Molécules Bioactives, équipe de Pharmacie Biogalénique, University of Strasbourg, 74 route du Rhin, 67401 Illkirch Cedex, France
e-mail: nanton@unistra.fr

F. Hallouard
DC2N Inserm U982, Laboratoire de Pharmacie Galénique Biopharmacie, Faculté de Médicine & Pharmacie, Université de Rouen, Rouen, France

Service de Pharmacie, Centre Hospitalier Asselin Hédelin, Yvetot, France

M.F. Attia
CNRS UMR 7199, Laboratoire de Conception et Application de Molécules Bioactives, équipe de Pharmacie Biogalénique, University of Strasbourg, 74 route du Rhin, 67401 Illkirch Cedex, France

National Research Center, P.O. 12622, Cairo, Egypt

A. Prokop, V. Weissig (eds.), *Intracellular Delivery III*, Fundamental Biomedical Technologies 8, DOI 10.1007/978-3-319-43525-1_11

Keywords Nano-emulsion • Formulation • High-energy method • Low-energy method • Nanomedicines • Targeting • Surface functionalization • Biomedical imaging

Abbreviations

ALA	Aminolaevulinic acid
AK	Actinic keratosis
BF200 ALA	BF200 aminolaevulinic acid
CMC	Critical micelle concentration
CT	Computed tomography
DLS	Dynamic light scattering
DTPA- PE	Diethylenetriaminepentaacetic acid phosphoethanolamine
DTX	Docetaxel
EPR	Enhanced permeation and retention
HLB	Hydrophilic-lipophilic balance
IONPs	Iron oxide nanoparticles
LNP	Lipid nanoparticle
MAL	5-aminolaevulinic acid
MRI	Magnetic resonance imaging
NIR	Near infrared
NIRF	Near infrared fluorescence
NMR	Nuclear magnetic resonance
NPs	Nanoparticle
PAV	Prednisolone acetate valerate
PBS	Phosphate buffer saline
PET	Positron emission tomography
PDI	Polydispersity index
PEG	Polyethylene glycol
PFCs	Perfluorocarbons
PFPE	Perfluoropolyether
QDs	Quantum-dots
RES	Rethiculoendothelial system
SPECT	Single photon emission computed tomography
SPIO	Superparamagnetic iron oxide
TEM	Transmission electron microscopy
TXT	Taxotere®

11.1 Introduction

Conventional emulsions, also called macro-emulsions, are typically sizing around the micrometer or larger. This is advantageous as regards the formulation process since they can be simply generated with conventional mechanical methods, like rotor-stator devices, but disadvantageous as regards their stability. Size range of macro-emulsions originates many physical destabilization processes, mainly gravitational-based phenomena that favor droplet flocculation and ultimately coalescence. On the other hand, the term "nano-emulsions" refers also to emulsions with the same structure than macro-emulsions but with drastic differences in both their stability properties and formulation processes. First, nano-emulsions (also known as miniemulsions, fine-dispersed emulsions, submicron emulsions and so forth) can be formulated from 10 to 20 nm up to 200–300 nm. In general, emulsions are thermodynamically unstable systems because the free energy ΔG_f is greater than zero. The physical destabilization of emulsions is related to the spontaneous trend to reduce interfacial area between the two immiscible phases. The minimization of interfacial area is attained by two mechanisms: *(i)* flocculation generally followed by coalescence, and *(ii)* Ostwald ripening. The global expression $\Delta G_f = \gamma \Delta A - T\Delta S_f$ (with γ the water–oil interfacial tension, ΔA the water–oil interfacial area gained with emulsification, $T\Delta S_f$ the entropy of droplet formation), means that the emulsion instability only comes from ΔA.

In the case of nano-emulsions, the stability comes from two phenomena; their nano-scale size range inhibits the effect of gravitation to the benefit of Brownian motion, and induces the predominance of the steric stabilization between droplets. As a consequence, Ostwald ripening is the only process that destabilizes the droplet giving a typical stability over several months. Actually, this kinetic stability is the main typical characteristic of nano-emulsions, making them prime candidate for numerous applications from nanomedicine, pharmaceutics, to agro-food industries. In nano-emulsion systems, the very small size of droplets will prevent the droplets undergoing reversible processes like flocculation and creaming (or sedimentation), and thus preventing the coalescence. However, the stability of nano-emulsions actually appears as a paradox since ΔA of nano-emulsions is definitively higher than the one of macro-emulsions, thus resulting in higher ΔG_f. On the other hand, the inhibition of the destabilization processes predominates, resulting in very stable nano-emulsions. Stability of nano-emulsions is an experimental fact, and very easily obtained with simple formulation procedures.

As we saw above the main property of nano-emulsions is their stability, but another interesting point due to the Brownian properties of the droplets is their ability to diffuse in all the available volume to give a very homogeneous dispersion of droplets. Therefore after encapsulating a given active principle (*e.g.* drugs, contrast agents) in oil nano-droplets, these hydrophobic or lipophilic molecules are homogeneously dispersed in the aqueous phase. The nano-scale of nano-emulsions droplets is also an advantage for biological and medical applications since they have privileged interactions with living cells, tissues, organs of pathologies, and they are as well compatible with the parenteral administration. In that way nano-emulsions naturally found many applications in various fields and notably in nanomedicines as lipid nano-carriers.

11.2 Formulation Processes

11.2.1 Generalities

Emulsification consists of dispersing one fluid into another non-miscible one, creating interface through the increasing of ΔA. The properties of the generated emulsion, *i.e.* size distribution and stability, are closely linked not only to the composition, chemical nature of the phases and stabilizing agents, ratio of viscosities of the dispersed and bulk phases, but also on the formulation protocol, temperature and time of processing, shear rate, cooling time, and type of emulsification apparatus. In addition, besides these criteria, the emulsion stability is strongly related to the size distribution of the dispersion after formulation. In the case of nano-emulsion, monodispersity will have a direct impact on the inhibition of Ostwald ripening, since the difference of the pressure between the bigger and smaller droplets will be reduced. The formulation of emulsions is conditioned by the energy supplied by the formulation device. Actually, most of apparatuses currently used for the fabrication of macro-emulsions do not allow decreasing the droplet size below 1 μm because of low emulsification yields due to the dissipation of the mechanical energy in heat. It is for example the case of rotor/stator apparatus like UltraTurrax® and colloidal mills. Only some methods allow the fabrication of nano-emulsions. Table 11.1 gathers the main methods for the formulation of emulsions along with the range of the generated droplet sizes (with limiting $\phi_v^d < 30\,\%$ to avoid droplet recombination during processing).

It follows that the three main possibilities for the formulation of nano-emulsions are: *(i)* high-pressure homogenization, *(ii)* ultrasound based methods (using sonotrode), and *(iii)* low-energy emulsification. In the following sections, these

Table 11.1 Summary of the accessible sizes in function of the emulsification apparatuses

Emulsification method	Typical size
Mechanical stirring	1–15 μm
Colloidal Mills	10–50 μm
* High-pressure homogenization	50 nm–5 μm
Membrane	0.2–100 μm
* Microfluidizer®	10 nm–1 μm
* Ultrasonication	20 nm–1 μm
* Low-energy methods	10–200 nm

* the ones generally compatible with the formulation of nano-emulsions

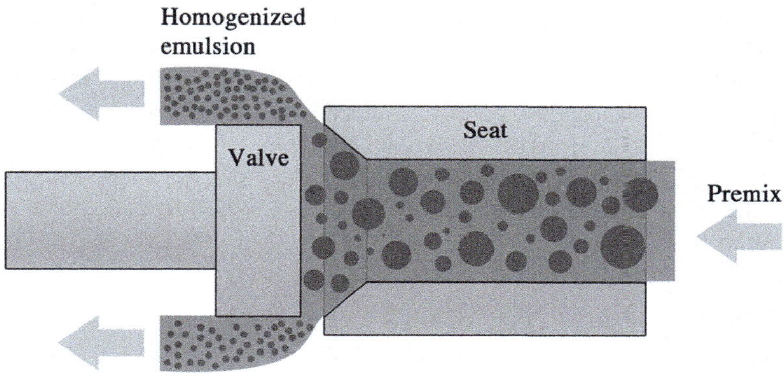

Fig. 11.1 Conceptual representation of high-pressure homogenizer chamber

processes will be presented and the conditions to reach nanometric-sized droplets will be discussed. It is to be noted that in the case of high-energy methods (high pressure or sonication), the formulation of nano-emulsions obligatory follows a preliminary step of pre-emulsification. The initial phase containing all components like emulsifiers, additives, and maintained at controlled temperature, undergoes a stage of dispersion that consist in creating a strong shearing that favors the breaking-up of the drops. Apparatuses indicated for this step are generally turbines type (*e.g.* Rushton type), that provide a high shearing along with an efficient recirculation of the liquid, giving rise to a premix emulsion between 10 and 100 μm. Then, the second step, homogenization step, will decrease the droplet size below 500 nm, narrow the size distribution, eventually improving the stability.

11.2.2 High-Pressure Methods

High-pressure homogenization is a very common industrial process, notably in the formulation of agri-food liquid or semi-liquid products. The premix emulsion is projected under high pressure from 30 to 1'000 bar through a homogenization head of particular geometry (see Fig. 11.1), and undergoes a combination of elongation and shear flows, impacts, and cavitations. The resulting size distributions are reproducible with a mean size ranging from 50 nm to 5 μm.

High-pressure emulsification arises from the competition between the droplets break-up die to deformation enhanced by the high-speed flow, and their recombination due to coalescence enhanced by collisions. The main parameters influencing the size and dispersity of the nano-emulsions are *(i)* the number of passages in the homogenizer, *(ii)* the value of the volume fraction of dispersed phase φ_v^d, *(iii)* the value of the pressure applied, and *(iv)* the concentration *C* of stabilizing agent like surfactants, proteins, polymers. We can note that the stabilizing agent concentration *C* is a critical parameter, since, if *C* is lower than *cmc*/10 (*cmc* is the critical micelle concentration), the droplet size does not decrease lower than 300 nm. This is due to the predominance of the coalescence against the stabilization of the fragmented droplets. On the other hand, when C is typically higher than $cmc \times 10$, the surfactants molecules are efficient to stabilize the droplets formed after fragmentation, allowing to decrease the size up to 50 nm. Of course, a further increase of C can reach the interface saturation that cannot allow a further decrease of the size. In addition, as easily understandable, this is only in this regime of surfactant concentration that the effect of the applied pressure is effective, actually the size follow a power law (Taisne et al. 1996).

11.2.3 Ultrasounds-Based Methods

This formulation method consists in applying a strong ultrasonic energy in a small volume sample under circulation. This is performed with a specific sonotrode that concentrate the energy in a surrounding volume (see Fig. 11.2). Ultrasonic emulsification is thus only efficient in this small volume around the sonotrode, and thus the fluid circulation in the sample is still necessary for optimizing the process, and as well involves a minimum time before stabilization of the droplet sizes. The frequencies of ultrasound are from 16 kHz to 1 MHz, and generated by a plane surface vibrating on a sinusoidal way with amplitude between 1 and 200 μm. Cavitation bubble alternatively undergo contraction and dilatations, and the micrometric globules of the premix emulsions are broken-up by the successive implosions of the cavitation bubble. The droplet fragmentation of the droplets gradual decreases their average diameter. This is illustrated in Fig. 11.2 (middle) that shows the nano-emulsion mean size describes an exponential decay up to stabilization, for a simple system composed of medium chain triglycerides (oily core), nonionic surfactant (stabilizing agent) and water as continuous phase (Delmas et al. 2011). The figure shows the follow-up of the mean size and polydispersity indexes (PDI). PDI reflects the quality and monodispersity of the dispersion, such that a suspension is considered of good monodispersity (good quality) if the value of PDI is below 0.2–0.15, and of very good monodispersity (very good quality) if the PDI is below 0.1. The exponential profiles reported, for a similar system can be "stretched" if the energy supplied is decreased, as illustrated in Fig. 11.2 (bottom). This result is actually important for understanding the process, as well as its design and optimization. If ultrasound-based technologies are compared with high-pressure homogenizers, the properties of the resulting emulsions are generally similar. However, high-pressure homogenizers are widely used in industrial processes, whereas ultrasonifiers are more adapted for the laboratory scale, research and development stages.

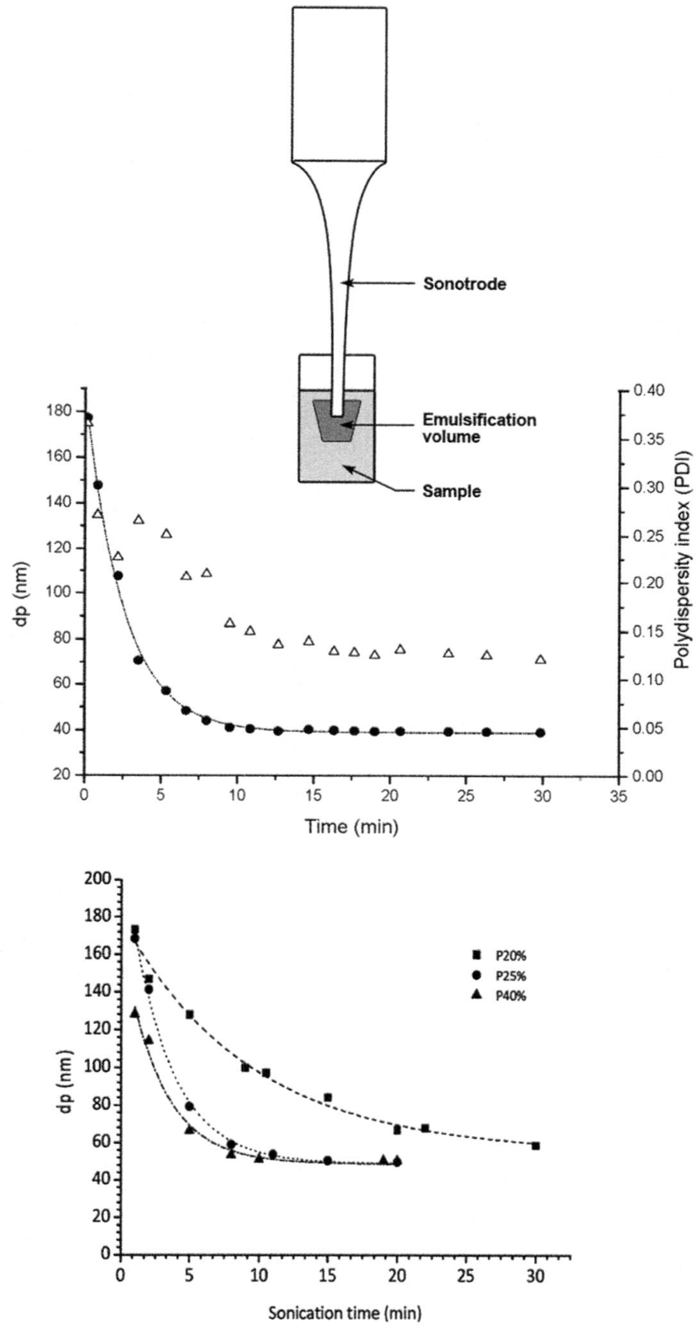

Fig. 11.2 (Reprinted adapted with permission from (Delmas et al. 2011)). Schematic of a sonication-based nano-emulsification process. *Top:* evolution of droplet size distribution along the sonication process: particle mean diameter (black disks), and polydispersity index (open triangles). *Bottom:* evolution of droplet size distribution along the sonication process in function of the power of the sonicator

11.2.4 Low-Energy Methods

Low-energy methods allow the formulation of nano-emulsions similar to those described above (with high pressure and sonication methods), but without mechanical energy. In this case, the increase of ΔA is obtained by taking benefit of the intrinsic physicochemical properties of the phases and stabilizing agents.

Low-energy methods are simple, cost-effective formulation processes and are very attractive. However, the low-energy methods are very studied but almost only on the fundamental point of view, and much less used in industrial processes than high-energy ones. This is likely due to several points, like the fact *(1)* that these methods only allow nano-emulsification and not homogenization of premix (*e.g.* not suitable for homogenizing milk), or *(2)* that the physicochemical properties of the dispersion cannot always be finely controlled, or finally *(3)* that they involve a non-negligible amount of surfactants that is not always compatible with all the specifications (impact on the taste in agrifood industries, or induce a toxicity, etc.).

Low-energy methods in general are based on spontaneous emulsification, or self-emulsification. It is to be noted that other low-emulsifications methods are reported, like phase inversion temperature methods of solvent diffusion, but it has recently been shown that their principle is based on the spontaneous emulsification described below. Spontaneous emulsification occurs when two immiscible fluids are brought in contact and gently homogenized. In a few seconds a kinetically stable nano-emulsion is formed. In contrast with high-energy methods, spontaneous emulsification allows a similar gain in ΔA without energy input. Spontaneous emulsification takes benefit of the physicochemical properties of the water, oil, and surfactant for generating the nano-droplets through a spinodal decomposition-like process.

Numerous explanations aimed understanding of this phenomenon, however, it is only recently than a universal and simple mechanism was proposed (Anton and Vandamme 2009), thus including all the spontaneous emulsification processes. Illustrated in Fig. 11.3, the spontaneous emulsification was attributed to the penetration of the water phase in the oily one, resulting in breaking-up the oil phase at the nanometric scale. Let us consider the oil phase homogeneously mixed with surfactant molecules in specific conditions (*e.g.* temperature) that make them lipophilic (Fig. 11.3a). Then, once this surfactant/oil phase is in contact with water (Fig. 11.3b), aqueous phase penetrates the oil one solubilizing the surfactant molecules (Fig. 11.3c). Surfactants stabilize the newly-formed oil nano-droplets generating the nano-emulsions (Fig. 11.3d), generally sizing from 20 nm to 200–300 nm. In addition, spontaneous nano-emulsions give suspension with a very good monodispersity excepted when we reach the limit of the feasibility region, *e.g.* at low surfactant concentrations, where the polydispersity indexes suspensions can increase higher than 0.2–0.3. It follows that spontaneous emulsification is driven by the physicochemical properties of the surfactant and its solubility towards the oily and aqueous phases. On the other hand, when a spontaneous emulsification process is efficient to give nano-emulsions, its own size can be finely controlled by the surfactant concentration, namely the higher is the surfactant concentration, the lower is

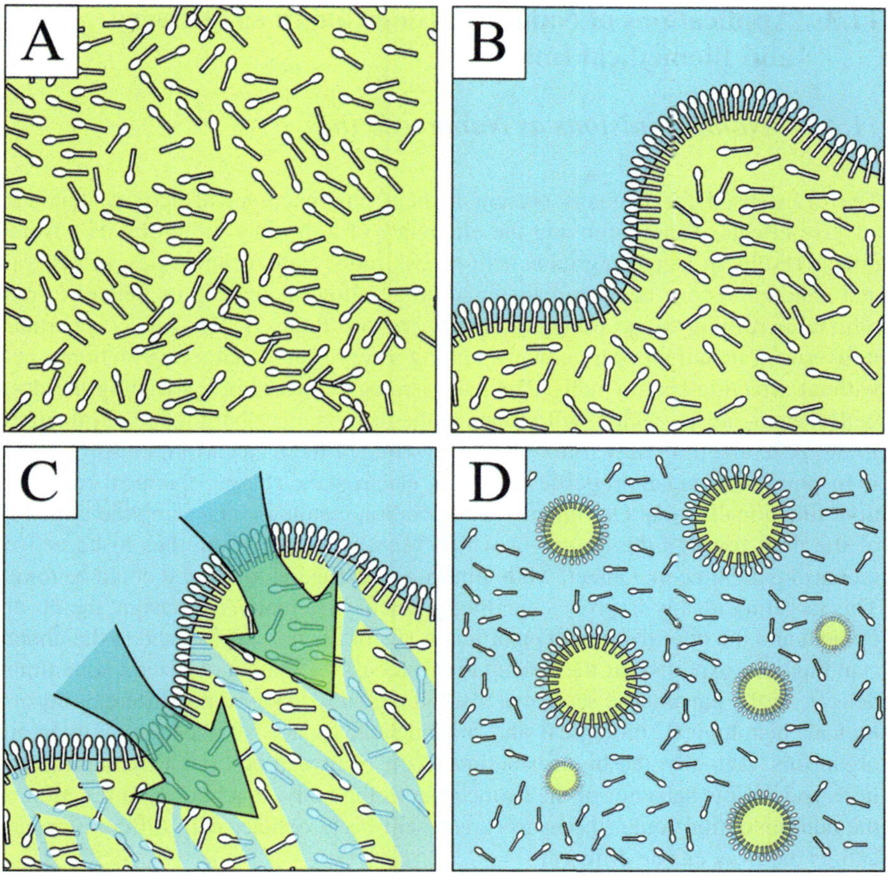

Fig. 11.3 (adapted from (Anton and Vandamme 2009)) Schematic representation of the mechanism driving the spontaneous emulsification

the average nano-emulsion size (Anton and Vandamme 2009). It is noteworthy that not all the surfactant/oil couples are compatible with the spontaneous nano-emulsification, and this is one of the main reasons why sometimes it is necessary to shift to high-energy methods. However, we can define some necessary conditions for the spontaneous emulsification, like the complete solubility of the surfactant to the oily phase. This affinity can be relative and be artificially improved with temperature, for example with nonionic surfactants that become lipophilic at high temperature (above their cloud point in general). Then with a sudden mixing with the aqueous phase at room temperature (*i.e.* colder water) the surfactant affinity for oil is suddenly decrease beneficially to the one of water, enhancing the water penetration. Another important factor that impact on the process is simply the chemical nature of each compound, not only the properties of nonionic surfactant but also the nature of oil, the nature of surfactant, and the additives in the aqueous phase.

11.3 Applications of Nano-emulsions for Drug Delivery and Biomedical Imaging

11.3.1 Nano-emulsions as Nanomedicines

The drugs used for Humans present a main pharmacological activity, but also adverse effects. Thus improving the efficiency of a drug by inhibiting its adverse effects results in increasing its benefit-to-risk ratio. Nanotechnologies, in general, and nano-emulsions being a realistic solution, allow increasing its benefit-to-risk ratio in modifying the becoming of the drug in the organism. This aspect is particularly crucial in antitumor treatments, aiming a very specific targeting on tumor cell without affecting healthy cells. The main strategy consists in associating the drug with a nano-carrier, which will influence the drug biodistribution and elimination. In this case, the drug may follow the *in vivo* fate of the nanocarrier, closely related to its physicochemical properties like size, electrostatic charge, chemical composition, functionalization. Optimizing this operation results in increasing the targeting of the drug towards the tissues in which we want this last one has to be active, increasing the time in contact, while limiting its diffusion to those it could be toxic. This targeting ideally improves the drug efficiency. Actually, the design, the development and the optimization of an efficient vector are not only related to the understanding and expertise in the formulation process of the nanocarriers, sometimes delicate at the nanometric scale, but also on the knowledge of the different physiological, histological, biological and biochemical aspects of the tissues in living organisms. Actually, the biodistribution of the drug will be driven by the interactions and *in vivo* behavior of the nanocarrier in the body. In addition, in function of the administration route, the nano-carrier will be in contact with different tissues and its pathway can be different.

As regards lipid nano-emulsions as drug delivery systems, and considering that the potential drugs are solubilized in oil cores without premature leakage, their targeting properties will be defined by the interactions of the lipid droplets in function of the administration route. Let us now focus on the different administration ways, and evaluate the different potentialities for targeting lipid nano-droplets.

Intravenous administration constitutes the main administration route adopted for research investigations or preclinical studies since blood is the ultimate location desired for a drug before targeting. However, beside physiological compatibilities like sterility, osmolarity and pH, the size of the nanocarrier is a main issue in their design. Once in the bloodstream, the nano-droplets will circulate in the blood pool, migrate to the heart by the vena cava, then in the lung to be back in heart by the pulmonary vena, and in the arterial circulation by the aorta. In this context, the risk of pulmonary emboli that arises above 5 μm is prevented by the size of nano-emulsions below 200–300 nm. Then, their elimination is performed by the macrophages of the reticuloendothelial system (RES) and particularly in liver and spleen. Küpfer cells in liver are particular macrophages very efficient for nanoparticles uptake; they are located in the border of hepatic sinuses, and constitute 25 % of the liver cells and 90 % of the macrophages in the body. Spleen is well an efficient trap for nanoparticles. Then, their elimination is performed through the bile by transcytosis within the gall bladder. On the other hand, when the particles are smaller than 5–10 nm they are filtrated by kidneys, and this happens in some applications with, *e.g.* micelles or small hydrophilic contrast agents.

It appears that the therapeutic efficiency of drugs can be improved by their potential targeting through nano-carriers, and thus depend on their time of circulation in the blood stream. The longer is this circulation time, the longer will be the contact with the target tissues. It follows that increasing the circulation time of the nano-carriers in bloodstream (*i.e.* making them stealth towards RES) is a key factor for improving the drug efficiency. Actually, before the macrophage uptake by phagocytosis as discussed above, the nanoparticles are recognized through opsonization, by fixation onto their surface of blood proteins called opsonin. Two parameters are known to inhibit this phenomenon, the size of the particles and the chemical nature of the surface. Indeed, the limit of 100 nm is generally considered as an efficient limitation of the RES recognition, as well as the decoration of the surface with neutral hydrophilic polymers like polyethylene glycol (PEG).

In the case of nano-emulsions, we have seen that the different formulation processes give rise to oil nano-droplet suspensions of similar physicochemical properties. Therefore the size and surface composition should be mainly related to the choice of stabilizing agent. Precisely, a large class of stabilizing agents, *i.e.* nonionic surfactants, fulfills such specifications and allows the simple formulation of small PEGylated nano-droplets. Nonionic surfactants are generally composed of a PEG chain as hydrophilic part connected to a more or less simple aliphatic chain. Owing to their high surfactant properties (HLB > 15–16) they allow to reach very small size below 200 or even 100 nm, and owing to the PEG part of the surfactant the resulting droplets are naturally protected from opsonization. Spontaneous emulsification has emerged as a very efficient method to produce stealth nano-emulsion droplets. Some examples are reported in Fig. 11.4, for different kinds of nanoparticles comparing nano-emulsions, polymeric nanocapsules and inorganic nanoparticles. Their morphologies are illustrated in TEM micrographs and the longitudinal quantification of their concentration in blood are reported for all nanoparticles.

In fact, these nanoparticles differ in their mean sizes, with nano-emulsions around 80 nm, and, inorganic nanoparticles or polymeric capsules around 200 nm. However we have to consider that the size distributions of these kinds of suspensions largely overlap, and thus the average size should not be considered as main factors influencing the blood clearance. Another basic factor that impacts on the blood pharmacokinetics, beyond the surface functionalization, is the nature of the particle core itself. In Fig. 11.4a, blood concentration of nano-droplets (*i.e.* proportional to X-ray contrast) are compared for droplets of similar size and made with the same surfactant (PEG-35 ricinoleate) but with different iodinated oils constituting their core. One can see a slight difference in the curve profiles, resulting in half-lives varying from 6 h to 9 h. However, when the surface is drastically modified with a polymeric capsule (Fig. 11.4b), the clearance is much more decreased up to 2.5 h, and finally reaches 1 h with inorganic (tungsten oxide crystals) (Fig. 11.4c).

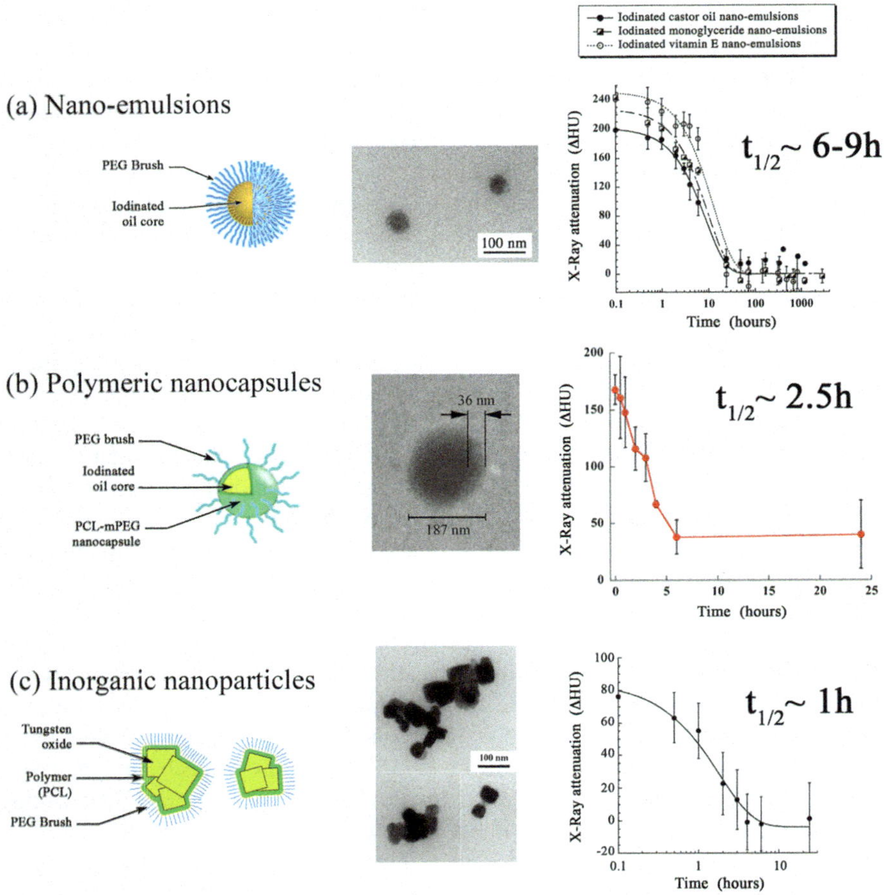

Fig. 11.4 (adapted from (Attia et al. 2014; Hallouard et al. 2013; Jakhmola et al. 2014)) Comparison of the residence time in bloodstream for different kinds of nanoparticles: (**a**) nano-emulsions, (**b**) polymeric nano-capsules, and (**c**) inorganic nanoparticle. The figure presents a schematic illustration of the nanoparticle structure (*left*), transmission electron micrograph (*middle*), and X-ray attenuation (proportional to the blood concentration of nanoparticles) in function of time post-intravenous administration in mice

Obviously the nature of the nanoparticles influence the opsonization process due to the surface charge, electronic density or protein attractions by other low-energy forces (Owens and Peppas 2006), but we can note as well that in the case of nano-emulsions the PEG coverage is actually optimal since any other material enters in the surface composition.

Besides the considerations on blood circulation time, the drug targeting process is associated to the nano-carriers accumulation into the aimed tissues during this blood circulation phase. This means that it is related to accumulation phenomena, specific to the nature of the nano-carrier and/or surface functionalization. This accumulation can be the direct consequence of the size, chemical nature or charge of the nano-carriers (so-called passive targeting), but can also be induced by the decoration of the surface by ligands specific to the receptors overexpressed in the target tissues (so-called active targeting). Sometimes the natural elimination pathways of the vectors towards the liver or spleen are used for the specific drug delivery in these organs, and can be also called passive targeting (of liver or spleen). In this case, we have disclosed the passive accumulation in these organs thanks to X-ray imaging that allows the longitudinal non-invasive *in vivo* quantification of the biodistribution. Figure 11.5 illustrates the biodistribution in liver, spleen and kidney of nano-emulsions with similar size and physicochemical properties, but different oil cores made with iodinated vitamin E, iodinated castor oil, and iodinated monoglyceride (as in Fig. 11.4a above). The quantification of their biodistribution with time emphasize drastic differences between the different organs, iodinated vitamin E is more accumulated in liver while iodinated castor oil prefers spleen, and on the contrary iodinated monoglyceride does not show an organ specificity and is simply more rapidly eliminated from the organism. It follows that the passive accumulation of nano-droplets through elimination pathway is a powerful tool for imaging pathologies actually located in these organs as described in (Almajdub et al. 2007; Martiniova et al. 2010; Aprahamian et al. 2011; Boll et al. 2011) in liver and spleen.

On the other hand, passive targeting includes the passive accumulation in tumors, for instance implanted subcutaneously, and considered to be undergone through the enhanced permeation and retention effect (EPR effect), targeting. EPR effect results from the passive accumulation of drug or nanocarriers due to their extravasation through leaky vasculature. It is well established that for certain pathologies like tumors, infarctus, inflammation areas, and under certain circumstances, the endothelial lining of the blood vessel wall becomes more permeable than in the normal state (Hobbs et al. 1998; Jain 1999). Thus, large molecules and nanoparticles ranging from 10 to 500 nm can leak from the vessel and accumulate in interstitial space, *i.e.* co-localize with the pathology. Actually, the efficiency of the passive EPR tumor accumulation is basically related to the time of the agent in bloodstream: the higher is the exposure, the higher is the accumulation. Thus, a prolonged time in blood is obtained with non-renal clearable entities (size > 5 nm) (Fang et al. 2001) decorated with well solvated and flexible polymer chain (PEG) was shown to slow down their opsonization and clearance by the RES (Klibanov et al. 1990; Torchilin and Trubetskoy 1995). The major examples reported in literature refer to liposomal formulations that accumulate in subcutaneous xenografted tumors (Karathanasis et al. 2009; Zheng et al. 2009; Anayama et al. 2013). The direct visualization of this accumulation phenomenon is provided by X-ray imaging techniques, and reveals the gradual enhancement of the contrast over time in the tumor surrounding regions. Some examples of EPR targeting with nano-emulsions were reported and have proved the compatibility of lipid droplets with EPR effect in general. For instance, in Ref. (Khalid et al. 2006), the authors report the formulation of nano-emulsions made with PEGylated surfactants (PEG 660-hydroxystearate) stabilizing medium chain triglyceride core. These lipid nano-droplets encapsulated anticancer mole-

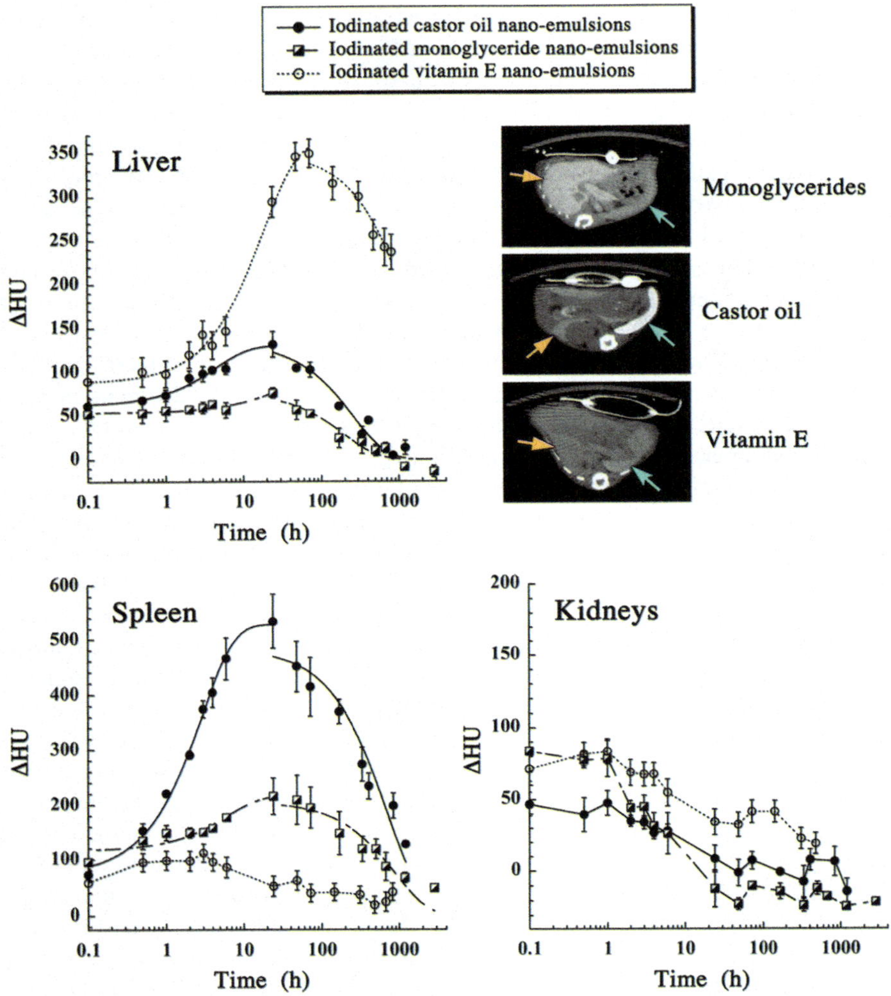

Fig. 11.5 (adapted from Ref. (Attia et al. 2014)) Quantification of the droplet concentration in liver, spleen and kidneys (through X-ray attenuation, ΔHU) after i.v. administration of iodinated nano-emulsions with similar size and physicochemical properties, but different oil cores made with iodinated vitamin E, iodinated castor oil, and iodinated monoglyceride. To illustrate the differences in contrast at 24 h post-injection, transverse slices through the liver and spleen are presented. Liver is indicated by orange arrows and spleen by blue arrows (n = 3 mice/group)

cule, docetaxel (DTX), at a concentration of 3 wt.% of the nano-droplet. The passive EPR accumulation the model tumor occurred thanks to the nano-carriers, studied and compared to a commercial control formulation (*i.e.* solubilized in micelles) of DTX (Taxotere®, TXT). C26 colon adenocarcinoma cells in growth medium were injected subcutaneously in three different locations on the back of each mouse, which produced three separate tumors. The formulations were admin-

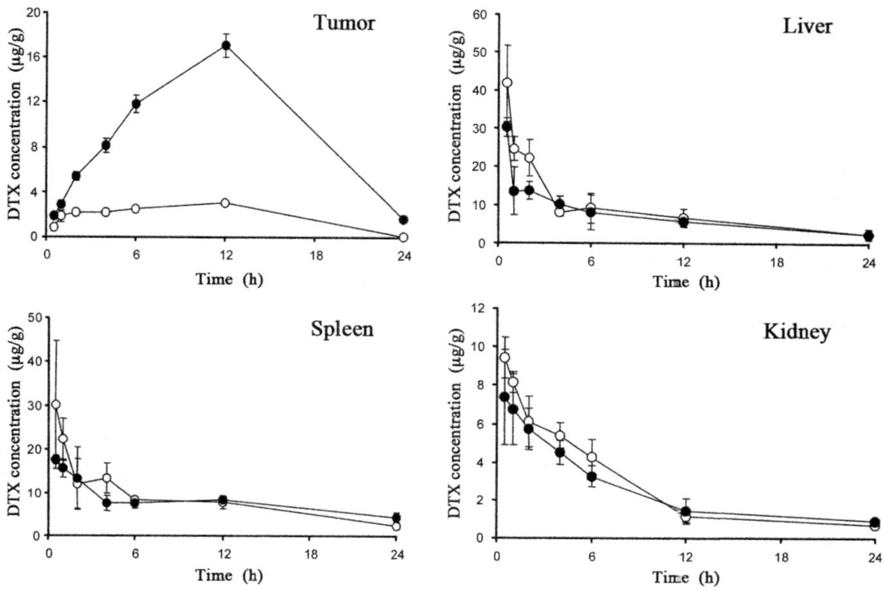

Fig. 11.6 (adapted from Ref. (Khalid et al. 2006)) Illustration of the passive targeting of nano-emulsions to xenografed subcutaneous tumor in mice after i.v. administration of nano-emulsions encapsulating DTX (*closed circles*) and of control injection of TXT (*open circles*). The drug dose level of 15 mg/kg and the concentration in the nanocarrier is 3 wt.% (n=4 mice/group)

istered when each tumor grew to a volume of about 20 mm^3, approximately 10 days after cell inoculation. These results are reported in Fig. 11.6, where the two above-described formulations were injected at similar doses of 15 mg/kg, comparing the pharmacokinetics and biodistributions of DTX in tumor. The results disclose a clear-cut and significant accumulation in tumor of the DTX encapsulated in nano-emulsions compared to the control roughly constant, while in liver, spleen and kidney the two curve profiles are similar. This clearly evidences the EPR effect induced by the lipid nano-carrier.

Actually in literature, EPR accumulation is generally illustrated with liposomes, for example in biomedical imaging (Karathanasis et al. 2009; Zheng et al. 2009; Anayama et al. 2013; Li et al. 2014). Herein for nano-emulsions, a similar size range and lipid nature gives rise to the similar efficiency in the passive accumulation of lipid nano-emulsion droplets in tumors, that can allow to draw a realistic parallel with the liposomes EPR accumulation. As a last remark, the active targeting of nano-emulsions, performed with post-insertion in nano-emulsion droplets of function lipids, remains anecdotal and not very efficient (Béduneau et al. 2008), likely due to the difficulty to increase the ligand concentration decorating the particle surface.

11.3.2 Clinical Applications of Nano-emulsions

Some nano-emulsions have already clinical applications and are authorized in different countries. We will present you in this part, different applications of such commercialized emulsions. Then, a description of current clinical assays on nano-emulsions will be performed. Mainly, nano-emulsions in commercialized formulations are used to improve drug solubility in water and therefore drug bioavailability. To illustrate these facts one can mention two examples. The first is Neoral® (Novartis Pharma, Rueil-Malmaison, France), an oral cyclosporine emulsion for transplantation (ANSM). The second is Propofol Lipuro® (B Braun AG, Melsungen, Germany), an emulsion of propofol in soy oil for intravenous administration (ANSM). This last is used for anesthesia in intensive care units. Another advantages of nano-emulsions are the possibility to inject oil intravenously and to improve emulsion stability. This is for example the case for Medialipide® (B Braun Medical, Boulogne, France), a commercialized oil-in-water emulsified lipid utilized for parenteral nutrition (ANSM). To be administrated, strictly speaking oil droplets size should be less than 5 microns but in practice the characterizing size is the average size of the lognormal distribution meaning that the population contains droplets much bigger than the average size. It follows that nano-emulsions to be compatible with parenteral administration have a size range below the micrometer.

We identified the current clinical assays on nano-emulsions using the Cochrane library. It is noteworthy that in this part, we will not present the assays using nano-emulsions as model for the evaluation of different treatments or drugs on hyperlipidemia. Experiments based on *ex vivo* model like explanted human tooth are likewise not considered.

Actually, these clinical assays could be classified in 3 applications: anesthesiology, dermatology and vaccines. One publication concerned a comparative study of emulsified propofol with Solutol® HS15 (BASF) in nanometric size and propofol in soybean oil emulsion for anesthesia during endoscopy. This phase II assay was performed on 150 patients (being 33–54 years old) (Rodrigues et al. 2012). The authors observed equivalence between emulsified propofol and propofol in nanometric emulsion form concerning its efficacy, safety and the side effects. However, it is worth to note that it was a lower incidence of pain during injection with propofol in nanometric emulsion form: 53.3 % of patients instead of 82.7 % with propofol in soybean oil emulsions. In additions, the incidence of nausea and vomiting was reduced with nano-emulsified propofol from 10.7 to 2.7 % of the patients. A lack of this work is the absence of comparative study with Propofol Lipuro®, propofol emulsion containing medium-chain triglycerides (Dubey and Kumar 2005). Indeed, this last formulation induced less injection pain than non-lipid forms of propofol. A hypothesis was a reduction or an absence of free propofol with Propofol Lipuro®. With a comparative study between Propofol Lipuro® and nano-emulsified propofol with Solutol® HS15 could be a good approach to verify this hypothesis.

Concerning dermatological applications, different applications of nanometric emulsion based drugs are clinically tested. A first clinical assay was conducted on

482 subjects for the development of drug against cold sores (Kircik et al. 2012). The developed active substance (NB-001) was formulated at different dosages (0.1, 0.3 or 0.5 wt.%) in nanometric emulsions. At the first signs or symptoms of a cold sore episode, this medicine was applied 5 times per day, approximately 3–4 h apart, for 4 days. Emulsions at a drug concentration of 0.3 wt.% showed the best efficiency by a 1.3 days improvement in the mean time to healing compared to placebo ($p=0.006$). This could be explained by a better skin penetration of the active substance at this concentration. Concerning drug safety, no serious side effects or dermal irritation at any concentration were observed. In additions, no active substance was detected in subject bloods. Compared to other medicine for cold sores, the observed efficacy was similar to oral nucleoside analogues and greater than topical formulations which present a reduction of healing time by only one half day. The originality of this new formulation is the lack of drug distribution in blood preventing most of side effects for an efficacy comparable to oral formulations.

The two next studies focused on drug evaluation for the treatment of distal subungual onychomycosis (Ijzerman et al. 2010a, b). The formulation consisted of emulsions with a mean droplet size of 180 nm containing a new active substance (NB-002). The first study corresponds to a phase II assay on 432 subjects having mild to moderate subungual onychomycosis of the toenails (Ijzerman et al. 2010a, b). The treatment was applied to all toenails and on 5 mm of adjacent skin for 42 weeks. The droplet size and emulsion composition allow selective uptake into the skin without irritation via hair follicles and skin pores. In additions, the droplet size of the formulation destabilized fungal hyphae and spores leading to a formulation efficiency of 84 % compared to 13 % with blank emulsion. No serious side effect was observed during this assay. Nevertheless, the most observed side effect was nail discoloration and concerned globally 10 subjects on 432. The second assay is a post-hoc analysis of the first study on 227 subject to assess the post-treatment mycologic cure and effective treatment rates of a 42-week treatment regimen of NB-002 compared to vehicle (Ijzerman et al. 2010a, b). At 4 and 8 weeks post-treatment, an evaluation of the treatment efficiency. At all emulsion concentration of NB-002, the mycologic cure rates 8 weeks post treatment was ranged from 4.2 to 16.9 % versus 5.3 % in blank emulsion. These assays showed clear antifungal activity with clinically significant nail clearing of NB-002. For more, the emulsion form had also an activity on treatment efficacy.

Several phase III assays concerned the use of nano-emulsions of BF200 5-aminolaevulinic acid (BF200 ALA) for actinic keratosis (AK) photo-treatment compared to a registered 5-aminolaevulinic acid cream (MAL) or placebo (Szeimies et al. 2010; Dirschka et al. 2012; Dirschka et al. 2013; Neittaanmäki-Perttu et al. 2014). BF-200 ALA (Biofrontera Bioscience GmbH) is a gel formulation of 5-aminolaevulinic acid (ALA) with nano-emulsion. This formulation was designed to prevent ALA instability and improve skin penetration (Szeimies et al. 2010; Dirschka et al. 2012). For the illumination, different devices were used 3 h after drug administration: narrow emission spectrum LED (590–670 nm; Aktilite CL128®) or broad emission spectrum LED (580–1400 nm; PhotoDyn 750®). These assays were performed on 13–663 patients presenting a light to mild AK on face and/or bald scalp. The protocol included one drug application followed by a LED luminescence. In cases of residual AK lesions corresponding mostly to grade II and III AK, a second treatment was performed 12 weeks after the first (Dirschka et al. 2013; Neittaanmäki-Perttu et al. 2014). Then, patients where followed during 3–12 months to determine influence of different treatment protocols on AK recurrence (Szeimies et al. 2010; Dirschka et al. 2013; Neittaanmäki-Perttu et al. 2014). Concerning the results, since 2010 and confirmed by all other works, the authors concluded that treatment efficacy depends on both drug design and luminescence technique. Narrow spectrum light sources presented the best results. For example, on 600 patients with MAL or BF200 ALA, this technique had a healing rate compared to broad spectrum lamps of 71.5 % versus 61.3 % and 84.8 % versus 65.7 % respectively (Dirschka et al. 2013). It was to note that only with placebo broad spectrum lamps had better results 21.6 % versus 12.8 % with narrow spectrum light sources. In point of view of used drug formulations, BF200 ALA presented the best efficacy compared to MAL and placebo in all assays. The healing rates on 600 patients for BF200 ALA compared to MAL were 78.2 % versus 64.2 % (p=0.05) with a remission rate at 3 months for BF200 ALA of 90.4 % (Dirschka et al. 2013). This difference could be explained by the observation on 13 patients of better histological clearance at 3 month with BF200 ALA compared to MAL: 61.5 % compared to 38.5 % (p=0.375) (Neittaanmäki-Perttu et al. 2014). At 12 months after last treatment, BF200 ALA showed the best remission rates at 47 % versus 36 % for MAL (Dirschka et al. 2013). Concerning the safety of BF200 ALA, the observed side effects were similar with MAL (Dirschka et al. 2013; Neittaanmäki-Perttu et al. 2014). All these results demonstrated that BF-200 ALA is a very effective and superior to a registered MAL. In additions efficacies and adverse events are also depended to the different light sources used.

Two publications report vaccines against influenza (Stanberry et al. 2012; Treanor et al. 2013). In both assays, emulsions had the role of adjuvant. In the first work, the authors used a glucopyranosyl lipid A, an emulsion designed to be an intramuscular vaccine (Treanor et al. 2013). They performed a phase I/II assay on 392 patients (being 18–49 years old) receiving two administrations at D1 and D21. Forty two days after the first vaccine administration, up to 82 % of the tested patients presented seropositivity to the different influenza stems. Concerning vaccine security, this study revealed that the half of patients presented light to mild side effects. In some cases, the authors determined serious side effects requiring the use of antihypertensive drugs. In the second work, the author used a new adjuvant W805EC corresponding to nano-emulsion. Indeed, nasal mucosa is the main entry for influenza virus (Stanberry et al. 2012). The authors performed a phase I assay on 199 patients receiving a single intra-nasal administration of vaccine for 3 different influenza strains. The innovation was to design a vaccine for an intra-nasal administration in order to induce both systemic and mucosal immunity. Despite immunization process with a single administration through an original administration route, more than 70 % of the volunteers were sero-protected for the 3 influenza strains through systemic and mucosal immunity. W805EC as adjuvant presented a good tolerance during this clinical assays: absence of serious side effects and observation of side effects similar with intra-nasal administration of PBS (Phosphate buffer saline).

11.3.3 Nano-emulsions in Biomedical Imaging

Another fundamental application of nano-emulsions concerns their use as contrast agents for biomedical imaging. The formulations strategies can be actually close to the formulations of medicines, with on the whole different purposes. Imaging is an important preliminary aspect of some treatments like cancer treatment, providing real-time monitoring with minimal invasiveness and tissue destruction. Biomedical imaging is often used for prediction, screening, staging, prognosis, biopsy guidance, therapy planning, and guidance. Computed X-ray tomography, magnetic resonance imaging, fluorescence and ultrasound have been traditional techniques of anatomical imaging. Ultrasound is commonly used as an external stimulus due to its accessibility, cost-effectiveness, and ability to be used in conjunction with multimodal systems. Besides, optical imaging tools, PET, SPECT, and photoacoustic imaging techniques, nano-emulsions have been widespread used in biomedical fields, daily consuming, particularly in cosmetics and even food industry owing to the huge stability, the encapsulation efficiency, and the high safety of these nano-emulsions.

As mentioned above, the key to their success hinges on their oily cores which act as reservoir allowing solublization and nano-encapsulation of hydrophobic molecules like drugs, contrast agents, organic fluorophores, and/or inorganic NPs at very high concentrations. Other fundamental point for their applications *in vivo* is the non-toxicity since there are lonely fabricated with parenteral-compatible materials. Herein we will shed more light on specific examples to show the importance of the excellent behaviors of nano-emulsions as contrast agent for imaging. And this could be categorized into integrated classes:

11.3.3.1 Nano-emulsions in X-ray Imaging

One of the earliest developed contrast agents-based nano-emulsions for micro-computed tomography (CT) are Fenestra VC® and Fenestra LC® in order to overcome the incompatibility of clinic products like small hydrophilic iodinated molecules. These latter undergo a rapid elimination from the kidneys causing renal failure and do not allow performing scans in preclinical studies.

Introducing the nanotechnology into the medicine made significant progress in micro-CT imaging. Over the last few years, our group has developed several efficient formulations of nano-emulsion-based contrast agents for preclinical X-ray imaging. The ideal solution to ensure a high encapsulation and prevent any release is the synthesis of oily molecules grafted with X-ray contrasting atoms like iodine, and then formulated as nano-emulsions. These droplets are coated with hydrophilic polymer (PEG) to reduce their opsonization that prolongs their residence time in the blood stream. For these applications, as discussed above, the size in the range is 50–250 nm, avoiding a fast clearance by the reticulo-endothelial system (Kulkarni and Feng 2013), but as well large enough to prevent a rapid elimination by kidneys (Choi et al. 2007). From these considerations, some examples of iodinated nano-emulsions were proposed as contrast agents for X-ray micro CT such as iodinated α-tocopherol nano-emulsions (Li et al. 2013), iodinated monoglyceride nano-emulsions (glyceryl monocaprylate), iodinated triglyceride nano-emulsions (castor oil) (Attia et al. 2014), and iodinated cholecalciferol nano-emulsions (Attia et al. 2015) and studied *in vivo*. The preclinical investigations, briefly discussed in the previous sections, concerned the pharmacokinetics, biodistributions, toxicity and contrasting efficiencies of nano-emulsions, eventually disclosing their outstanding efficacy for X-ray biomedical imaging.

11.3.3.2 Nano-emulsions in Magnetic Resonance Imaging (MRI)

It is approved that some moieties like iron oxide nanoparticles (IONPs), Gd^{3+} chelates, and/or perfluorocarbons PFCs in combination with MRI modality are able to enhance the imaging performance. We will restrain here on the nano-emulsion formulations loading those contrast agents, and it is as well to note that many other examples reports association of MRI contrast agent combined with imaging agents for other modalities.

Recently, a multifunctional and biodegradable nanocarrier system based on oil-in-water nano-emulsions with three different mean diameters (30, 60, and 95 nm) were developed, and used as tumor targeting agent through EPR effect (Jarzyna et al. 2009). The formulation is performed with iron oxide nanoparticles having a hydrophobic coating, dispersed in the soybean oil, solubilizing simultaneously a near infrared fluorophore. These nano-emulsions are injected in nude mice and their accumulation in subcutaneous tumors followed. The optimized quantity of iron oxide nanocrystals induces a remarkably high transverse relaxivity (R2), which is desirable for T2-weighted MRI imaging. The multimodal *in vivo* MRI and fluorescence imaging is shown in Fig. 11.7.

Another example of MRI imaging with nano-emulsions concerns the use of perfluorocarbons (PFCs) as oily phase. PFCs are synthetic organic compounds in which all or most of the hydrogen atoms have been replaced with fluorine atoms. The ^{19}F isotope of PFC's is biologically and chemically inert and thus provides excellent sensitivity *in vivo*. PFCs nano-emulsions have been used for ^{19}F MRI cell tracking (Nöth et al. 1997; Ahrens et al. 2005; Lanza et al. 2005; Partlow et al. 2007) as an alternative to superparamagnetic iron oxide (SPIO) agents, thanks to the high Nuclear magnetic resonance (NMR) sensitivity of the ^{19}F atom. As well, they have been shown to label *in vivo* the monocytes and macrophages, and give positive signals at sites of inflammation (Nöth et al. 1997; Flögel et al. 2008).

11.3.3.3 NEs as Fluorescent Probes in Fluorescent Imaging

Fluorescence is one of the most powerful and commonly used tools in biophysical studies of biomembrane structure and dynamics that can be applied on different levels, from lipid monolayers and bilayers to living cells, tissues, and whole

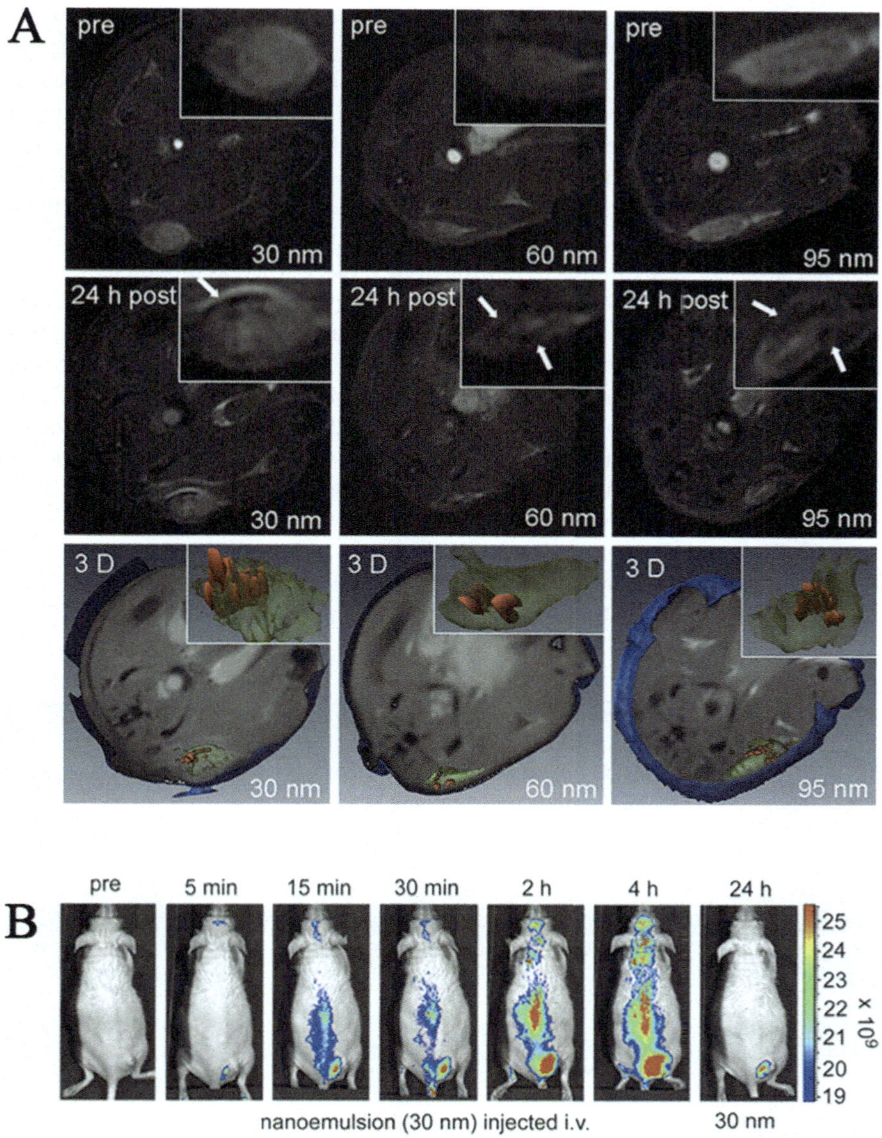

Fig. 11.7 (adapted from (Jarzyna et al. 2009)) MRI images of nude mice bearing subcutaneous EW7 tumors, injected with the three different nano-emulsions of 30 nm, 60 nm and 90 nm. (**a**) T2-weighted MRI images of the pre-scans,, after 24 h after intravenous injection, and 3D images that show the proton density MR image (*grey*), the whole tumor in 3D (*transparent green*) and the areas of iron oxide deposition in red. (**b**) *In vivo* near infrared (*NIR*) fluorescence imaging series showing the accumulation of the 30 nm nano-emulsion

animals. Nano-emulsions are also very suited for encapsulating fluorescent probes at high concentrations either organic dyes or inorganic quantum-dots (QDs), thus enabling a multimodal imaging as shown previously in Fig. 11.7. We have shown that the slight modification of classical dyes enable the increase of their concentration in oil without an excessive loss of quantum yield, resulting in ultrabright nanodroplets with no dye leakage (Klymchenko et al. 2012). This can be performed using classical dyes derivatives, like Nile Red. Actually, we disclosed that in living animal model (zebrafish), classical Nile Red dye encapsulated in nano-emulsions shows a strong release towards the surrounding tissues, whereas modified lipophilic Nile Red does not leak and allows accurate imaging of the blood. In 2010, Goutayer et al. (2010) were able to formulate a lipid nanoparticles (LNP) loading lipophilic fluorescent dye (DiD), then functionalized with the cRGD peptide binding to $\alpha v \beta_3$ integrin, a well-known angiogenesis biomarker, allowing their *in vivo* tracking using fluorescence imaging. *In vitro* study on HEK293(β3) cells over-expressing the $\alpha v \beta_3$ integrins demonstrates the functionalization, specific targeting, and internalization of cRGD-functionalized LNP in comparison with LNP-cRAD or LNP-OH used as negative controls. Following their intravenous injection in Nude mice, LNP-cRGD can accumulate actively in slow-growing HEK293(β3) cancer xenografts, leading to tumor over skin fluorescence ratio of 1.53 ± 0.07 (n=3) 24 h after injection. Figure 11.8 indicates the *in vivo* imaging in the study.

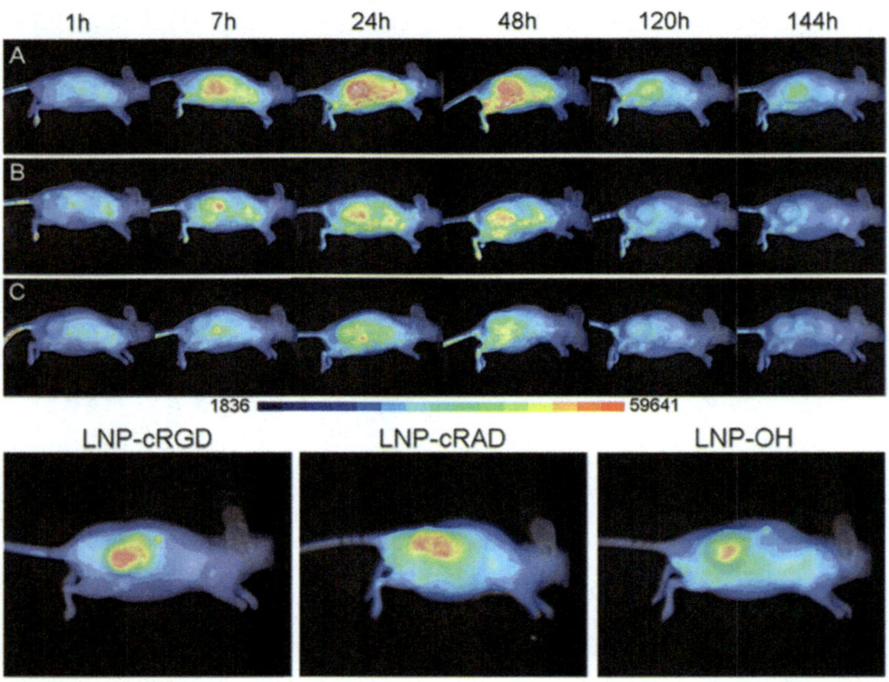

Fig. 11.8 (adapted from (Goutayer et al. 2010)) *In vivo* injection of LNP-cRGD (**a**), LNP-cRAD (**b**), or LNP-OH (**c**) in HEK293(β3) xenografted nude mice

11.3.3.4 Nano-emulsions as Multimodal Imaging Probes and Theragnostics

The main characteristic of nano-emulsions is their cargo structure that allows solubilizing and dispersing lipophilic compounds in the droplet's core, even if the molecules are very different. This property allows encapsulating different contrast agents for different imaging modalities, but also contrasts agents and drugs, so-called theragnostics agents.

A study has been performed by Jiali Ding et al. (2013) to design CT/fluorescence dual-modal nano-emulsion platform for investigating atherosclerotic plaques. Hydrophobic QDs were embedded in iodinated oil, which subsequently were dispersed in water to form multimodal contrast agent. The simultaneous X-ray contrasting properties and fluorescence properties were proved. These authors show that the nano-emulsion droplets entered murine macrophage cells and human liver cells, conferring them bimodal imaging properties. *In vivo* evaluation on atherosclerotic rabbits showed that bimodal nano-emulsions are detected to target specifically macrophages and allows visualizing atherosclerotic plaques.

On the other hand, the co-encapsulation of contrast agent and drug, *i.e.* theragnostics, has seen a real emergence in the last decade since it opens a real chance to monitor the actual dosage and amount of drug delivery in real time. This is even more interesting when non-invasive imaging techniques are more and more efficient, thus offering the possibility to follow quantitatively the drug targeting within a living organism. An interesting example reports a multimodal theragnostics nano-emulsion, containing MRI and fluorescent contrast agent (IONPs and Cy7, respectively), along with a glucocorticoid prednisolone acetate valerate (PAV) used for cancer treatment (Gianella et al. 2011). The droplets were functionalized with $\alpha v\beta_3$-specific RGD peptides, and administrated in subcutaneous tumors-bearing mice (at doses of 30 mg of FeO/kg and 10 mg of PAV/kg). The results, reported in Fig. 11.9 show a significant accumulation of the contrast agents in the tumor region, while tumor growth profiles revealed a potent inhibitory effect in all of the PAV containing nano-emulsion (compared to the control without PAV).

Another study (Tiwari et al. 2006) was reported for fabrication of DTPA-PE-Gd^{3+} (diethylenetriaminepentaacetic acid phosphoethanolamine) complexes nano-emulsions encapsulating paclitaxel hydrophobic drug and their T1 relaxation times measured. Based on NMR and MRI results, it was concluded that this system could act as a contrast agent. *In vitro* cell experiments also confirmed the drug delivery efficacy of the nano-emulsions, since they penetrated the cell membrane and killed the cancerous cells. O'Hanlon *et al.* (O'Hanlon et al. 2012) have synthesized the NIR-labeled perfluoropolyether (PFPE) nano-emulsions for drug delivery and imaging. In brief, PFPE-tyramide was successfully formulated into a non-steroidal anti-inflammatory drug, celecoxib (0.2 mg/mL)-carrying nano-emulsions with dual imaging modalities, ^{19}F MR and NIR. The results obtained by Dynamic light scattering (DLS), ^{19}F NMR, NIR fluorescence microscopy, and biological studies indicate that the nano-emulsion formulation may be useful for parenteral administration of celecoxib.

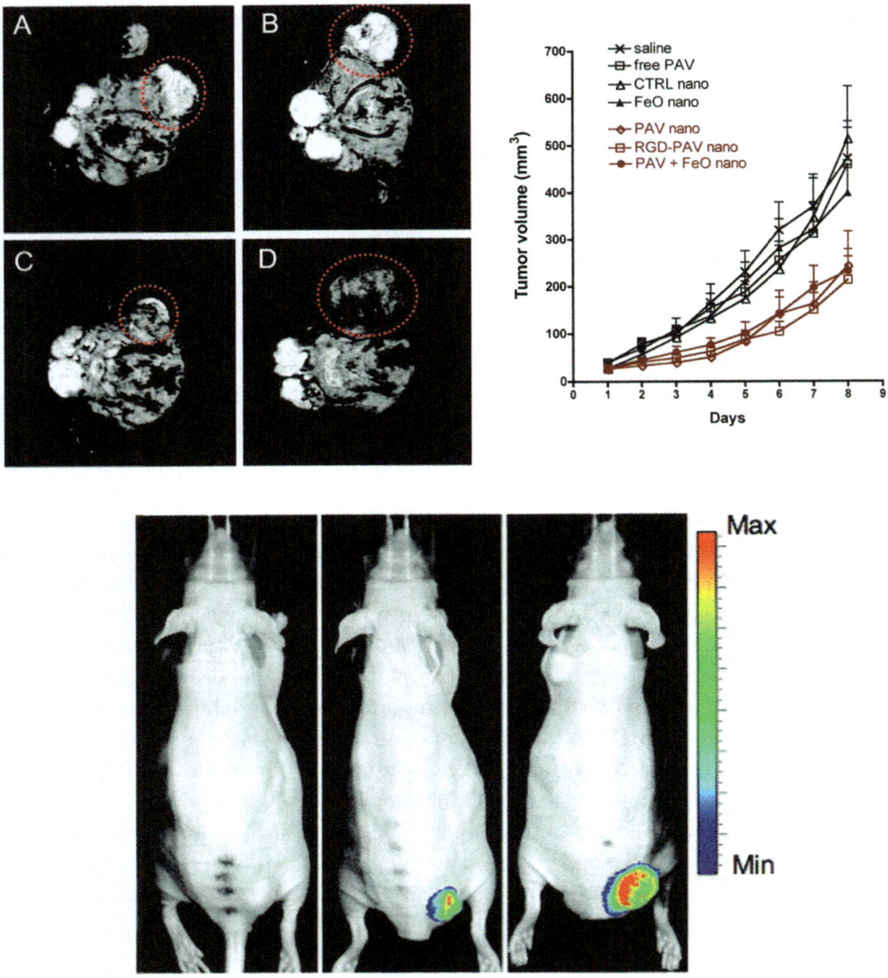

Fig. 11.9 (**a-d**) *In vivo* MRI imaging of xenografted tumor in mice (transverse sections), *red circles* indicate the tumors; (**a**) Selected MR images of PAV nano-emulsions and (**b**) control nano-emulsions. (**c**) MR images of PAV + FeO nano-emulsion and (**d**) FeO NE injected mice. *Red circles* indicate the tumors. In (**a**) and (**b**) tumors appeared bright compared to surrounding muscle tissue, and in (**c**) and (**d**), tumor areas appeared hypo-intense, indicative of FeO accumulation. Starting at day 6, tumor growth profiles showed statistically significant tumor growth inhibition ($P<0.001$) in all of the PAV nano-emulsions treated groups compared to saline, control nano-emulsions, FeO nano-emulsions, and free PAV injected groups. *Bottom: In vivo* NIRF images of a mouse injected with unlabeled NE (*left*) and mice (two different sized tumors) injected with Cy7 NE (*middle* and *right*) at the end of the study. (adapted from.(Ding et al. 2013))

11.4 Conclusion

To summarize, we can say that nano-emulsions are stable nano-cargo with important loading capability. These unique features open the doors of mixing the nature of encapsulating materials, enable a wide range of applications from biomedical imaging to therapies. In addition, any application of nanoparticulate systems is conditioned by the toxicity, biodistribution, pharmacokinetics and potentialities to target specific sites, and this chapter has been precisely focused on these different points to show the extent in which nano-emulsions are promising in many fields related to nanomedicines and imaging.

References

Ahrens ET, Flores R, Xu H, Morel PA (2005) In vivo imaging platform for tracking immunotherapeutic cells. Nat Biotechnol 23(8):983–987

Almajdub M, Nejjari M, Poncet G, Magnier L, Chereul E, Roche C, Manier M (2007) In vivo high-resolution X-ray microtomographyfor liver and spleen tumor assessment in mice. Contrast Media Mol Imaging 2:88–93

Anayama T, Nakajima T, Dunne M, Zheng J, Allen C, Driscoll B, Vines D, Keshavjee S, Jaffray D, Yasufuku K (2013) A novel minimally invasive technique to create a rabbit VX2 lung tumor model for nano-sized image contrast and interventional studies. PLoS One 8, e67355

ANSM. Résumé des caractéristiques du produit – MEDIALIPIDE 20 POUR CENT, émulsion pour perfusion - Base de données publique des médicaments.

ANSM. Résumé des caractéristiques du produit – NEORAL 100 mg/ml, solution buvable - Base de données publique des médicaments.

ANSM. Résumé des caractéristiques du produit – PROPOFOL LIPURO 1 % (10 mg/ml), émulsion injectable ou pour perfusion – Base de données publique des médicaments.

Anton N, Vandamme TF (2009) The universality of low-energy nano-emulsification. Int J Pharm 377:142–147

Aprahamian M, Bour G, Akladios CY, Fylaktakidou K, Greferath R, Soler L, Marescaux J, Egly JM, Lehn JM, Nicolau C (2011) Myo-InositolTrisPyroPhosphate treatment leads to HIF-1α suppression and eradication of early hepatoma tumors in rats. ChemBioChem 12(5):777–783

Attia M, Anton N, Chiper MC, Akasov R, Anton H, Messaddeq N, Fournel S, Klymchenko A, Mély Y, Vandamme TF (2014) Biodistribution of X-ray iodinated contrast agent in nano-emulsions is controlled by the chemical nature of the oily core. ACS Nano 8(10):10537–10550

Attia MF, Anton N, Akasov R, Chiper M, Markvicheva E, Vandamme TF (2015) Biodistribution and toxicity of X-ray iodinated contrast agent in nano-emulsions in function of their size. Pharm Res 33(3):603–614

Béduneau A, Hindré F, Clavreul A, Leroux JC, Saulnier P, Benoit JP (2008) Brain targeting using novel lipid nanovectors. J Control Release 126(1):44–49

Boll H, Nittka S, Doyon F, Neumaier M, Marx A, Kramer M, Groden C, Brockmann MA (2011) Micro-CT based experimental liver imaging using a nanoparticulate contrast agent: a longitudinal study in mice. PLoS One 6:e25692–e25697

Choi HS, Liu W, Misra P, Tanaka E, Zimmer JP, Ipe BI, Bawendi MG, Frangioni JV (2007) Renal clearance of nanoparticles. Nat Biotechnol 25:1165–1170

Delmas T, Piraux H, Couffin AC, Texier I, Vinet F, Poulin P, Cates ME, Bibette J (2011) How to prepare and stabilize very small nanoemulsions. Langmuir 27:1683–1692

Ding J, Wang Y, Ma M, Zhang Y, Lu S, Jiang Y, Qi C, Luo S, Dong G, Wen S, An Y, Gu N (2013) CT/fluorescence dual-modal nanoemulsion platform for investigating atherosclerotic plaques. Biomaterials 34:209–216

Dirschka T, Radny P, Dominicus R, Mensing H, Brüning H, Jenne L, Karl L, Sebastian M, Oster-Schmidt C, Klövekorn W, Reinhold U, Tanner M, Gröne D, Deichmann M, Simon M, Hübinger F, Hofbauer G, Krähn-Senftleben G, Borrosch F, Reich K, Berking C, Wolf P, Lehmann P, Moers-Carpi M, Hönigsmann H, Wernicke-Panten K, Helwig C, Foguet M, Schmitz B, Lübbert H, Szeimies R-M, Group A-CS (2012) Photodynamic therapy with BF-200 ALA for the treatment of actinic keratosis: results of a multicentre, randomized, observer-blind phase III study in comparison with a registered methyl-5-aminolaevulinate cream and placebo. Br J Dermatol 166:137–146

Dirschka T, Radny P, Dominicus R, Mensing H, Brüning H, Jenne L, Karl L, Sebastian M, Oster-Schmidt C, Klövekorn W, Reinhold U, Tanner M, Gröne D, Deichmann M, Simon M, Hübinger F, Hofbauer G, Krähn-Senftleben G, Borrosch F, Reich K, Berking C, Wolf P, Lehmann P, Moers-Carpi M, Hönigsmann H, Wernicke-Panten K, Hahn S, Pabst G, Voss D, Foguet M, Schmitz B, Lübbert H, Szeimies R-M, Group A-CS, Group A-CS (2013) Long-term (6 and 12 months) follow-up of two prospective, randomized, controlled phase III trials of photodynamic therapy with BF-200 ALA and methyl aminolaevulinate for the treatment of actinic keratosis. Br J Dermatol 168:825–836

Dubey PK, Kumar A (2005) Pain on injection of lipid-free propofol and propofol emulsion containing medium-chain triglyceride: a comparative study. Anesth Analg 101(4):1060–1062

Fang J, Nakamura H, Maeda H (2001) The EPR effect: unique features of tumor blood vessels for drug delivery, factors involved, and limitations and augmentation of the effect. Adv Drug Deliv Rev 63:136–151

Flögel U, Ding Z, Hardung H, Jander S, Reichmann G, Jacoby C, Schubert R, Schrader J (2008) In vivo monitoring of inflammation after cardiac and cerebral ischemia by fluorine magnetic resonance imaging. Circulation 118(2):140–148

Gianella A, Jarzyna PA, Mani V, Ramachandran S, Calcagno C, Tang J, Kann B, Dijk WJR, Thijssen VL, Griffioen AW, Storm G, Fayad ZA, Mulder WJM (2011) Multifunctional nanoemulsion platform for imaging guided therapy evaluated in experimental cancer. ACS Nano 5(6):4422–4433

Goutayer M, Dufort S, Josserand V, Royère A, Heinrich E, Vinet F, Bibette J, Coll JL, Texier I (2010) Tumor targeting of functionalized lipid nanoparticles: assessment by in vivo fluorescence imaging. Eur J Pharm Biopharm 75:137–147

Hallouard F, Briançon S, Anton N, Li X, Vandamme TF, Fessi H (2013) Poly(ethylene glycol)-poly(ε-caprolactone) iodinated nanocapsules as contrast agents for X-ray imaging. Pharm Res 30:2023–2035

Hobbs SK, Monsky WL, Yuan F, Roberts WG, Griffith L, Torchilin VP, Jain RK (1998) Regulation of transport pathways in tumor vessels: role of tumor type and microenvironment. Proc Natl Acad Sci U S A 95:4607–4612

Ijzerman M, Backer J, Flack M, Robinson P (2010a) Efficacy of topical nanoemulsion (NB-002) for the treatment of distal subungual onychomycosis: a randomized, double-blind, vehicle-controlled trial. J Am Acad Dermatol 62:AB76

Ijzerman M, Backer J, Flack M, Robinson P (2010b) Forty-two–week safety study of topical nanoemulsion (NB-002) for the treatment of mild to moderate distal subungual onychomycosis: a randomized, double-blind, vehicle-controlled trial. J Am Acad Dermatol 62:AB77

Jain RK (1999) Transport of molecules, particles, and cells in solid tumors. Annu Rev Biomed Eng 1:241–263

Jakhmola A, Anton N, Anton H, Messaddeq N, Hallouard F, Klymchenko A, Mely Y, Vandamme TF (2014) Poly-ε-caprolactone tungsten oxide nanoparticles as a contrast agent for X-ray computed tomography. Biomaterials 35(9):2981–2986

Jarzyna PA, Skajaa T, Gianella A, Cormode DP, Samber DD, Dickson SD, Chen W, Griffioen AW, Fayad ZA, Mulder WJM (2009) Iron oxide core oil-in-water emulsions as a multifunctional nanoparticle platform for tumor targeting and imaging. Biomaterials 30:6947–6954
Karathanasis E, Suryanarayanan S, Balusu SR, McNeeley K, Sechopoulos I, Karellas A, Annapragada AV, Bellamkonda RV (2009) Imaging nanoprobe for prediction of outcome of nanoparticle chemotherapy by using mammography. Radiology 250(2009):398–406
Khalid MN, Simard P, Hoarau D, Dragomir A, Leroux JC (2006) Long circulating poly(ethylene glycol)-decorated lipid nanocapsules deliver docetaxel to solid tumors. Pharm Res 23(4):752–758
Kircik L, Jones TM, Jarratt M, Flack MR, Ijzerman M, Ciotti S, Sutcliffe J, Boivin G, Stanberry LR, Baker JR, N.-S. Group (2012) Treatment with a novel topical nanoemulsion (NB-001) speeds time to healing of recurrent cold sores. J Drugs Dermatol 11:970–977
Klibanov AL, Maruyama K, Torchilin VP, Huang L (1990) Amphipathic polyethyleneglycols effectively prolong the circulation time of liposomes. FEBS Lett 268(1):235–237
Klymchenko AS, Roger E, Anton N, Anton H, Shulov I, Vermot J, Mély Y, Vandamme TF (2012) Highly lipophilic fluorescent dyes in nano-emulsions: towards bright non-leaking nano-droplets. RSC Adv 2:11876–11886
Kulkarni SA, Feng SS (2013) Effects of particle size and surface modification on cellular uptake and biodistribution of polymeric nanoparticles for drug delivery. Pharm Res 30:2512–2522
Lanza GM, Winter PM, Neubauer AM, Caruthers SD, Hockett FD, Wickline SA (2005) 1H/19 F magnetic resonance molecular imaging with perfluorocarbon nanoparticles. Curr Top Dev Biol 70:57–76
Li X, Anton N, Zubera G, Zhao M, Messaddeq N, Hallouard F, Fessie H, Vandamme TF (2013) Iodinated α-tocopherol nano-emulsions as non-toxic contrast agents for preclinical X-ray imaging. Biomaterials 34:481–491
Li X, Anton N, Zuber G, Vandamme TF (2014) Contrast agents for preclinical targeted X-ray imaging. Adv Drug Deliv Rev 76:116–133
Martiniova L, Schimel D, Lai EW, Limpuanfthip A, Kvetnansky R, Pacak K (2010) In vivo micro CT imaging for liver lesions in small animal models. Methods 50:20–25
Neittaanmäki-Perttu N, Karppinen T t, Grönroos M, Tani T t, Snellman E (2014) Daylight photodynamic therapy for actinic keratoses: a randomized double-blinded nonsponsored prospective study comparing 5-aminolaevulinic acid nanoemulsion (BF-200) with methyl-5-aminolaevulinate. Br J Dermatol 171:1172–1180
Nöth U, Morrissey SP, Deichmann R, Jung S, Adolf H, Haase A, Lutz J (1997) Perfluoro-15-crown-5-ether labelled macrophages in adoptive transfer experimental allergic encephalomyelitis. Artif Cells Blood Substit Immobil Biotechnol 25(3):243–254
O'Hanlon CE, Amede KG, O'Hear MR, Janjic JM (2012) NIR-labeled perfluoropolyether nanoemulsions for drug delivery and imaging. J Fluor Chem 137:27–33
Owens DE, Peppas NA (2006) Opsonization, biodistribution and pharmacokinetics of polymeric nanoparticles. Int J Pharm 307:93–102
Partlow KC, Chen J, Brant JA, Neubauer AM, Meyerrose TE, Creer MH, Nolta JA, Caruthers SD, Lanza GM, Wickline SA (2007) 19F magnetic resonance imaging for stem/progenitor cell tracking with multiple unique perfluorocarbon nanobeacons. FASEB J 21(8):1647–1654
Rodrigues TA, Alexandrino RA, Kanczuk ME, Gozzani JL, Mathias LA, Da ST (2012) A comparative study of non-lipid nanoemulsion of propofol with solutol and propofol emulsion with lecithin. Rev Bras Anestesiol 62(3):325–334
Stanberry LR, Simon JK, Johnson C, Robinson PL, Morry J, Flack MR, Gracon S, Myc A, Hamouda T, Baker JR Jr (2012) Safety and immunogenicity of a novel nanoemulsion mucosal adjuvant W805EC combined with approved seasonal influenza antigens. Vaccine 30:307–316
Szeimies R-M, Radny P, Sebastian M, Borrosch F, Dirschka T, Krähn-Senftleben G, Reich K, Pabst G, Voss D, Foguet M, Gahlmann R, Lübbert H, Reinhold U (2010) Photodynamic therapy with BF-200 ALA for the treatment of actinic keratosis: results of a prospective, randomized, double-blind, placebo-controlled phase III study. Br J Dermatol 163:386–394

Taisne L, walstra B, Cabane B (1996) Transfer of oil between emulsion droplets. J Colloid Interface Sci 184:378–390

Tiwari S, Tan YM, Amiji M (2006) Preparation and in vitro characterization of multifunctional nanoemulsions for simultaneous MR imaging and targeted drug delivery. J Biomed Nanotechnol 2:1–8

Torchilin V-P, Trubetskoy V-S (1995) Wich polymers can make nanoparticulates drug carriers long circulating? Adv Drug Deliv Rev 16(2–3):141–155

Treanor JJ, Essink B, Hull S, Reed S, Izikson R, Patriarca P, Goldenthal KL, Kohberger R, Dunkle LM (2013) Evaluation of safety and immunogenicity of recombinant influenza hemagglutinin (H5/Indonesia/05/2005) formulated with and without a stable oil-in-water emulsion containing glucopyranosyl-lipid A (SE+GLA) adjuvant. Vaccine 31:5760–5765

Zheng J, Jaffray D, Allen C (2009) Quantitative CT imaging of the spatial and temporal distribution of liposomes in a rabbit tumor model. Mol Pharm 6(2):571–580

Chapter 12
The Tumor Microenvironment in Nanoparticle Delivery and the Role of Imaging to Navigate Roadblocks and Pathways

Dmitri Artemov and Zaver M. Bhujwalla

Abstract Effective delivery and distribution of nanoparticles (NPs) are at the core of successful applications of the exciting new advances being made in nanomedicine to achieve precision medicine in cancer. The tumor microenvironment (TME) can pose a barrier but can also be exploited for delivery and distribution. Here we have outlined the characteristics of the TME and the influence that different aspects of the TME such as vascularization, permeability, the extracellular matrix (ECM), and stromal cells can have on NP delivery and distribution. We also discuss TME modification strategies to improve delivery and distribution of NPs.

Keywords Nanoparticle delivery • Tumor microenvironment • Molecular imaging

Abbreviations

CAFs	Cancer associated fibroblasts
EPR	Enhanced permeability and retention
ECM	Extracellular matrix
H&E	Hematoxylin and Eosin
HT	Hyperthermia
MPLSM	Multiphoton laser scanning microscopy
NPs	Nanoparticles
NIR	Near infrared

D. Artemov (✉)
Division of Cancer Imaging Research, Department of Radiology, The Johns Hopkins University School of Medicine, Rm 217, Traylor Bldg., 720, Rutland Avenue, Baltimore, MD 21205, USA
e-mail: dartemo2@jhmi.edu

Z.M. Bhujwalla
Division of Cancer Imaging Research, Department of Radiology, The Johns Hopkins University School of Medicine, Rm 208C, Traylor Bldg., 720, Rutland Avenue, Baltimore, MD 21205, USA
e-mail: zaver@mri.jhu.edu

A. Prokop, V. Weissig (eds.), *Intracellular Delivery III*, Fundamental Biomedical Technologies 8, DOI 10.1007/978-3-319-43525-1_12

PEG	Polyethylene-glycol
SHG	Second harmonic generation
TAMs	Tumor-associated macrophages
TME	Tumor microenvironment

12.1 Introduction

The promise of nanomedicine for precision medicine in cancer has resulted in remarkably innovative strategies in the design of nanoparticles (NPs) and approaches to deliver these with a high degree of specificity to cancer specific targets. The unique tumor microenvironment (TME) has presented barriers as well as opportunities in the delivery and targeting of NPs in cancer. Here we have provided examples of NPs that are currently available together with advantages and disadvantages. Aspects of the TME and their modification in the delivery of NPs are discussed. We also discuss the role of molecular imaging in NP delivery and in cancer theranostics.

12.2 Nanoplatforms

Successful delivery of nanoparticles (NPs) to the target of interest is one of the prerequisites of nanomedicine and an important component of precision medicine. Despite exquisitely specific targeting abilities in culture, this specificity rapidly diminishes once NPs encounter the complexities of the TME, highlighting the importance of noninvasive imaging to visualize their movement *in vivo*. Several NP platforms with a typical size of 5–400 nm are currently available with the patent advantages of a high payload capacity that enables carrying imaging reporters as well as therapeutic cargo. The plasma pharmacokinetics of NPs can be optimized for long circulation and minimal clearance by the reticuloendothelial system. NPs can be functionalized for high affinity binding to molecular targets expressed in cancer, as well as to respond to unique physiological signals present in the TME to activate various physicochemical processes, such as changing the surface charge, shedding a protective sheath and exposing functional moieties, and decomposing to release the cargo. NPs can be directly used for therapeutic applications using their unique optical or magnetic properties to damage target cells or environments. Typical matrix materials used to synthesize NPs and their properties are summarized in Table 12.1.

Table 12.1 Nanoparticle types

Type	Shape	Size	Loading capacity	Inherent functionality
Iron oxide NP	Spherical	20–1000s nm	Low, primarily in the surface layer	MRI contrast; Magnetism for MPI*; Hyperthermia in AMF**; Mechanical moment in AMF
Gold nanoshells	Spherical	1.5–200 nm	Low, surface conjugation	Light absorption by SPR***; produce hyperthermia
Gold nanorods	Needle-shape	Diameter 10–50 nm; length 20–250 nm	Low, surface conjugation	Light absorption by SPR; produce hyperthermia
Liquid-metal core-shell nanospheres (Lu et al. 2015)	Spherical	~100 nm	Low to moderate, shell surface loading via conjugated cyclodextrin	Biodegradable GaIn core under mild acidic conditions
Polymer NP	Various	1–1000 nm	High, chemical linkage or encapsulation in the matrix	
Dendrimers	Spherical	2 (G1)- 15 (G10) nm, monodisperse	High, multiple surface functional groups	
Carbon nanotubes	Needle-shape	Single wall: Diameter 1–1.5 nm; length >10 μm Multiwall: Diameter 10–30 nm; length 1–5 μm	High	
Micelles	Spherical	5–100 nm	High for lipophilic agents	
Liposomes	Spherical	Unilamellar: 20–400 nm	High for both lipophilic and hydrophilic agents	

* – MPI – magnetic particle imaging Gleich et al. (2005) and Goodwill et al. (2012)
** – AMF – alternating magnetic field
*** – SPR – surface plasmon resonance Bedford et al. (2012)

12.3 Tumor Microenvironment

The TME is a complex milieu consisting of stromal cell populations intertwined within a complex extracellular matrix (ECM). Spatial and temporal heterogeneities in vascularization, lymphatics, oxygenation, acidic extracellular pH, stromal cell trafficking, cytokines, and proteolytic enzymes make the TME uniquely different from the environment of normal tissues, and underline the importance of imaging in following these heterogeneities.

12.3.1 Tumor Vasculature

The TME is characterized by disorganized vasculature with immature capillaries that are leaky and tortuous and cannot provide sufficient oxygen and nutrients to cells within the tumor resulting in hypoxia, acidic extracellular pH, and necrosis. A comparison between a regularly arranged vasculature in the muscle tissue and the chaotic vascular network in tumors is shown in Fig. 12.1.

When a cluster of cancer cells exceeds a few millimeters in size, nutrients supplied by diffusion become insufficient and the cancer cells induce neovascularization. Cancer cells induce neovascularization by co-opting existing vasculature and remodeling it *via* intussusceptive angiogenesis (Burri et al. 2004) by stimulating angiogenesis and inducing sprouting of new blood capillaries from existing ones (Gerhardt et al. 2003; Kalluri 2003), or by vasculogenesis, which is formation of *de novo* capillaries from endothelial progenitor cells (angioblasts) recruited from the host (Jin and Patterson 2009; Lancrin et al. 2009). Tumor angiogenesis occurs as a stress response to low oxygen and nutrients resulting in the overproduction of pro-angiogenic cytokines, such as VEGF, by cancer cells and by stromal cells

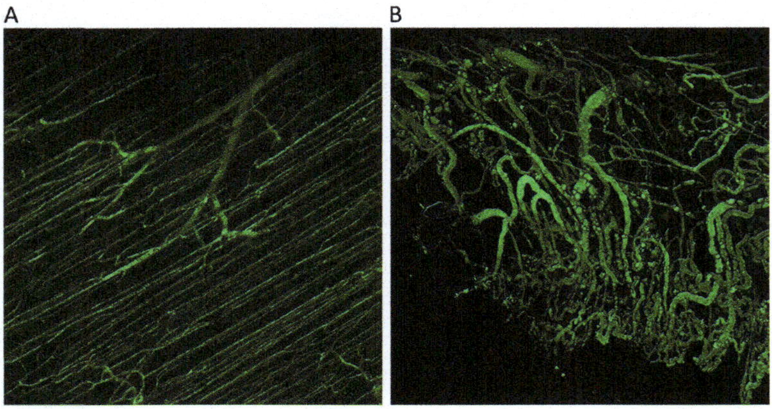

Fig. 12.1 Multiphoton intravital microscopy of vasculature in (**a**) muscle tissue and (**b**) in a human xenograft grown in a nude mouse. A fluorescent dextran probe was used as a vascular marker

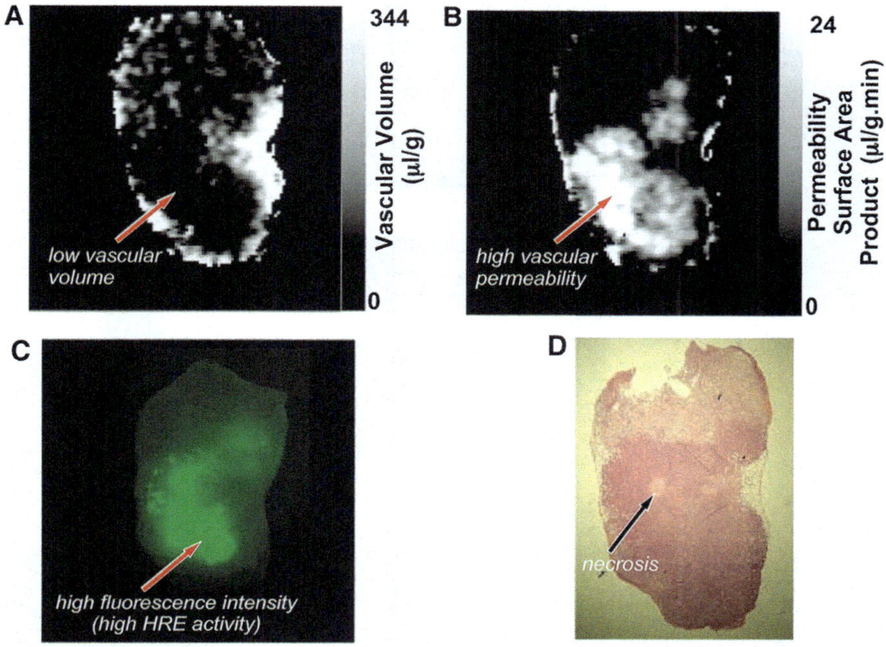

Fig. 12.2 Maps of (**a**) vascular volume and (**b**) permeability-surface area product obtained from a central slice of an HRE-EGFP PC-3 tumor. (**c**) fluorescent microscopy (1×) of a fresh tissue slice obtained from the imaged slice, using a Nikon TS100-F microscope with a wavelength of 512 nm. (**d**) H&E-stained section from the central MR-imaged slice. The region exhibiting EGFP consisted of viable cells. The small necrotic area of dying cells is marked by *black arrow* (From Raman et al. 2006)

present in the tumor (Claesson-Welsh and Welsh 2013; Semenza 2001). The high interstitial concentration of cytokines is one reason for disregulation of the normal angiogenic cascade that is driven by the interplay between VEGF receptors VEGFR1, VEGFR2, and the neuropilin receptor NPR1 (Imoukhuede and Popel 2011; Vempati et al. 2014). As a result there is multiple sprouting of endothelial cells and the formation of a disorganized capillary network. These immature highly permeable blood vessels are leaky to macromolecules, have reduced blood flow, and are unable to support oxygen and nutrient supply for the bulk of the tumor, causing formation of hypoxia and necrosis (Dudley 2012). The immature vasculature is also susceptible to collapse from the high interstitial pressure that exists in tumors resulting in diffusion limited hypoxia and necrosis as well as acute hypoxia due to vascular collapse (Stylianopoulos et al. 2013). Intermittent hypoxia has also been observed in tumors (Toffoli and Michiels 2008).

The spatial relationship between tumor vascular volume, vascular permeability, hypoxia, and tissue viability is demonstrated in Fig. 12.2 in a study that combined fluorescence detection of hypoxia with MRI characterization of vascular volume and permeability of albumin-GdDTPA (Raman et al. 2006). Hypoxic tumor areas detected by green fluorescence protein expression under control of hypoxic response

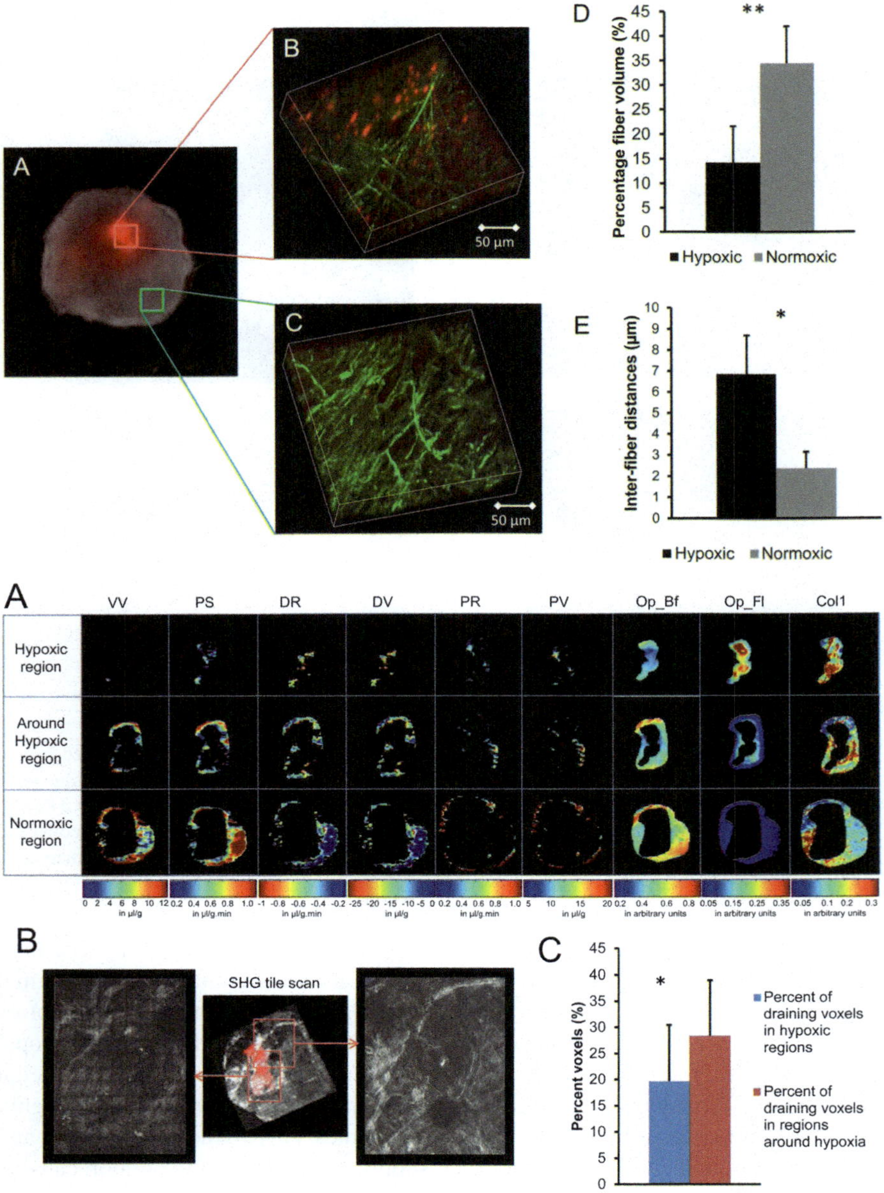

Fig. 12.3 *Panel 1* (**a**) *Ex vivo* 16 bright-field and fluorescence image overlay showing the locations of hypoxic ROIs on the tumor section. 3D visualization of (**b**) hypoxic and (**c**) normoxic FOVs; hypoxic regions are displayed in *red* and Col1 fibers in *green*. The FOV image size was $335 \times 335 \times 60$ μm^3 with a voxel size of 0.44 μm^3. (**d**) Hypoxic FOVs showed a significantly lower fiber volume as compared to normoxic FOVs. (**e**) Hypoxic FOVs have a significantly larger inter-fiber distance as compared to normoxic FOVs. *Panel 2* (**a**) Displays of the distribution of vascular parameters (*VV* vascular volume, *PS* permeability surface area product), transport parameters (*DR* draining rate, *DV* draining volume, *PR* pooling rate, *PV* pooling volume) and optical imaging parameters (*OP_Bf* optical bright field, *OP_Fl* optical fluorescence field, *Col1* collagen 1 fibers) in hypoxic regions (top panel), ~100 μm around hypoxic regions (*middle panel*), and in normoxic

elements have low vascular volume and high vascular permeability, whereas normoxic tumor regions have high vascular volume but low permeability. These data suggest that relatively low interstitial delivery of macromolecular agents will occur in well vascularized viable tumor regions (Raman et al. 2006).

12.3.2 Tumor Lymphatics

Most tumors do not have a functional lymphatic system resulting in impaired clearance of interstitial fluid with reduced or absent macroscopic flow of interstitial fluid (Jain 1997). Functional lymphatics are missing in solid tumors despite an increased expression of VEGFR-C, the factor that stimulates proliferation of lymphatic endothelium and is a risk factor for lymphatic tumor dissemination (Ciobanu et al. 2013; Decio et al. 2014; Gogineni et al. 2013). This results in the buildup of the interstitial hydrostatic pressure preventing vascular delivery of macromolecules such as NPs that depend on the gradient between the vasculature and interstitial pressure rather than on pure diffusion for interstitial movement (Jain and Stylianopoulos 2010). On the other hand, non-functional lymphatics result in the prolonged retention of macromolecules and NPs in the tumor interstitium because of slow or absent lymphatic drain.

12.3.3 The Extracellular Matrix

The extracellular matrix (ECM) is composed of fiber proteins and glycosaminoglycans. Glycosaminoglycans are carbohydrate proteins, such as proteoglycans (heparan sulfate, chondroitin sulfate, keratin sulfate) and hyaluronic acid that play a role in cell signaling and in sequestering and releasing multiple growth factors and cytokines in the tumor interstitium (Multhaupt et al. 2016). Fiber proteins consist of collagen (Type I – XIV), fibronectin, and elastin that provide structure to the tumor (Friedl and Brocker 2000). These proteins form an interlocking mesh that theoretically should prevent diffusion of NPs through the tumor interstitium. On the other hand, recent work suggests that collagen 1 fibers can in fact mediate macromolecular transport (Kakkad et al. 2013). A similar effect of augmented transport of dislodged cancer cells along the fibers can explain the association between collagen I fiber density and lymph node metastasis observed in breast cancer (Kakkad et al. 2013).

Significantly reduced transport of high-molecular weight contrast agent in the hypoxic tumor areas, which are also characterized by a lower density of the collagen 1 fibers measured by second harmonic generation (SHG) multiphoton microscopy, is shown in Fig. 12.3.

Fig. 12.3 (continued) regions (*bottom panel*) of a representative tumor section from a breast cancer xenograft. (**b**) Tile scan of SHG microscopy of Col1 fibers of the same section, showing Col1 fiber patterns in the magnified inserts. (**c**) Quantifications of draining voxels in percent showing a significant increase in percent draining voxels in areas surrounding hypoxic regions. Adapted from Kakkad et al. (Kakkad et al. 2013)

12.3.4 *Stromal Cells*

Stromal cells are frequently co-opted by cancer cells and the TME is characterized by the presence of several types of cancer associated stromal cells, such as tumor-associated macrophages (TAMs) (Komohara et al. 2016; Vinogradov et al. 2014) and cancer-associated fibroblasts (CAF) (Shiga et al. 2015; Jung et al. 2016). These cell populations play a role in NP delivery by expressing remodeling cytokines. The potential phagocytosis of NPs by TAMs can influence the distribution and availability of NPs for cancer cells (Chen et al. 2014).

12.4 NP Delivery in Tumors

The delivery of NPs can be active or passive (Sharma et al. 2013). Passive delivery is attributed to increased accumulation of nanomaterials in the tumor due to the enhanced permeability and retention (EPR) effect (Maeda 2001; Nehoff et al. 2014) that depends on the enhanced permeability of the tumor neovasculature and reduced lymphatic drain and is most pronounced for hydrophilic nanoparticles in the range of 100 nm. The penetration of NPs into the tumor interstitium is a passive process in which the NP circulates in the blood stream, translocates across the endothelial lining of the tumor blood vessels, and finally diffuses to the target site.

Spherical NPs that have molecular sizes optimal for EPR delivery have negligible diffusion within the tumor interstitium (Du et al. 2015). Therefore, their delivery is typically limited to the perivascular space and the delivery pattern forms "hot spots" with very little amount of nanomaterial in the rest of the tumor mass. An example of extravasation patterns of a nonspecific mAb and 50 nm microspheres in tumor sections 6 h after injection is shown in Fig. 12.4 (Nakahara et al. 2006). However, as discussed in the subsequent section, elongated NPs seem to traverse the ECM and attach to targets on the cancer cell more efficiently despite sizes of ~200 nm (Chen et al. 2016).

There are several important consequences of this non-uniform pattern of nanodelivery. Firstly, in nanomedicine applications, the therapeutic NP should either induce a strong bystander effect by affecting not just the neighboring cells but also those at a distance of several hundred microns, or upon extravasation into the tumor, release small molecular weight agents that can effectively diffuse into the tumor. Secondly, many activation-based delivery strategies that are discussed in the next section rely on the unique properties of the TME such as acidic pH, low oxygen, or high concentrations of proteolytic enzymes. These important characteristics of TME may not be present in the "hot spot" areas of the tumor that are in fast exchange with plasma through the leaky walls of tumor associated blood capillaries. For instance, serum is a potent inhibitor of multiple proteases and may significantly reduce the activation efficiency of NP designed to be sensitive to these enzymes.

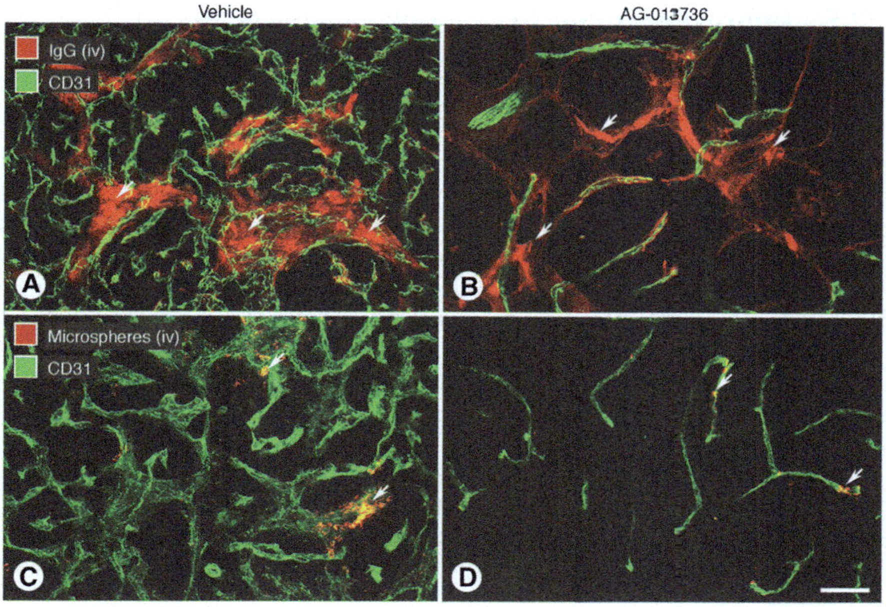

Fig. 12.4 Confocal micrographs showing the distribution of extravasated nonspecific IgG and 50-nm microspheres at 6 h after injection in RIP-Tag2 tumors. Extravasated IgG is shown under baseline conditions (**a**) and after treatment with AG-013736 for 7 days (**b**). Extravasated IgG (*arrows*) had a patchy distribution with only vague association with CD31 stained blood vessels (**a**). After AG-013736, the number of tumor vessels was markedly reduced (**b**) but extravasated IgG still had a patchy distribution (**b**, *arrows*). Some IgG was associated with tumor vessels and some was not. Extravasated microspheres were closely associated with focal regions of tumor vessels both under baseline conditions (**c**, arrows) and after AG-013736 (**d**, arrows). Fewer tumor vessels were present after treatment (**d**). Microsphere extravasation was reduced more than the vasculature (**d**). Bar, 50 μm (From Nakahara et al. 2006)

Target-specific NP delivery is a valuable strategy for specific imaging to detect expression of the target. This approach has been explored in several reports for specific imaging of tumors using targeted NPs (Bai et al. 2016; Bakhtiary et al. 2015; Weissleder et al. 1990, 2000). The mechanisms for nanomaterial retention at the target site can be passive or active. Passive retention is facilitated by dysfunctional lymphatics, or by changes in the polarity and hydrophilicity of the particle. Active delivery is achieved by trapping the NP in the tumor by a specific process such as binding to a molecular target, internalization into the target cell, or the disassembly of the NP and release of the cargo in specific environments.

12.5 Improving Tumor Delivery of Nanoparticles

12.5.1 Optimization of Circulation Time

Prolonged systemic circulation time is a prerequisite for efficient tumor delivery of NPs. To achieve a slow systemic clearance the NP size should be within a range of approximately 5 nm – the normal endothelium pore size and cutoff size for renal clearance by glomerular filtration (Longmire et al. 2008), to several hundred nanometers to minimize hepatic clearance by Kupffer cells and hepatocytes. Particle charge also plays a major role in determining the clearance rate; charged particles rapidly adsorb serum protein inducing rapid opsonization and hepatic clearance. Modification of the NP surface with highly hydrophilic neutral polyethylene-glycol (PEG) chains was shown to be an efficient means to prevent serum protein adsorption, opsonization and clearance (Choi et al. 2007). On the other hand, PEGylation suppresses internalization of NPs by target cells. Therefore, if endocytosis is a desirable feature of the NP design, PEGylation can hinder this process and this creates a so-called "PEG dilemma" (Hama et al. 2015; Hatakeyama et al. 2007; Mishra et al. 2004).

12.5.2 NP Extravasation

The walls of mature functional blood vessels are comprised of several layers of endothelial cells, basement membrane, and blood vessel-associated pericytes. Although in tumors many angiogenic blood capillaries lack this complex structure and are simply formed by a relatively leaky single layer of endothelial cells, the uniform delivery of NPs can require improved transcapillary transport. As shown in Fig. 12.5, using a specific antibody to target caveolae in the tumor endothelium can provide an effective mean to pump NPs across the endothelial layer from the blood into the tumor interstitium (Oh et al. 2014).

12.5.3 NP Shape and Size

While spherical NPs have severely limited diffusion and movement through the tumor interstitium, significantly improved delivery and diffusion properties have been observed for elongated nanoconstructs such as nanorods, nanotubes, or linear polymer chains due to easier extravasation because of their 'string-like' structures and potentially faster diffusion through the ECM (Raphael et al. 2015). Increased tumor targeting by elongated nanoworms compared to spherical NP was

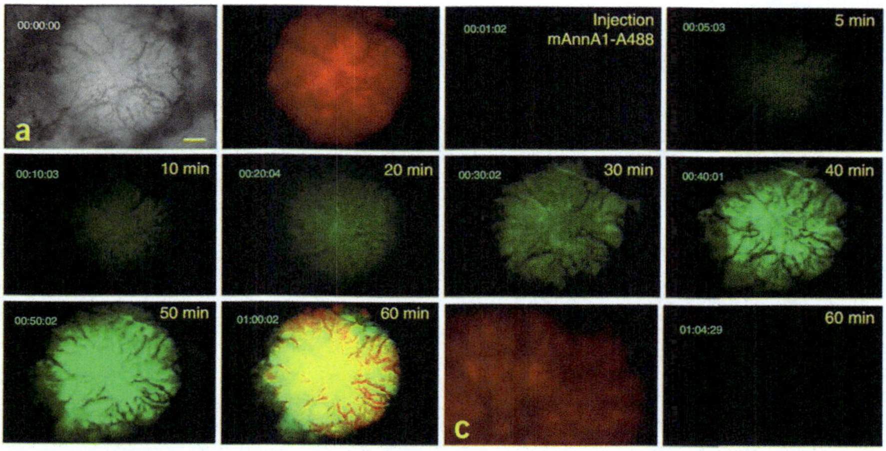

Fig. 12.5 Fluorescence intravital microscopy imaging of H2B-mCherry–expressing N202 mammary tumors (*red*) in mice after intravenous injection of tumor endothelial caveolae targeting antibody, mAnnA1, shows rapid and pervasive tumor penetration. Time series imaged after tail vein injection of 4 μg of mAnnA1 (**a**) and control isotype-matched mouse IgG (mIgG) (**c**). The static images shown in (**a**) and (**c**) were captured in the *red* or *green* fluorescence channel at the indicated times. The overlap of *red*- and *green*-channel images (tumor and mAnnA1 probe, respectively) is shown at 60 min in (**a**) (From Oh et al. 2014)

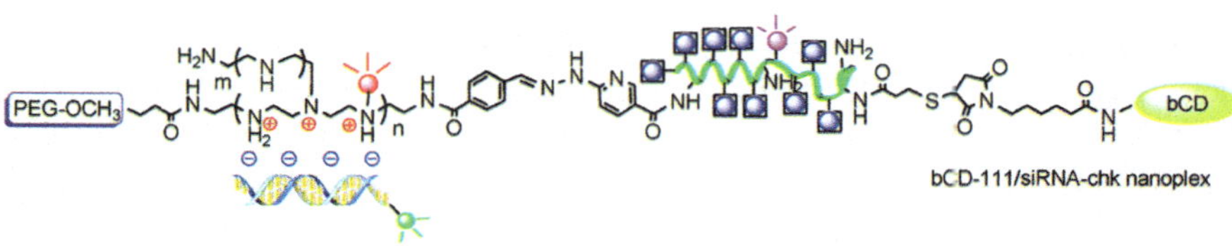

Fig. 12.6 Schematic structure of the nanoplex (From Li et al. 2010)

reported by Park et al. that was also attributed to the prolonged circulation time and multivalent interactions between homing moieties and their *in vivo* targets (Park et al. 2009). String-like NPs with a poly-l-lysine backbone, called nanoplexes, showed an efficient delivery despite having a molecular weight of 355 kDa and an average size of 55 nm in a preclinical model of human breast and prostate carcinomas, starting 4 h after systemic injection (Chen et al. 2016; Li et al. 2008, 2010). These nanoplexes (Fig. 12.6) were used to deliver, under image-guidance, the prodrug enzyme bacterial cytosine deaminase (bCD) together with small interfering siRNA (siRNA) to downregulate choline kinase (chk) for specific molecular targeting (Fig. 12.7).

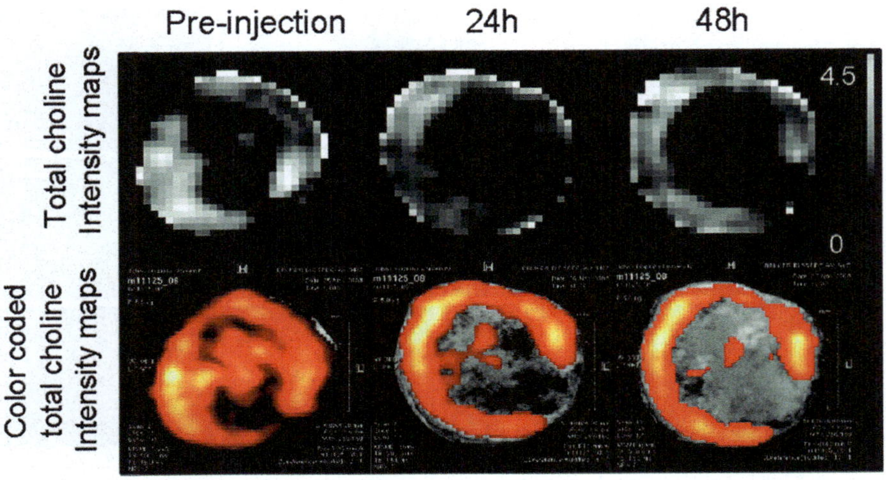

Fig. 12.7 Representative in vivo tCho maps and color coded tCho intensity maps overlaid on corresponding T1-weighted images of a tumor before and at 24 and 48 h after NP injection (300 mg/kg, i.v.) (From Li et al. 2010)

12.5.4 TME Modification Strategies

Although the ECM may present a barrier for diffusion of NPs, strategies are being developed to improve NP diffusion using enzymatically-induced degradation of ECM components such as hyaluronic acid with hyaluronidase (Eikenes et al. 2010; Scodeller et al. 2013; Zhou et al. 2016) or collagenase (Netti et al. 2000). A significant increase in the transport of 54 nm fluorescent NP was detected after pre-infusion of a rat brain with hyaluronidase (Neeves et al. 2007). In addition, therapeutic targeting of CAFs can also help to remodel the tumor ECM and potentially improve nanodelivery (Narunsky et al. 2014).

A novel strategy to improve extravasation and delivery of NP to tumors is applying mild hyperthermia. Hyperthermia has multiple biological functions and is known to significantly enhance the permeability of the vasculature (Hildebrandt et al. 2002; Emami and Song 1984). The putative mechanism of this effect is nonspecific shrinkage of endothelial cells lining the capillaries due to a rapid rearrangement of the actin cytoskeleton of the endothelial cells subjected to the hyperthermic shock (Hildebrandt et al. 2002; Bagley et al. 2015; Clark et al. 1983). Other biological response mechanisms that can potentially contribute to hyperthermia-induced permeability of the tumor vasculature may include modulation of the tumor blood flow, paracrine effects of cytokines such as histamine, bradykinin, and TNF-α, and modification of the tumor ECM (Hildebrandt et al. 2002). Thermal exposure of tumors to 41.5^0 C for 30 min induced a significant increase in the uptake of vaccinia virus (an average diameter of 250–300 nm) (Chang et al. 2005). A significant increase in the accumulation of NPs with diameters in the range of 100–400 nm in SKOV-3 tumors was observed within the first 15–30 min after exposure of the tumor to mild HT at 42 °C (Kong et al. 2000). Interestingly, the uptake of smaller NPs with a diameter of 100 nm was significantly higher than for NPs with diameters of 200 and 400 nm, suggesting a "sieve" effect of the pore size in the tumor endothelium (Kong et al. 2000, 2001). Increased permeability persisted for at least 4 h after the end of hyperthermia, and the vascular permeability was completely restored to the basal state after 6 h (Kong et al. 2001).

Hyperthermia can be induced by various means including direct heating of the tissue, high intensity focused ultrasound, as well as using unique inherent properties of nanomaterials. For instance, superparamagnetic NPs can generate heat when exposed to alternating magnetic fields due to hysteresis losses in the magnetic material (Ivkov 2013). Special gold NPs called "nanoshells" are capable of producing significant heat through absorption of NIR irradiation by the surface plasmon resonance (SPR) (Zhao et al. 2014). NIR-absorbing PEGylated gold nanorods were used as "activators" of the delivery of doxorubicin-loaded liposomes *in vivo* (Park et al. 2010). These nanorods provided efficient heating of the tumor to a surface temperature of 45 °C (NIR light, 810 nm at 0.75 W/cm^2 for 30 min), which significantly increased tumor uptake of NPs such as superparamagnetic nanoworms and therapeutic liposomes, with a corresponding increase in the therapeutic efficacy (Park et al. 2010). Unfortunately, HT-induced nanodelivery is complicated by the effect of thermotolerance, which manifests itself as reduced sensitivity of endothelial cells to repetitive cycles of hyperthermia. Physiologically, thermotolerance prevents extravasation of nanoparticles from the tumor vasculature that previously experienced HT (Bagley et al. 2015; Kong et al. 2001).

12.6 TME Targeting Strategies

Specific targeting can significantly modify pharmacokinetics of NPs within the tumor, resulting in increased tumor accumulation. For an extensive review of varies strategies for development of NP specifically targeted TME we refer the readers to the review by Du et al. (2015). Table 12.2 presents a brief summary of the available methods and related properties of the nanomaterials. Generally, molecular targets expressed on the surface of cancer cells and in the TME are used for high-affinity binding of functionalized NPs. NP internalization is activated, or tumor environmental conditions induce structural changes in NPs causing their trapping and prolonged retention in the tumor. In addition, an important concept in targeting of the TME is development of specific agents, which recognize unique targets expressed by the tumor microvasculature. These targets are exposed to circulating blood. As a

Table 12.2 Mechanisms of targeting, activation, and retention of nanomaterials

Target	Targeting principle	Mechanisms of activation and/or retention
Acidic extracellular pH_e in TME	(i) Hydrophilic-hydrophobic transition	(i) Dissociation of micelles
	(ii) Electrostatic associating between polymer chains	(ii) On-demand ligand exposure on the surface of NP
	(iii) pH catalyzed bond cleavage	(iii) Internalization of NP undergoing transition form neutral or negative to positively charged
	(iv) peptides with membrane insertion sequences	(iv) Shedding PEG layer from NP by cleavage of the linker
		(v) Insertion of the protonated C-terminal of a peptide chain into the cell across the plasma membrane
Hypoxia or low pO_2	Hypoxia responsive electron acceptor, such as nitroimidazole	(i) Photoactivated drug release
		(ii) Disassembly of nitroimidazole derivate carboxymethyl-dextran micelles
		(iii) Detaching PEG sheath from NP
High activity of proteolytic enzymes	Cleavage of MMP-specific peptide chains	(i) Release of chemically linked moieties
		(ii) Removal of protective caps from porous NP releasing the encapsulated cargo
Matrix metalloproteinases (MMPs), phospholipase A2 (PLA2)		(iii) Detaching PEG protective layer from NP
		(iv) Exposing cell-penetrating peptides, which induce NP internalization
		(v) Enzymatic degradation of gelatin nanogels
Cancer cell surface receptors (HER2, PSMA, EGFR, TfR, CD44)	Specific high affinity binding of an antibody, peptide, or chemical ligand to the specific receptor	Immobilization of the NP on the cell surface followed by rapid internalization
Cancer associated stromal cells (CAF, TAM)	(i) High affinity ligands for FAPα expressed on CAF	Internalization of the NP by the stromal cells
	(ii) Phagocytosis of NP by CAF Ernsting et al. (2015)	
	(iii) Mannosylated NP to target TAM	
Cancer associated neovasculature ($\alpha_V\beta_3$, VEGFR, PSMA)	High affinity ligands such as mAb or peptide and proteins	Attachment of circulating NP to endothelial cells with subsequent internalization and/or translocation across the tumor endothelium

result systemically injected NPs do not have to traverse any barriers to reach and bind the target epitope. On the other hand, vascular targets are localized to a two-dimensional surface made up by blood vessel walls within a tumor volume. For a typical blood volume of 5 %, the concentration of vascular targets is at least an order of magnitude lower compared to cancer cell targets with a similar number of molecules expressed per cell. Direct applications of this approach can either be diagnostic imaging of cancer associated vascular markers (Zhu et al. 2010; Pan et al. 2011; Carbary-Ganz et al. 2014; Anderson et al. 2010), or delivery of a therapeutic cargo to tumor capillaries in the context of antiangiogenic therapy (Blankenberg et al. 2011; Rusckowski et al. 2016).

Another important aspect of cell surface epitope targeting is possible formation of a "binding site barrier" in the vasculature or in the tumor periphery and perivascular space that can further retard penetration of targeted NPs into the interstitial space (Saga et al. 1995; Juweid et al. 1992). This effect can explain somewhat paradoxical results obtained when comparing the delivery of vascular-receptor VEGFR2 targeted liposomes to untargeted ones (Kato et al. 2015). In these studies a reduced accumulation of magnetically labeled liposomes in the tumors was detected when the surface was modified with single-chain VEGF constructs to provide high-affinity binding to VEGFR2 expressed in the tumor neovasculature (Imoukhuede and Popel 2014).

12.7 Molecular Imaging in Optimizing NP Delivery and in Theranostics

The spatial and temporal heterogeneities that exist in the TME call for the use of molecular and functional imaging to visualize NP targeting and determine the functional effects of NPs. Clinically available imaging technologies such as MRI, PET, and ultrasound that transition easily from benchtop to bedside are making significant inroads in such applications (Glunde and Bhujwalla 2011; Glunde et al. 2007; Stasinopoulos et al. 2011). Nuclear imaging technologies such as PET and SPECT provide quantitative measurements of tracers *in vivo* in the pM range but with relatively low spatial resolutions in the millimeter range. MRI provides a higher resolution of about 50 μm in preclinical models using high-field scanners, however sensitivity is an issue. Typically MRI molecular imaging agents have to be used at concentrations in the 1–100 μM range that far exceed tracer concentration limits and result in high target occupancy by the probe (Penet et al. 2013). Similarly, *in vivo* fluorescence and bioluminescence imaging is an important tool for high sensitivity preclinical imaging, however true 3D localization and quantitation, as well as depth of imaging limitations present significant problems for these modalities.

Targeted contrast-enhanced ultrasound (US) is gaining significant ground as an imaging modality that is finding increasing applications in molecular imaging of tumor targets. Traditional US uses unique properties of micro- or nanobubbles to generate an acoustic signature. Nanobubbles can be prepared with the average diameter of 0.5–1 μm (Cai et al. 2015) and can be functionalized with high affinity probes for specific binding to molecular targets (Yeh et al. 2015). A relatively large diameter practically restricts applications of microbubbles contrast to US imaging of vascular targets (Yan et al. 2015; Payen et al. 2015). More recently, hybrid technologies such as photoacoustic imaging are attracting attention for biological and translational applications (Liu et al. 2016). The major advantages of photoacoustic imaging are the ability to use light-absorbing probes with optimized molecular size and structure to generate an acoustic signature that can be detected with high spatial and temporal resolution (Chen et al. 2015), a relatively large penetration depth for whole-body preclinical imaging in small animals (Kimbrough et al. 2015), and the potential for multiplexing to detect more than one NP within the region of interest. This technology was recently extended to clinical applications, where the combination of high specificity and sensitivity of detection can potentially provide a noninvasive imaging technology for cancer detection (Chen et al. 2015).

Macroscopic imaging provides important information to develop and optimize NP delivery. Important parameters such as kinetics of accumulation at the target site, background specific and nonspecific uptake, and clearance from the systemic circulation can be readily measured longitudinally in the same animal, which can also be used as its own control. On the other hand, information regarding the finer microstructural details of the interaction of functionalized NPs with their targets at the microscopic level is often required and is usually beyond the capability of the whole-body imaging technologies.

Intravital microscopy provides the ability to directly visualize fluorescent NPs *in vivo* with a minimal effect on the biological microenvironment. Traditional methods of *in vivo* microscopy include the use of optically transparent windows, such as dorsal skin-fold chambers (Koehl et al. 2009; Dunphy et al. 2009) or intracranial glioma models with a cranial window (Masamoto et al. 2013; Sckell and Klenke 2009). While these models provide unique information about the fundamental properties of the TME, they cannot fully recapitulate the TME without direct contact with a coverslip window (Koehl et al. 2009). Intravital multiphoton laser scanning microscopy (MPLSM) can be performed using minimally invasive procedures, such as opening a skin flap or inserting needle-shaped microprobe objectives to generate images at the depth of up to a millimeter below the surface. Examples of intravital MPLSM shown for liver accumulation of positively and negatively charged mesoporous silica nanoparticles (MSNs) (Fig. 12.8) and VEGFR2 targeted peptides in MDA-MB-231 human triple negative breast tumor models (Fig. 12.9) demonstrate the use of this technology in tracking NPs in the vasculature and the TME. In combination with SHG microscopy detection of collagen 1 fibers, the technology can be used to understand the role of fibers in the ECM to facilitate or pose a barrier to NP delivery and distribution.

Another exciting development in the combination of imaging and therapy with NP delivery is theranostic imaging. This rapidly expanding field has resulted in the development of several innovative classes of NPs for theranostics of cancer cells

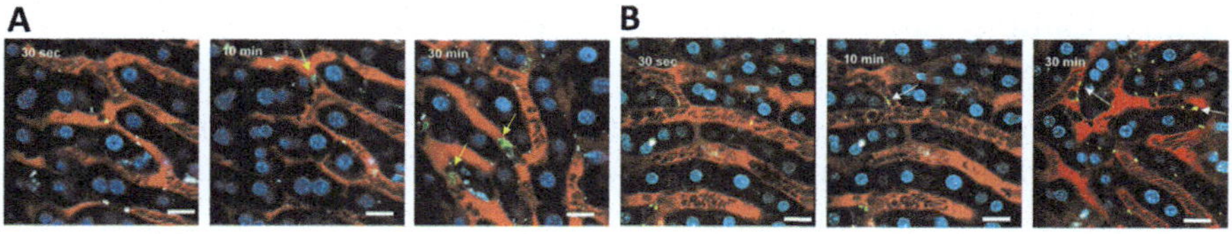

Fig. 12.8 Real-time hepatic visualization of negatively and positively charged FITC-labeled MSNs. Time-lapsed multiphoton images acquired approximately 30 μm below the capsule. Time (in minutes) indicates period postinjection of (**a**) FITC-MSN (*yellow arrows*) and (**b**) FITC-MSN-3X (*white arrows*) (scale bars: 50 μm). Images were merged from three channels. *Red*: rhodamine/dextran R6G(MW70 000) staining of sinusoids. *Green*: fluorescence of FITC-MSNs. *Blue*: hepatocyte nuclei labeled with Hoechst 33342 (From Chen et al. 2012)

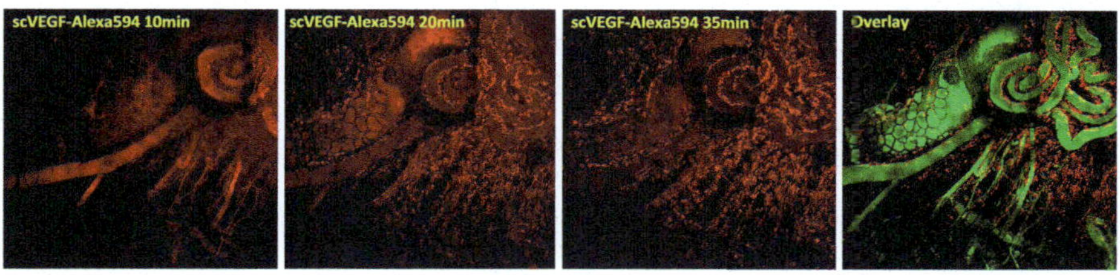

Fig. 12.9 Examples of labeling of blood capillaries (visualized by the blood pool dextran marker) with a VEGFR2-specific fluorescent probe. Complete clearance of the probe from circulation was detected after 30 min of perfusion

and the TME (Stasinopoulos et al. 2011). There are several possible modes of theranostic applications of NPs. Physical phenomena (often similar to those used for imaging) can be used for activating therapeutic properties of nanomaterials. Photothermal therapy can be used in combination with photoacoustic imaging (Liu et al. 2016; Lyu et al. 2016). Ultrasound cavitation of microbubbles can induce cell membrane or vascular (Nejad et al. 2016; Kiessling et al. 2014) or US-triggered drug delivery (Min et al. 2015; Paefgen et al. 2015). Similarly, fluorescent imaging can be combined with photoimmunotherapy (Mitsunaga et al. 2011; Maawy et al. 2015) or NIR hyperthermia and NIR-driven drug release (Jabeen et al. 2014; Zhao et al. 2015). Iron oxide NPs are important platforms for the development of high-sensitivity MRI-detectable agents (Schleich et al. 2015) that can also be used to generate AMF-induced hyperthermia using magnetic NPs that specifically accumulate in the tumor (Ivkov 2013; Attaluri et al. 2015).

Specific delivery of biologically active nanostructures under image guidance for prodrug activation therapy (Li et al. 2008, 2010; Chen et al. 2012), or multistep formation of cross-linked nanocarriers that have improved cell targeting and internalization properties via *in situ* chemical ligation (Hapuarachchige et al. 2014; Zhu et al. 2007), are other examples of theranostic application of NPs in cancer therapy.

12.8 Conclusion

Nanomedicine using NPs is the new frontier in medicine, providing a virtually limitless scope of highly specific targeting for precision medicine and early detection. Future developments in nanomedicine will also include preventive nanomedicine where reporter systems can alert a person on changes in the tissue even prior to the development of malignant cells. At the core of successful nanomedince is effective delivery. Tumor microenvironments with their continually changing landscapes and heterogeneities continue to present a challenge in NP delivery. The incorporation of molecular and functional imaging with NP delivery will advance the effective applications of nanomedicine in cancer through detection of NPs, in the evaluation of strategies to modify the TME to improve NP delivery, and in the rapidly expanding field of theranostic imaging.

Acknowledgements Support from P50 CA103175, P30 CA006973, R21 CA198243, R01 CA193365, R01 CA136576, R01 CA138515, R01 CA82337, R01 CA73850, and RO1 CA154738 is gratefully acknowledged.

References

Anderson CR, Rychak JJ, Backer M et al (2010) scVEGF microbubble ultrasound contrast agents: a novel probe for ultrasound molecular imaging of tumor angiogenesis. Invest Radiol 45(10):579–585

Attaluri A, Kandala SK, Wabler M et al (2015) Magnetic nanoparticle hyperthermia enhances radiation therapy: a study in mouse models of human prostate cancer. Int J Hyperther 31(4):359–374

Bagley AF, Scherz-Shouval R, Galie PA et al (2015) Endothelial thermotolerance impairs nanoparticle transport in tumors. Cancer Res 75(16):3255–3267

Bai J, Wang JT, Rubio N et al (2016) Triple-modal imaging of magnetically-targeted nanocapsules in solid tumours in vivo. Theranostics 6(3):342–356

Bakhtiary Z, Saei AA, Hajipour MJ et al (2015) Targeted superparamagnetic iron oxide nanoparticles for early detection of cancer: possibilities and challenges. Nanomedicine 12(2):287–307

Bedford EE, Spadavecchia J, Pradier CM et al (2012) Surface plasmon resonance biosensors incorporating gold nanoparticles. Macromol Biosci 12(6):724–739

Blankenberg FG, Levashova Z, Goris MG et al (2011) Targeted systemic radiotherapy with scVEGF/177Lu leads to sustained disruption of the tumor vasculature and intratumoral apoptosis. J Nucl Med 52(10):1630–1637

Burri PH, Hlushchuk R, Djonov V (2004) Intussusceptive angiogenesis: its emergence, its characteristics, and its significance. Dev Dyn 231(3):474–488

Cai WB, Yang HL, Zhang J et al (2015) The optimized fabrication of nanobubbles as ultrasound contrast agents for tumor imaging. Sci Rep 5:13725

Carbary-Ganz JL, Barton JK, Utzinger U (2014) Quantum dots targeted to vascular endothelial growth factor receptor 2 as a contrast agent for the detection of colorectal cancer. J Biomed Opt 19(8):086003

Chang E, Chalikonda S, Friedl J et al (2005) Targeting vaccinia to solid tumors with local hyperthermia. Hum Gene Ther 16(4):435–444

Chen Z, Penet MF, Nimmagadda S et al (2012) PSMA-targeted theranostic nanoplex for prostate cancer therapy. ACS Nano 6(9):7752–7762
Chen P, Zhang X, Jia L et al (2014) Optimal structural design of mannosylated nanocarriers for macrophage targeting. J Control Release 194:341–349
Chen H, Yuan Z, Wu C (2015) Nanoparticle probes for structural and functional photoacoustic molecular tomography. BioMed Res Int 2015:757101
Chen Z, Penet MF, Krishnamachary B et al (2016) PSMA-specific theranostic nanoplex for combination of TRAIL gene and 5-FC prodrug therapy of prostate cancer. Biomaterials 80:57–67
Choi HS, Liu W, Misra P et al (2007) Renal clearance of quantum dots. Nat Biotechnol 25(10):1165–1170
Ciobanu M, Eremia IA, Craitoiu S et al (2013) Lymphatic microvessels density, VEGF-C, and VEGFR-3 expression in 25 cases of breast invasive lobular carcinoma. Rom J Morphol Embryol 54(4):925–934
Claesson-Welsh L, Welsh M (2013) VEGFA and tumour angiogenesis. J Intern Med 273(2):114–127
Clark AW, Robins HI, Vorpahl JW et al (1983) Structural changes in murine cancer associated with hyperthermia and lidocaine. Cancer Res 43(4):1716–1723
Decio A, Taraboletti G, Patton V et al (2014) Vascular endothelial growth factor c promotes ovarian carcinoma progression through paracrine and autocrine mechanisms. Am J Pathol 184(4):1050–1061
Du J, Lane LA, Nie S (2015) Stimuli-responsive nanoparticles for targeting the tumor microenvironment. J Control Release 219:205–214
Dudley AC (2012) Tumor endothelial cells. Cold Spring Harb Perspect Med 2(3):a006536
Dunphy MP, Entenberg D, Toledo-Crow R et al (2009) In vivo microcartography and subcellular imaging of tumor angiogenesis: a novel platform for translational angiogenesis research. Microvasc Res 78(1):51–56
Eikenes L, Tufto I, Schnell EA et al (2010) Effect of collagenase and hyaluronidase on free and anomalous diffusion in multicellular spheroids and xenografts. Anticancer Res 30(2):359–368
Emami B, Song CW (1984) Physiological mechanisms in hyperthermia: a review. Int J Radiat Oncol Biol Phys 10(2):289–295
Ernsting MJ, Hoang B, Lohse I et al (2015) Targeting of metastasis-promoting tumor-associated fibroblasts and modulation of pancreatic tumor-associated stroma with a carboxymethylcellulose-docetaxel nanoparticle. J Control Release 206:122–130
Friedl P, Brocker EB (2000) The biology of cell locomotion within three-dimensional extracellular matrix. Cell Mol Life Sci 57(1):41–64
Gerhardt H, Golding M, Fruttiger M et al (2003) VEGF guides angiogenic sprouting utilizing endothelial tip cell filopodia. J Cell Biol 161(6):1163–1177
Gleich B, Weizenecker J (2005) Tomographic imaging using the nonlinear response of magnetic particles. Nature 435(7046):1214–1217
Glunde K, Bhujwalla ZM (2011) Metabolic tumor imaging using magnetic resonance spectroscopy. Semin Oncol 38(1):26–41
Glunde K, Pathak AP, Bhujwalla ZM (2007) Molecular-functional imaging of cancer: to image and imagine. Trends Mol Med 13(7):287–297
Gogineni A, Caunt M, Crow A et al (2013) Inhibition of VEGF-C modulates distal lymphatic remodeling and secondary metastasis. PLoS One 8(7):e68755
Goodwill PW, Saritas EU, Croft LR et al (2012) X-space MPI: magnetic nanoparticles for safe medical imaging. Adv Mater 24(28):3870–3877
Hama S, Itakura S, Nakai M et al (2015) Overcoming the polyethylene glycol dilemma via pathological environment-sensitive change of the surface property of nanoparticles for cellular entry. J Control Release 206:67–74

Hapuarachchige S, Zhu W, Kato Y et al (2014) Bioorthogonal, two-component delivery systems based on antibody and drug-loaded nanocarriers for enhanced internalization of nanotherapeutics. Biomaterials 35(7):2346–2354
Hatakeyama H, Akita H, Kogure K et al (2007) Development of a novel systemic gene delivery system for cancer therapy with a tumor-specific cleavable PEG-lipid. Gene Ther 14(1):68–77
Hildebrandt B, Wust P, Ahlers O et al (2002) The cellular and molecular basis of hyperthermia. Crit Rev Oncol Hematol 43(1):33–56
Imoukhuede PI, Popel AS (2011) Quantification and cell-to-cell variation of vascular endothelial growth factor receptors. Exp Cell Res 317(7):955–965
Imoukhuede PI, Popel AS (2014) Quantitative fluorescent profiling of VEGFRs reveals tumor cell and endothelial cell heterogeneity in breast cancer xenografts. Cancer Med 3(2):225–244
Ivkov R (2013) Magnetic nanoparticle hyperthermia: a new frontier in biology and medicine? Int J Hyperther 29(8):703–705
Jabeen F, Najam-ul-Haq M, Javeed R et al (2014) Au-nanomaterials as a superior choice for near-infrared photothermal therapy. Molecules 19(12):20580–20593
Jain RK (1997) Delivery of molecular and cellular medicine to solid tumors. Adv Drug Deliv Rev 26(2–3):71–90
Jain RK, Stylianopoulos T (2010) Delivering nanomedicine to solid tumors. Nat Rev Clin Oncol 7(11):653–664
Jin SW, Patterson C (2009) The opening act: vasculogenesis and the origins of circulation. Arterioscler Thromb Vasc Biol 29(5):623–629
Jung YY, Kim HM, Koo JS (2016) The role of cancer-associated fibroblasts in breast cancer pathobiology. Histol Histopathol 31(4):371–378
Juweid M, Neumann R, Paik C et al (1992) Micropharmacology of monoclonal antibodies in solid tumors: direct experimental evidence for a binding site barrier. Cancer Res 52(19):5144–5153
Kakkad SM, Penet MF, Akhbardeh A et al (2013) Hypoxic tumor environments exhibit disrupted collagen I fibers and low macromolecular transport. PLoS One 8(12):e81869
Kalluri R (2003) Basement membranes: structure, assembly and role in tumour angiogenesis. Nat Rev Cancer 3(6):422–433
Kato Y, Zhu W, Backer MV et al (2015) Noninvasive imaging of liposomal delivery of superparamagnetic iron oxide nanoparticles to orthotopic human breast tumor in mice. Pharm Res 32(11):3746–3755
Kiessling F, Fokong S, Bzyl J et al (2014) Recent advances in molecular, multimodal and theranostic ultrasound imaging. Adv Drug Deliv Rev 72:15–27
Kimbrough CW, Khanal A, Zeiderman M et al (2015) Targeting acidity in pancreatic adenocarcinoma: multispectral optoacoustic tomography detects pH-low insertion peptide probes in vivo. Clin Cancer Res 21(20):4576–4585
Koehl GE, Gaumann A, Geissler EK (2009) Intravital microscopy of tumor angiogenesis and regression in the dorsal skin fold chamber: mechanistic insights and preclinical testing of therapeutic strategies. Clin Exp Metastasis 26(4):329–344
Komohara Y, Fujiwara Y, Ohnishi K et al (2016) Tumor-associated macrophages: potential therapeutic targets for anti-cancer therapy. Adv Drug Deliv Rev 99(Pt B):180–185
Kong G, Braun RD, Dewhirst MW (2000) Hyperthermia enables tumor-specific nanoparticle delivery: effect of particle size. Cancer Res 60(16):4440–4445
Kong G, Braun RD, Dewhirst MW (2001) Characterization of the effect of hyperthermia on nanoparticle extravasation from tumor vasculature. Cancer Res 61(7):3027–3032
Lancrin C, Sroczynska P, Stephenson C et al (2009) The haemangioblast generates haematopoietic cells through a haemogenic endothelium stage. Nature 457(7231):892–895
Li C, Penet MF, Winnard P Jr et al (2008) Image-guided enzyme/prodrug cancer therapy. Clin Cancer Res 14(2):515–522
Li C, Penet MF, Wildes F et al (2010) Nanoplex delivery of siRNA and prodrug enzyme for multimodality image-guided molecular pathway targeted cancer therapy. ACS Nano 4(11):6707–6716
Liu Y, Nie L, Chen X (2016) Photoacoustic molecular imaging: from multiscale biomedical applications towards early-stage theranostics. Trends Biotechnol 34(5):420–433

Longmire M, Choyke PL, Kobayashi H (2008) Clearance properties of nano-sized particles and molecules as imaging agents: considerations and caveats. Nanomedicine 3(5):703–717
Lu L, Hu Q, Lin Y et al (2015) Transformable liquid-metal nanomedicine Nat Commun 6:10066
Lyu Y, Fang Y, Miao Q et al (2016) Intraparticle molecular orbital engineering of semiconducting polymer nanoparticles as amplified theranostics for in vivo photoacoustic imaging and photothermal therapy. ACS Nano 10(4):4472–4481
Maawy AA, Hiroshima Y, Zhang Y et al (2015) Near infra-red photoimmunotherapy with anti-CEA-IR700 results in extensive tumor lysis and a significant decrease in tumor burden in orthotopic mouse models of pancreatic cancer. PLoS One 10(3):e0121989
Maeda H (2001) The enhanced permeability and retention (EPR) effect in tumor vasculature: the key role of tumor-selective macromolecular drug targeting. Adv Enzym Regul 41:189–207
Masamoto K, Takuwa H, Tomita Y et al (2013) Hypoxia-induced cerebral angiogenesis in mouse cortex with two-photon microscopy. Adv Exp Med Biol 789:15–20
Min HS, You DG, Son S et al (2015) Echogenic glycol chitosan nanoparticles for ultrasound-triggered cancer theranostics. Theranostics 5(12):1402–1418
Mishra S, Webster P, Davis ME (2004) PEGylation significantly affects cellular uptake and intracellular trafficking of non-viral gene delivery particles. Eur J Cell Biol 83(3):97–111
Mitsunaga M, Ogawa M, Kosaka N et al (2011) Cancer cell-selective in vivo near infrared photoimmunotherapy targeting specific membrane molecules. Nat Med 17(12):1685–1691
Multhaupt HA, Leitinger B, Gullberg D et al (2016) Extracellular matrix component signaling in cancer. Adv Drug Deliv Rev 97:28–40
Nakahara T, Norberg SM, Shalinsky DR et al (2006) Effect of inhibition of vascular endothelial growth factor signaling on distribution of extravasated antibodies in tumors. Cancer Res 66(3):1434–1445
Narunsky L, Oren R, Bochner F et al (2014) Imaging aspects of the tumor stroma with therapeutic implications. Pharmacol Ther 141(2):192–208
Neeves KB, Sawyer AJ, Foley CP et al (2007) Dilation and degradation of the brain extracellular matrix enhances penetration of infused polymer nanoparticles. Brain Res 1180:121–132
Nehoff H, Parayath NN, Domanovitch L et al (2014) Nanomedicine for drug targeting: strategies beyond the enhanced permeability and retention effect. Int J Nanomedicine 9:2539–2555
Nejad SM, Hosseini H, Akiyama H et al (2016) Reparable cell sonoporation in suspension: theranostic potential of microbubble. Theranostics 6(4):446–455
Netti PA, Berk DA, Swartz MA et al (2000) Role of extracellular matrix assembly in interstitial transport in solid tumors. Cancer Res 60(9):2497–2503
Oh P, Testa JE, Borgstrom P et al (2014) In vivo proteomic imaging analysis of caveolae reveals pumping system to penetrate solid tumors. Nat Med 20(9):1062–1068
Paefgen V, Doleschel D, Kiessling F (2015) Evolution of contrast agents for ultrasound imaging and ultrasound-mediated drug delivery. Front Pharmacol 6:197
Pan D, Pramanik M, Senpan A et al (2011) Molecular photoacoustic imaging of angiogenesis with integrin-targeted gold nanobeacons. FASEB J 25(3):875–882
Park JH, von Maltzahn G, Zhang L et al (2009) Systematic surface engineering of magnetic nanoworms for in vivo tumor targeting. Small 5(6):694–700
Park JH, von Maltzahn G, Xu MJ et al (2010) Cooperative nanomaterial system to sensitize, target, and treat tumors. Proc Natl Acad Sci U S A 107(3):981–986
Payen T, Dizeux A, Baldini C et al (2015) VEGFR2-targeted contrast-enhanced ultrasound to distinguish between two anti-angiogenic treatments. Ultrasound Med Biol 41(8):2202–2211
Penet MF, Artemov D, Farahani K et al (2013) MR – eyes for cancer: looking within an impenetrable disease. NMR Biomed 26(7):745–755
Raman V, Artemov D, Pathak AP et al (2006) Characterizing vascular parameters in hypoxic regions: a combined magnetic resonance and optical imaging study of a human prostate cancer model. Cancer Res 66(20):9929–9936
Raphael AP, Sisney JP, Liu DC et al (2015) Enhanced delivery of nano- and submicron particles using elongated microparticles. Curr Drug Deliv 12(1):78–85

Rusckowski M, Wang Y, Blankenberg FG et al (2016) Targeted scVEGF/(177)Lu radiopharmaceutical inhibits growth of metastases and can be effectively combined with chemotherapy. EJNMMI Res 6(1):4
Saga T, Neumann RD, Heya T et al (1995) Targeting cancer micrometastases with monoclonal antibodies: a binding-site barrier. Proc Natl Acad Sci U S A 92(19):8999–9003
Schleich N, Danhier F, Preat V (2015) Iron oxide-loaded nanotheranostics: major obstacles to in vivo studies and clinical translation. J Control Release 198:35–54
Sckell A, Klenke FM (2009) The cranial bone window model: studying angiogenesis of primary and secondary bone tumors by intravital microscopy. Methods Mol Biol 467:343–355
Scodeller P, Catalano PN, Salguero N et al (2013) Hyaluronan degrading silica nanoparticles for skin cancer therapy. Nanoscale 5(20):9690–9698
Semenza GL (2001) HIF-1 and mechanisms of hypoxia sensing. Curr Opin Cell Biol 13(2):167–171
Sharma A, Jain N, Sareen R (2013) Nanocarriers for diagnosis and targeting of breast cancer. BioMed Res Int 2013:960821
Shiga K, Hara M, Nagasaki T et al (2015) Cancer-associated fibroblasts: their characteristics and their roles in tumor growth. Cancers (Basel) 7(4):2443–2458
Stasinopoulos I, Penet MF, Chen Z et al (2011) Exploiting the tumor microenvironment for theranostic imaging. NMR Biomed 24(6):636–647
Stylianopoulos T, Martin JD, Snuderl M et al (2013) Coevolution of solid stress and interstitial fluid pressure in tumors during progression: implications for vascular collapse. Cancer Res 73(13):3833–3841
Toffoli S, Michiels C (2008) Intermittent hypoxia is a key regulator of cancer cell and endothelial cell interplay in tumours. FEBS J 275(12):2991–3002
Vempati P, Popel AS, Mac Gabhann F (2014) Extracellular regulation of VEGF: isoforms, proteolysis, and vascular patterning. Cytokine Growth Factor Rev 25(1):1–19
Vinogradov S, Warren G, Wei X (2014) Macrophages associated with tumors as potential targets and therapeutic intermediates. Nanomedicine 9(5):695–707
Weissleder R, Reimer P, Lee AS et al (1990) MR receptor imaging: ultrasmall iron oxide particles targeted to asialoglycoprotein receptors. AJR Am J Roentgenol 155(6):1161–1167
Weissleder R, Moore A, Mahmood U et al (2000) In vivo magnetic resonance imaging of transgene expression. Nat Med 6(3):351–355
Yan F, Xu X, Chen Y et al (2015) A lipopeptide-based alphavbeta(3) integrin-targeted ultrasound contrast agent for molecular imaging of tumor angiogenesis. Ultrasound Med Biol 41(10):2765–2773
Yeh JS, Sennoga CA, McConnell E et al (2015) A targeting microbubble for ultrasound molecular imaging. PLoS One 10(7):e0129681
Zhao J, Wallace M, Melancon MP (2014) Cancer theranostics with gold nanoshells. Nanomedicine 9(13):2041–2057
Zhao P, Zheng M, Luo Z et al (2015) NIR-driven smart theranostic nanomedicine for on-demand drug release and synergistic antitumour therapy. Sci Rep 5:14258
Zhou H, Fan Z, Deng J et al (2016) Hyaluronidase embedded in nanocarrier PEG shell for enhanced tumor penetration and highly efficient antitumor efficacy. Nano Lett 16(5):3268–3277
Zhu W, Okollie B, Artemov D (2007) Controlled internalization of Her-2/neu receptors by cross-linking for targeted delivery. Cancer Biol Ther 6(12):1960–1966
Zhu L, Niu G, Fang X et al (2010) Preclinical molecular imaging of tumor angiogenesis. Q J Nucl Med Mol Imaging 54(3):291–308

Chapter 13
Microscopic Mass Spectrometry for the Precise Design of Drug Delivery Systems

Yasuhiro Matsumura and Masahiro Yasunaga

Abstract To bring drugs formulated in drug delivery systems (DDS) successfully into the clinic, preclinical studies have to be conducted which are aimed at obtaining pharmacological data relevant to the clinical application of such drugs. For such preclinical studies high-performance liquid chromatography (HPLC) or liquid chromatography mass spectrometry (LC-MS) is generally used. However, these methods do not generate data about the drug distribution in a specific target area, although obtained data allow optimizing the drug design in order to achieve a more efficient targeted delivery.

Microscopic mass spectrometry (MMS), in which a microscope is coupled to an atmospheric pressure matrix-assisted laser desorption/ionization (MALDI) and quadruple ion trap time-of-flight (TOF) analyzer has been developed for the investigation of the distribution of molecules such as small peptide metabolites and low-molecular weight drugs. The matrix-coated drug sample is ionized and then separated based on its mass-to-charge ratio (m/z). Images are acquired from imaging mass spectrometry or tandem mass spectrometry (MS/MS) data, respectively.

Here we introduce a drug imaging system with enhanced resolution and sensitivity which is based on using MMS. In our analysis, MS and MS/MS were used for quantification and validation, respectively. Our short review describes how the use of MMS allows the analysis of the precise distribution of a DDS drug complex including both, active and passive targeting systems. Notably, we successfully visualized and quantified the distribution of a non-radiolabeled and non-chemically modified drug in various frozen tissue slices microscopically.

In conclusion, MMS may provide a new strategy for facilitating the design of DDS with incorporated low-molecular weight drugs.

Keywords Microscopic mass spectrometry • DDS • Active targeting • Passive targeting • EPR effect

---

Y. Matsumura (✉) • M. Yasunaga
Division of Developmental Therapeutics, Exploratory Oncology Research & Clinical Trial Center, National Cancer Center, 6-5-1 Kashiwanoha, 277-8577 Kashiwa, Japan
e-mail: yhmatsum@east.ncc.go.jp

A. Prokop, V. Weissig (eds.), *Intracellular Delivery III*, Fundamental Biomedical Technologies 8, DOI 10.1007/978-3-319-43525-1_13

Abbreviations

DDS	Drug delivery system
MALDI	Matrix-assisted laser desorption/ionization
MMS	Microscopic mass spectrometry
EPR	Enhanced permeability and retention
RES	Reticuloendothelial system
ADC	Antibody drug conjugate
ACA	Anticancer agent
MMAE	Monomethyl auristatin E
PTX	Paclitaxel
CHCA	α-cyano-4-hydroxycinnamic acid
DHB	2,5-Dihydroxybenzoic acid

13.1 Introduction

Tumor-targeted delivery of therapeutic agents has been a longstanding pharmacological goal to improve the treatment selectivity and the therapeutic index. The 'passive' targeting afforded by the "Enhanced Permeability and Retention (EPR) effect" provides a versatile and non-saturable approach for tumor-selective delivery (Matsumura and Maeda 1986; Maeda et al. 2000). Small molecules easily leak from normal blood vessels in the body, which gives small molecules a short plasma half-life. On the other hand, macromolecules have a long plasma half-life because they are too large to pass through the normal vessel walls, unless they are trapped by the reticuloendothelial system (RES) in various organs. In solid tumor tissues, it was found that solid tumors generally possess several pathophysiological characteristics: hypervasculature, secretion of vascular permeability factors stimulating extravasation of macromolecules within the cancer, and absence of effective lymphatic drainage from tumors that impedes the efficient clearance of macromolecules accumulated in solid tumor tissues. These characteristics of solid tumors are the basis of the EPR effect (Fig. 13.1). Several techniques have been developed to maximally utilize the EPR effect such as modification of drug structures and development of drug carriers. Polymeric micelle-based anticancer drugs were originally developed by Kataoka et al. in the late 1980s or early 1990s (Yokoyama et al. 1990, 1991; Kataoka et al. 1993). Polymeric micelles were expected to increase the accumulation of drugs in tumor tissues by utilizing the EPR effect as well as to incorporate various kinds of drugs into their inner core with relatively high stability by chemical conjugation or physical entrapment. Also, the size of micelles can be controlled within the diameter range of 20–100 nm to ensure that they do not permeate normal vessel walls. With this development, it is expected that the incidence of drug-induced side effects may be decreased due to the reduced drug distribution within normal tissues. Polymeric micelles are ideally suited to exploit the EPR effect, and are now under clinical evaluations (Fig. 13.2) (Matsumura et al. 2014).

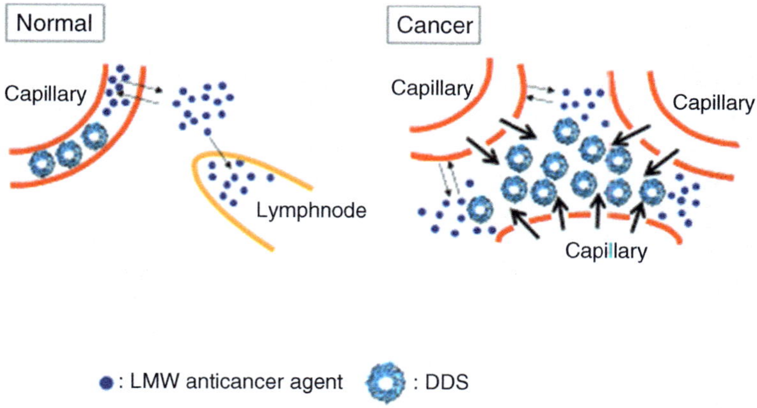

Fig. 13.1 Diagram of the EPR effect. Solid tumors generally possess several pathophysiological characteristics, namely, hypervasculature, secretion of vascular permeability factors stimulating extravasation of macromolecules within the cancer, and absence of effective lymphatic drainage from tumors that impedes the efficient clearance of macromolecules accumulated in solid tumor tissues. These characteristics of solid tumors are the basis of the EPR effect

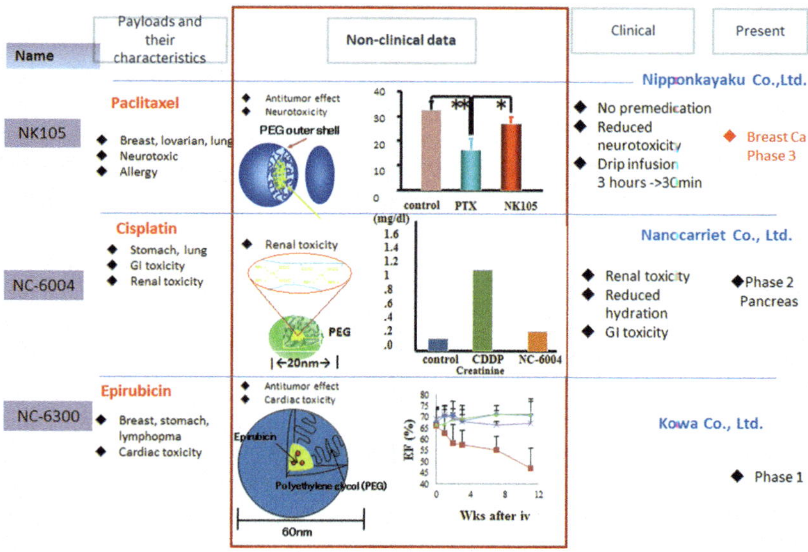

Fig. 13.2 Preclinical and clinical study of ACA incorporating micelles. NK105 is a PTX incorporating micelle. The phase 3 trial of NK105 is now underway in patients with metastatic breast cancer. NC-6004 is a CDDP incorporating micelles. The phase 3 trial of NC-6004 is now underway in patients with metastatic pancreatic cancer. NC-6300 is an epirubicin incorporating micelles. The phase 1 trial is now underway

Recent developments of antibody drug conjugate (ADC) for active tumor targeting are promising. ADCs as high molecular weight proteins (Matsumura and Maeda 1986) passively accumulate in solid tumor tissue due to the EPR effect and in addition actively accumulate in targeted malignant cells due to specific antibody-antigen binding. This ADC strategy should be confined, however only to highly toxic anticancer agents (ACA) such as monomethyl auristatin E (MMAE) but not to ordinary ACA such as paclitaxel (PTX) or Adriamycin due to the limit loading capacity of antibodies. Generally less than three ACA molecules should be conjugated to a single mAb in order to preserve the antibody's affinity. Too many ACA molecules conjugated to a single mAb may severely diminish its biological activity, i.e. affinity to its corresponding antigen (Junutula et al. 2008). Subsequently, an unrealistically high dose of ADC would have to be administered if ordinary ACA are conjugated to mAbs. As alternative, ordinary ACA should be formulated into nanoparticle or nanovesicle-based DDS, both of which allow for a much higher payload in comparison to using mAb as DDS. In both passive and active targeting, it is pivotal to know whether DDS formulations can deliver their payloads to cancer cells in the cancer tissues.

Pharmacokinetics (PK) and pharmacodynamics (PD) studies are very important not only for evaluation of the efficacy and toxicity of such drugs but also for optimal drug design. For these purposes, tissue homogenate samples are generally used and analyzed by high-performance liquid chromatography (HPLC) or liquid chromatography mass spectrometry (LC-MS) (Garrett and Workman 1999).

Although many HPLC or LC-MS-based studies have indicated that drugs formulated into DDS accumulate selectively in tumor tissue in comparison to the free drug, the question whether DDS could deliver the drug to cancer-cell clusters within the tumor tissue remained open. Solid tumor tissue is heterogeneous and consists not only of cancer cells but also of abundant tumor stroma, which in turn can severely impede the diffusion of macromolecules, including micelles and ADCs, within the tumor tissue (Matsumura 2012).

Another name for MMS is Matrix-assisted laser desorption ionization imaging mass spectrometry (MALDI-IMS) which has been developed for analyzing the biodistribution of molecules such as small peptides, drugs, and their metabolites (Cornett et al. 2007; Schwamborn and Caprioli 2010; Castellino et al. 2011; Saito et al. 2011; Lorenz et al. 2013; Römpp and Spengler 2013). Moreover, this method can be used to assess the distribution of numerous molecules in a single measurement without having to employ any specialized probes. Therefore, this method allows the direct observation of a drug as well as its metabolites within tissue samples. We have developed a mass microscopy method in which a microscope is coupled to a high-resolution atmospheric pressure-laser desorption/ionization and quadruple ion trap time-of-flight (TOF) analyzer, which results in an enhanced tissue resolution of MALDI-IMS. We investigated the ability of our mass microscopy technique to visualize the tissue distribution of ACA released from micellar formulation (Yasunaga et al. 2013) or ADCs (Fujiwara et al. 2016) and we obtained precise regional information about the drug distribution in a specific anatomical area.

13.2 Principle of Microscopic Mass Spectrometry (MMS)

Figure 13.3 shows a schematic representation of our MMS-based drug imaging system. All imaging data were acquired using MMS. In the analysis, mass spectrometry (MS) and tandem mass spectrometry (MS/MS) were used for quantification and validation, respectively (Fig. 13.3).

Although the delivered drug content per weight of tissue can be determined by conventional HPLC or LC-MS, the detailed drug distribution within the tumor and normal tissue cannot. Therefore, we used our drug imaging system to assess the distribution of the drug within the target tissue.

Recent progress in MMS analysis, including the new features present in our instrument, has achieved a MMS resolution of 10 μm or less, which is advantageous for analyzing the drug distribution in specific cells or areas of interest within tissues (Castellino et al. 2011; Saito et al. 2011; Lorenz et al. 2013; Römpp and Spengler 2013). The improved resolution also allows an MMS image to be overlaid on an optical image of the same sample. In fact, we were able to distinguish the nerve component from the surrounding tissue and evaluate the specific distribution of free drug in the region (Yasunaga et al. 2013). Tissue samples should be frozen in the absence of any liquids to avoid diffusion or loss of the drug from the tissue. For efficient ionization of the drug molecules to be analyzed, the sample was spray-

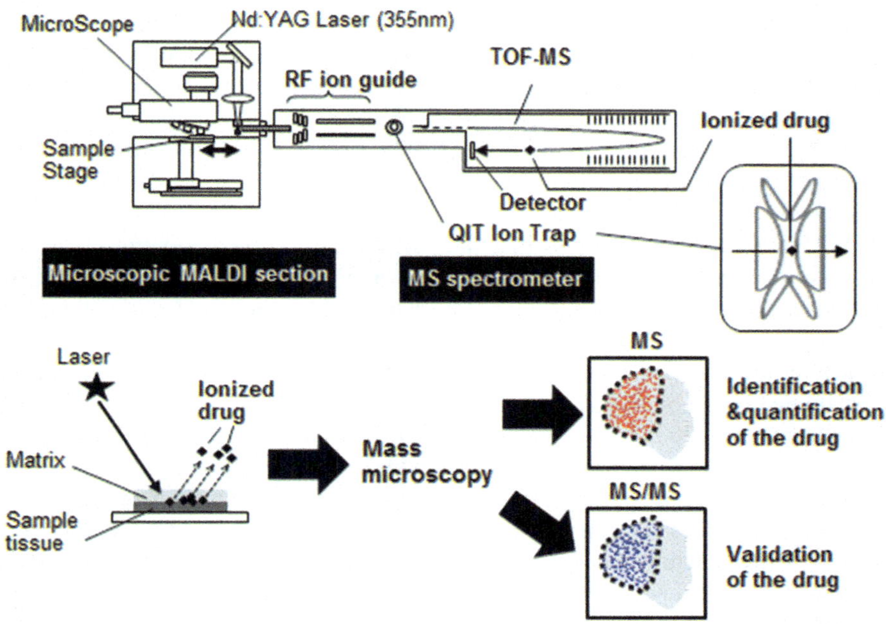

Fig. 13.3 Diagram of the microscopic mass spectrometry. A schematic illustration of the drug imaging system. The matrix-coated drug sample is ionised and then separated on the basis of its m/z. Images from MS or MS/MS analysis are recorded

coated with a sufficient quantity of matrix. An optimal matrix should be selected to facilitate the efficient ionization of the drug. We are currently attempting the use of several matrix materials to enhance the sensitivity of our MMS technique. Moreover, we also used a combination of MS and MS/MS for the imaging analysis. In the MS analysis, accurate quantification of the drug under investigation was demonstrated in vivo. In the MS/MS analysis, the presence of that drug was validated by a fragment-specific signal, which does not overlap with any other signals. The combination of MS and MS/MS thus facilitates the accurate identification of the drug-originated signal by distinguishing the drug signal from endogenous metabolites with a similar m/z (Yasunaga et al. 2013).

13.3 MMS for Analysis of Drugs Delivered Via Active Targeting

Antibody-drug conjugates (ADCs) are a class of immunotherapeutic agents that enable the delivery of cytotoxic drugs specifically to malignant cells. ADCs passively accumulate in solid tumor tissue due to the EPR effect (Matsumura and Maeda 1986) and subsequently target malignant cells based on the specific antibody-antigen binding (Hellstrom et al. 2013). The proper selections of a suitable monoclonal antibody (mAb), an anticancer agent (ACA) as well as a linker have been proven to be crucial for the clinical success of ADCs (Sievers and Senter 2013). The ADC strategy should be confined to highly toxic ACAs and not applied to ordinary ACAs, such as taxane, adriamycin, and others, because fewer than four ACA molecules should be conjugated to the mAb to prevent a decrease in the affinity of the mAb which would be the case if too many ACA molecules were attached to a single antibody macromolecule (Junutula et al. 2008). MMAE is one of the most useful and potent ACAs for the clinical development of novel ADCs (Zolot et al. 2013). MMAE inhibits cell division by blocking the polymerization of tubulin.

In our laboratory, we have developed an anti-human tissue factor (TF) mAb which is attached to valine-citrulline (Val-Cit)-MMAE (human TF ADC) and we have reported its antitumor effect against xenografts of a human pancreatic cancer cell line, BxPC-3 (Koga et al. 2015). The Val-Cit-MMAE has been designed for maximum serum stability and efficient release into the tumor environment (Sanderson et al. 2005). Once human TF ADC binds to the target malignant cells, it is internalized via receptor-mediated endocytosis, and subsequently MMAE is supposed to be released into the tumor environment via hydrolysis of the linker by. lysosomal enzymes. TF is a transmembrane glycoprotein involved in the initiation of the extrinsic pathway of blood coagulation (Mackman 2009), it is expressed in various types of cancer, and it plays a role in cancer progression, angiogenesis, tumor growth, and metastasis (van den Berg et al. 2012). Because pancreatic cancer tissue expresses high levels of TF, it is a useful target antigen for this condition (Flossel et al. 1994; Nitori et al. 2005; Khorana et al. 2007).

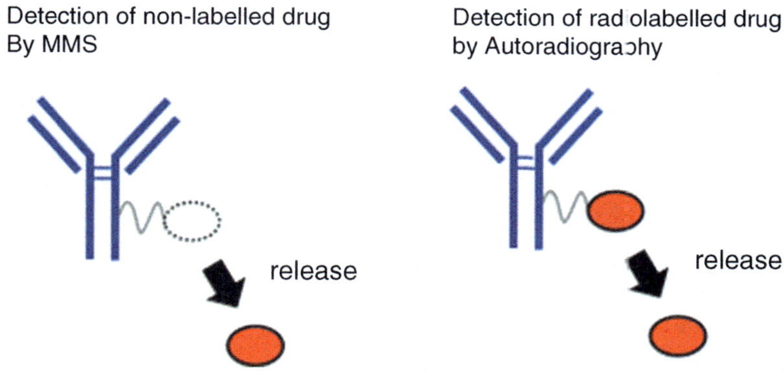

Fig. 13.4 Difference between MMS and Autoradiography. The MMS can distinguish released drug from drug conjugated to an ADC. On the other hand, Autoradiography cannot distinguish free radiolabelled drug from the radiolabelled drug conjugated to an ADC

To assess the efficacy of an ADC against TF-positive solid tumors, preclinical pharmacological studies of the ADC have to be performed to determine whether the human TF ADC has been designed optimally. With respect to their antitumor effects, ACAs must efficiently penetrate the tumor tissue and be retained there at a high and biologically active concentration (Minchinton and Tannock 2006; Kim et al. 2013) For such analyses, high-performance liquid chromatography (HPLC) or liquid chromatography mass spectrometry (LC-MS) are generally used. However, these techniques do not provide information about the drug distribution in a specific target area, although they allow optimization of the drug design to some extent, which might result in a more efficiently targeted delivery. While autoradiography can be used to examine the tissue distribution of radiolabelled small molecules (Solon 2012), this method cannot distinguish between a radiolabelled drug conjugated to an ADC and free radiolabelled drug released from the ADCs (Fig. 13.4) (Xie et al. 2004).

There are two types of MMS: One type can detect larger molecules (Rowinsky et al. 1993), although it is currently difficult to directly ionize high-molecular weight proteins, such as mAbs. The other type, which was used in our study, is specifically designed for low-molecular weight substances, such as ACAs.

Although it is difficult to determine the distribution of ADCs from a technical standpoint, MMS is a useful analytical tool for verifying whether the ADCs release their cytotoxic agents at a specific target area within the tumor tissues as designed. Mass spectrometry (MS) and tandem MS (MS/MS) do not require labelling reagents, and MMS can provide accurate maps of the target molecules in tissue specimens directly (Yasunaga et al. 2013).

For the application of MALDI to MMAE analysis, α-cyano-4-hydroxycinnamic acid (CHCA) was selected as a matrix. The chemical formula of MMAE is C39H67N5O7, and its monoisotopic mass is 717.504. Three positive-ion peaks derived from MMAE were observed by MS analysis: single-charge hydrogen, and the sodium and potassium adducts, denoted as [M+H]+, [M+Na]+, and [M+K]+,

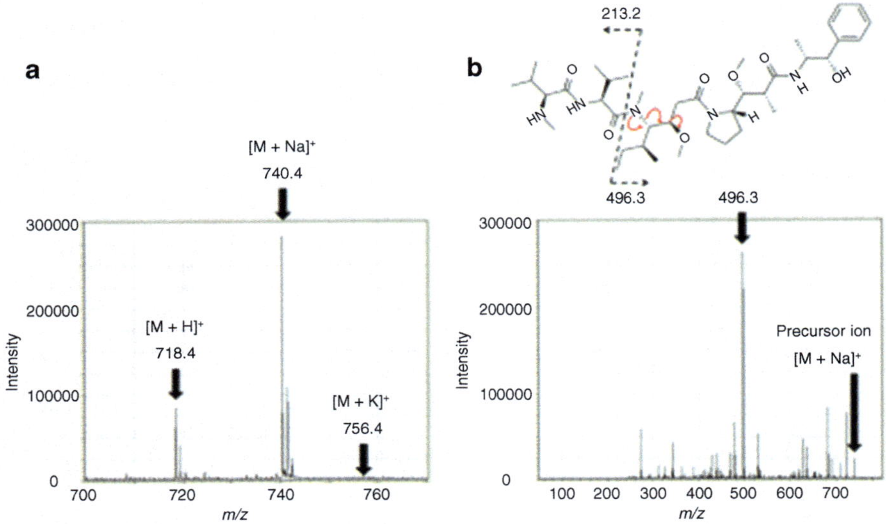

Fig. 13.5 MS and MS/MS of MMAE. (**a**) To define the MS and MS/MS analyses of MMAE, the matrix, CHCA in 75% acetonitrile, 0.02% trifluoroacetic acid, 2.0-mM sodium acetate and a 1/1000 dilution of aniline (10 mg/mL, 1.0 μL) was used for the ionization of the MMAE standard (1.0 μM, 1.0 μL). [M+H]+, [M+Na]+, and [M+K]+were m/z 718.4, 740.4, and 756.4, respectively. (**b**) The MMAE-specific MS/MS fragment, m/z 496.3, was determined when [M+Na]+was used as a precursor ion in the MS/MS analysis

respectively (Fig. 13.5a). The main fragment at m/z 496.3 was observed when m/z 740.4 was used as a precursor ion in the MS/MS analysis (Fig. 13.5b). MMAE was semi-quantitatively detected at different concentrations with MMS (Fig. 13.6a). Then, 0.2-mg/kg MMAE alone was injected intravenously into the tail vein of mice bearing a BxPC-3 xenograft. The 5.0-mg/mL CHCA solution was applied by pin-point spray gun to the tissue sections. The signals originating from the MMAE were detected in the tumor tissues at 3, 24, and 72 h after the administration of MMAE alone. The MMAE signal decreased in a time-dependent manner (Fig. 13.6b). Tissue sections serial to those used for MMS were also quantified using LC-MS/MS (Fig. 13.6c), and the results were consistent with the IMS data from MMAE.

Anti-TF ADC and its control ADC, which did not bind the TF antigen, were prepared with a drug-to-antibody ratio (DAR) of the ADCs of 2–4 (Fig. 13.7a). The ADCs were crystallized with the CHCA solution. MMS was performed on the ADC sample and on MMAE alone in MS mode (Fig. 13.7b–c). The signal intensity of MMAE increased in a concentration-dependent manner, and the signal intensity obtained from 1.0 μL of 1.0-μM MMAE alone was significantly higher than that from 1.0 μL of 1.0-μM human TF ADC and the control ADC (Tukey-Kramer, $P<0.01$).

The MALDI analysis was capable of distinguishing between MMAE alone and ADCs conjugated with MMAE. The signals originating from the MMAE released

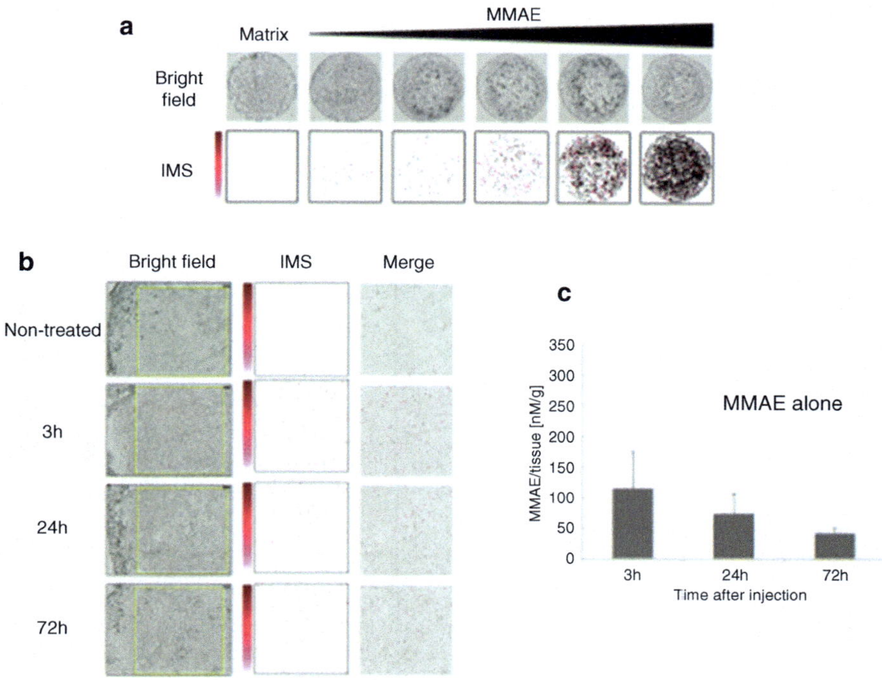

Fig. 13.6 Semi-quantitative analysis of MMAE using MMS. (**a**) Only CHCA was used as a negative control. From *left* to *right*, the concentrations of MMAE were 0.10, 1.0, 10, 100, and 1000 nM. The CHCA solution (10 mg/mL, 1.0 μL) was used for the ionization of each MMAE sample (1.0 μL). MMAE was semi-quantitatively detected in a concentration-dependent manner using MALDI-IMS. (**b**) The *yellow rectangles* on the bright field show the measurement area (2500×2500 μm). Pixel size, 20 μm; original magnification, ×5; scale bar, 200 μm. The images obtained from m/z 496.3 using MALDI-IMS for MMAE detection in tumor tissues are shown as a pseudo-colour image (*Red-Purple*). The 5.0-mg/mL CHCA solution was applied with a pinpoint spray gun. Minimum intensity, 4; maximum intensity, 10. Each merged image superposed the IMS image on the measurement area image obtained 3, 24, or 72 h after the administration of MMAE alone. (**c**) LC-MS/MS analysis quantified the MMAE concentration in the tumors 3, 24, and 72 h after the administration of MMAE alone. The contents of MMAE in the tumor tissues 3, 24, and 72 h after administration were 1155±609, 736±319, and 413±88 fmol/g (mean±SD, N=5 for each group), respectively

from ADCs were detected in the tumor tissues at 3, 24, and 72 h after the administration of the ADCs. The MMAE signal detected following the accumulation of the human TF ADC in the tumor tissue was greatest at 24 h after administration, compared with the control ADC at the same time (Tukey-Kramer, $P<0.01$, Fig. 13.8a–b). Tissue sections serial to those for MMS were also quantified with LC-MS/MS (Fig. 13.8c), and the results were consistent with the IMS data from MMAE.

These data indicate that MMS can be a powerful tool for optimizing the design of ADCs in the preclinical stage.

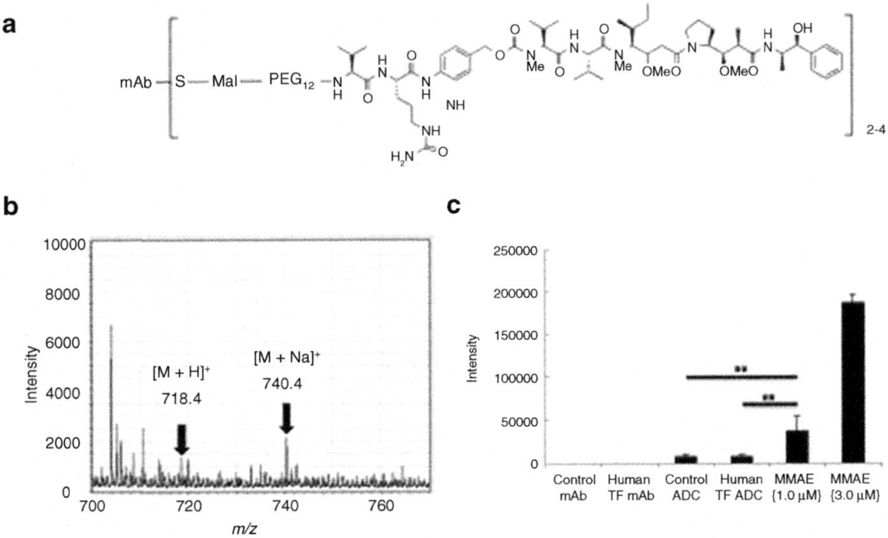

Fig. 13.7 Selective detection of MMAE alone over MMAE conjugated to mAb using MMS. (**a**) The DAR of the ADCs was 2–4. A CHCA solution (10 mg/mL, 1.0 μL) was used for the ionization of mAbs (1.0 μM, 1.0 μL), ADCs (1.0 μM, 1.0 μL) and MMAE (1.0 and 3.0 μM, 1.0 μL). (**b**) The mass spectrum of human TF ADC (1.0 μM, 1.0 μL) is indicated. (**c**) The intensities (m/z 740.4) of the mAbs, ADCs, and each MMAE sample were measured with MS (Tukey-Kramer, ** $P<0.01$, $N=3$ for each group). The intensity was expressed as the mean ± SD. Pixel size, 20 μm; measurement area, 200 × 200 μm

13.4 MMS for a Passive Targeting

PTX is a mitotic inhibitor and an ACA that is used to treat various cancers. However, PTX is associated with peripheral neuropathy, a serious adverse effect (Rowinsky et al. 1993). NK105, a PTX-incorporating micelle, was developed to address this limitation of PTX14–17. Due to the EPR effect (Matsumura and Maeda 1986; Maeda et al. 2000), NK105 preferentially accumulates in a solid tumor, which in turn results in both, an enhanced antitumor effect and a reduction of adverse effects including peripheral neuropathy. The high efficacy and low toxicity of NK105 have been demonstrated in both preclinical and clinical studies (Hamaguchi et al. 2005; Hamaguchi et al. 2007; Kato et al. 2012; Matsumura and Kataoka 2009). Although the administered drug content per tissue weight was determined by conventional HPLC or LC-MS, the detailed drug distribution within the tumor and normal tissue has not yet been examined. Therefore, we used our drug imaging system to evaluate the difference in the distribution of NK105 and free PTX within tumor and normal peripheral neuronal tissue (Yasunaga et al. 2013). To achieve efficient ionization, the sample was spray-coated with a sufficient amount of 2,5-Dihydroxybenzoic acid (DHB) as the chosen matrix. (Yasunaga et al. 2013).

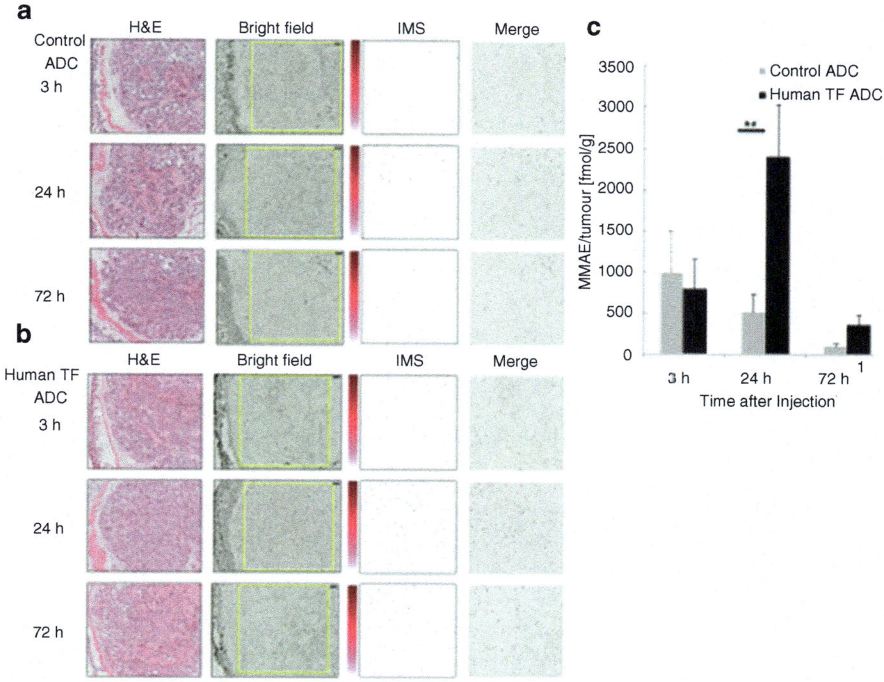

Fig. 13.8 Visualization of MMAE released from ADCs in tumor tissues using MMS. (**a**) H&E staining is shown in the *left column* (original magnification, ×5). The *yellow rectangles* on the bright field show the measurement area (2500×2500 μm). Pixel size, 20 μm; scale bar, 200 μm. (**a–b**) The images obtained from m/z 496.3 using MALDI-IMS for detection of MMAE in tumor tissues are shown as pseudo-colour images (*Red-Purple*). The 5.0-mg/mL CHCA solution was applied with a pinpoint spray gun. Minimum intensity, 4; maximum intensity, 10. Each merged image was superposed with the IMS image on the measurement area image obtained 3, 24, and 72 h after the administration of the ADCs. At 24 h after the administration of the ADCs, the accumulation of MMAE released from human TF ADC was visibly higher than that of the control ADC. (**c**) The contents of MMAE in the tumor tissues 3, 24, and 72 h after the administration of the control ADC were 985±500, 504±222, and 92±31 fmol/g (mean±SD, N=5 for each group), respectively. The contents of MMAE in the tumor tissues 3, 24, and 72 h after the administration of the human TF ADC were 785±354, 2399±637, and 353±125 fmol/g (mean±SD, N=5 for each group), respectively. There was a significant difference between the contents of the control ADC and human TF ADC 24 h after administration (Tukey-Kramer, ** $P<0.01$, N=5 for each group)

NK105 or free PTX was administered at a PTX equivalent dose of 50 mg/kg/day to mice bearing BxPC3 pancreatic cancer xenografts on days 0, 4, and 8. NK105 showed significantly higher antitumor activity than the control (saline) and free PTX (Fig. 13.9a). To confirm the correlation of the distribution with the antitumor effect, the corresponding tumor sections were subjected to MMS. A drug signal originating from PTX was detected in the tumors at 15 min and 1 h after the administration of PTX, but this signal decreased at 6 h and was below the limit of detection

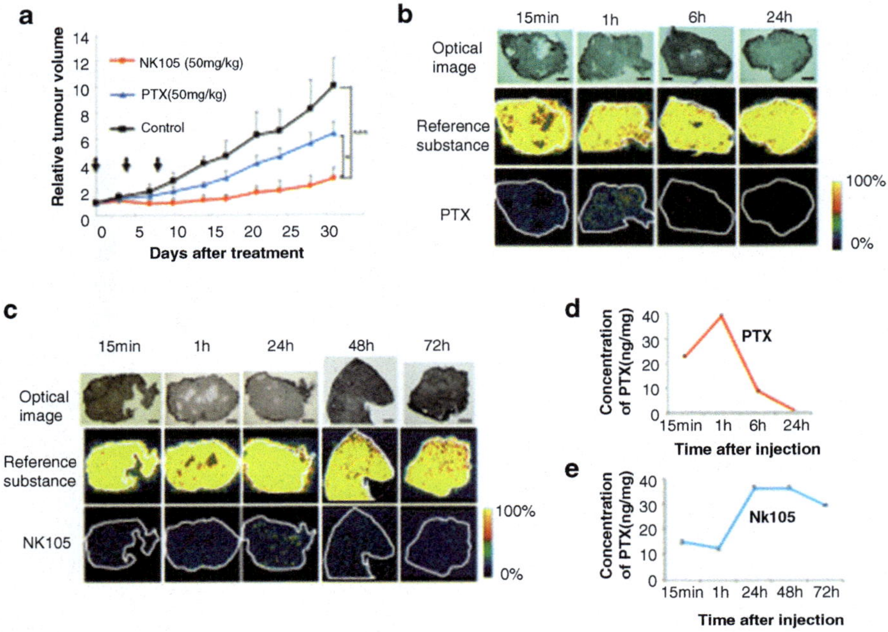

Fig. 13.9 Antitumor activity and visualisation of PTX and NK105 distribution with MS analysis. (**a**) Antitumor activity was examined in an animal model with BXPC3 xenografts. NK105, PTX, or saline (as a control) was administered at a PTX equivalent dose of 50 mg/kg on days 0, 4, and 8. *P<0.05 (PTX vs. NK105), ***P<0.001 (saline vs. NK105). Bar=SD. (**b**)(**c**) Imaging of PTX within the tumor was performed after PTX (**a**) or NK105 (**b**) administration at a dose of 100 mg/kg. The *upper*, *middle*, and *lower* columns display the optical images, reference substance (an arbitrary signal of m/z 824.6), and PTX (specific signal of m/z 892.3 [M+K]+), respectively. Bar, 1 mm. (**d**) (**e**) LC-MS analysis of the PTX concentration in the tumors treated by PTX (**d**) or NK105 (**e**). Tissue sections serial to those shown in (**b**) and (**c**)

after 24 h (Fig. 13.9b). By contrast, the signal originating from the PTX released from NK105 (rPTX) following the accumulation of NK105 in the tumor was detected at 15 min as well as at 1, 24, 48, and 72 h after the administration of NK105. The signal intensity was greatest at 24 h (Fig. 13.9c). Tissue sections serial to those for MMS were also quantified by LC-MS (Fig. 13.9d, e), the results of which correlated with the drug imaging results (Fig. 13.9b–e). The data did not contradict previous data obtained by HPLC (Hamaguchi et al. 2005). The results of the MMS analysis demonstrate that significant levels of PTX were present in the tumor clusters, including within the center of the tumor tissue.

A mechanical stress test that measured the degree of peripheral neurotoxicity demonstrated that the mice in the PTX treatment group exhibited a significantly stronger hypersensitive reaction to the mechanical stress test than those in the control and NK105 treatment groups (Fig. 13.10a). To confirm the correlation of the distribution with the abnormal neurological reaction, we also applied MMS and examined the distribution of PTX in peripheral neural tissue at 30 min, 1 h, and 24 h

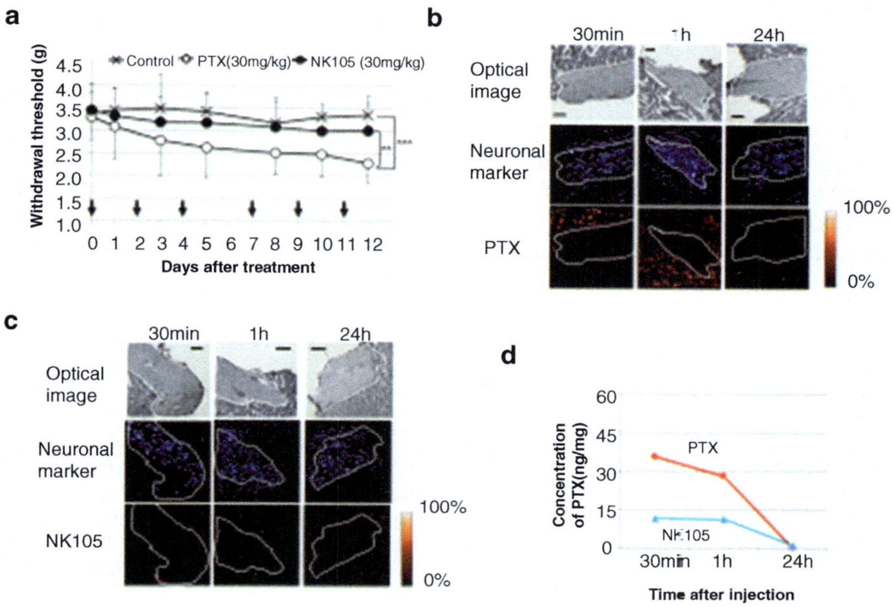

Fig. 13.10 Peripheral neurotoxicity and visualisation of PTX and NK105 distribution by MS analysis. (**a**) Mechanical sensory stress was assayed in an animal model of PTX-induced peripheral neuropathy. NK105, PTX, or saline was administered at 30 mg/kg on days 0, 2, 4, 7, 9, and 11. **P<0.01 (PTX vs. NK105), ***P<0.001 (saline vs. PTX). Bar = SD. (**b**) (**c**) PTX within neuronal tissue was imaged after PTX (**b**) or NK105 (**c**) administration at a dose of 50 mg/kg. The *upper*, *middle*, and *lower* columns show the optical images, a neuronal marker (sphingomyelin-specific signal of 851.6 m/z), and PTX (specific signal of m/z 892.3 [M+K]+), respectively. The neuronal area is delineated by a white line. Bar, 200 μm. (**d**) Analysis of the PTX concentration by LC-MS. Tissue sections serial to those shown in (**b**) and (**c**)

after administration. The signals surrounding and inside the nerve were lower after NK105 injection than after PTX injection (Fig. 13.10b,c). LC-MS analysis of the neural samples revealed that the concentration of rPTX after NK105 injection was also lower than that after PTX injection (Fig. 13.10d). Significant levels of PTX, even in the core of the tumor tissue, were observed following NK105 administration, and the NK105 was retained for a long period of time.

Low molecular weight (LMW) ACAs, including molecular targeting agents, can easily extravasate from normal blood vessels and cause various adverse effects. DDS drugs such as NK105, which exhibit low short-term accumulation in normal tissues that lack the EPR effect, can minimize this drug toxicity. Our data clearly demonstrate that the distribution of released PTX from NK105 in the peripheral nerve and surrounding tissues was quite low compared with PTX alone. These observations support the low incidence of observed peripheral neuropathy when PTX is administered as NK105.

13.5 Conclusion and Future Prospect

We successfully visualized and quantified the distribution of a non-radiolabeled and non-chemically modified drug in various frozen tissue slices microscopically. In addition to MMAE and PTX, we also have successfully visualized other anticancer agents (data not shown). Our achieved success indicates that the MMS technique can be applied to clinical biopsy specimens or surgically resected tissues after neo-adjuvant chemotherapy. In addition, the data obtained by MMS can be utilized to facilitate the design of drugs formulated into a wide variety of DDS.

References

Castellino S, Groseclose MR, Wagner D (2011) MALDI imaging mass spectrometry: bridging biology and chemistry in drug development. Bioanalysis 3:2427–2441

Cornett DS, Reyzer ML, Chaurand P et al (2007) MALDI imaging mass spectrometry: molecular snapshots of biochemical systems. Nat Methods 4:828–833

Flossel C, Luther T, Muller M et al (1994) Immunohistochemical detection of tissue factor (TF) on paraffin sections of routinely fixed human tissue. Histochemistry 101:449–453

Fujiwara Y, Furuta M, Manabe S et al (2016) Imaging mass spectrometry for the precise design of antibody-drug conjugates. Sci Rep 6:24954

Garrett MD, Workman P (1999) Discovering novel chemotherapeutic drugs for the third millennium. Eur J Cancer 35:2010–2030

Hamaguchi T, Matsumura Y, Suzuki M et al (2005) NK105, a paclitaxel-incorporating micellar nanoparticle formulation, can extend in vivo antitumour activity and reduce the neurotoxicity of paclitaxel. Br J Cancer 92:1240–1246

Hamaguchi T, Kato K, Yasui H et al (2007) A phase I and pharmacokinetic study of NK105, a paclitaxel-incorporating micellar nanoparticle formulation. Br J Cancer 97:170–176

Hellstrom KE, Hellstrom I, Brown JP (2013) Human tumor-associated antigens identified by monoclonal antibodies. Springer Semin Immunopathol 5:127–146

Junutula JR, Raab H, Clark S et al (2008) Site-specific conjugation of a cytotoxic drug to an antibody improves the therapeutic index. Nat Biotechnol 26:925–932

Kataoka K, Kwon GS, Yokoyama M et al (1993) Block copolymer micelles as vehicles for drug delivery. J Control Release 24:119–132

Kato K, Chin K, Yoshikawa K et al (2012) Phase II study of NK105, a paclitaxel-incorporating micellar nanoparticle, for previously treated advanced or recurrent gastric cancer. Investig New Drugs 30:1621–1627

Khorana AA, Ahrendt SA, Ryan CK et al (2007) Tissue factor expression, angiogenesis, and thrombosis in pancreatic cancer. Clin Cancer Res 13:2870–2875

Kim M, Gillies RJ, Rejniak KA (2013) Current advances in mathematical modeling of anti-cancer drug penetration into tumor tissues. Front oncol 3:278

Koga Y, Manabe S, Aihara Y et al (2015) Antitumor effect of anti-tissue factor antibody-MMAE conjugate in human pancreatic tumor xenografts. Int J Cancer 137:1257–1466

Lorenz M, Ovchinnikova OS, Kertesz V et al (2013) Laser microdissection and atmospheric pressure chemical ionisation mass spectrometry coupled for multimodal imaging. Rapid Commun Mass Spectrom 27:1429–1436

Mackman N (2009) The many faces of tissue factor. J Thromb Haemost 7:136–139

Maeda H, Wu J, Sawa T et al (2000) Tumor vascular permeability and the EPR effect in macromolecular therapeutics: a review. J Control Release 65:271–284

Matsumura Y (2012) Cancer stromal targeting (CAST) therapy. Adv Drug Deliv Rev 64:710–719
Matsumura Y (2014) The drug discovery by NanoMedicine and its clinical experience. Jpn J Clin Oncol 44:515–525
Matsumura Y, Kataoka K (2009) Preclinical and clinical studies of anticancer agent-incorporating polymer micelles. Cancer Sci 100:572–579
Matsumura Y, Maeda H (1986) A new concept for macromolecular therapeutics in cancer chemotherapy: mechanism of tumoritropic accumulation of proteins and the antitumor agent smancs. Cancer Res 46:6387–6392
Minchinton AI, Tannock IF (2006) Drug penetration in solid tumours. Nat Rev Cancer 6:583–592
Nitori N, Ino Y, Nakanishi Y et al (2005) Prognostic significance of tissue factor in pancreatic ductal adenocarcinoma. Clin Cancer Res 11:2531–2539
Römpp A, Spengler B (2013) Mass spectrometry imaging with high resolution in mass and space. Histochem Cell Biol 139:759–783
Rowinsky EK, Chaudhry V, Forastiere AA et al (1993) Phase I and pharmacologic study of paclitaxel and cisplatin with granulocyte colony-stimulating factor: neuromuscular toxicity is dose-limiting. J Clin Oncol 11:2010–2020
Saito Y, Waki M, Hameed S et al (2011) Development of imaging mass spectrometry. Biol Pharm Bull 35:1417–1424
Sanderson RJ, Hering MA, James SF et al (2005) In vivo drug-linker stability of an anti-CD30 dipeptide-linked auristatin immunoconjugate. Clin Cancer Res 11:843–852
Schwamborn K, Caprioli RM (2010) Molecular imaging by mass spectrometry – looking beyond classical histology. Nat Rev Cancer 10:639–646
Sievers EL, Senter PD (2013) Antibody-drug conjugates in cancer therapy. Annu Rev Med 64:15–29
Solon EG (2012) Use of radioactive compounds and autoradiography to determine drug tissue distribution. Chem Res Toxicol 25:543–555
van den Berg YW, Osanto S, Reitsma PH et al (2012) The relationship between tissue factor and cancer progression: insights from bench and bedside. Blood 119:924–932
Xie H, Audette C, Hoffee M et al (2004) Pharmacokinetics and biodistribution of the antitumor immunoconjugate, cantuzumab mertansine (huC242-DM1), and its two components in mice. J Pharmacol Exp Ther 308:1073–1082
Yasunaga M, Furuta M, Ogata K et al (2013) The significance of microscopic mass spectrometry with high resolution in the visualisation of drug distribution. Sci Rep 3 3050
Yokoyama M, Miyauchi M, Yamada N et al (1990) Polymer micelles as novel drug carrier: adriamycin-conjugted poly(ethylene glycol)-poly(aspartic acid) block copolymer. J Control Release 11:269–278
Yokoyama M, Okano T, Sakurai Y et al (1991) Toxicity and antitumor activity against solid tumors of micelle-forming polymeric anticancer drug and its extremely long circulation in blood. Cancer Res 51:3229–3236
Zolot RS, Basu S, Million RP (2013) Antibody-drug conjugates. Nat Rev Drug Discov 12:259–260

Part VII
Quantitative PK Treatment, Systems Biology and Drug Discovery

Chapter 14
Pharmacokinetics and Pharmacodynamics of Nano-Drug Delivery Systems

Amit Ranjan Maity and David Stepensky

Abstract Incorporation of drugs into nano-drug delivery systems (nano-DDSs) leads to profound changes in their disposition (distribution and elimination) and pharmacological activity (magnitude of desired and adverse effects). Nano-DDSs of different types (liposomes, particles, drug conjugates, etc.) and of different composition are intensively investigated nowadays in pre-clinical and clinical settings. Specifically, the relationships between the nano-DDSs composition, efficiency of their targeting to the intended site of action, and the magnitude of the desired vs. adverse effects are examined. In this chapter, the pathways of nano-DDSs disposition at the systemic, local, and intracellular levels are described, and the formulation properties that affect the drug targeting efficiency at each of these levels are discussed. Complexity of the nano-DDSs disposition pathways and limitations of drug targeting approaches are illustrated using specific examples: (1) delivery of antigenic peptides to the endoplasmic reticulum (ER) of the antigen presenting cells using PLGA-based nano-DDSs for the purpose of anti-cancer vaccination, (2) delivery of analgesic peptides to the brain using bolavesicle-based nano-DDSs.

Keywords Nano-drug delivery systems • Disposition • Targeting efficiency • Pharmacokinetics • Pharmacodynamics

Abbreviations

AChE	Acetylcholine esterase
BBB	Blood–brain barrier
B-CSF-B	Blood-cerebrospinal fluid barrier
CNS	Central nervous system
DC	Dendritic cell
DDS	Drug delivery system

A.R. Maity • D. Stepensky (✉)
Department of Clinical Biochemistry and Pharmacology, Faculty of Health Sciences, Ben-Gurion University of the Negev, POB 653, 84105 Beer-Sheva, Israel
e-mail: davidst@bgu.ac.il

A. Prokop, V. Weissig (eds.), *Intracellular Delivery III*, Fundamental Biomedical Technologies 8, DOI 10.1007/978-3-319-43525-1_14

EPR	Enhanced permeability and retention effect
ER	Endoplasmic reticulum
PD	Pharmacodynamics
PK	Pharmacokinetics

14.1 Introduction

Pharmacological effects of drugs are mediated by their interactions with specific targets (receptors, channels, soluble proteins, lipids, etc.) located in different organs, tissues, or body fluids. Effective and safe drug treatment requires efficient interaction of the drug with the targets located at the intended site of action and lack or limited interactions with the targets located elsewhere. Many pharmacological agents distribute to different parts of the body following their administration and interact with the desired, as well with the undesired targets, leading to the unfavorable balance of intended vs. adverse effects. In this case, targeted delivery of the drug to its intended site of action can enhance its interaction with the intended targets and diminish the undesired effects. Such targeted delivery is possible using specialized nano-drug delivery systems (nano-DDSs) that mask the drugs' physicochemical properties and alter its disposition (distribution and elimination) pathways in the body following administration. To this end, the drug can be either encapsulated or chemically modified to generate different types of nano-DDSs that can be classified according to their structure and composition into nanoparticles, liposomes, micelles, polymer conjugates, and other formulation types (see Table 14.1).

During the last decades, targeted drug delivery using different nano-DDSs types was the topic of extensive investigation by numerous research groups. Especial emphasis was set on the nano-DDSs (the formulations with dimensions in the 1–1000 or 100–1000 nm range; there is no consensus among the researchers regarding the exact definition of the nano-size range (Keck and Muller 2013; Naahidi et al. 2013)) which are in the focus of this chapter. Specifically, different formulations, formulation components, external targeting signals, and self-triggered targeting approaches were proposed to enhance the nano-DDSs targeting efficiency (see Table 14.1). For the detailed description of the different formulations and targeting approaches, we refer the readers to the several excellent recent reviews on this topic (Lammers et al. 2008; Steichen et al. 2013; Babu et al. 2014).

Results of numerous investigations report development, characterization, and *in vitro* and *in vivo* assessment of nano-DDSs (Weissig et al. 2014; Weissig and Guzman-Villanueva 2015). On a weekly and even on a daily basis, new promising experimental results are reported, especially in the 'hot' research fields, such as delivery of anticancer agents. Usually, these publications contain claims regarding efficient drug targeting using different types of nano-DDSs, and expectations of their future applications for diagnosis and therapy. However, the current clinical impact of nano-DDSs is limited and does not match the level of expectations from this technology on behalf of researchers and non-scientists. It appears that only a

Table 14.1 The major approaches for nano-DDSs targeting to the intended site of action

Formulation types	Administration routes	Approaches for self-triggered targeting	DDS formulation components that can affect targeting efficiency	Approaches for externally-activated targeting
Nanoparticles Liposomes Micelles Bolavesicles Polymer conjugates Dendrimers Carbon nanotubes Pasty polymers Hydrogels	Intravenous Intraarterial Intraperitoneal Intranasal Local (focal) injection	Nano-DDS size Decoration with targeting residues that can interact with molecules overexpressed at the intended site of action pH-dependent drug release Enzyme-dependent drug release	Magnetic or paramagnetic components Sheddable PEGylation Multistage drug release (e.g., nano-DDS disintegration into smaller particles) Promoter drugs (vascular damage, apoptosis induction, TGF-beta blockade, etc.) Permeation enhancers	Magnetic field Ultrasound Local heating Light-induced chemical activation

small fraction of the administered drug is capable of exerting the pharmacological effect at the intended site of action in the body, and the overall targeting efficiency of the nano-DDSs is low, leading to their limited clinical effectiveness and safety.

This limited clinical impact of nano-DDSs stems from the fact that, despite the scientific advances, the knowledge on the relationship between the nano-DDSs formulation properties and their pharmacokinetics and pharmacodynamics is still limited. Different types of nano-DDSs (see Table 14.1) vary in their properties that affect their disposition (Danhier et al. 2010; Kanapathipillai et al. 2014; Wicki et al. 2015): (a) size and composition, (b) ability to encapsulate different drugs (e.g., hydrophilic and lipophilic small molecular weight drugs, nucleic acids, therapeutic proteins, etc.), (c) drug loading efficiency and release kinetics, (d) stability in different locations in the body (in the body fluids and inside cells), and ability to control it. Specifically, efficiency of drug delivery using nano-DDSs is highly dependent on the surface properties and drug release kinetics of the specific formulation. Indeed, nano-DDSs surface is the major site of interaction between the nano-DDSs and biological systems, and the surface properties of a specific nano-DDS have important effect on its disposition pathways (see below), such as endocytosis (by the endothelial, phagocytic and target cells), interaction with the soluble factors (activation of complement, formation of 'corona' made of endogenous proteins), etc. In the next sections of this chapter, we will describe in a detailed fashion the major pharmacokinetic and pharmacodynamic properties of nano-DDSs at different levels (in the systemic circulation, in the tissues/organs/body fluids, and inside the cells), and will illustrate using specific examples the complexities of nano-DDSs targeting to the intended site of action.

14.2 The Pharmacokinetics and Pharmacodynamics of Systemically-Administered Nano-DDSs

Pharmacokinetics (PK) describes the time course of drug concentrations that are attained in different locations in the body following drug administration. Pharmacodynamics (PD) refers to the pharmacological effects elicited by the drug, including the mechanisms of these effects at different locations in the body and concentration-effect relationships. Nano-DDSs possess substantially different pharmacokinetic properties, as compared to the 'classical' drugs (small-molecular drug compounds), which also affect their pharmacodynamics. Specifically, many nano-DDSs do not pass efficiently the biological barriers, which necessitates their parenteral administration (usually *via* intravenous route).

In the bloodstream, the major disposition pathways of the 'classical' drugs (e.g., permeation into the extracellular fluids and into the cells, liver and kidney elimination) are less relevant to the pharmacokinetics of nano-DDSs, which are exposed to other distribution and elimination processes instead. These processes operate in different ways in different locations in the body, and in this chapter we will describe the three levels of disposition that are typical for many nano-DDSs and for the pharmacological effects induced by them (see Fig. 14.1). The first level takes place in the systemic circulation, where preferential nano-DDS distribution (targeting) to a specific organ/tissue of interest can be beneficial for the treatment efficiency. Subsequently, at the organ/tissue level, nano-DDSs targeting to the specific cells may be desirable. Ultimately, for the drugs that act on the intracellular targets (such

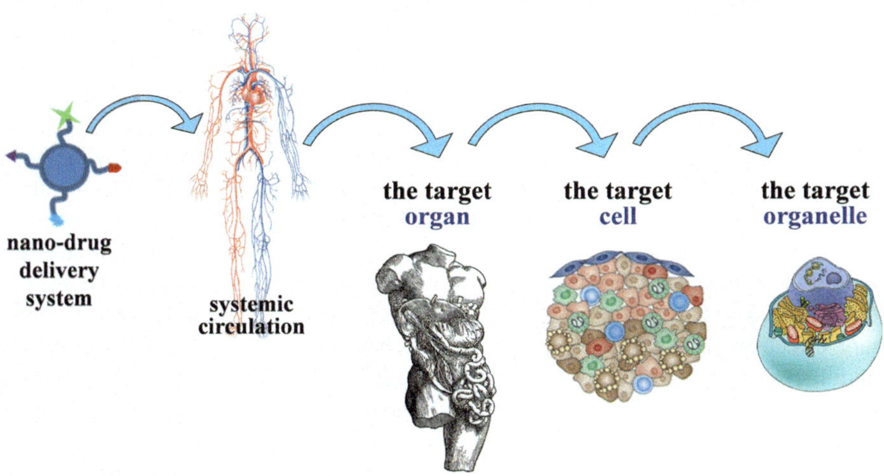

Fig. 14.1 The 3 levels of drug targeting using nano-DDSs. Parts of this figure were adapted from (Holzel et al. 2013) by permission from Macmillan Publishers Ltd, and from ASU Ask A Biologist, Arizona Board of Regents (https://askabiologist.asu.edu/content/cell-parts)

Table 14.2 The major pathways of nano-DDSs disposition following systemic administration

Pathway
Release of the encapsulated drug
Interaction with the endogenous compounds (formation of 'corona')
Aggregation in the bloodstream
Degradation in the systemic circulation
Elimination via the liver and/or the kidneys
Uptake by mononuclear phagocyte system (MPS) cells
Uptake by the endothelial cells
Uptake by the red blood cells
Permeation to the target organ
Permeation to other organs and tissues

as nucleic acids, many anti-cancer agents, etc.) nano-DDSs targeting to the specific organelle may lead to enhanced efficiency on the subcellular level.

It should be noted that at each one of the above-mentioned levels, both the nano-DDS and the drug exist in multiple forms, with each one of those exposed to a different extent to the individual disposition pathways. For instance, the nano-DDS-encapsulated drug can be released to the body fluids (plasma, extracellular or intracellular fluids), bind to proteins or other endogenous components, undergo metabolism, return back to the nano-DDS (via surface absorption or reloading processes) (Das et al. 2016), etc. These processes complicate the understanding the concentration-effect relationships for nano-DDSs, and may introduce bias in the results and conclusions of pharmacokinetic studies that apply analytical methods that do not discriminate between the different drug/nano-DDSs forms.

In the next sections, we will describe the major disposition pathways of nano-DDSs at the individual levels, and the obstacles for efficient targeting to the intended site of action on each one of them (see Fig. 14.1).

14.2.1 PK and PD of Nano-DDSs in the Systemic Circulation

Following systemic administration and permeation into the bloodstream, nano-DDSs are exposed to numerous disposition pathways (see Table 14.2), only one of which mediates their permeation to the intended site of action. Specifically, nano-DDSs can release the encapsulated drugs (decapsulate) in the systemic circulation prior to reaching the target tissue, aggregate in the bloodstream (with subsequent deposition in capillaries or uptake by the phagocytic cells in the bloodstream), undergo elimination via hepatic and renal elimination pathways, endocytosis and metabolism by the cells that are circulating in the blood or lining the blood vessels, etc. For detailed analysis of the factors that affect the disposition pathways of free and nano-DDS-encapsulated drugs, the readers are referred to several excellent recent reviews (Ruenraroengsak et al. 2010; Moghimi et al. 2012).

It can be seen that the chances of the individual nano-DDS to reach the target tissue are low and drug targeting to the intended site of action requires blocking or avoiding the above-mentioned interfering pathways. Efficiency of the individual disposition pathways (see Table 14.2) can be substantially affected by the dimensions (size and shape) and the surface properties (charge, surface residues) of the nano-DDSs. For instance, the cells of the mononuclear phagocyte system (MPS) cells have their 'preferences' for the size, shape, and charge of the nano-DDSs and will preferably endocytose charged (with positive or negative surface charge) nano-DDSs with diameter bigger than 200 nm (Decuzzi et al. 2010; Moghimi et al. 2012). An addition, sinusoidal endothelium of selected organs and tissues (liver, spleen, bone marrow, endocrine organs), in combination with presence of specific cells (e.g., sinusoidal endothelial cells and Kupffer cells in the liver) capture and eliminate from the bloodstream nano-DDSs of certain size ranges. For instance, liver efficiently captures and eliminates nano-DDSs 10–20 nm diameter in size. On the other hand, urinary excretion is rapid and efficient for nano-DDSs with hydrodynamic radius smaller than ~5 nm (Longmire et al. 2008). Nano-DDSs targeting to the liver, therefore, is not a very challenging task, and systemically-administered formulations with the diameter of 10–20 nm will tend to accumulate in the liver (and other above-mentioned MPS organs). On the other hand, formulations with the diameter of 20–200 nm may have higher chances to arrive to the intended site of action that is located elsewhere (not in the liver or other MPS organs).

During their presence in the systemic circulation, nano-DDSs can interact with and adsorb endogenous compounds such as albumin, transferrin, lipoproteins, and other plasma components (Moghimi et al. 2012; Dufort et al. 2012; Bigdeli et al. 2016; Palchetti et al. 2016). As a result of this interaction, coating or 'corona' is formed on the surface of nano-DDSs which affects their surface properties and efficiency of their disposition pathways (see Table 14.2), and can further reduce the chances for efficient drug targeting to the intended site of action. A popular strategy to block these pathways and to avoid 'corona' formation is nano-DDSs decoration (*via* non-covalent adsorption or covalent conjugation) with polyethylene glycol (PEG) residues. PEGylated nano-DDSs are often referred to as 'stealth' formulations, and PEGylation technology that forms 'inert' steric barrier on the nano-DDSs' surface can indeed be used to prolong their residence in the bloodstream (Parveen and Sahoo 2006; Barenholz 2012; Perry et al. 2012). However, repeated administration of PEGylated nano-DDSs can provoke immune reactions of the host against the PEG residues, *via* formation of anti-PEG IgM antibodies and complement activation, that can increase the nano-DDSs clearance and diminish their accumulation at the target site (Ishida and Kiwada 2008; Schellekens et al. 2013). These immunogenic aspects of PEGylation require more detailed investigation, and combination of PEGylation with 'active' targeting approaches may lead to enhanced efficiency of nano-DDSs targeting to the intended site of action.

Systemic nano-DDSs disposition can be substantially altered in some disease states. For instance, formation of gaps or openings in the vascular endothelium will lead to enhanced extravasation of nano-DDSs at these locations, at expense of other disposition pathways (see Table 14.2). Such gaps or openings can form in some solid cancers or in inflammatory diseases (leading to the so-called enhanced permeability and retention effect, EPR) (Maeda 2012; Maeda et al. 2013; Nichols and Bae 2014), in the brain tissue (due to the physical trauma, stroke, severe and prolonged epileptic attacks, etc.) and in other conditions (Azzopardi et al. 2013; Boyd et al. 2015). It is plausible that nano-DDSs disposition can be altered due to perfusion changes (in congestive heart failure, edema, burns), but these alterations have not been characterized in sufficient detail yet.

Efficient drug targeting at the systemic level (i.e., efficient or preferential drug delivery to the specific organ/tissue/body fluid) is claimed in many experimental studies with nano-DDSs. However, few of these studies report quantitative data on the drug targeting efficiency. Due to the complexity of the disposition pathways (see Table 14.2), only a small fraction of the administered nano-DDSs apparently accumulates in the target organ/tissue. For instance, only in rare cases >5–6 % of the administered nano-DDSs distribute to the solid tumors following intravenous administration to the human patients, even in rapidly growing tumors with prominent EPR effect (Ruenraroengsak et al. 2010; Lammers et al. 2012; Taurin et al. 2012; Nichols and Bae 2014). The targeting efficiency can be higher in the animal models of cancer disease, due to the exaggerated EPR effect, which can be seldom attained in the clinical settings.

The brain-targeting ability of the currently available nano-DDSs is also limited, and in many cases the brain tissue is exposed to lower amounts of drug, as compared to the internal organs and tissues (Kozlovskaya and Stepensky 2013). Efficiency of brain drug delivery depends on the drug formulation: intravenously-administered nanoparticle-based nano-DDSs appear to be more efficient than the liposome- or conjugate-based nano-DDSs (Kozlovskaya and Stepensky 2013). Incorporation of specialized reagents, such as PEG residues and surface-active agents, decoration with specific targeting residues, use of external targeting signals apparently can increase the efficiency of nano-DDS delivery to the brain and can lead to brain-targeted drug delivery. Moreover, nano-DDSs administration *via* a nasal route can also enhance their brain targeting efficiency due to existence of direct nose-to-brain traffic routes (via olfactory and trigeminal cranial nerves) (Kozlovskaya et al. 2014). A more detailed quantitative analysis of the nano-DDS disposition and its mechanisms following systemic and nasal administration is required to identify the limiting factors for drug brain delivery and the drug formulations that are the most suitable for this purpose (Kozlovskaya and Stepensky 2013; Kozlovskaya et al. 2014).

Table 14.3 Pathways of nano-DDS disposition at the intended site of action and in other organs/tissues/body fluids

Pathway
Adsorption to the capillary wall
Extravasation (can be enhanced due to the inflammation or in solid tumors – the enhanced permeability and retention effect)
Diffusion
Convection
Binding to the cells and/or extracellular matrix
Uptake by the cells
Metabolism outside and/or inside the cells
Lymphatic elimination
Venous elimination

14.2.2 PK and PD of Nano-DDSs in the Tissues

Following arrival to the target or non-target tissue/organ/body fluid, nano-DDSs are exposed to several disposition pathways, which are summarized in Table 14.3. Due to these multiple disposition pathways, nano-DDSs (and the drug that is released from them) are become distributed in a highly non-homogeneous fashion on the organ/tissue level. For instance, the majority of the extravasated nano-DDSs accumulate in close proximity to the blood vessels, while penetration to the more 'deep' layers is limited (Stepensky 2014; Zhang et al. 2013), leading to a gradient of drug/nano-DDS concentrations in the specific organ or tissue.

Usually, it can be assumed that the desired pharmacological effects are mediated by specific targets expressed by a selected type of cells (e.g., the cancer cells in the tumor, the neurons in the brain, etc.; see Fig. 14.1). Therefore, many of the pathways listed in Table 14.3, such as adsorption to the capillary wall, local metabolism or elimination with the venous blood or with the lymph, prevent or interfere with effectiveness of nano-DDSs. On the other hand, diffusion, convection, binding and uptake processes, depending on their location and direction, can either contribute to the desired effects, or interfere with them. For instance, overexpression of efflux pumps (e.g., Pgp, BCRP and others) and of xenobiotic metabolizing enzymes, can reduce the drug accumulation inside the target cells (e.g., the cancer cells in the tumor tissue) and diminish the treatment efficiency (Rochat 2009). On the other hand, overexpression of specific surface receptors (e.g., HER-2, EGFR, etc.) by the target cells can serve for 'active targeting' of the nano-DDSs to these cells using specific targeting residues (Li et al. 2012; van der Meel et al. 2013). Such 'active targeting' seldom succeeds since it can limit the nano-DDSs permeability to the 'deep' parts of the tissue and reduce the treatment efficiency (Lammers et al. 2012; Popilski and Stepensky 2015), or since the targeting residues on the nano-DDSs surface can be degraded or masked by the 'corona'.

For a more detailed discussion of the factors that affect nano-DDSs disposition at the tissue level, especially for the purpose of targeted drug delivery to the solid tumors, the readers are referred to several recent reviews of this topic (Minchinton and Tannock 2006; Ruenraroengsak et al. 2010; Danquah et al. 2011; Nichols and Bae 2012; Ernsting et al. 2013; Stepensky 2014; Popilski and Stepensky 2015).

14.2.3 Intracellular PK and PD of Nano-DDSs

Many drugs act on intracellular targets within specific organelles, and efficient penetration of these drugs to specific intracellular site of action is a pre-requisite for its pharmacological activity. In order to reach the intracellular targets, nano-DDSs (that encapsulate an active drug) usually need to undergo endocytosis and to release the drug in its free form inside the cell. The resulting efficiency of drug interaction with its target can be low due to a high degree of compartmentalization of the cells' interior (organelles separated by biological membranes) and complexity of the intracellular trafficking pathways. Specifically, sequence of the following intracellular pharmacokinetic processes is needed for mounting the desired effects of intracellularly-acting drugs: nano-DDS endocytosis by the target cells, endosomal escape, trafficking to (or in vicinity of) the specific organelle of interest, release of the encapsulated drug, and drug-target interaction in the organelle of interest. Each one of these processes can pose a barrier for drug accumulation in the target organelle (Stepensky 2010). For instance, many types of cells possess low endocytic behavior that precludes the subsequent targeting stages. Moreover, nano-DDSs which enter cells via endocytosis tend to trap inside endosomes and undergo degradation during their maturation into late endosomes and lysosomes, which prevents nano-DDS permeation to the cytosol and their accumulation in the target organelle (Paillard et al. 2010).

Nano-DDSs formulation properties can affect intracellular drug pharmacokinetics at multiple stages. It is known, for instance, that the nano-DDS formulation properties (formulation type, size, charge, type and density of surface residues, their stability) affect the efficiency of the nano-DDS uptake and intracellular trafficking mechanisms. The best investigated process appears to be the nano-DDS endocytosis by some types of cells where some important discoveries regarding the effects of nano-DDS size/shape/charge have been made recently (Decuzzi et al. 2010; Moghimi et al. 2012; Ernsting et al. 2013). On the other hand, efficiency of nano-DDS endosomal escape, spontaneous or mediated by specific endosome-destabilizing formulation components, is much less characterized. Moreover, it is not clear how the intracellular trafficking mechanisms can be modulated for preferential delivery of nano-DDSs to the organelle of interest (i.e., for the purpose of subcellullarly-targeted drug delivery), which types of surface targeting residues are best suited for this, and to which extent the efficiency of this subcellular targeting differs between cells of different origin (Maity and Stepensky 2015, 2016b).

During the recent years, targeting of nano-DDSs to the specific intracellular organelles (i.e., subcellular targeting) was investigated in numerous publications, but targeting efficiency of these systems was seldom reported (Maity and Stepensky

2016b). It appears that specialized nano-DDSs decorated with specific targeting residues or active groups can alter the 'passive' pathways of drug intracellular disposition (distribution and elimination) and lead to 'active' targeting of the drug to specific organelle (i.e., leading to higher drug concentration in the target organelle, as compared to its concentrations in other organelles). Nucleus and mitochondrion were applied more frequently, as compared to other organelles, as intracellular targets of drug delivery, and discussion of nano-DDS designed to target these organelles appears in several recent reviews (Torchilin 2006; D'Souza and Weissig 2009; Duncan and Richardson 2012).

Different chemical approaches can be used to decorate nano-DDSs surface with specific subcellular targeting residues (Yu et al. 2012), and intracellular disposition of nano-DDSs can be studied using a variety of biochemical and imaging-based approaches (such as subcellular fractionation based on density gradient centrifugation, colocalization analysis of fluorescently-labeled nano-DDSs and fluorescent organelle markers, etc.) (Torchilin 2005; Seib et al. 2006; Sneh-Edri and Stepensky 2011; Gilleron et al. 2013). Despite the technological advances that allow quantitative analysis of nano-DDSs formulation properties, their endocytosis and intracellular disposition, our understanding of relationships between nano-DDSs properties and their biofate is only partial (Breunig et al. 2008; D'Souza and Weissig 2009; Stepensky 2010; Fu et al. 2014). Extensive research is needed to clarify these issues, and to identify the formulation properties that lead to efficient subcellular targeting. This research should involve detailed investigation of nano-DDS uptake and intracellular disposition in the target and non-target cells, in different experimental settings. Otherwise, the expectations regarding targeted drug/nano-DDS delivery can turn out to be exaggerated.

Overall, it appears that conjugation of specific targeting residues can affect intracellular fate of nano-DDSs and result in preferential drug accumulation within an organelle of interest. Studies that quantitatively assess the mechanisms, barriers and efficiencies of subcellularly-targeted drug delivery and of the associated toxic effects are required to determine the therapeutic potential of subcellular nano-DDSs targeting, and to translate the research results into clinical applications.

14.3 Targeted Delivery of Analgesic Peptides to the Brain Using Bolavesicle-Based Nano-DDSs

Analgesic peptides (such as leu-enkephalin and kyotorphin) act on the opioid receptors that are involved in pain transmission and control in many regions of the central nervous system, including spinal cord, midbrain and thalamus (Snyder and Pasternak 2003; Law et al. 2000). Similar to other biopharmaceuticals, analgesic peptides possess limited stability in the body fluids and tissues, and inefficient permeability *via* the biological barriers, including the blood–brain and blood-cerebrospinal fluid barriers (the BBB and the B-CSF-B). Targeted drug delivery to the site of action in the brain can overcome these pharmacokinetic limitations of the analgesic peptides, and

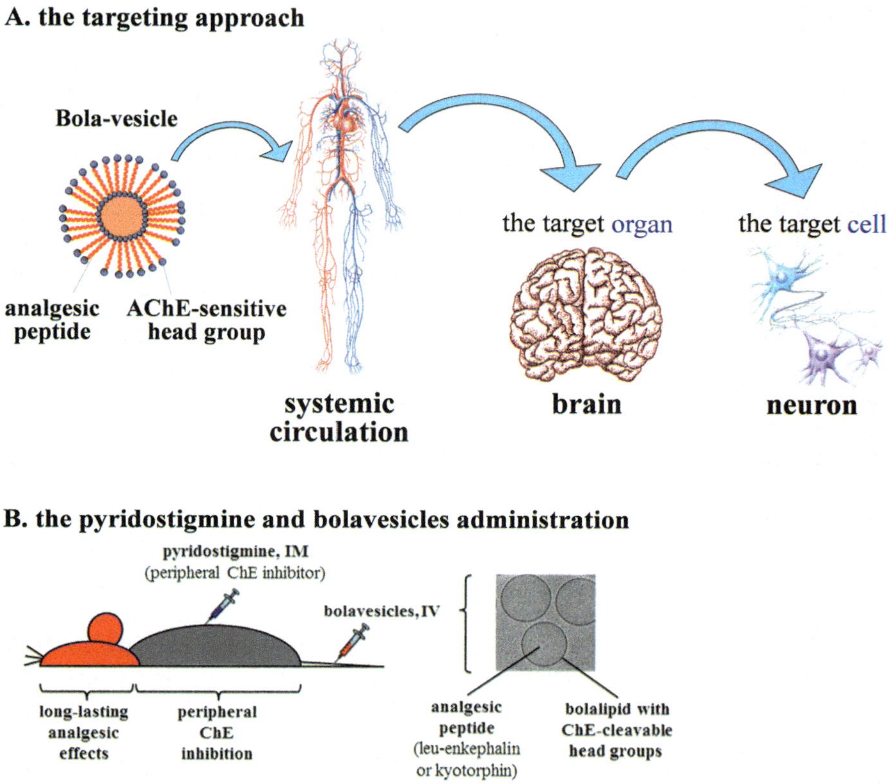

Fig. 14.2 Approach for brain delivery of analgesic peptides using bolavesicle-based nano-DDSs. (Part of this figure is reprinted from (Popov et al. 2013), with permission from Elsevier)

enhance the efficiency of their analgesic effects. For this purpose, targeted delivery to the selected parts of the brain is desired (i.e., targeting on the organ and cell levels; see Fig. 14.2a). The third intracellular targeting levels is not necessary in this case, since analgesic peptides do not need to permeate the cells in order to exert their pharmacological effects (which are mediated by interaction with the extracellular parts of the opioid receptors that belong to the G-proteins family).

For the purpose of brain delivery of analgesic peptides, we used bolavesicle-based nano-DDSs (Grinberg et al. 2005; Grinberg et al. 2008; Popov et al. 2010). Bolavesicles are monolayer membrane vesicles that are assembled from synthetic bolalipids containing two hydrophilic head groups at each end of a hydrophobic alkyl chain (see Fig. 14.2a). Bolalipids form vesicles that possess high chemical and physical stability that can be useful for their permeation via the BBB and B-CSF-B. On the other hand, bolavesicles can be designed to be destabilized in a controlled fashion – enzymatic cleavage of the headgroups can trigger rapid disruption of the vesicular structure, leading to preferential release of the encapsulated material at the desired locations, including the brain (Gupta et al. 2015).

In our studies, we used bolavesicles made of the GLH-20 bolalipid that is a substrate of acetylcholine esterase (AChE) enzyme (Dakwar et al. 2012). This enzyme can be efficiently inhibited in the bloodstream and in the different organs/tissues/body fluids by pyridostigmine, which is charged at the physiological pH and does not permeate to the brain tissue. Thus, pyridostigmine pretreatment can stabilize the GLH-20 bolavesicles in the systemic circulation, but allows their decapsulation upon permeation into the central nervous system, enhancing the delivery of bolavesicle-encapsulated drugs to the brain (see Fig. 14.2b). We investigated the tissue distribution and the pharmacological effects induced by bolavesicles loaded with fluorescent markers and/or analgesic peptides (see Fig. 14.2) (Dakwar et al. 2012; Popov et al. 2013). Apart from these components, the bolavesicle-based nano-DDSs contained stabilizing agents (cholesterol) and in some cases were also decorated with chitosan moieties that enhance bolavesicles permeation *via* the BBB and other biological barriers.

The experimental results indicated that the AChE-sensitive bolavesicles efficiently distributed into various tissues with marked accumulation in the brain, whereas non-encapsulated (free) BSA-FITC possessed much lower permeability, especially to the brain tissues (see Fig. 14.3a). The investigated bolavesicles delivered analgesic peptides to the brain in quantities sufficient to elicit efficient and prolonged analgesic activity, while the non-encapsulated peptides induced much lower pharmacological effects (see Fig. 14.3b). Based on these data, we conclude that bolavesicle-based nano-DDSs, in combination with the pyridostigmine, alter substantially the pharmacokinetics of the analgesic peptides and enhance the efficiency of their pharmacological effects.

The studied formulations possessed limited targeting efficiency on the systemic level (lack of preferential accumulation in the brain vs. other tissues; see Fig. 14.3a), and apparently also on the organ/tissue level (lack of preferential accumulation in the parts of the brain that mediate analgesic activity). Nevertheless, encapsulation of the investigated analgesic peptides into the bolavesicle-based nano-DDSs altered their pharmacokinetics to a substantial extent, and enhanced and prolonged their analgesic activity. We and our collaborators successfully applied bolavesicle-based nano-DDSs for brain delivery of other pharmacological agents, including small molecular weight drugs, biopharmaceuticals, and marker compounds, as described in detailed fashion in (Shechter et al. 2010; Dakwar et al. 2012; Philosof-Mazor et al. 2013; Popov et al. 2013; Gupta et al. 2015).

14.4 Targeted Delivery of Antigenic Peptides to the Endoplasmic Reticulum of the Antigen-Presenting Cells for Anti-Cancer Vaccination

Tumor antigens are antigenic substances (peptides, glycopeptides, etc.) produced in tumor cells that are able to trigger immune response in the host. Mounting efficient anticancer vaccination can be possible by administration of tumor-associated

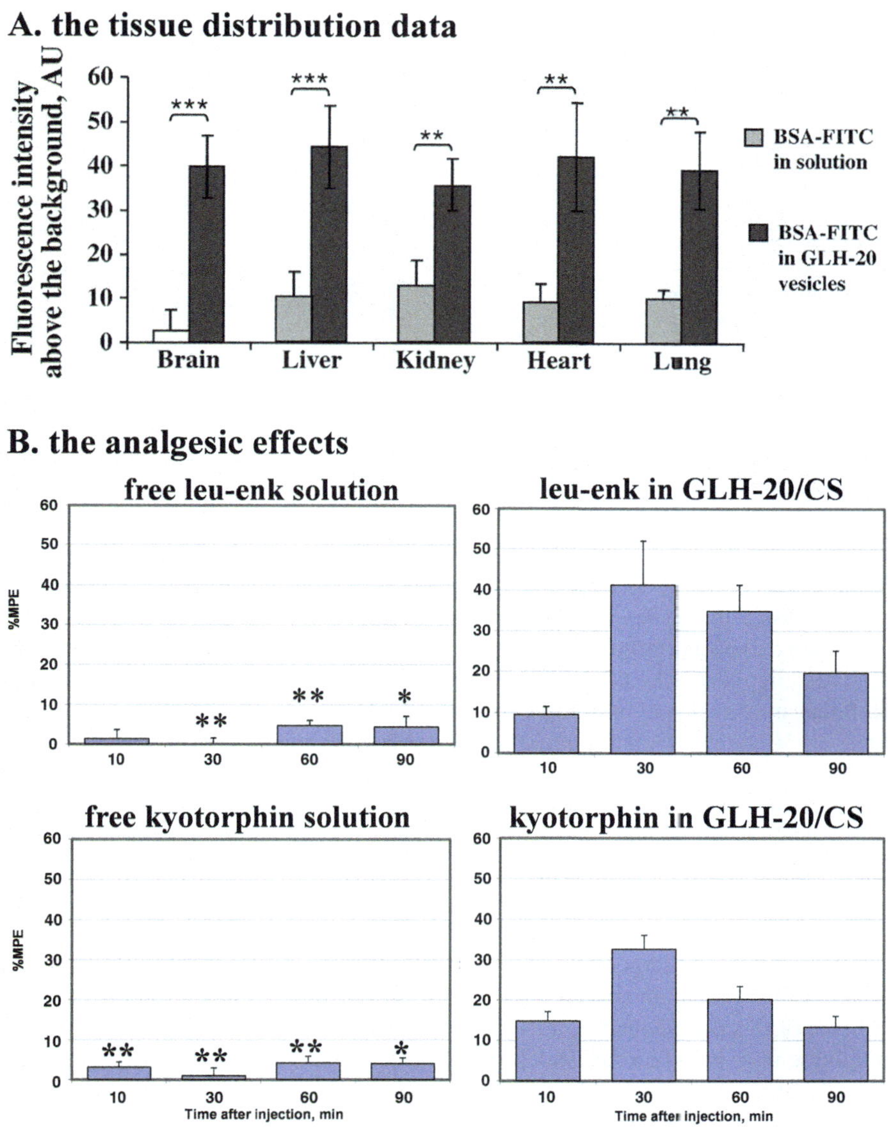

Fig. 14.3 Results of brain delivery of analgesic peptides using bolavesicle-based nano-DDSs. (Reprinted from (Dakwar et al. 2012) and from (Popov et al. 2013), with permission from Elsevier)

antigens to the host (in the form of antigenic peptides, proteins, or DNA that encodes for them). For efficient anticancer vaccination, antigenic peptides should reach the intracellular organelles within the antigen-presenting cells which are capable of antigen cross-presentation process (dendritic cells, DCs) (Amigorena and Savina 2010; Segura and Villadangos 2011; Mantegazza et al. 2013). As a result of this

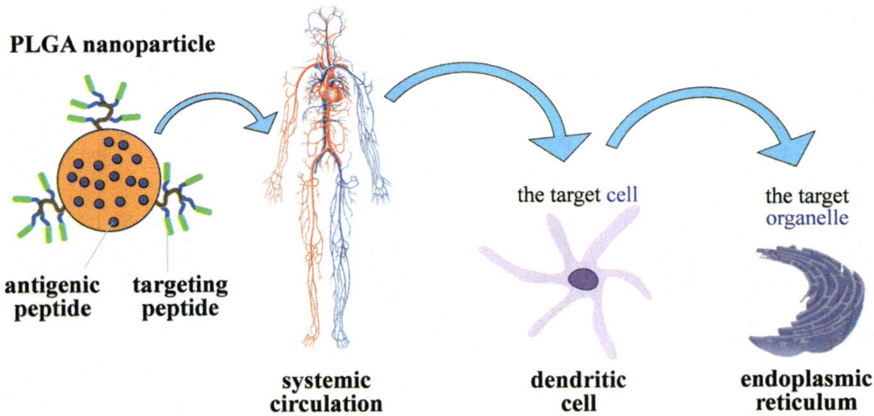

Fig. 14.4 Approach for targeted delivery of antigenic peptides to the endoplasmic reticulum (ER) of the antigen-presenting cells for anti-cancer vaccination using decorated PLGA nanoparticles

process, that takes place predominantly in the endoplasmic reticulum (ER) and endosomal compartments, DCs process and present tumor-associated antigens with MHC class I molecules to CD8+ T cells (cytotoxic T cells), which can become activated and exert cytotoxic anti-tumor responses.

DCs circulate in the bloodstream, are phagocytic, and possess specialized intracellular antigen processing and cross-presentation machinery. Therefore, targeted delivery of antigenic material to the DCs is possible by encapsulating it into nano-DDSs that undergo phagocytosis by the DCs (in the central circulation following intravenous administration, or in the tissues following injection, e.g., by subcutaneous route). Thus, for the purpose of anti-cancer vaccination, targeting on the cell and intracellular (subcellular) levels is desired to deliver efficiently the antigenic peptides to the cross-presenting organelles within the DCs (see Fig. 14.4), while the organ/tissue level of targeting is less important or is not required.

We investigated targeting of the specific antigenic peptide (SIINFEKL) to the endoplasmic reticulum (ER) of the DCs for the purpose of anti-cancer vaccination. Choice of this peptide and of the *in vitro* experimental system for assessment of its targeting were based on the availability of specific and highly-sensitive reagents and experimental systems (antibodies and assays) for quantitative assessment of its intracellular fate and pharmacological effect (i.e., cross-presentation efficiency). SIINFEKL peptide and fluorescent marker were encapsulated into poly(lactic-co-glycolic) acid (PLGA)-based nano-DDSs to generate nanoparticles with the required size (~200–500 nm range) which undergo efficient phagocytosis by the DCs (Bachmann and Jennings 2010; Foged et al. 2005), are loaded with sufficient amount of antigenic peptide, and gradually release it over the desired time period (3–5 days) (Sneh-Edri et al. 2011).

We decorated the PLGA-based nano-DDSs with peptidic targeting residues using a 3-stage decoration approach based on carbodiimide and Click (azide-alkyne Huisgen cycloaddition using copper (I) catalysis) reactions (Sneh-Edri et al. 2011;

Kaplun and Stepensky 2014). As a targeting signal, we used a peptide containing specific ER-targeting moieties (KKXX signal) that was previously shown to target intracellular proteins to the ER (Andersson et al. 1999; Teasdale and Jackson 1996), or control (scrambled) peptide. We choose to assess targeting to the ER organelle since it is a major site of peptide loading on the MHC class I molecules, and efficient delivery of exogenous antigenic peptide to this organelle can enhance dramatically its cross-presentation efficiency.

We characterized the size, morphology, surface properties and decoration efficiency of the developed nano-DDS, and analyzed their endocytosis and intracellular trafficking in the DCs *in vitro* and the resulting cross-presentation efficiency of the antigenic peptide (Sneh-Edri et al. 2011). Decoration of the nanoparticles with peptidic residues profoundly affected their uptake and intracellular trafficking in the DCs (see Fig. 14.5), indicating substantial differences in the surface properties between the different formulations. Unexpectedly, following their endocytosis, nanoparticles decorated with the ER-targeting peptide accumulated to a lower extent in the ER as compared to the control peptide-conjugated nanoparticles. We attributed this finding to the effect of control and targeting peptides on efficiency of the individual endocytosis pathways of the studied nano-DDSs and on their subsequent intracellular distribution.

We concluded that decoration of nano-DDSs with specific targeting residues affects their endocytosis and intracellular trafficking in the DCs, and can be beneficial for anti-cancer vaccination (Sneh-Edri et al. 2011). Further studies that quantitatively assess the mechanisms, barriers, and efficiency of subcellular drug delivery are required to identify potent targeting residues and their density on the nano-DDSs surface that is required for efficient subcellular targeting to the individual target organelles (Maity and Stepensky 2016a). Subcellularly-targeted drug delivery is a promising new approach for enhancing and controlling the drug pharmacological activities for the purpose of anti-cancer vaccination (see Fig. 14.4) and for other therapeutic applications (Torchilin 2006; D'Souza and Weissig 2009), albeit the subcellular targeting efficiency of the currently available nano-DDSs appears to be low (Stepensky 2010; Maity and Stepensky 2015; Maity and Stepensky 2016b).

14.5 Rational Design of Nano-DDSs

Design of nano-DDSs with the desired balance of the disposition pathways on the different levels (the systemic circulation, the tissue and intracellular levels; see Fig. 14.1) is not an easy task due to a limited understanding of the relationship between the nano-DDS formulation and the individual pathways efficiency. Numerous parameters that originate from the physico-chemical properties of the specific drug, the formulation properties of the nano-DDS and the physiological and pathophysiological factors affect the nano-DDSs disposition on each one of the above-mentioned levels (see Table 14.4). Moreover, the drug and the nano-DDSs exist in multiple forms in the body (e.g., free drug, nano-DDS-encapsulated drug,

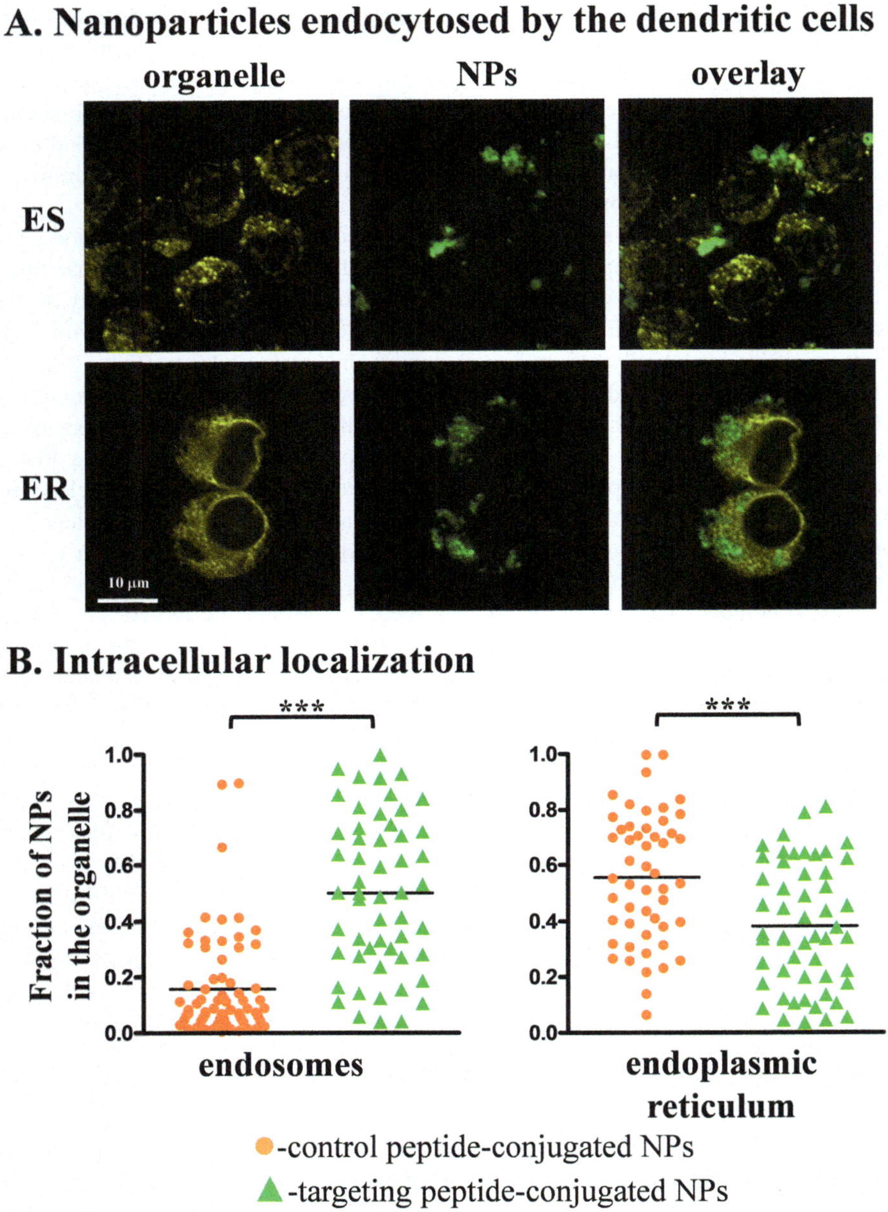

Fig. 14.5 Results of targeted delivery of antigenic peptides to the endoplasmic reticulum (ER) of the antigen-presenting cells for anti-cancer vaccination using decorated PLGA nanoparticles. (Part of this figure is reprinted from (Sneh-Edri and Stepensky 2011) with permission from American Chemical Society)

Table 14.4 The major parameters that govern the nano-DDS disposition and affect the efficiency of drug targeting to the intended site of action

Group	Parameter
Drug properties	Molecular weight
	Water solubility
	Lipophilicity (logP, $logD_{7.4}$)
Nano-DDS properties	Type and composition
	Size and shape
	Surface properties of the nano-DDS (ζ-potential, decoration with targeting residues, PEGylation, etc.)
	Drug encapsulation efficiency
	Rate of drug release from the nano-DDS
Properties of the intended site of action and of other organs/tissues/body fluids	Cells (types and properties: growth, renewal, endocytic behavior, metabolic activity, etc.)
	Extracellular matrix (composition, density)
	Vascularization (type of capillaries, perfusion and its heterogeneity, etc.)
	Extracellular fluid flow (efficiency and directions of convective flow)
	Lymphatic drainage (efficiency and direction towards specific draining lymph nodes)
	Expression of targets for drug action (e.g., specific receptors)
	Presence of drug and nano-DDS-metabolizing enzymes
	Sensitivity to the drug

protein-bound drug, uncoated and 'corona'-coated nano-DDS, etc.) and there are analytical difficulties with the quantitative analysis of nano-DDSs (and of the drug released from them) in the different locations.

Formerly, drug concentrations in the individual organs and tissues following administration of nano-DDSs was measured mostly based on radioactivity, fluorescence, or UV absorbance of tissue homogenates (with or without chromatography-based separation of the samples). Imaging-based approaches (e.g., fluorescent imaging of excised organs) are more commonly applied for the same purpose. Unfortunately, these approaches do not detect separately the and un-encapsulated drug. As a result, the analytical assays applied for quantitative analysis of nano-DDSs disposition in many studies appear to have limited robustness and reliability and may lead to artefacts and erroneous conclusions. For instance, absorption and scattering of the fluorescent signal in the individual tissues can significantly interfere with the fluorescence-based measurement outcomes (Liu et al. 2012). Fluorophore self-quenching and pH sensitivity of the fluorescence intensity can also limit the robustness of the fluorescence-based analytical methods (North 2006; Waters 2009), and lead to erroneous drug/nano-DDS concentration/amount measurements. Analysis of nano-DDSs disposition at the intracellular level is prone to all the above-described problems, and also suffers from the limited resolution (especially in the depth dimension) of the imaging-based analytical techniques. Therefore, outcomes of fluorescence-based quantitative analysis should be interpreted with caution, and preferably referred to as qualitative and not quantitative measurements.

Part of the above-described problems with investigation and analysis of nano-DDSs disposition can be solved using mathematical (*in silico*) models. To this end, the major factors that govern the nano-DDS disposition (e.g., on the organ/tissue level, see Table 14.3) and the resulting pharmacological effects are encoded in the form of mathematical equations (see the review of such models for analysis of intratumoral nano-DDSs disposition in (Popilski and Stepensky 2015)). Subsequent investigation using specialized software packages allows to describe and analyze the experimental data (e.g., the extravasation and convection contribution to the local nano-DDS disposition), and to predict the impact of the individual factors (e.g., increased expression of the efflux pumps or metabolizing enzymes) on nano-DDS disposition and effectiveness. It is expected that the imaging and the mathematical modeling approaches will be increasingly used for analysis of the nano-DDSs disposition, and that they will contribute in the future to design of nano-DDSs with more efficient targeting efficiency on the systemic, organ/tissue, and intracellular levels.

14.6 Conclusion

Drugs can be encapsulated into nano-DDSs to mask their physicochemical properties, alter their pharmacokinetics and pharmacodynamics, and enhance treatment efficiency. Nano-DDSs are exposed to complex pathways of disposition in the different locations in the body, including the central circulation, organs/tissues/body fluids and inside the cells. More detailed and mechanistic understanding of the nano-DDSs disposition pathways and their modulation by the nano-DDSs formulation properties is needed. More advanced analytical methods should be used for this purpose for precise and robust quantification of the different forms of the nano-DDSs and of the released drugs at these locations. These methods, combined with the mathematical modeling techniques, will allow rational design of nano-DDSs for the purpose of targeted drug delivery to the intended site of action (on the tissue, cell, and intracellular levels), and their efficient translation for different therapeutic applications.

Acknowledgements We want to thank our collaborators in the bolavesicle project: Prof. Eli Heldman, Dr. Sarina Grinberg, and Prof. Charles Linder from the Ben-Gurion University of the Negev, Beer-Sheva, Israel. The experimental results presented in this chapter were generated with support of research grants from the Israel Science Foundation, the USA-Israel Binational Science Foundation, Prof. Yannai Tabb Cancer Research Foundation, and the Planning and Budgeting Committee of the Israeli Council for Higher Education for Outstanding Post-doctoral Fellows from China and India.

References

Amigorena S, Savina A (2010) Intracellular mechanisms of antigen cross presentation in dendritic cells. Curr Opin Immunol 22:109–117

Andersson H, Kappeler F, Hauri HP (1999) Protein targeting to endoplasmic reticulum by dilysine signals involves direct retention in addition to retrieval. J Biol Chem 274:15080–15084

Azzopardi EA, Ferguson EL, Thomas DW (2013) The enhanced permeability retention effect: a new paradigm for drug targeting in infection. J Antimicrob Chemother 68:257–274

Babu A, Templeton AK, Munshi A et al (2014) Nanodrug delivery systems: a promising technology for detection, diagnosis, and treatment of cancer. AAPS PharmSciTech 15:709–721

Bachmann MF, Jennings GT (2010) Vaccine delivery: a matter of size, geometry, kinetics and molecular patterns. Nat Rev Immunol 10:787–796

Barenholz Y (2012) Doxil – the first FDA-approved nano-drug: lessons learned. J Control Release 160:117–134

Bigdeli A, Palchetti S, Pozzi D et al (2016) Exploring cellular interactions of liposomes using protein corona fingerprints and physicochemical properties. ACS Nano 10:3723–3737

Boyd BJ, Galle A, Daglas M et al (2015) Traumatic brain injury opens blood-brain barrier to stealth liposomes via an enhanced permeability and retention (EPR)-like effect. J Drug Target 23:847–853

Breunig M, Bauer S, Goepferich A (2008) Polymers and nanoparticles: intelligent tools for intracellular targeting? Eur J Pharm Biopharm 68:112–128

Dakwar GR, Abu Hammad I, Popov M et al (2012) Delivery of proteins to the brain by bolaamphiphilic nano-sized vesicles. J Control Release 160:315–321

Danhier F, Feron O, Preat V (2010) To exploit the tumor microenvironment: passive and active tumor targeting of nanocarriers for anti-cancer drug delivery. J Control Release 148:135–146

Danquah MK, Zhang XA, Mahato RI (2011) Extravasation of polymeric nanomedicines across tumor vasculature. Adv Drug Deliv Rev 63:623–639

Das A, Adhikari C, Nayak D et al (2016) First evidence of the liposome-mediated deintercalation of anticancer drug doxorubicin from the drug-DNA complex: a spectroscopic approach. Langmuir 32:159–170

Decuzzi P, Godin B, Tanaka T et al (2010) Size and shape effects in the biodistribution of intravascularly injected particles. J Control Release 141:320–327

D'Souza GG, Weissig V (2009) Subcellular targeting: a new frontier for drug-loaded pharmaceutical nanocarriers and the concept of the magic bullet. Expert Opin Drug Deliv 6:1135–1148

Dufort S, Sancey L, Coll J-L (2012) Physico-chemical parameters that govern nanoparticles fate also dictate rules for their molecular evolution. Adv Drug Deliv Rev 64:179–189

Duncan R, Richardson SC (2012) Endocytosis and intracellular trafficking as gateways for nanomedicine delivery: opportunities and challenges. Mol Pharm 9:2380–2402

Ernsting MJ, Murakami M, Roy A et al (2013) Factors controlling the pharmacokinetics, biodistribution and intratumoral penetration of nanoparticles. J Control Release 172:782–794

Foged C, Brodin B, Frokjaer S et al (2005) Particle size and surface charge affect particle uptake by human dendritic cells in an in vitro model. Int J Pharm 298:315–322

Fu A, Tang R, Hardie J et al (2014) Promises and pitfalls of intracellular delivery of proteins. Bioconjug Chem 25:1602–1608

Gilleron J, Querbes W, Zeigerer A et al (2013) Image-based analysis of lipid nanoparticle-mediated siRNA delivery, intracellular trafficking and endosomal escape. Nat Biotechnol 31:638–646

Grinberg S, Linder C, Kolot V et al (2005) Novel cationic amphiphilic derivatives from vernonia oil: synthesis and self-aggregation into bilayer vesicles, nanoparticles, and DNA complexants. Langmuir 21:7638–7645

Grinberg S, Kolot V, Linder C et al (2008) Synthesis of novel cationic bolaamphiphiles from vernonia oil and their aggregated structures. Chem Phys Lipids 153:85–97

Gupta K, Afonin KA, Viard M et al (2015) Bolaamphiphiles as carriers for siRNA delivery: from chemical syntheses to practical applications. J Control Release 213:142–151

Holzel M, Bovier A, Tuting T (2013) Plasticity of tumour and immune cells: a source of heterogeneity and a cause for therapy resistance? Nat Rev Cancer 13:365–376
Ishida T, Kiwada H (2008) Accelerated blood clearance (ABC) phenomenon upon repeated injection of PEGylated liposomes. Int J Pharm 354:56–62
Kanapathipillai M, Brock A, Ingber DE (2014) Nanoparticle targeting of anti-cancer drugs that alter intracellular signaling or influence the tumor microenvironment. Adv Drug Deliv Rev 79–80:107–118
Kaplun V, Stepensky D (2014) Efficient decoration of nanoparticles intended for intracellular drug targeting with targeting residues, as revealed by a new indirect analytical approach. Mol Pharm 11:2906–2914
Keck CM, Muller RH (2013) Nanotoxicological classification system (NCS) – a guide for the risk-benefit assessment of nanoparticulate drug delivery systems. Eur J Pharm Biopharm 84:445–448
Kozlovskaya L, Stepensky D (2013) Quantitative analysis of the brain-targeted delivery of drugs and model compounds using nano-delivery systems. J Control Release 171:17–23
Kozlovskaya L, Abou-Kaoud M, Stepensky D (2014) Quantitative analysis of drug delivery to the brain via nasal route. J Control Release 189:133–140
Lammers T, Hennink WE, Storm G (2008) Tumour-targeted nanomedicines: principles and practice. Br J Cancer 99:392–397
Lammers T, Kiessling F, Hennink WE et al (2012) Drug targeting to tumors: principles, pitfalls and (pre-) clinical progress. J Control Release 161:175–187
Law PY, Wong YH, Loh HH (2000) Molecular mechanisms and regulation of opioid receptor signaling. Annu Rev Pharmacol Toxicol 40:389–430
Li Y, Wang J, Wientjes MG et al (2012) Delivery of nanomedicines to extracellular and intracellular compartments of a solid tumor. Adv Drug Deliv Rev 64:29–39
Liu Y, Tseng Y-C, Huang L (2012) Biodistribution studies of nanoparticles using fluorescence imaging: a qualitative or quantitative method? Pharm Res 29:3273–3277
Longmire M, Choyke PL, Kobayashi H (2008) Clearance properties of nano-sized particles and molecules as imaging agents: considerations and caveats. Nanomedicine (Lond) 3:703–717
Maeda H (2012) Macromolecular therapeutics in cancer treatment: the EPR effect and beyond. J Control Release 164:138–144
Maeda H, Nakamura H, Fang J (2013) The EPR effect for macromolecular drug delivery to solid tumors: improvement of tumor uptake, lowering of systemic toxicity, and distinct tumor imaging in vivo. Adv Drug Deliv Rev 65:71–79
Maity AR, Stepensky D (2015) Delivery of drugs to intracellular organelles using drug delivery systems: analysis of research trends and targeting efficiencies. Int J Pharm 496:268–274
Maity AR, Stepensky D (2016a) Efficient subcellular targeting to the cell nucleus of quantum dots densely decorated with a nuclear localization sequence peptide. ACS Appl Mater Interfaces 8:2001–2009
Maity AR, Stepensky D (2016b) Limited efficiency of drug delivery to specific intracellular organelles using subcellularly "targeted" drug delivery systems. Mol Pharm 13:1–7
Mantegazza AR, Magalhaes JG, Amigorena S et al (2013) Presentation of phagocytosed antigens by MHC class I and II. Traffic 14:135–152
Minchinton AI, Tannock IF (2006) Drug penetration in solid tumours. Nat Rev Cancer 6:583–592
Moghimi SM, Hunter AC, Andresen TL (2012) Factors controlling nanoparticle pharmacokinetics: an integrated analysis and perspective. Annu Rev Pharmacol Toxicol 52:481–503
Naahidi S, Jafari M, Edalat F et al (2013) Biocompatibility of engineered nanoparticles for drug delivery. J Control Release 166:182–194
Nichols JW, Bae YH (2012) Odyssey of a cancer nanoparticle: from injection site to site of action. Nano Today 7:606–618
Nichols JW, Bae YH (2014) EPR: evidence and fallacy. J Control Release 190:451–464

North AJ (2006) Seeing is believing? A beginners' guide to practical pitfalls in image acquisition. J Cell Biol 172:9–18

Paillard A, Hindre F, Vignes-Colombeix C et al (2010) The importance of endo-lysosomal escape with lipid nanocapsules for drug subcellular bioavailability. Biomaterials 31:7542–7554

Palchetti S, Colapicchioni V, Digiacomo L et al (2016) The protein corona of circulating PEGylated liposomes. Biochim Biophys Acta 1858:189–196

Parveen S, Sahoo SK (2006) Nanomedicine: clinical applications of polyethylene glycol conjugated proteins and drugs. Clin Pharmacokinet 45:965–988

Perry JL, Reuter KG, Kai MP et al (2012) PEGylated PRINT nanoparticles: the impact of PEG density on protein binding, macrophage association, biodistribution, and pharmacokinetics. Nano Lett 12:5304–5310

Philosof-Mazor L, Dakwar GR, Popov M et al (2013) Bolaamphiphilic vesicles encapsulating iron oxide nanoparticles: new vehicles for magnetically targeted drug delivery. Int J Pharm 450:241–249

Popilski H, Stepensky D (2015) Mathematical modeling analysis of intratumoral disposition of anticancer agents and drug delivery systems. Expert Opin Drug Metab Toxicol 11:767–784

Popov M, Linder C, Deckelbaum RJ et al (2010) Cationic vesicles from novel bolaamphiphilic compounds. J Liposome Res 20:147–159

Popov M, Abu Hammad I, Bachar T et al (2013) Delivery of analgesic peptides to the brain by nano-sized bolaamphiphilic vesicles made of monolayer membranes. Eur J Pharm Biopharm 85:381–389

Rochat B (2009) Importance of influx and efflux systems and xenobiotic metabolizing enzymes in intratumoral disposition of anticancer agents. Curr Cancer Drug Targets 9:652–674

Ruenraroengsak P, Cook JM, Florence AT (2010) Nanosystem drug targeting: facing up to complex realities. J Control Release 141:265–276

Schellekens H, Hennink WE, Brinks V (2013) The immunogenicity of polyethylene glycol: facts and fiction. Pharm Res 30:1729–1734

Segura E, Villadangos JA (2011) A modular and combinatorial view of the antigen cross-presentation pathway in dendritic cells. Traffic 12:1677–1685

Seib FP, Jones AT, Duncan R (2006) Establishment of subcellular fractionation techniques to monitor the intracellular fate of polymer therapeutics I. Differential centrifugation fractionation B16F10 cells and use to study the intracellular fate of HPMA copolymer – doxorubicin. J Drug Target 14:375–390

Shechter Y, Heldman E, Sasson K et al (2010) Delivery of neuropeptides from the periphery to the brain: studies with enkephalin. ACS Chem Neurosci 1:399–406

Sneh-Edri H, Stepensky D (2011) 'IntraCell' plugin for assessment of intracellular localization of nano-delivery systems and their targeting to the individual organelles. Biochem Biophys Res Commun 405:228–233

Sneh-Edri H, Likhtenshtein D, Stepensky D (2011) Intracellular targeting of PLGA nanoparticles encapsulating antigenic peptide to the endoplasmic reticulum of dendritic cells and its effect on antigen cross-presentation in vitro. Mol Pharm 8:1266–1275

Snyder SH, Pasternak GW (2003) Historical review: opioid receptors. Trends Pharmacol Sci 24:198–205

Steichen SD, Caldorera-Moore M, Peppas NA (2013) A review of current nanoparticle and targeting moieties for the delivery of cancer therapeutics. Eur J Pharm Sci 48:416–427

Stepensky D (2010) Quantitative aspects of intracellularly-targeted drug delivery. Pharm Res 27:2776–2780

Stepensky D (2014) Pharmacokinetic and pharmacodynamic aspects of focal and targeted delivery of drugs. In: Domb AJ, Khan W (eds) Focal controlled drug delivery. Springer, New York

Taurin S, Nehoff H, Greish K (2012) Anticancer nanomedicine and tumor vascular permeability; where is the missing link? J Control Release 164:265–275

Teasdale RD, Jackson MR (1996) Signal-mediated sorting of membrane proteins between the endoplasmic reticulum and the golgi apparatus. Annu Rev Cell Dev Biol 12:27–54

Torchilin VP (2005) Fluorescence microscopy to follow the targeting of liposomes and micelles to cells and their intracellular fate. Adv Drug Deliv Rev 57:95–109

Torchilin VP (2006) Recent approaches to intracellular delivery of drugs and DNA and organelle targeting. Annu Rev Biomed Eng 8:343–375

van der Meel R, Vehmeijer LJ, Kok RJ et al (2013) Ligand-targeted particulate nanomedicines undergoing clinical evaluation: current status. Adv Drug Deliv Rev 65:1284–1298

Waters JC (2009) Accuracy and precision in quantitative fluorescence microscopy. J Cell Biol 185:1135–1148

Weissig V, Guzman-Villanueva D (2015) Nanopharmaceuticals (part 2): products in the pipeline. Int J Nanomedicine 10:1245–1257

Weissig V, Pettinger TK, Murdock N (2014) Nanopharmaceuticals (part 1): products on the market. Int J Nanomedicine 9:4357–4373

Wicki A, Witzigmann D, Balasubramanian V et al (2015) Nanomedicine in cancer therapy: challenges, opportunities, and clinical applications. J Control Release 200:138–157

Yu MK, Park J, Jon S (2012) Targeting strategies for multifunctional nanoparticles in cancer imaging and therapy. Theranostics 2:3–44

Zhang L, Wang Y, Cao Y et al (2013) Transport barriers and strategies of antitumor nanocarriers delivery system. J Biomed Mater Res A 101:3661–3669

Chapter 15
PBPK Modelling of Intracellular Drug Delivery Through Active and Passive Transport Processes

Lars Kuepfer, Christoph Niederalt, Thomas Wendl, Jan-Frederik Schlender, Michael Block, Thomas Eissing, and Donato Teutonico

Abstract Physiologically based pharmacokinetic (PBPK) models describe adsorption, distribution, metabolisation and excretion (ADME) of drugs in the body of an organism based on a large amount of prior anatomical and physiological knowledge. In contrast to compartmental pharmacokinetic modeling which uses rather empirical model structures, PBPK models aim for a detailed mechanistic representation of physiological processes underlying drug pharmacokinetics within the body. That means that the relevant organs or tissues of an organism are explicitly represented in a PBPK model. Organs in PBPK models are usually divided in sub-compartments such as plasma, interstitial space, intracellular space and red blood cells. Mass transfer between the different compartments which ultimately governs intracellular drug delivery is quantified either by so-called distribution models for the calculation of organ-plasma partition coefficients or by permeability-surface area products quantifying passive diffusion, respectively. Notably, both types of calculation methods are parameterized based upon the physicochemical properties of the investigated drug, respectively. These properties include amongst others lipophilicity and the molecular weight of the compound. Additional physiological information ranging from the whole body level (e.g. organ volumes, blood flow rates, tissue composition) to relative tissue-specific protein abundance is explicitly provided in the model. PBPK models are nowadays routinely used to analyze pharmacokinetics in drug development Due to the large amount of mechanistic information which is implicitly provided in the structural equations, PBPK models are in particular well-suited for both in-depth analyses of ADME processes underlying drug pharmacokinetics as well as for extrapolation to novel indications, patient populations or treatment designs. In this review we will present and discuss calculation methods used in PBPK model to describe and to quantify intracellular drug delivery.

L. Kuepfer (✉) • C. Niederalt • T. Wendl • J-F. Schlender • M. Block • T. Eissing
D. Teutonico
Complex Systems Modelling, Bayer Technology Service GmbH,
51368 Leverkusen, Germany
e-mail: lars.kuepfer@bayer.com

A. Prokop, V. Weissig (eds.), *Intracellular Delivery III*, Fundamental Biomedical Technologies 8, DOI 10.1007/978-3-319-43525-1_15

Keywords Physiologically-based pharmacokinetic modelling • PBPK • Cross-species extrapolation • Drug ADME • Pharmacokinetic modelling

Abbreviations

ABCB11	ATP-binding cassette sub-family B member 11
ADME	Adsorption distribution, metabolisation and excretion
ABCG2	ATP-binding cassette sub-family G member 2
P×SA	Permeability surface area products
PBPK	Physiologically-based pharmacokinetic modelling
PD	Pharmacodynamic
PK	Pharmacokinetic
Vmax	catalytic activity [μmol/l/min]
kcat	catalytic efficiency [1/min] and the total concentration of the catalyzing protein
E_0 $\mathrm{E_0}$	total concentration of the catalyzing protein [μmol/l] at the organism level
$e_{rel,j}$	relative protein expression in in tissue *j*
SLC22A8	Solute carrier family 22 member 8e carrier family 22 member 8
OATP1B3	organic anion-transporting polypeptide 1B3

15.1 An Introduction to Physiologically Based Pharmacokinetic Modelling

In the last decades, pharmacokinetic (PK) compartmental models have been widely used as guidance during the different steps of the drug development process. In such models, compartments do not have an explicit physiological meaning but represent simply volumes with homogenous composition. Parameters as well are not necessarily related to any physical or biochemical meaning, but they can be transformed to provide clinically relevant parameters such as clearance or volume of distribution.

Alternatively, physiologically-based pharmacokinetic (PBPK) modelling aims for a detailed mechanistic representation of human or animal physiology at the whole-body level, in which prior information available on the biological system and on the drug are combined together to predict *a priori* the expected PK. Concerning the parameters included in the model, on the one hand PBPK models are based on large-scale collections of physiological parameters such as organ volumes or blood flows which build up the physiological database or organism properties for each species like mouse, rat, dog or human. On the other hand, physicochemical properties of the drug such as lipophilicity and molecular weight are used to parametrize the distribution model describing the steady state tissue concentrations as well as the corresponding permeation rates. Hence, even though PBPK models may

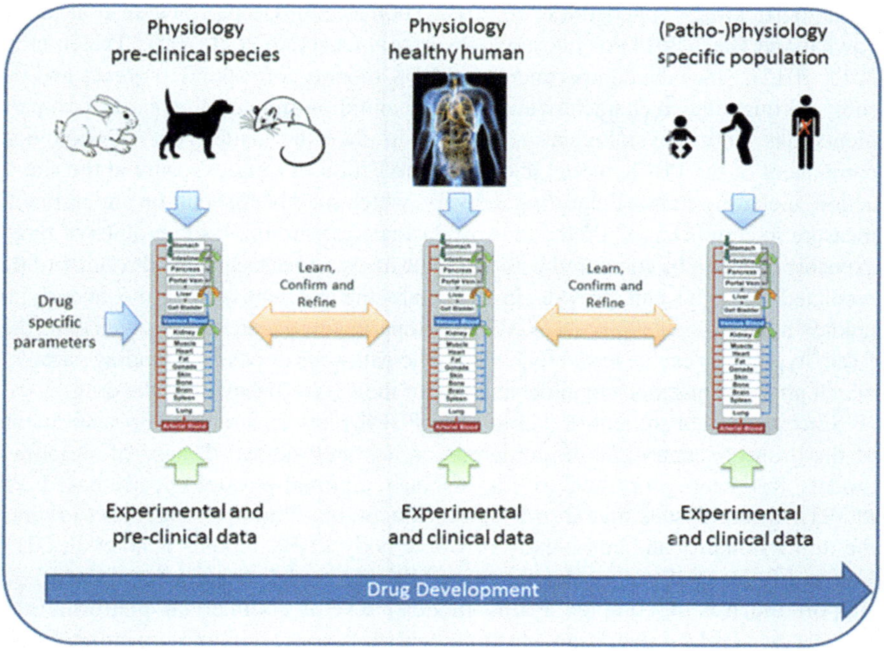

Fig. 15.1 Generic workflow for the establishment of a PBPK modelling along the drug development process. The drug specific parameters are used as input and combined with species-specific parametrization describing the organism physiology. Such predictions are then compared with experimental data and additional compound-specific parameters, e.g. on measured or assumed biochemical processes, can then be included in the model to explain possible deviations from experimental data. Once established, the model can also be scaled from preclinical species to human as well as from healthy to specific populations, such as diseased patients, or pediatric and elderly senior populations, providing insight and support along the drug development process. At the different stages, model predictions can be performed as individual simulations, for instance considering average physiological values, as well as population simulations, taking also into account assumed physiological the variability associated to the population in the predictions

comprise several hundreds of ordinary differential equations, the number of independent model parameters is usually small (less than 5 per compound in most of the cases), due to the large degree of prior physiological information contained in the models.

A generic whole-body PBPK model (Fig. 15.1) can be built by combining individual physiological models of the muscle, fat, skin, brain, liver, kidney and other tissues relevant for the drug distribution (Peters 2012). These models usually contain the most relevant organs and tissues from the physiological and pharmacological point of view, i.e. heart, lung, brain, stomach, spleen, pancreas, gut, liver, kidney, thymus, adipose, muscle, bone and skin. Each organ is characterized by an associated blood flow rate, organ volume, permeability and tissue partition coefficient and they are linked together by arterial and venous blood compartments. In some cases, if required, a more detailed mechanistic description of some organs may also be included in the model, for instance to describe brain distribution (Liu et al. 2005), liver (Hoehme et al. 2010), kidney (Niederalt et al. 2012; Thomas 2009), lung (Tawhai et al. 2011; Tawhai and Bates 2011) or intestinal absorption (Agoram et al. 2001; Thelen et al. 2011, 2012). Since the compartments in PBPK models correspond to organs and tissues, the quantitative characterization of concentration-time profiles in the compartments has high physiological relevance. In fact the underlying physiological component of the PBPK model allows the prediction of drug exposure at the site of action, including intracellular drug delivery, which may be difficult or impossible to measure in certain cases. Predictions of plasma concentration-time profiles or target exposure can then be compared with available measured data and any deviation of the predicted outcome can provide insights into the mechanisms of the underlying unknown physiological processes. When comparing model predictions with data, different hypothesis can be tested (e.g. metabolic pathways or possible binding partners) which provides insights and understanding of the *in vivo* behavior of the drug.

Since the establishment of whole body PBPK platforms requires measurements of the drug concentrations in numerous organs and tissues, the use of simplified models has been proposed, as for instance minimal-physiologically-based PK models, where tissues that shows similar kinetic are "lumped" together to reduce the dimensionality and complexity of whole body PBPK models (Cao et al. 2013; Cao and Jusko 2012) or mechanism-based PK-PD models (Ploeger et al. 2009). To support the use of complex PBPK models, several commercial platforms also became available, integrating physiological databases and implementing PBPK modeling approaches, such as GastroPlus™ (Simulations Plus, Lancaster, USA), SimCyp® (SimCyp, Sheffield, UK) and PK-Sim® and MOBI® (Bayer Technology Services, Leverkusen, Germany).

These PBPK modelling platforms provide a generic model structure and parameterization for the physiology of predefined populations. The PBPK platforms then combine such physiological data with compound specific information as input and translate then such information to a full-blown model. The general structure of a typical PBPK model is composed of different type of information, which can be divided into three different building blocks: organism properties, drug properties and study protocol and formulation properties.

The *organism properties* include all the information related to the living organism considered in the model, i.e. mostly *physiological and anatomical parameters*. Examples of organism properties are organ volumes, organ composition, blood flows or surface areas. Such properties are related to the species considered as well as to the specific condition of the organism. For instance in the case of special populations, these organism properties would take into account the physiological differences of such populations (e.g. diseased or pediatric populations) with respect to the healthy adult population.

Drug properties represent the properties of the studied compounds, and these properties can in principle be defined at the beginning of the drug development process and carry-forward along the different preclinical and clinical development programs. Drug properties include the *physico-chemical properties* (such as compound lipophilicity, solubility or molecular weight), which are fully independent from the organism physiology, as well as *drug biological properties* (such as fraction unbound, or tissue-plasma partition coefficient) which are also drug specific properties but they are defined by an interaction between the drug and the organism, and as consequence they also depend on the organism considered (e.g. animal species or special population).

By combining the drug properties with the organism properties, it is possible to estimate the passive processes involved in the distribution of the drug in the body. Examples of such processes are the diffusion processes dependent on the physicochemical properties of the drug. On the other side, active processes, as for instance metabolic processes, active binding or active transporters have also a physiological location within the organism but they are heavily dependent on the specific drug, and for this reason they need to be specified on a drug-by-drug basis. Passive and active processes will be discussed in more details in the next sections.

Finally, information on the *study protocol* and *formulation properties* are needed to define a PBPK simulation. Considering typical applications of a PBPK modelling exercise, where the active ingredient (the drug) is already selected, these last components of the model contains most of the aspects which can be actually optimized during a preclinical and clinical development program. For instance typical application of a PBPK model for a single drug could be the optimization of the formulation, the evaluation of different routes of administration or the prediction of PK concentration after different doses or different administration protocols.

15.2 Intracellular Drug Delivery in PBPK Modelling

Organs in PBPK are generally differentiated into four subcompartments representing the blood plasma, red blood cells, interstitium and the intracellular space (Fig. 15.2). Mass transfer between the different compartments is generally enabled by passive transport though diffusion or active protein-mediated transport. In the following the various processes underlying drug distribution in PBPK modelling are discussed.

15.2.1 Organ-Plasma Partitioning

Partitioning between blood plasma and the surrounding tissue largely governs drug distribution within the body. The quantification of these processes, however, is experimentally very labor-intensive (Blakey et al. 1997; Jones et al. 2009). In PBPK modelling, so-called distribution models allow the calculation of organ/plasma partition coefficient to estimate the concentration ratio between the blood plasma and the adjacent tissue. Their calculation from easily accessible *in vitro* parameters such as lipophilicity or plasma protein binding is a major improvement of PBPK modelling. Different concepts for mechanistic correlations are nowadays available for the *in silico* prediction of organ/plasma partition coefficients. Binding to proteins and

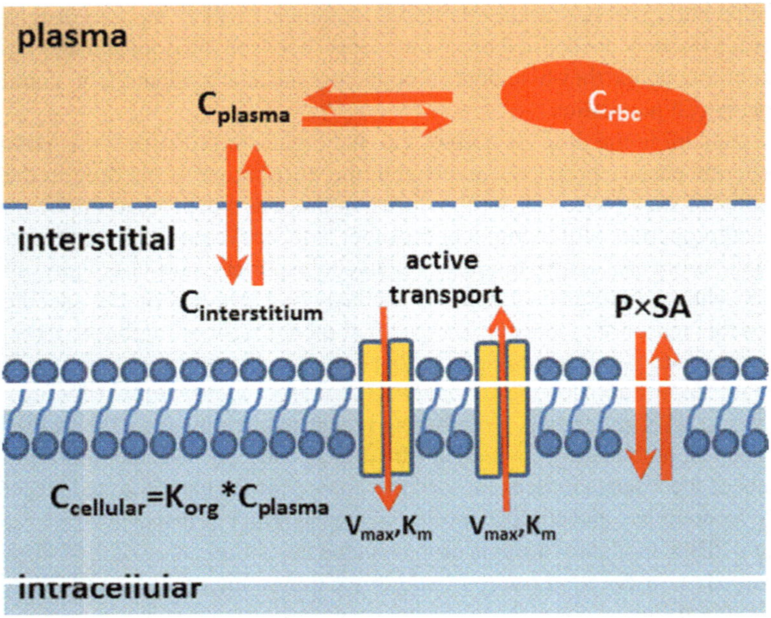

Fig. 15.2 Transport processes in PBPK modelling governing intracellular drug delivery adapted from (Willmann et al. 2003). Organs are generally divided into plasma, red blood cells (*rbc*) and interstitial and cellular space, respectively. Organ/plasma partition coefficients K_{org}, together with permeation surface area (P×SA) products determine passive distribution within each organ. Notably, both parameters can be calculated from physicochemical properties of a compound. In addition, active drug transport contributes to uptake and secretion of drugs into and out of the cell

lipids as the two major tissue constituents is generally taken into account. The calculation methods deviate with respect to the kind of parameters used for the calculations. All known partition coefficient models include partition coefficients for water/protein and water/lipid and they all require specification of tissue composition. The overall organ/plasma partition component is calculated as a weighted sum of the partition coefficients multiplied by the volume fractions of each tissue component. The first distribution model was developed by Poulin and co-workers (Poulin et al. 2001; Poulin and Theil 2000) who assumed that the lipo-hydrophilicity of organic tissue is calculated as a mixture of neutral lipids and phospholipds, respectively. Notably, drug binding is unspecific to neutral lipids yet specific (fu_p) for phospholipids. Poulin et al. (2001) suggest olive oil as a surrogate for lipids in adipose tissue, and n-octanol in all other organs. Additionally, Poulin et al. use volumetric tissue composition, and drug pKa for calculation of organ-plasma partition coefficients. Rodgers and co-workers extended the concepts of Poulin et al. to the binding of ionized compounds to acidic phospholipids at physiological pH (Rodgers et al. 2005; Rodgers and Rowland 2006). Their model explicitly considers partitioning of unionized drugs between neutral lipids and neutral phospholipids while ionized fraction of a drug interacts with acidic tissue phospholipids. Berezhkovskiy

(Berezhkovskiy 2004) proposed another extension of the concepts by Poulin et al., accounting for peripheral drug elimination. Likewise, Willmann and coworkers (Willmann et al. 2005) extended the concept by Poulin et al. by the consideration of proteins as tissue components. Their model for the calculation of plasma partition coefficients use membrane affinity (Loidl-Stahlhofen et al. 2001) to quantify partitioning between water and an artificial cellular membrane. Finally, Schmitt (2008) calculated the organ/plasma partition coefficient for organic compounds, by further compartmentalizing tissue as water, neutral lipids, neutral phospholipids, acidic phospholipids and proteins. It should be noted that the different calculation methods for organ/plasma partition constituents lead to structurally different model equations, although they are conceptually very similar. An iterative testing of different distribution models during model development model hence provides a powerful tool to improve the predictive accuracy of a specific PBPK model.

15.2.2 Passive Transport: Permeability-Surface Area Products

In addition to the organ-plasma partition coefficients, permeability surface area (P×SA) products are the other key parameter determining passive diffusion for each organ which largely governs intracellular drug delivery. P×SA products quantify the permeation across the cell membranes (interstitial-cell barrier) and are composed out of a compound specific term (permeability) and a species- or physiology-specific term (surface area). P×SA products have been determined by estimating the organ specific kinetics of a cyclosporin A derivative by means of a rat PBPK model (Kawai et al. 1994).

Scaling of permeability can be obtained from the physicochemical properties of the drug through proprietary calculation methods (Willmann et al. 2003). In general, the permeability P is considered to be proportional to the drugs' permeability through the phospholipid bilayer as such representing cellular membranes. Small or lipophilic drugs pass more easily through the endothelial wall than larger or hydrophilic drugs. All lipid membranes in the body are usually assumed to have the same permeability for a given substance in PBPK modelling.

15.2.3 Active Transport: First Order Kinetics and Michaelis-Menten Kinetics

In addition to passive diffusion across membranes, drugs can also be imported or secreted by active drug transporters. In PBPK modelling it is possible to quantify the relative abundance of transporters at the whole-body level by using relative tissue-specific gene or protein expression as a surrogate for protein abundance (Meyer et al. 2012). In brief, the catalytic activity of each active process is quantified by *Vmax* (μmol/l/min), which is in turn the product of the catalytic efficiency *kcat* (1/min) and the total concentration of the catalyzing protein E_0 (μmol/l) at the organism level:

$$Vmax = kcat * E_0$$

The relative protein expression $e_{rel,j}$ can be used to estimate the catalytic activity $Vmax_j$ of a specific protein in tissue *j*:

$$Vmax_j = kcat * E_{0,j} = kcat * E_0 * e_{rel,j} = kcat^* * e_{rel,j}$$

Note that *kcat** is a global model parameter which needs to be adjusted during model development, whereas the $e_{rel,j}$ can be obtained from different expression databases (Meyer et al. 2012). For example, tissue-specific expression of OATP1B1, OAT3 or MRP1 has been taken into account in PBPK models for pravastatin or simvastatin (Lippert et al. 2012; Meyer et al. 2012). In these studies, carefully validated models have been used to analyze the effect of a single nucleotide polymorphism in *SLCO1B1*, a gene encoding for the hepatic uptake transporter OATP1B1, on statin pharmacokinetics. In particular, it could be shown that plasma concentration profiles in heterozygous subgroups of patients could be predicted from OATP1B1 transporter activities identified in both homozygous subgroups of patients (Lippert et al. 2012).

Given the availability of adequate measurements both transcriptomics or proteomics data can directly be incorporated in the PBPK model. Notably, relative protein abundance in different tissues is a physiological property which is highly specific for the individual organism. This has important implications for extrapolation between different patient subgroups (Maharaj and Edginton 2014) or even species (Thiel et al. 2015), thereby strongly supporting one of the key applications of PBPK modelling (Fig. 15.1).

15.3 Cross-Species Extrapolation

In a recent study analyzing feasibility of cross-species extrapolation for ten marketed drugs, it could be shown, that neither enzyme or transporter kinetics nor tissue-specific expression profiles alone support a significant improvement in PBPK-based PK simulations (Thiel et al. 2015). It is only through the simultaneous consideration of kinetics and expression data, that an adequate agreement between experimentally measured drug plasma profile and simulated time-concentration curves could be achieved (Thiel et al. 2015). This is not unexpected since transporter kinetics can be seen as a way of contextualizing expression data at the organism level. Notably, such expression profiles may be highly specific for each species and even more for each protein. For example, a comparison of relative and normalized gene-expression data for humans (Meyer et al. 2012) and mice (Thiel et al. 2015) shows only a modest agreement for the transporter gene *ABCB11* ($r=0.51$,

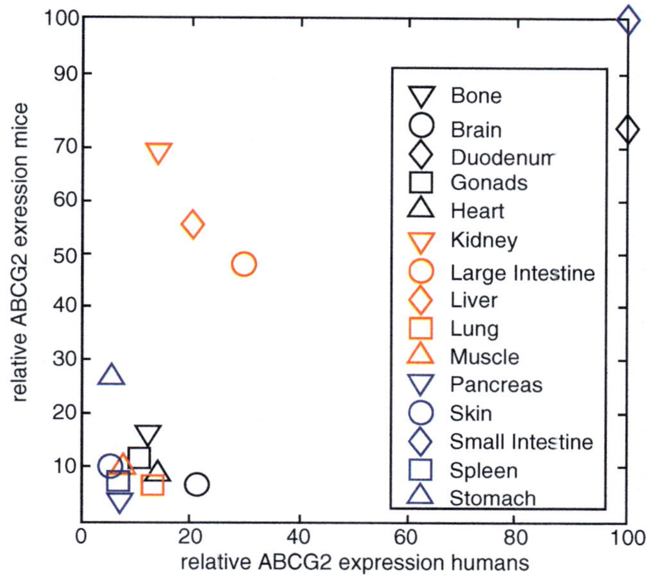

Fig. 15.3 Relative tissue-specific expression of the gene ABCG2 also known as BCRP (Breast Cancer Resistance Protein) in humans and mice (Meyer et al. 2012; Thiel et al. 2015). The expression is represented is normalized to the tissue or organ with the highest expression

$p<0.05$; ATP-binding cassette, sub-family B member 11) but in turn a high agreement for the transporter genes *SLC22A8* ($r=0.99$, $p<1e\text{-}5$; solute carrier family 22 member 8e carrier family 22 member 8), *OATP1B3* ($r=0.93$, $p<1e\text{-}5$; organic anion-transporting polypeptide 1B3) and *ABCG2* ($p=0.7$, $r<1e\text{-}3$; ATP-binding cassette sub-family G member 2) (Fig. 15.3). These examples show that due to an unforeseeable number and functional equivalence of orthologues species-specific genetics rather than physiology may be a limiting factor in PBPK-based cross-species extrapolation.

15.4 Discussion

PBPK models are nowadays routinely used in pharmaceutical development and their application is increasingly supported by regulatory agencies (Jones and Rowland-Yeo 2013; Jones et al. 2009; Sager et al. 2015; Zhao et al. 2012). A major advance in PBPK is the possibility to calculate model parameters from physicochemical drug parameters such as molecular weight, fraction unbound or lipophilicity, which are otherwise difficult to determine experimentally. Due to the various calculation methods available in PBPK software packages, the parametrization of PBPK models is hence comparatively easy, even though such models may comprise hundreds of parameters and an equally large systems of ordinary differential equations (Eissing et al. 2011). Calculation of organ-plasma partition coefficients and of permeability surface area (P × SA) products are the two key calculation methods in PBPK modelling (Fig. 15.2). They allow in particular to calculate passive drug distribution in organs. In addition to the passive processes, active drug transport also contributes to intracellular drug delivery. The relative activity for each organ can be quantified by using omics measurements such as gene expression data as a surrogate for estimation of tissue-specific protein abundance (Meyer et al. 2012). Organs in PBPK modelling are usually subdivided in plasma, red blood cells, interstitium or cellular space (Fig. 15.2). Diffusion from plasma to interstitial space across the endothelial wall or in turn diffusion across cellular membranes from the interstitium to the cellular space determines how fast drug distribution occurs. Notably, this detailed representation of organ physiology enables the quantification or identification of single steps during cellular drug delivery as such allowing the comprehensive differentiation between perfusion limited and permeability limited drug kinetics.

The large level of physiological detail which is implicitly included in the PBPK model structure and significantly supports mechanistic analyses of processes governing adsorption, distribution, metabolisation and excretion of a particular drug. Moreover, specific clinical subgroups such as cirrhotic patients (Edginton and Willmann 2008), different genotypes (Lippert et al. 2012) or age classes (Maharaj and Edginton 2014) can be explicitly considered. In addition to that, since the general model structure in PBPK models supports a rigorous differentiation between physicochemical properties of the drug on the one hand and physiological properties of the organism on the other, PBPK modelling also greatly supports extrapolation to different clinical cases from an established reference benchmark model.

An example for PBPK-based translation is cross-species extrapolation (Thiel et al. 2015): In a systematic consideration of species-specific physiology it could be shown that species-specific physiology is the most important information when using PBPK for the design of a first-in-man study. However, it also became obvious that ADME genetics may be difficult to translate across species due to an unknown number and function of gene orthologues (Fig. 15.3).

Pediatric extrapolation (Maharaj and Edginton 2014) and the model based design of pediatric investigation plans is another prominent application for PBPK-based translation. Here, switching between different age groups requires the collection and integration of age-specific physiological information which needs to be assembled for example from literature and public databases. The medical benefit of applying computational modelling in the design of clinical research programs, improving efficacy and safety of clinical trials, is so high that pediatric extrapolation will certainly become a prominent use case of PBPK modelling in the future. Here and in many other applications of PBPK modelling, the detailed description of the various physiological processes governing intracellular drug delivery provides the platform which enable further mechanistic analyses in pharmaceutical development programs.

References

Agoram B, Woltosz WS, Bolger MB (2001) Predicting the impact of physiological and biochemical processes on oral drug bioavailability. Adv Drug Deliv Rev 50(Suppl 1):S41–S67

Berezhkovskiy LM (2004) Volume of distribution at steady state for a linear pharmacokinetic system with peripheral elimination. J Pharm Sci 93(6):1628–1640. doi:10.1002/jps.20073

Blakey GE, Nestorov IA, Arundel PA, Aarons LJ, Rowland M (1997) Quantitative structure-pharmacokinetics relationships: I. Development of a whole-body physiologically based model to characterize changes in pharmacokinetics across a homologous series of barbiturates in the rat. J Pharmacokinet Biopharm 25(3):277–312

Cao Y, Jusko WJ (2012) Applications of minimal physiologically-based pharmacokinetic models. J Pharmacokinet Pharmacodyn 39(6):711–723. doi:10.1007/s10928-012-9280-2

Cao Y, Balthasar JP, Jusko WJ (2013) Second-generation minimal physiologically-based pharmacokinetic model for monoclonal antibodies. J Pharmacokinet Pharmacodyn 40(5):597–607. doi:10.1007/s10928-013-9332-2

Edginton AN, Willmann S (2008) Physiology-based simulations of a pathological condition: prediction of pharmacokinetics in patients with liver cirrhosis. Clin Pharmacokinet 47(11):743–752. doi:10.2165/00003088-200847110-00005

Eissing T, Kuepfer L, Becker C et al (2011) A computational systems biology software platform for multiscale modeling and simulation: integrating whole-body physiology, disease biology, and molecular reaction networks. Front Physiol 2:4. doi:10.3389/fphys.2011.00004

Hoehme S, Brulport M, Bauer A et al (2010) Prediction and validation of cell alignment along microvessels as order principle to restore tissue architecture in liver regeneration. Proc Natl Acad Sci U S A 107(23):10371–10376. doi:10.1073/pnas.0909374107

Jones H, Rowland-Yeo K (2013) Basic concepts in physiologically based pharmacokinetic modeling in drug discovery and development. CPT Pharmacometrics Syst Pharmacol 2, e63. doi:10.1038/psp.2013.41

Jones HM, Gardner IB, Watson KJ (2009) Modelling and PBPK simulation in drug discovery. AAPS J 11(1):155–166. doi:10.1208/s12248-009-9088-1

Kawai R, Lemaire M, Steimer JL, Bruelisauer A, Niederberger W, Rowland M (1994) Physiologically based pharmacokinetic study on a cyclosporin derivative, SDZ IMM 125. J Pharmacokinet Biopharm 22(5):327–365

Lippert J, Brosch M, von Kampen O et al (2012) A mechanistic, model-based approach to safety assessment in clinical development. CPT Pharmacometrics Syst Pharmacol 1, e13. doi:10.1038/psp.2012.14

Liu X, Smith BJ, Chen C et al (2005) Use of a physiologically based pharmacokinetic model to study the time to reach brain equilibrium: an experimental analysis of the role of blood–brain barrier permeability, plasma protein binding, and brain tissue binding. J Pharmacol Exp Ther 313(3):1254–1262. doi:10.1124/jpet.104.079319

Loidl-Stahlhofen A, Hartmann T, Schottner M et al (2001) Multilamellar liposomes and solid-supported lipid membranes (TRANSIL): screening of lipid-water partitioning toward a high-throughput scale. Pharm Res 18(12):1782–1788

Maharaj AR, Edginton AN (2014) Physiologically based pharmacokinetic modeling and simulation in pediatric drug development. CPT Pharmacometrics Syst Pharmacol 3, e150. doi:10.1038/psp.2014.45

Meyer M, Schneckener S, Ludewig B, Kuepfer L, Lippert J (2012) Using expression data for quantification of active processes in physiologically based pharmacokinetic modeling. Drug Metab Dispos 40(5):892–901. doi:10.1124/dmd.111.043174

Niederalt C, Wendl T, Kuepfer L et al (2012) Development of a physiologically based computational kidney model to describe the renal excretion of hydrophilic agents in rats. Front Physiol 3:494. doi:10.3389/fphys.2012.00494

Peters SA (2012) Physiologically-Based Pharmacokinetic (PBPK) modeling and simulations: principles, methods, and applications in the pharmaceutical industry. Wiley, Hoboken

Ploeger BA, van der Graaf PH, Danhof M (2009) Incorporating receptor theory in mechanism-based pharmacokinetic-pharmacodynamic (PK-PD) modeling. Drug Metab Pharmacokinet 24(1):3–15

Poulin P, Theil FP (2000) A priori prediction of tissue:plasma partition coefficients of drugs to facilitate the use of physiologically-based pharmacokinetic models in drug discovery. J Pharm Sci 89(1):16–35. doi:10.1002/(SICI)1520-6017(200001)89:1<16::AID-JPS3>3.0.CO;2-E

Poulin P, Schoenlein K, Theil FP (2001) Prediction of adipose tissue: plasma partition coefficients for structurally unrelated drugs. J Pharm Sci 90(4):436–447

Rodgers T, Rowland M (2006) Physiologically based pharmacokinetic modelling 2: predicting the tissue distribution of acids, very weak bases, neutrals and zwitterions. J Pharm Sci 95(6):1238–1257. doi:10.1002/jps.20502

Rodgers T, Leahy D, Rowland M (2005) Physiologically based pharmacokinetic modeling 1: predicting the tissue distribution of moderate-to-strong bases. J Pharm Sci 94(6):1259–1276. doi:10.1002/jps.20322

Sager JE, Yu J, Raguenau-Majlessi I, Isoherranen N (2015) Physiologically Based Pharmacokinetic (PBPK) modeling and simulation approaches: a systematic review of published models, applications and model verification. Drug Metab Dispos. doi:10.1124/dmd.115.065920

Schmitt W (2008) General approach for the calculation of tissue to plasma partition coefficients. Toxicol In Vitro 22(2):457–467. doi:10.1016/j.tiv.2007.09.010

Tawhai MH, Bates JH (2011) Multi-scale lung modeling. J Appl Physiol (1985) 110(5):1466–72 doi:10.1152/japplphysiol.01289.2010

Tawhai M, Clark A, Donovan G, Burrowes K (2011) Computational modeling of airway and pulmonary vascular structure and function: development of a "lung physiome". Crit Rev Biomed Eng 39(4):319–336

Thelen K, Coboeken K, Willmann S, Burghaus R, Dressman JB, Lippert J (2011) Evolution of a detailed physiological model to simulate the gastrointestinal transit and absorption process in humans, part 1: oral solutions. J Pharm Sci 100(12):5324–5345. doi:10.1002/jps.22726

Thelen K, Coboeken K, Willmann S, Dressman JB, Lippert J (2012) Evolution of a detailed physiological model to simulate the gastrointestinal transit and absorption process in humans, part II: extension to describe performance of solid dosage forms. J Pharm Sci 101(3):1267–1280. doi:10.1002/jps.22825

Thiel C, Schneckener S, Krauss M et al (2015) A systematic evaluation of the use of physiologically based pharmacokinetic modeling for cross-species extrapolation. J Pharm Sci 104(1):191–206. doi:10.1002/jps.24214

Thomas SR (2009) Kidney modeling and systems physiology. Wiley Interdiscip Rev Syst Biol Med 1(2):172–190. doi:10.1002/wsbm.14

Willmann S, Lippert J, Sevestre M, Solodenko J, Fois F, Schmitt W (2003) PK-Sim®: a physiologically based pharmacokinetic 'whole-body' model. BioSilico 1(4):121–124

Willmann S, Lippert J, Schmitt W (2005) From physicochemistry to absorption and distribution: predictive mechanistic modelling and computational tools. Expert Opin Drug Metab Toxicol 1(1):159–168. doi:10.1517/17425255.1.1.159

Zhao P, Rowland M, Huang SM (2012) Best practice in the use of physiologically based pharmacokinetic modeling and simulation to address clinical pharmacology regulatory questions. Clin Pharmacol Ther 92(1):17–20. doi:10.1038/clpt.2012.68

Chapter 16
Exploiting Nanocarriers for Combination Cancer Therapy

Yi Wen Kong*, **Erik C. Dreaden***, **Paula T. Hammond, and Michael B. Yaffe**

Abstract Combination chemotherapy has vastly improved patient outcomes following treatment for cancer. Combining multiple drugs with non-overlapping mechanisms of action has been shown to forestall the development of drug resistance, leading to increased efficacy. Emerging insights into cancer pathophysiology from tumor genomics, metabolomics, and proteomics now present us with unprecedented opportunities to combine targeted molecular therapies together, or to combine molecular therapies with cytotoxic chemotherapy in a rationally designed manner based on unique molecular signatures. However, the clinical implementation of these improved drug combinations is frequently limited by overlapping drug toxicities. By using new nanotechnology platforms to enhance tumor targeting, and provide precise spatial and temporal control of drug delivery for each agent within a multi-drug regimen, it should be possible to mitigate these toxicity limitations and treat tumors with increasing safety, efficacy and durability. This chapter discusses recent efforts in developing nanoparticles to deliver multiple types of drugs for temporally-sequenced concurrent or sequential combination chemotherapy.

Keywords Nanomedicine • Systems pharmacology • Nanotechnology • Chemotherapy • Rational drug combinations • Synergy • Synthetic lethality • Cancer treatment • Mechanism of action • Molecular targeting • Silica nanoparticles • Micelles • Toxicity profile

---

*Authors contributed equally to this work.

Y.W. Kong
Koch Institute for Integrative Cancer Research, Massachusetts Institute of Technology, 77 Massachusetts Avenue, Cambridge, MA 02139, USA

Department of Biology, Massachusetts Institute of Technology, 77 Massachusetts Avenue, Cambridge, MA 02139, USA

E.C. Dreaden
Koch Institute for Integrative Cancer Research, Massachusetts Institute of Technology, 77 Massachusetts Avenue, Cambridge, MA 02139, USA

Department of Chemical Engineering, Massachusetts Institute of Technology, 77 Massachusetts Avenue, Cambridge, MA 02139, USA

A. Prokop, V. Weissig (eds.), *Intracellular Delivery III*, Fundamental Biomedical Technologies 8, DOI 10.1007/978-3-319-43525-1_16

Abbreviations

17-AAG	17-allylamino-17-demethoxygeldanamycin
5-FU	5-fluorouracil
6MP	6-mercaptopurine (6MP)
AlClPc	aluminum chloride phthalocyanine
dsRNA	double-stranded RNA
EGFR	epidermal growth factor receptor
EPR	enhanced permeability and retention
HMSNs	hollow mesoporous silica nanoparticles
LbL	Layer-by-Layer
miRNA	microRNA
mRNA	messenger RNA
MMs	macromonomers
MOMP	nitrogen mustard with vincristine, methotrexate, and prednisone
MOPP	nitrogen mustard with vincristine, procarbazine, and prednisone
MPS	mononuclear phagocyte system
MSNs	mesoporous silica nanoparticles
MTD	maximum-tolerated-dose
NSAIDSs	nonsteroidal anti-inflammatory drugs
PCL	poly caprolactone
PDT	photodynamic therapy
PEG	poly(ethylene glycol)
PGLA	poly(D,L-lactide-co-glycolide)
PLA	polylactic acid
RNAi	RNA interference

P.T. Hammond (✉)
Koch Institute for Integrative Cancer Research, Massachusetts Institute of Technology, 77 Massachusetts Avenue, Cambridge, MA 02139, USA

Department of Chemical Engineering, Massachusetts Institute of Technology, 77 Massachusetts Avenue, Cambridge, MA 02139, USA

Institute for Soldier Nanotechnologies, Massachusetts Institute of Technology, 77 Massachusetts Avenue, Cambridge, MA 02139, USA
e-mail: hammond@mit.edu

M.B. Yaffe (✉)
Koch Institute for Integrative Cancer Research, Massachusetts Institute of Technology, 77 Massachusetts Avenue, Cambridge, MA 02139, USA

Department of Biology, Massachusetts Institute of Technology, 77 Massachusetts Avenue, Cambridge, MA 02139, USA

Department of Biological Engineering, Massachusetts Institute of Technology, 77 Massachusetts Avenue, Cambridge, MA 02139, USA

Divisions of Acute Care Surgery, Trauma, and Critical Care and Surgical Oncology, Department of Surgery, Beth Israel Deaconess Medical Center, Harvard Medical School, Boston, MA 02115, USA
e-mail: myaffe@mit.edu

siRNA	small interfering RNA
SSRIs	serotonin reuptake inhibitors
TNBC	triple-negative breast cancer
UCNs	upconversion nanoparticles
UV	ultraviolet
VAMP	vincristine, amethopterin, 6MP, and prednisone

16.1 Introduction

Cancer is a complex collection of diseases that through deregulation of myriad cellular pathways achieves unchecked growth, metabolism, and migration, thus making it the second leading cause of death in the United States. Solid tumors in particular have complex cyto-architecture and diverse microenvironments, making single-agent therapy largely ineffective due to sub-populations of cells that are resistant to a given agent (Burrell et al. 2013; Hainaut and Plymoth 2013; Hanahan and Weinberg 2000, 2011). Efforts have been taken to better understand the genetic alterations and complex molecular mechanisms in cancer to identify more effective therapies.

Due to the past failure of mono-therapies, most cancer patients now receive some form of combination therapy as the standard of care treatment for their tumor and this approach has improved patient outcomes. However, for the most part, current regimens were not "designed", rather, they were empirically determined to be combinations of drugs with non-overlapping toxicities, so that each drug could be administered at near-maximal dosage. Since these drugs generally have different mechanisms of action, there tends to be minimal cross-resistance, decreasing the emergence of drug resistant tumors (Mayer and Janoff 2007; Harasym et al. 2007; Ramsey 2005; Zoli et al. 2001). Importantly, however, this does not mean that current combinations are the best cocktail of drugs to achieve lasting remissions in patients. Rather they are the best cocktail of drugs identified to date that avoid unmanageable toxicity. Recent advances in nano-scale drug carriers that can target tumors preferentially and limit systemic toxicity are now revolutionizing the way we approach combination therapy.

16.2 Drug Combinations for Cancer Treatment

16.2.1 A Brief History of Combination Cancer Therapy

16.2.1.1 Combinations of Independently Active Drugs

Following the discovery of **cytotoxic chemotherapy** by Goodman et al. (1946; Gilman and Philips 1946) in 1943 and Farber et al. (1948) in 1947, researchers and clinicians began focusing on strategies to prolong cancer remissions in patients and

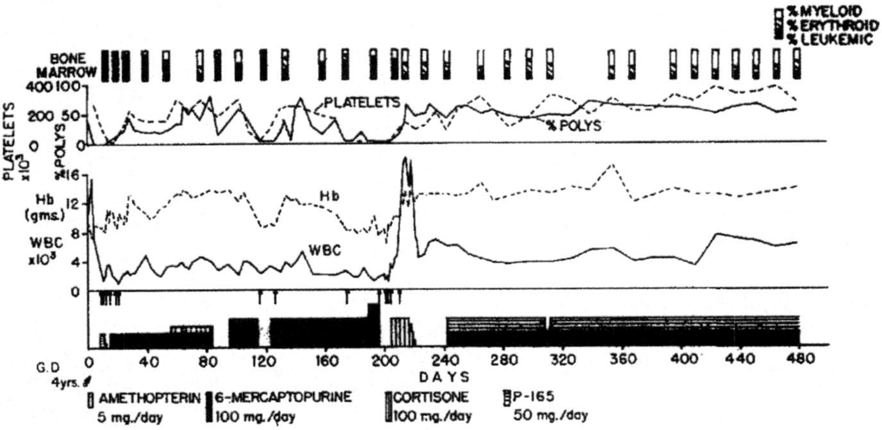

Fig. 16.1 An early example of sequential combination cancer therapy c.1954. Treatment regimen and corresponding blood counts for a 4 year old male with 6-mercaptopurine-resistent acute leukemia receiving amethopterin (antifolate), 6-mercaptopurine (purine antagonist), cortisone (steroid), and azaserine/P-165 (purine antagonist) at the Sloan-Kettering Institute (Reproduced with permission from Burchenal et al. (1954). Copyright 1954 John Wiley and Sons)

to delay drug resistance. Inspired by the observation of synergism between 6-mercaptopurine (6MP) and azaserine (O-diazoacetyl-L-serine), as well as 6MP and antifolates (Skipper et al. 1954) in preclinical mouse models of leukemia, Burchenal and colleagues conducted pilot studies (Burchenal et al. 1954) and later a randomized trial (Heyn et al. 1960) in children with acute leukemia receiving sequential combinations of 6MP, azaserine, and steroids (Fig. 16.1). While widespread improvements in overall survival were not observed, Frei et al. later demonstrated greatly enhanced remission rates in children with acute leukemia treated with 6MP and antifolates in 1961 – noting that "advantage derives from the two drugs acting independently on the patient" (Frei et al. 1961). With the discovery of antineoplastic activity from plant alkaloids from *Vinca rosea* at Eli Lilly, Freireich et al. (1964) quickly developed a "quadruple combination therapy" known as VAMP (vincristine, amethopterin, 6MP, and prednisone) which augmented remission rates in patients with acute leukemia treated in 1964. VAMP soon gave way to MOMP (Devita et al. 1965) (nitrogen mustard with vincristine, methotrexate, and prednisone) and then to MOPP (Devita et al. 1970) (nitrogen mustard with vincristine, procarbazine, and prednisone), the latter of which demonstrated startling activity in patients with Hodgkin's lymphoma – nearly 80 % displayed complete remission, and of these, 60 % demonstrated disease free survival. Throughout the advent of MOPP, the rationale for combination chemotherapy relied heavily on two fundamental underlying concepts: independent drug action and non-overlapping toxicity. As stated by Vincent DeVita: "*differing mechanisms of action and various dose-limiting toxicities could presumably overcome [low therapeutic index]*" (Devita et al. 1970).

16.2.1.2 Rational Drug Combinations Targeting a Shared Mechanism of Action

In 1982, *in vitro* studies demonstrated that pretreatment of cells with leucovorin (folinic acid), an innocuous nontoxic agent, to cancer cells could enhance the biochemical effects of the nucleoside analog, 5-fluorouracil (5-FU). Researchers found that leucovorin augmented inhibition of thymidylate synthase, the canonical target of 5-FU, depleting cellular nucleotide levels and inducing apoptosis (Pritchard et al. 2013; Longley et al. 2003). Preclinical studies in tumor xenograft-bearing mice, and later pilot studies in patients (Machover et al. 1982) showed clear benefits in response rates from the combination therapy versus that seen after treatment with 5-FU alone, although the improvements in overall survival were more moderate. The combination, still applied today as part of FOLFOX, FOLFIRI, and FOLFIRINOX treatment regimens, marked a significant departure from the strategy originally established by Frei, Freireich, and Zubrod. These drug combinations came about from "trial and error"/empirical testing in patients and are used not because they are the most effective treatment for tumor killing, but because they are capable of achieving the maximal tumor reduction within the maximal toxicity tolerated by patients. With the advancement of targeted drug delivery, "rational drug combinations" can now be used. Rather than combining drugs with independent activity or dose-limiting toxic effects, *rational drug combinations* could act cooperatively on overlapping molecular targets to selectively to kill cancer cells.

16.2.1.3 Molecularly Targeted Therapies

The discovery that multidrug transporter proteins could play a key role in the development of *adaptive resistance* to chemotherapy (Rothenberg and Ling 1989), led to the initiation of clinical trials in the late 1980s to concurrently inhibit promiscuous drug efflux pumps during the administration of cytotoxic anti-tumor agents. Here, competitive substrates of the multidrug transporters – already approved to treat other conditions (e.g. cyclosporine and verapamil) – were co-administered in patients with refractory disease. Although these trials, and those that followed over multiple generations of inhibitors, were largely disappointing (Fletcher et al. 2010; Kaye 1993), the general strategy of identifying and preemptively blocking the biological mechanisms responsible for adaptive resistance continued to guide a significant portion of subsequent work.

During this same period in the late 1980s, researchers were also beginning to understand structure-activity relationships for a novel small molecule contraceptive agent that showed unusually selective antitumor activity in breast cancer patients – tamoxifen (Jordan 2003). The drug was found to selectively inhibit the estrogen receptor, its 'target' protein, in breast tumors. These findings led to a shift in the focus of commercial anti-cancer drug discovery away from enhanced non-specific cell killing towards rational target inhibition. So-called 'molecularly **targeted therapies**' against the BCR-ABL fusion protein (i.e. imatinib/Gleevec) (Capdeville et al. 2002), **monoclonal antibody** therapies (e.g. trastuzumab/Herceptin, and related molecules targeting EGFR family members) (Hudis 2007), and **recombinant proteins** (e.g. interleukin-2) (Dranoff 2004) followed soon afterward. Rational drug combinations incorporating these molecularly-targeted drugs added further complexity to prior combination therapeutic approaches.

16.2.1.4 Large-Scale Screens, Nucleic Acid Therapies, and Beyond

Later, with the discovery of RNA interference in mammalian cells in 2001 (Elbashir et al. 2001), large scale loss-of-function screens were used to identify a subset of new molecular targets for cancer therapy (Ngo et al. 2006; Luo et al. 2008), as well as new gene combinations whose pairwise loss blocked cancer cell survival, resulting in '*synthetic lethality*' (Kaelin 2005; Luo et al. 2009). A number of these early discoveries were complicated by poor reproducibility, despite improvements in **small interfering RNA** (siRNA) and **messenger RNA** (mRNA) delivery (Kormann et al. 2011). Improvements in **CRISPR-Cas9** technology have overcome many of these difficulties (Platt et al. 2014), further expanding this toolkit to include amplified or synthetic protein expression, in addition to genetic loss of function. Although it is currently unclear to what extent **epigenetic modifiers** (e.g. chromatin regulators) (Floyd et al. 2013; Keung et al. 2014), **immune checkpoint antagonists** (Mahoney et al. 2015), or **chimeric proteins/receptors** (Kalos et al. 2011; Morsut et al. 2016) will contribute to future multiplexed combination therapies, the number of possible pairwise combinations of the above drug classes alone provides ample opportunities for the creation of new and highly potent therapies with durable treatment responses.

16.2.2 Challenges in Delivering Drug Combinations to Tumors

16.2.2.1 Co-delivery

Implementation of drug combinations generally requires co-localization of each agent within the malignant cells for efficient tumor cell killing. This presents a variety of complex intrinsic challenges due to the unique physiochemical properties of each drug such as size, charge, hydrophobicity, and stability, among others. For example, current frontline two-agent therapy for ovarian cancer requires co-delivery of cisplatin and paclitaxel. Although both drugs are roughly neutral at physiological pH, paclitaxel is more than double the molecular weight of cisplatin and its relative hydrophobicity (octanol:water partition coefficient) is more than five logs higher than cisplatin. Cellular colocalization in the complex tumor microenvironment is thus severely constrained. Further challenges include colocalization of either drug with bevacizumab – a monoclonal antibody directed against the pro-angiogenic cytokine VEGF-A, which is 1000 times larger in molecular weight, as part of platinum-sensitive disease therapy, or liposomal doxorubicin, a topoisomerase inhibitor whose nanoparticle is almost 10,000 times larger than the free small molecule paclitaxel – as part of platinum-insensitive disease therapy. Interestingly, Wittrup and coworkers (Schmidt and Wittrup 2009) have modeled tumor uptake data for biomolecules of varying size and affinity and found that intermediate-sized targeting agents (ca. 25 kDa, 5.3 nm dia) exhibit the lowest tumor uptake, whereas higher tumor uptake levels are observed for either smaller agents (e.g. peptides and small molecules) and larger agents (e.g. IgG, liposomes). Bawendi, Jain, and Fukumura (Stylianopoulos et al. 2012) have similarly examined size-dependent penetration of nanoscale particles into the interstitium of tumor xenografts and found that drug size is inversely correlated with tumor penetration (Fig. 16.2a). Because intratumoral distribution profiles of drugs are often heterogeneous, enhanced or diminished cell killing can thus occur in a spatially dependent manner. Using inorganic colloids as model drug carriers, Chan and coworkers also found that 20 nm particles efficiently penetrate and are retained in the tumor interstitium at significantly greater depths than comparable 40–100 nm particles (Perrault et al. 2009).

16.2.2.2 Stoichiometry/Ratiometric Dosing

Historically, combinations of free drugs are administered at their respective maximum-tolerated-doses (MTDs); however, it is now widely acknowledged that drug combinations can act synergistically at specific drug ratios, as well as additively or even antagonistically at other drug stoichiometries. For example, Dreaden et al. found that MEK and PI3K inhibitors, when co-administered, exhibit stoichiometry-dependent drug synergy *in vitro* (Fig. 16.2b). Likewise, optimally synergistic pairwise combinations of irinotecan/floxuridine (Batist et al. 2009), cytarabine/daunorubicin (Tardi et al. 2009a), irinotecan/cisplatin (Tardi et al. 2009b), paclitaxel/tanespimycin (17-AAG) (Katragadda et al. 2013), and quercetin/vincristine (Wong et al. 2010) have also been identified.

16.2.2.3 Drug Sequence and Timing

Cellular responses to perturbations occur in a time-dependent manner and drug combinations that exploit these response networks can often maximize therapeutic potential through sequence- and time-staggered delivery. Although conventional delivery methods such as intravenous or intraperitoneal infusion can be staged manually, poor drug colocalization and unfavorable drug stoichiometry within tumors can abrogate the therapeutic potential of even the most potent rational combination therapies. Lee et al. (2012), for example, recently employed a systems-based approach to understand how targeted cancer therapies rewire oncogenic cell signaling networks, and methods by which this 'dynamic rewiring' can be exploited to improve tumor cell killing (Fig. 16.2c). Interesting, they found that time-staggered

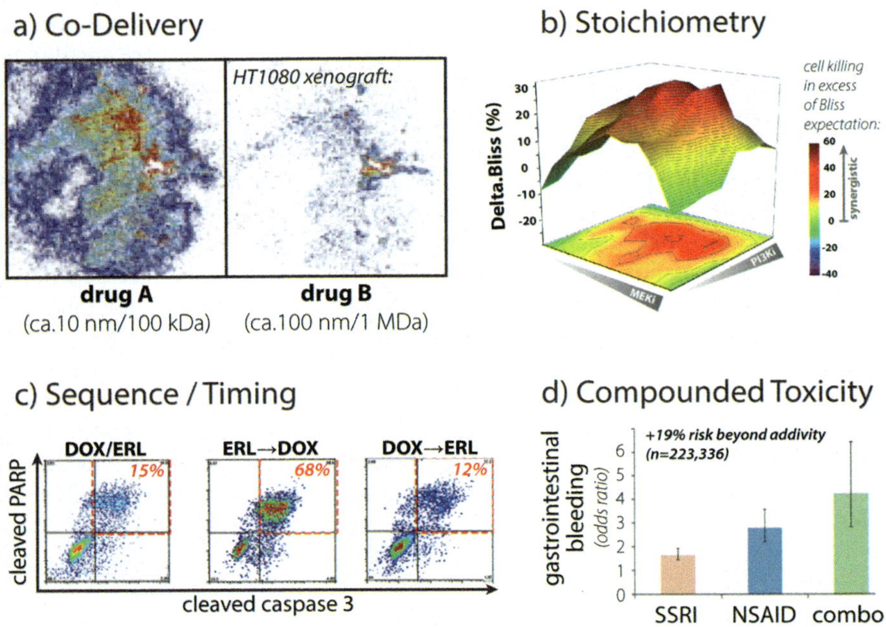

Fig. 16.2 Challenges in delivering drug combinations. (**a**) Drugs with differing targets often display disparate physiochemical properties that, here, affect intratumoral distribution, but also pharmacokinetics, pharmacodynamics, and tissue disposition. Model drug A is 10-fold smaller than B and penetrates more deeply into HT1080 tumor xenografts at 6 h as measured by fluorescence microscopy of hitological tissue sections. (**b**) In many canses, multiplexed combinations exhibit optimally synergistic drug stoichiometry that is difficult to recapitulate using free drug compounds. Here, small molecule inhibitors of Mek and PI3K are synergistic toxic only over a narrow range of drug ratios *in vitro* as measured by CellTiter Glo. (**c**) The sequence and timing with which drugs modulate (rewire) complex cell signaling networks also determines capacity for cell killing. Here, combinations of the cytotoxic chemotherapeutic, doxorubicin, and erlotinib, an inhibitor of EGFR, optimally induce apoptosis (double positive flow cytometry) in a sequence-dependent manner *in vitro*. (**d**) Dose-limiting toxic effects can also act synergistically, here increasing the relative risk of gastrointestinal bleeding 19 % beyond additivity when selective serotonin reuptake inhibitors (SSRIs) and nonsteroidal anti-inflammatory drugs (NSAIDSs) are combined in patients (n = 223,336) (Reproduced with permission from (**a**) Stylianopoulos et al. (2012), (**b**) Dreaden et al. (2015), (**c**) Lee et al. (2012), (d) Anglin et al. (2014). Copyright (**a**) 2012 Elsevier, (**b**) 2015 American Association of Cancer Research, (**c**) 2012 Cell Press, (**d**) 2014 Nature Publishing Group)

inhibition of epidermal growth factor receptor (EGFR), but not simultaneous co-administration, could sensitize a subset of triple negative breast cancer cells to DNA damaging chemotherapy. Not only was sequence a critical determinant for enhanced cell death, but a time delay of as little as 4 h could dramatically alter apoptotic response *in vitro*. In a related approach, Seino et al. (2016) recently found that JNK pathway may play a key role in adaptive resistance of ovarian cancer cells towards frontline platinum/taxane therapy. Although concurrent targeted inhibition of JNK induced contrasting effects on cisplatin (enhanced cytotoxicity) and paclitaxel (decreased cytotoxicity), time-staggered inhibition of JNK prior to chemotherapy greatly augmented *in vitro* cell killing in a time-dependent manner. These findings also suggest that basal JNK activity may correlate with drug resistance in ovarian cancer. Sengupta and coworkers (Goldman et al. 2015) have also examined drug-induced tumor cell perturbations and found that taxane therapy induces a transient cell state characterized by Src family kinase (SFK)/Hck pathway activation and suppression of apoptosis. By studying the *in vivo* induction kinetics of this transient state, the authors could preemptively block anti-apoptotic responses *in vivo* through targeted inhibition of SFK/Hck staggered approximately 6 days after taxane therapy.

16.2.2.4 Compounding and Overlapping Toxicity

As the focus of combination cancer therapies shifts towards synergy and network rewiring, compounded dose-limiting toxic effects again present a significant challenge to multi-drug treatment regimens. Drug interactions are often complex to predict, and dose-limiting toxic effects – like therapeutic effects – can occur synergistically. For example, combined administration of common therapeutics such as nonsteroidal anti-inflammatory drugs (NSAIDSs) with selective serotonin reuptake inhibitors (SSRIs) increases the relative risk of gastrointestinal bleeding 19 % beyond the additive expectation (Anglin et al. 2014) (Fig. 16.2d). Similarly, doxocubicin and trastuzumab (anti-HER2/ErbB2; Herceptin) – commonly co-administered in breast cancer patients – are known to induce Type I and Type II cardiotoxicity, respectively, effects which compound the risk of dose-limiting cardiotoxic events in patients (Cardinale et al. 2010). More recently, combined blockade of MAPK and PI3K pathway signaling has been shown to synergistically kill a variety of solid tumors *and* pre-emptively block resistance-associated signaling in preclinical animal models (Engelman et al. 2008). A retrospective Phase I clinical study of patients with advanced solid tumors receiving small molecule inhibitors of MAPK or PI3K pathway signaling, alone or in combination, found that simultaneous blockade significantly decreased average tumor burden relative to monotherapy (Shimizu et al. 2012); however, these favorable outcomes were accompanied by a 2.0-fold increase in the prevalence of dose-limiting toxicity and a 3.0-fold increase in the prevalence of drug-related high grade (>III) adverse events, primarily hepatic-related.

16.3 Nanoparticle Formulations to Optimize Anti-cancer Combination Therapies

Nanomaterials can be used to co-deliver multimodal cancer therapeutic agents to achieve maximum therapeutic effects (Hu and Zhang 2012). Some of the nanomaterials used to study combination therapy include liposomes, polymer-drug conjugates, dendrimers, and polymeric micelles (Peer et al. 2007). We will discuss a few notable examples in detail below.

16.3.1 Mesoporous Silica Nanoparticles

Mesoporous silica nanoparticles (MSNs) exhibit a range of features that are amenable to drug delivery (Baek et al. 2015). They have high surface area, which allows for large amounts of drug loading, tunable porosity and size, structural diversity, easily modifiable chemistry and suitability for functionalization, and are biocompatible. MSNs have been used extensively as multifunctional nanocarrier systems by combination or hybridization with biomolecules, drugs, and other nanoparticles, and can be stimulated by signals such as pH, optical signal, redox reaction, òr electric and magnetic fields (Baek et al. 2015).

Lipid-coated MSN have recently been used to delivery synergistic gemcitabine and paclitaxel to both subcutaneous and orthotopic pancreatic tumors in mice. Mice with both subcutaneous and orthotopic tumors receiving systemic gemcitabine and paclitaxel loaded MSN achieved more effective tumor shrinkage than those receiving individual or free drug. The authors also observed elimination of metastatic foci without evidence of local or systemic toxicity (Meng et al. 2015).

Another way to use MSNs is for the combination of photodynamic and chemotherapies. One such example is the use of MSNs loaded with aluminum chloride phthalocyanine (AlClPc) and cisplatin for cancer treatment. Vivero-Escoto and colleagues showed that these MSNs are taken up by HeLa cells, and upon light exposure, the AlClPc-cisplatin-MSN combination was more cytotoxic then the AlClPc-MSN and cisplatin-MSN controls. These data suggest that there is great potential for the use of MSN platforms as nanocarriers for combination photodynamic and chemotherapies to treat cancer (Vivero-Escoto and Elnagheeb 2015). In another example, Zhang and colleagues synthesized a polymeric prodrug (doxorubicin)-coated hollow mesoporous silica nanoparticles (HMSNs) with an NIR absorbing dye IR825 loaded into the hollow cavity of the HMSN to form a multifunctional hybrid HMSNs-DOX/IR825 (Zhang et al. 2016). Cancer cells efficiently took up the hybrid nanoparticle, and the conjugated doxorubcin was successfully released in the cellular environment. *In vitro* cytotoxicity study showed that anticancer activity of HMSNs-DOX/IR825 was significantly improved by the NIR irradiation, suggesting that the hybrid nanoparticle could potentially be used for combined photothermal-chemotherapy of cancer (Zhang et al. 2016).

16.3.2 Self-Assembly Copolymer Carriers – Micelles

Many self-assembling molecules are amphiphilic, comprising of both hydrophobic and hydrophilic domains. Amphiphilic copolymers can self-assemble into micelles, vesicles, and molecular gels composed of tubules, fibrils, and fibers (Giddi et al. 2007; Nishiyama and Kataoka 2006; Rösler et al. 2001).

Micelles are amphiphilic molecules that self-assemble into a spherical structure with a hydrophobic core and hydrophilic exterior making it suitable for encapsulating hydrophobic cancer drugs (Jhaveri and Torchilin 2014). It is estimated that about 40 % of marketed drugs and up to 75 % compounds under development are poorly water soluble (Jhaveri and Torchilin 2014; Di et al. 2009; Williams et al. 2013). Polymeric micelles can be used for combination therapy by loading multiple anticancer agents in polymeric micelles in a one-step drug-loading process without chemical modification of drugs (Shin et al. 2011). Multi-drug release may occur by simple hydrolysis or be triggered by an acidic pH and/or lysosomal enzymes, and can be tuned by chemical linkage for concurrent or sequential delivery (Duncan 2006; Greco and Vicent 2009).

Shin and colleagues showed that they could encapsulate paclitaxel, rapamycin, and 17-allylamino-17-demethoxygeldanamycin (17-AAG) in PEG-*b*-PLA micelles without changing the pharmacokinetics of each drug at low doses. The pharmacokinetic profiles were however altered when drugs are delivered at higher doses (Shin et al. 2012). Rapamycin and 17-AAG acts concurrently to inhibit the PI3K/Akt/mTOR and Ras/Raf/MEK/ERK signaling pathways, enhancing cancer cell killing by paclitaxel. Bae and colleagues directly conjugated doxorubicin and 17-hydroxy ethylamino-17-demethoxygeldanamycin (GDM-OH) to a poly(ethylene glycol)-poly(aspartate hydrazide) block copolymers through acid-labile hydrazone bonds. The pH-sensitive micelles were combined and appeared to minimize a schedule-dependent change in combined drug efficacy when compared to the free drug combination (Bae et al. 2010). Karaca and colleagues, reported the use of methoxy poly(ethyleneglycol)-block-ploy(2-methyl-2-carboxyl-propylenecarbonate)-graft-dodecanol (mPEG-b-PCC-g-DC) copolymer to conjugate gemcitabine and encapsulate a Hedgehog inhibitor, Vismodegib (GDC-0449) into its hydrophobic core for the treatment of pancreatic ductal adenocarcinoma (PDAC). The *in vivo* stability of gemcitabine increased significantly after conjugation, and the drug combination, when administered to athymic nude mice bearing subcutaneous tumors generated using MIA PaCa-2 cells, efficiently inhibited tumor growth (Karaca et al. 2016).

16.3.3 Nanotechnology Approaches to Enhance Co-delivery

By physically confining drug combinations within a single carrier, pharmacokinetics for multiple drugs can be unified, ensuring all particle-treated cells receive a pairwise combination of drugs – maximizing therapeutic potential. One key challenge in this area involves combining three general categories of drugs, all of which display differing combinations of physiochemical properties: (i) large and hydrophilic proteins, (ii) small and hydrophobic small molecules, and (iii) moderately sized, hydrophilic, and highly anionic nucleic acids. Targeted, nanoscale delivery of therapeutic proteins remains an area currently underexplored in cancer therapy – particularly immunotherapy. Fahmy and coworkers recently addressed this challenge in the development of a combination cancer immunotherapy that reverts immunosuppressive tumor microenvironments (Fig. 16.3a) (Park et al. 2012). Using the cytokine, IL-2, which stimulates NK cell and cytotoxic T lymphocyte activity,

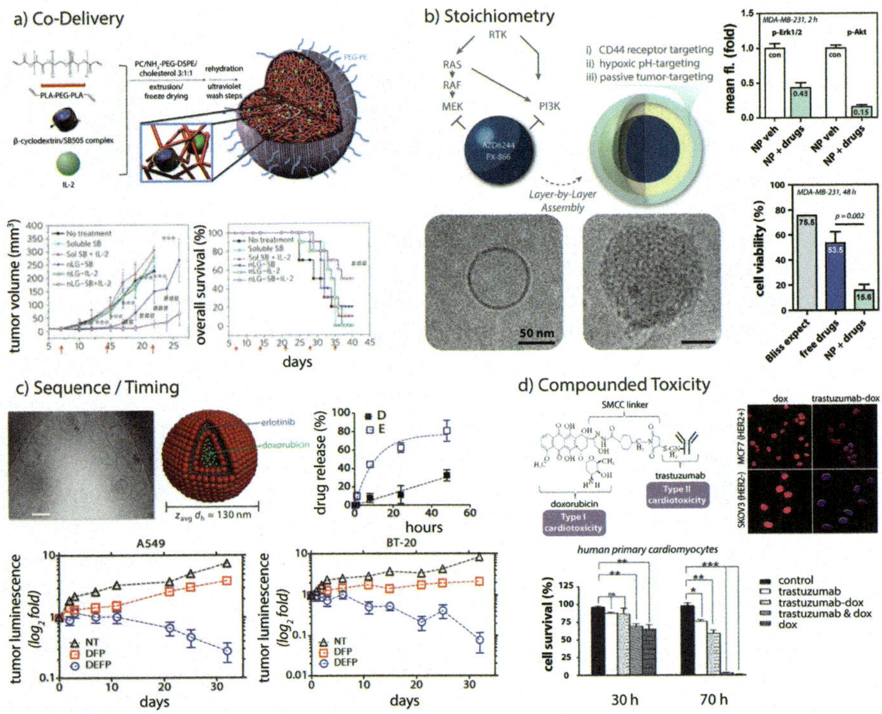

Fig. 16.3 Nanomedicines Overcome Challenges in Combination Drug Delivery. (**a**) Bulky and hydrophilic proteins are notably difficult to co-deliver with small, hydrophobic molecular inhibitors. Here, IL-2 is entrapped in a photo-polymerizable hydrogel matrix composed of poly(ethylene glycol), polylactide, and a small molecule/β-cyclodextrin inclusion complex. The hydrogel core is encapsulated by a liposomal shell and photocured under ultraviolet light. Systemically administered 'nanolipogels' decreased tumor burden and improved overall survival better than single drug-loaded particles or intratumoral injection of free drugs in immunocompetent, subcutaneous tumor xenograft mouse models (B16-F10). IL-2 promotes a hostile tumor microenvironment by stimulating NK cell and cytotoxic T lymphocyte activity, while a TGF-β signal blockade (SB505124) does the same while additionally depleting tumor-promoting regulatory T lymphocytes (Tregs). (**b**) The delivery of synergistic drug ratios can also be 'pre-programmed' into nanoscale drug carriers. Here, a hydrophobic inhibitor of MAPK pathway signaling is co-encapsulated with a hydrophilic inhibitor of the PI3K pathway at a pre-defined synergistic drug ratio. Layer-by-Layer (LbL) polymer assembly around the drug-containing liposomal core facilitated both affinity-based and microenvironment-responsive tumor targeting, while simultaneously blocking both pathways and enhancing drug synergy beyond that of the free drug combination *in vitro* and *in vivo*. (**c**) Time-staggared delivery can be achieved through structural partitioning of drugs within nanoscale drug carriers. Here, a hydrophobic inhibitor of EGFR (erlotinib) partitions into the lipid bilayer of a nanoscale lipid vesicle, while a hydrophilic genotoxic agent, doxorubicin, is loaded into the aqueous particle core. Initial release of erlotinib 'dynamically rewires' cell signaling in lung and breast tumors in a manner that enhances subsequent cell killing by doxorubicin in a time-dependent manner. Systemically administered, folate receptor-targeting constructs induce a partial response in lung (A549) and breast (BT-20) tumor xenograft-bearing mice, while mice receiving single drug-loaded particles (doxorubicin) exhibit progressive disease. (**d**) Nanomedicines can also mitigate dose-limiting side effects of drugs with overlapping toxicity profiles. Although, doxocubicin and

in combination with TGF-β signal blockade, which does the same while additionally depleting tumor-promoting regulatory T lymphocytes (Tregs), the authors were able to develop a systemically administerable drug carrier that decreased tumor burden and improved overall survival when compared to single drug-loaded particles or to intratumoral injection of free drugs in immunocompetent, subcutaneous tumor xenograft mouse models (B16-F10). To achieve this, IL-2 was entrapped in a photo-polymerizable hydrogel matrix composed of poly(ethylene glycol) (PEG), polylactide, and a small molecule/β-cyclodextrin inclusion complex. The hydrogel core is encapsulated by a liposomal shell and photocured under ultraviolet light to obtain PEG-stabilized 'nanolipogels'. A variety of other novel co-delivered drug combinations are listed in Table 16.1.

16.3.4 Nanotechnology Solutions: Stoichiometry/Ratiometric Dosing

Through ratiometric drug loading into nanoscale carriers, intracellular drug concentrations can be 'pre-programmed' for drug synergy, thus maximizing therapeutic potential and avoiding possible antagonistic interactions resulting from spatially and temporally heterogeneous delivery of free drug compounds. For example, Dreaden et al 2015. co-encapsulated a hydrophobic inhibitor of MAPK pathway signaling with a hydrophilic inhibitor of the PI3K pathway at a pre-defined synergistic drug ratio (Fig. 16.3b) (Dreaden et al. 2015). Layer-by-Layer (LbL) polymer assembly around the drug-containing liposomal core facilitated both affinity-based and microenvironment-responsive tumor targeting, while simultaneously blocking both pathways and enhancing drug synergy beyond that of the free drug combination *in vitro* and *in vivo*. A related therapeutic in clinical development, liposomal irinotecan/floxuridine (CPX-1, 1:1 mol; Celator), co-encapsulates drugs at a previously identified synergistic drug ratio and maintains this drug stoichiometry both in plasma and in the tumor bed (Batist et al. 2009). Interestingly, efficacy of the liposomal formulation in tumor-bearing mice was superior to both free drugs dosed at their respective MTDs and the additive expectation from both singly loaded

Fig. 16.3 (continued) trastuzumab (anti-HER2/ErbB2; Herceptin) – alone – induce Type I and Type II cardiotoxicity, respectively, the two are often co-administered in breast cancer patients. Engineering of an trastuzumab-doxorubicin antibody drug conjugate (ADC) rescues doxorubicin toxicity towards human primary cardiomyocytes by limiting cellular uptake to HER2-expressing cells. Here, a heterobifunctional linker (SMCC) containing an amine-reactive NHS ester and a thiol-reactive malimide crosslinks doxorubicin and IgG, respectively (Reprinted with permission from (a) Park et al. 2012, (b) Dreaden et al. 2015, (c) Morton et al. 2014, and (d) Zhang et al. 2013). Copyright (**a**) 2012 Nature Publishing Group, (**b**) 2015 American Association of Cancer Research, (**c**) 2014 American Association for the Advancement of Science, and (**d**) 2013 original authors under the Creative Commons Attribution License)

Table 16.1 Selected examples of nanotechnology-enabled combination cancer therapies

Material	Cytotoxic	Small molecule	Protein	Nucleic acid	Indication	References
Liposome	5:1 cytarabine and daunorubicin				Acute myeloid leukemia	Tardi et al. (2009a)
Liposome	1:1 irinotecan and floxuridine				Colorectal cancer	Batist et al. (2009)
Liposome	7:1 irinotecan and cisplatin				Small-cell lung cancer	Tardi et al. (2009b)
Liposome	6-mercaptopurine and daunorubicin				Acute lymphocytic leukemia	Agrawal et al. (2005)
Liposome	1:2 quercetin and vincristine				Breast cancer	Wong and Chiu (2010)
Polymer nanoparticle	Doxorubicin and docetaxel				Prostate cancer	Zhang et al. (2007)
Polymer conjugate	Platinum, doxorubicin, and camptothecin				Ovarian cancer	Liao et al. (2014)
Polymer conjugate	Gemcitabine and doxorubicin				Prostate cancer	Lammers et al. (2009)
Polymersome (LbL)	5-fluorouracil, irinotecan, oxaliplatin				Pancreatic cancer	Li et al. (2015)
Drug self-assembly	Irinotecan and chlorambucil				Breast cancer	Huang et al. (2014)
Liposome/silica hybrid	Doxorubicin, 5-fluorouracil, and cisplatin				Hepatocellular carcinoma	Ashley et al. (2011)
Silica	Paclitaxel and gemcitabine				Pancreatic cancer	Meng et al. (2015)
Polymer/liposome hybrid	Combretastatin and doxorubicin				Lung carcinoma, melanoma	Sengupta et al. (2005)

Emulsion	Paclitaxel	Curcumin			Ovarian cancer	Ganta and Amiji (2009)
Liposome	Doxorubicin	Verapamil			Leukemia	Wu et al. (2007)
Liposome (LbL)	Doxorubicin	Erlotinib			Breast, lung cancer	Morton et al. (2014)
Polymer nanoparticle	Paclitaxel	Ceramide			Ovarian cancer	van Vlerken et al. (2010)
Polymer nanoparticle	Paclitaxel	Tariquidar			Various cancers	Patil et al. (2009)
Polymer nanoparticle	Vincristine	Verapamil			Breast cancer	Song et al. (2009)
Polymer nanoparticle	Doxorubicin	Cyclosporin A			Various cancers	Emilienne Soma et al. (2000)
Polymer conjugate	Doxorubicin	Wortmannin			Breast cancer	Bae et al. (2007)
Dendrimer	Methotrexate	All-trans retinoic acid			Leukemia	Tekade et al. (2008)
Dendrimer	Methotrexate	All-trans retinoic acid			Leukemia	Tekade et al. (2009)
Dendrimer	Paclitaxel	Alendronate			Bone metastases	Clementi et al. (2011)
Liposome (LbL)		Selumetinib and PX-866			Breast, lung cancer	Dreaden et al. (2015)
Liposome		SB505124	IL-2		Melanoma	Park et al. (2012)
Liposome	Doxorubicin		TRAIL		Breast cancer	Jiang et al. (2014)
Polymer nanoparticle	Paclitaxel		TRAIL		Breast cancer	Lee et al. (2011)
Polymer nanoparticle	Paclitaxel		Anti-EGFR		Lung cancer	Karra et al. (2013)

(continued)

Table 16.1 (continued)

Material	Cytotoxic	Small molecule	Protein	Nucleic acid	Indication	References
Polymer conjugate	Camptothecin		Anti-HER2		Breast cancer	Han and Davis (2013)
Polymer nanoparticle	Doxorubicin		Anti-HER2		Breast cancer	Shi et al. (2009)
Polymer nanoparticle	Docetaxel and cisplatin		Anti-HER2		Breast cancer	Mi et al. (2013)
Nanocapsule	Paclitaxel		Caspase-3		Cervical cancer	Kim et al. (2015)
Graphene oxide	Doxorubicin		TRAIL		Lung, colorectal cancer	Jiang et al. (2015)
Polymer nanoparticle		Sorafenib		Survivin shRNA	Hepatocellular carcinoma	Shen et al. (2014)
Liposome (LbL)	Doxorubicin			MRP-1 and Bcl-2 siRNA	Lung cancer	Saad et al. (2008)
Polymer nanoparticle	Paclitaxel			Bcl-2 siRNA	Breast cancer	Wang et al. (2006)
Polymer nanoparticle	Paclitaxel			VEGF siRNA	Prostate cancer	Zhu et al. (2010)
Polymer nanoparticle	Doxorubicin			Bcl-xL shRNA	Prostate cancer	Kim et al. (2010)
Polymer nanoparticle	Doxorubicin			MDR-1 siRNA	Breast cancer	Misra et al. (2014)
Dendrimer	Doxorubicin			Luc siRNA	Glioblastoma	Kaneshiro and Lu (2009)
Dendrimer	Paclitaxel			Akt siRNA	Ovarian cancer	Kala et al. (2014)
Silica	Doxorubicin			P-gp siRNA	Breast cancer	Meng et al. (2013)
Silica	Doxorubicin			Bcl-2 siRNA	Ovarian cancer	Chen et al. (2009a)

liposomes. Strikingly, a liposomal formulation encapsulating a previously identified antagonistic drug ratio (10:1 mol) was less effective than its singly loaded counterpart, suggesting a putative role for heterogeneous combination drug delivery in promoting resistance to therapy. Similar approaches employing cytarabine/daunorubicin (CPX-351, 5:1 mol) (Tardi et al. 2009a), irinotecan/cisplatin (CPX-571, 7:1 mol) (Tardi et al. 2009b), and paclitaxel/tanespimycin (17-AAG) (Katragadda et al. 2013), quercetin/vincristine (1:2 mol), and doxorubicin/camptothecin/Pt (1:2.5:3.6 mol) (Liao et al. 2014) are currently under investigation.

16.3.5 Nanotechnology Approaches to Tailor Drug Combination Timing and Sequence

Another powerful property of nanoscale drug carriers is their capacity to not only spatially regulate drug release in the body, but also to temporally control the sequence and kinetics of therapeutics released. To exploit the observation by Lee et al. that time-staggered inhibition of EGFR could sensitize breast cancer cells to DNA damaging chemotherapy, Yaffe, Hammond, and coworkers engineered a nanoscale drug carrier which achieved staged drug delivery through structural partitioning of drugs within a liposomal vesicle (Fig. 16.3c) (Morton et al. 2014). Here, a hydrophobic inhibitor of EGFR, erlotinib, partitions into the lipid bilayer of the vesicle, while a hydrophilic genotoxic agent, doxorubicin, is loaded into the aqueous particle core. The initial release of erlotinib from these nanoparticles 'dynamically rewired' cell signaling in lung and breast tumors in a manner that recapitulates optimally staggered delivery kinetics seen with free drug administration, enhancing subsequent cell killing by doxorubicin. Systemically administered, folate receptor-targeted Erlotinib/doxorubicin nanoparticles were shown to induce a partial response in both lung (A549) and breast (BT-20) tumor xenografts in nude mice, while mice receiving only single drug-loaded nanoparticles (doxorubicin) exhibited progressive disease.

Gnanasammandhan and colleagues described a noninvasive method to deliver drugs that allow for a high degree of spatial and temporal control. Upconversion nanoparticles (UCNs) were used to convert deeply penetrating near-infrared (NIR) light to UV-visible wavelengths that match the absorption spectrum of photosensitive therapeutics. This allowed for the use of deep-penetrating and biologically friendly NIR light for photoactivation (Gnanasammandhan et al. 2016). The UCNs were used for photodynamic therapy (PDT) and photoactivated control of gene expression. For PDT, the UCNs are coated with polyethylene glycol (PEG) for stabilization and folic acid for tumor targeting and then loaded with photosensitizers that would be expected to kill cells by singlet oxygen production, whereas for the photoactivated control of gene expression, knockdown of essential tumor genes is achieved using UCNs loaded with caged nucleic acid (Gnanasammandhan et al. 2016).

To achieve controlled release of drugs, Liao and colleagues recently used two novel macromonomers (MMs) and a novel cross-linker as building blocks for the construction of a multi-drug-loaded nanoparticle. **CPT-MM** and **DOX-MM** are branched MMs that release unmodified CPT and DOX in response to cell culture media and long-wavelength ultraviolet (UV) light, respectively (Liao et al. 2014).

16.3.6 Nanotechnology Approaches to Limit Compounding and Overlapping Toxicity

Combination drug carriers can overcome challenges from overlapping toxicity profiles by biasing tissue disposition away from off-target tissues or by decelerating bolus drug release in vital organs. To address cardiotoxicity from doxorubicin and trastuzumab (anti-HER2/ErbB2; Herceptin) combination therapy, Zhang et al. engineered an antibody-drug conjugate from the pair, thereby limiting cytotoxic doxorubicin delivery to cells expressing high levels of HER2, while blocking the compounded toxicity towards human primary cardiomyocytes (Zhang et al. 2013) (Fig. 16.3d). To address dose-limiting hepatotoxic effects from combined MAPK and PI3K pathway inhibition, Dreaden et al. engineered LbL nanoparticles that biased tissue disposition towards solid tumors and rescued both hepatic and renal tissue damage while improving antitumor efficacy *in vivo*. Similarly, Farokhzad and coworkers have found that aptamer-targeted PLGA nanoparticles can rescue the nephrotoxic effects of platinum chemotherapeutics while maintaining equivalent antitumor activity *in vivo* (Dhar et al. 2011; Kolishetti et al. 2010) and also accommodating the chemotherapeutic, docetaxel, in polylactide containing particles (Xu et al. 2013).

16.3.7 Combining Nucleic Acid Therapies with Other Drug Combinations

Since the discovery of RNA interference (RNAi) in 1997, there has been great interest in harnessing RNAi for the treatment of disease. RNAi is activated by double-stranded RNA (dsRNA), which includes short interfering RNA (siRNA) and microRNA (miRNA) and utilizes the endogenous RNAi pathway for the post-transcriptional silencing of gene expression. MicroRNAs form central nodal points in cancer development pathways and exert their effects by targeting various oncogenes and tumor suppressors (Kong et al. 2012; Zhang et al. 2007), while siRNAs can be used to efficiently silence the expression of any gene with high specificity. These include targets that are considered to be difficult to drug. Here we describe some platforms used to deliver RNAi-drug combinations.

One particular tumor type, for example, that could greatly benefit from RNAi therapy is triple-negative breast cancer (TNBC), which is characterized by the lack of progesterone, estrogen and HER2 receptors. It is non-responsive to conventional hormonal therapy (such as tamoxifen or aromatase inhibitors) or therapies that target HER2 receptors, such as Herceptin (trastuzumab) (Foulkes et al. 2010). RNAi-based approaches can therefore be beneficial for the treatment of TNBC. One way to accomplish this is by modulating endogenous miRNA levels in TNBC. miR-221 and miR-205 have been shown to be up and down regulated in TNBC respectively (Nassirpour et al. 2013; Piovan et al. 2012). Conde and colleagues recently showed that they could deliver a miR-205 mimic and an antagomiRNA (miRNA inhibitor) using a self-assembled RNA-triple-helix structure which is conjugated to dendrimers to form stable triplex nanoparticles that can achieve nearly 90 % tumor shrinkage 2 weeks post-gel implantation in a triple-negative breast cancer mouse model (Conde et al. 2016).

A promising approach made available by nanotechnology is to combine siRNA and chemotherapeutics in a single platform. Deng and colleagues used a controlled layer-by-layer process to co-deliver siRNA against a drug-resistance pathway (multidrug resistance protein 1) and a chemotherapy drug (doxorubicin) to challenge a highly aggressive form of triple-negative breast cancer resulting in an 8-fold decrease in tumor volume compared to control treatments with no toxicity observed (Deng et al. 2013). Xu and colleagues combined siRNA against targets involved in error-prone translesion DNA synthesis pathway (REV1 and REV3L) with conventional DNA-damaging chemotherapy (cisplatin prodrug) through self-assembly of a biodegradable poly(lactide-coglycolide)-b-poly(ethylene glycol) diblock copolymer and a self-synthesized cationic lipid. This nanoparticle formulation had a synergistic effect on tumor inhibition in a xenograft mouse model of human lymph node carcinoma of the prostate that was noticeably more effective than platinum monotherapy (Xu et al. 2013).

Other carriers that have been used to deliver the siRNA based combinations include liposomes (Gabizon et al. 1994; Chen et al. 2009a, b, 2010a, b; Li et al. 2008), micelles (Zheng et al. 2013; Shim et al. 2011; Zhu et al. 2010), polymers poly (D,L-lactide-co-glycolide) (PLGA) (Li et al. 2001; Fonseca et al. 2002), poly lactic acid (PLA) (Tobío et al. 1998; Dong and Feng 2004), polycapro lactone (PCL) (Yang et al. 2006), dendrimers (Biswas et al. 2013; Kaneshiro and Lu 2009; Kulhari et al. 2011), natural chitosan polymeric nanoparticles (Wei et al. 2013; Nagpal et al. 2010), silica (Santra et al. 2001; Qhobosheane et al. 2001; Kneuer et al. 2000) and other inorganic nanoparticles e.g calcium, gold, quantum dots, etc. (Sokolova and Epple 2008).

16.4 Limitations to Developing Combination Chemotherapeutics, Tumor-Specific Targeting, and Enabling Approaches

Like Frei, Freireich, and Zubrod, drug discovery has historically focused on the development of compounds with independent antitumor activity – those intended for use as monotherapies. Modern approaches to combination development have, in contrast, been largely limited to off-patent cytotoxic drugs. To address the challenge of integrating patent-protected targeted therapies in combination approaches,

Merck and AstraZeneca initiated a seminal partnership in 2009 to share the costs of developing combination candidates, for example, AstraZeneca's MEK inhibitor (AZD6244) and Merck's protein kinase B inhibitor (MK-22060). Merck and Sanofi later followed with a similar agreement to investigate Merck's MEK inhibitor, MSC1936369B, in combination with Sanofi's PI3K/mTOR inhibitor, SAR245409, and class I PI3K inhibitor, SAR245408 . Bristol-Myers Squibb and Roche have likewise established agreements to combine Roche's vemurafenib (Zelboraf) with BMS's ipilimumab (Yervoy) for BRAF mutant metastatic melanoma. While highly promising, dose-limiting hepatotoxicity from the latter two combination approaches in Phase I clinical trials (Xu et al. 2013) highlights a key weakness of this approach: by neglecting combination effects during the discovery phase, more safe and/or effective combination candidates may be discarded simply as a result of exhibiting less potent independent antitumor activity. Although small molecule targeted therapies, in the past, have provided less incentive for early combination development (due to their single agent efficacy), the recent resurgence of cancer immunotherapy development will likely accelerate the integration of combination approaches earlier in the development pipeline, providing opportunities for the investigation of combinations with weak independent activity, but potent and safe combined therapeutic effects in the future.

A major but as-yet incompletely realized opportunity for nanoparticle therapeutics is the potential for tumor-specific targeting. Compounded organ-specific toxic effects such as those described above highlight a potential intrinsic advantage of – and challenge to – nanoscale drug delivery, whereby dose-limiting toxicities could be mitigated through nanoparticle-altered pharmacokinetics combined with favorable tumor tissue targeting profiles. The latter phenomenon can occur through size-dependent 'passive' tumor targeting or the 'active' targeting of tissues via stimuli-responsive behavior or affinity directed accumulation. While a number of recent publications seek to revisit the importance and prevalence of passive targeting in tumor delivery (Prabhakar et al. 2013; Park 2013) – largely attributed to the so-called enhanced permeability and retention (EPR) effect (Matsumura and Maeda 1986; Matsumoto et al. 2016) – nanoparticles are well known to preferentially accumulate in organs of the mononuclear phagocyte system (MPS), namely the liver and spleen, and to augment the accumulation of renally excreted drugs (i.e. <10 nm) in solid tumors. This property can be advantageous when designing treatments for hepatocellular carcinoma or immunotherapies, respectively; however, affinity directed targeting of tumor tissues remains an integral and underexplored challenge to the field. Chan and coworkers (Wilhelm et al. 2016), highlight this disparity in a recent retrospective literature analysis, noting only modest (0.5-fold) improvements in median tumor accumulation afforded by active targeting across multiple studies. In contrast, relatively smaller antibody-drug conjugates (ADCs) (Vaklavas and Forero-Torres 2012) and molecular polyconjugates (Rozema et al. 2007), while less prevalent, have demonstrated notably reproducible in vivo targeting capabilities.

Future research investigating specific targeting ligands including small molecules (e.g. folic acid, bisphosphonate, carbohydrates), peptides (e.g. GE11, RGD, knottin), and proteins (e.g. IgG, Fab fragments, Centyrins), is expected to improve both the effectiveness and reproducibility of affinity targeting strategies employed by the field, as well as subsequent treatment outcomes from rational drug combinations.

16.5 Outlook and Conclusions

The intersection between cancer biology and nanotechnology is an exciting and emerging area of current academic and industrial research. With ongoing efforts in both fields, combination anticancer treatments are continuing to evolve, raising hopes for unprecedented antitumor responses and reduced toxicity. The emergence of newly engineered combination drug delivery platforms should allow us to combine different classes of drugs into a single nanoparticle with tunable functionality over local or temporal control of drug delivery increasing safety, efficacy, and durability. The co-delivery of different cancer therapeutic agents provides promising options to overcome chemoresistance. Recent reports provide strong evidence that combining different drugs using nanoparticles improves tumor killing compared to single agent therapy. While these approaches hold great promise, there still remain key limitations in their proof of concept. Most nanocarrier studies are currently performed in preclinical models, and desperately need to be translated into human clinical trials, particularly since the biodistribution, localization, and release profiles of these drugs may differ in humans. It is also pertinent for the safety profiles of the various carriers used for the delivery of these therapeutic agents to be further studied, with special focus on their toxicity and immune response Given the progress that has been made in the field during the past 5 years, the future of rationally designed and personalized combination therapy using customizable nanoparticle delivery platforms looks promising.

Acknowledgments We wish to thank Dr Ian Cannell for reading and commenting on the chapter.

Financial Support This work was supported by the Ovarian Cancer Research Foundation (MBY, PTH; Program Project Development Grant), the Breast Cancer Alliance (MBY, PTH; Exceptional Project Award), the Department of Defense (PTH; CDMRP Ovarian Cancer Research Program Teal Innovator Award), the National Institutes of Health (MBY; R01-ES015339, R01-GM104047, ECD; NIBIB 1F32EB017614-02), and the Misrock Foundation (YWK). Support was provided in part by the Koch Institute Support Grant (P30-CA14051) from the National Cancer Institute.

Conflict of Interest Statement The authors declare no conflict of interest.

References

Agrawal V, Paul MK, Mukhopadhyay AK (2005) 6-mercaptopurine and daunorubicin double drug liposomes—preparation, drug-drug interaction and characterization. J Liposome Res 15:141–155

Anglin R et al (2014) Risk of upper gastrointestinal bleeding with selective serotonin reuptake inhibitors with or without concurrent NonSteroidal anti-inflammatory use: a systematic review and meta-analysis. Am J Gastroenterol 109(6):811–819

Ashley CE et al (2011) The targeted delivery of multicomponent cargos to cancer cells by nanoporous particle-supported lipid bilayers. Nat Mater 10(5):389–397

Bae Y et al (2007) Mixed polymeric micelles for combination cancer chemotherapy through the concurrent delivery of multiple chemotherapeutic agents. J Control Release 122(3):324–330

Bae Y et al (2010) Mixed pH-sensitive polymeric micelles for combination drug delivery. Pharm Res 27(11):2421–2432

Baek S et al (2015) Smart multifunctional drug delivery towards anticancer therapy harmonized in mesoporous nanoparticles. Nanoscale 7(34):14191–14216

Batist G et al (2009) Safety, pharmacokinetics, and efficacy of CPX-1 liposome injection in patients with advanced solid tumors. Clin Cancer Res 15(2):692–700

Biswas S, Deshpande PP, Navarro G, Dodwadkar NS, Torchilin VP (2013) Lipid modified triblock PAMAM-based nanocarriers for siRNA drug co-delivery. Biomaterials 34(4):1289–1301, http://doi.org/10.1016/j.biomaterials.2012.10.024

Burchenal JH et al (1954) Clinical studies on 6-mercaptopurine. Ann N Y Acad Sci 60(2):359–368

Burrell RA et al (2013) The causes and consequences of genetic heterogeneity in cancer evolution. Nature 501(7467):338–345

Capdeville R et al (2002) Glivec (STI571, imatinib), a rationally developed, targeted anticancer drug. Nat Rev Drug Discov 1(7):493–502

Cardinale D et al (2010) Trastuzumab-induced cardiotoxicity: clinical and prognostic implications of troponin I evaluation. J Clin Oncol 28(25):3910–3916

Chen AM et al (2009a) Co-delivery of doxorubicin and Bcl-2 siRNA by mesoporous silica nanoparticles enhances the efficacy of chemotherapy in multidrug-resistant cancer cells. Small 5(23):2673–2677

Chen Y et al (2009b) Novel cationic lipid that delivers siRNA and enhances therapeutic effect in lung cancer cells. Mol Pharm 6(3):696–705

Chen Y, Bathula SR et al (2010a) Multifunctional nanoparticles delivering small interfering RNA and doxorubicin overcome drug resistance in cancer. J Biol Chem 285(29):22639–22650

Chen Y, Wu JJ, Huang L (2010b) Nanoparticles targeted with NGR motif deliver c-myc siRNA and doxorubicin for anticancer therapy. Mol Ther 18(4):828–834

Clementi C et al (2011) Dendritic poly(ethylene glycol) bearing paclitaxel and alendronate for targeting bone neoplasms. Mol Pharm 8(4):1063–1072

Conde J et al (2016) Self-assembled RNA-triple-helix hydrogel scaffold for microRNA modulation in the tumour microenvironment. Nat Mater 15(3):353–363

Deng ZJ, Morton SW, Ben-Akiva E, Dreaden EC, Shopsowitz KE, Hammond PT (2013) Layer-by-layer nanoparticles for systemic codelivery of an anticancer drug and siRNA for potential triple-negative breast cancer treatment. ACS Nano 7(11):9571–9584, http://doi.org/10.1021/nn4047925

Devita VT, Moxley JH, Brace K, Frei E III (1965) Intensive combination chemotherapy and X-irradiation in the treatment of Hodgkin's disease. Ant Intern Med 6:881–895

Devita VT, Serpick AA, Carbone PP (1970) Combination chemotherapy in the treatment of advanced Hodgkin's disease. Ann Intern Med 73(6):881–895, http://doi.org/10.7326/0003-4819-73-6-881

Di L, Kerns EH, Carter GT (2009) Drug-like property concepts in pharmaceutical design. Curr Pharm Des 15(19):2184–2194

Dhar S, Kolishetti N, Lippard SJ, Farokhzad OC (2011) Targeted delivery of a cisplatin prodrug for safer and more effective prostate cancer therapy in vivo. Proc Natl Acad Sci U S A 108(5):1850–1855, http://doi.org/10.1073/pnas.1011379108

Dong Y, Feng S-SS-S (2004) Methoxy poly(ethylene glycol)-poly(lactide) (MPEG-PLA) nanoparticles for controlled delivery of anticancer drugs. Biomaterials 25(14):2843–2849

Dranoff G (2004) Cytokines in cancer pathogenesis and cancer therapy. Nat Rev Cancer 4(1):11–22

Dreaden EC et al (2015) Tumor-targeted synergistic blockade of MAPK and PI3K from a layer-by-layer nanoparticle. Clin Cancer Res 21(19):4410–4419

Duncan R (2006) Polymer conjugates as anticancer nanomedicines. Nat Rev Cancer 6(9):688–701

Elbashir SM et al (2001) Duplexes of 21-nucleotide RNAs mediate RNA interference in cultured mammalian cells. Nature 411(6836):494–498

Emilienne Soma C et al (2000) Reversion of multidrug resistance by co-encapsulation of doxorubicin and cyclosporin A in polyalkylcyanoacrylate nanoparticles. Biomaterials 21(1):1–7

Engelman JA et al (2008) Effective use of PI3K and MEK inhibitors to treat mutant Kras G12D and PIK3CA H1047R murine lung cancers. Nat Med 14(12):1351–1356

Farber S et al (1948) Temporary remissions in acute leukemia in children produced by folic acid antagonist, 4-aminopteroyl-glutamic acid (aminopterin). New England J Med 238(23):787–793

Fletcher JI et al (2010) ABC transporters in cancer: more than just drug efflux pumps. Nat Rev Cancer 10(2):147–156

Floyd SR et al (2013) The bromodomain protein Brd4 insulates chromatin from DNA damage signalling. Nature 498(7453):246–250

Fonseca C, Simões S, Gaspar R (2002) Paclitaxel-loaded PLGA nanoparticles: preparation, physicochemical characterization and in vitro anti-tumoral activity. J Control Release 83(2):273–286

Foulkes WD, Smith IE, Reis-Filho JS (2010) Triple-negative breast cancer. New England J Med 363(20):1938–1948

Frei E, Freireich EJ, Gehan E, Pinkel D, Holland JF, Selawry O, Haurani F, Spurr CL, Hayes DM, James GW, Rothberg H, Sodee DB, Rundles RW, Schroeder LR, Hoogstraten B, Wolman IJ, Traggis DG, Cooper T, Gendel BR, Ebaugh F, Taylor R (1961) Studies of sequential and combination antimetabolite therapy in acute leukemia: 6-mercaptopurine and methotrexate. Blood 18(4):431–454

Freireich EJ, Karon M, Frei E III (1964) Quadruple combination therapy (VAMP) for acute lymphocytic leukemia of childhood. Proc Am Ass Cancer Res 5:20

Gabizon A et al (1994) Prolonged circulation time and enhanced accumulation in malignant exudates of doxorubicin encapsulated in polyethylene-glycol coated liposomes. Cancer Res 54(4):987–992

Ganta S, Amiji M (2009) Coadministration of paclitaxel and curcumin in nanoemulsion formulations to overcome multidrug resistance in tumor cells. Mol Pharm 6(3):928–939

Giddi HS, Arunagirinathan MA, Bellare JR (2007) Self-assembled surfactant nano-structures important in drug delivery: a review. Indian J Exp Biol 45(2):133–159

Gilman A, Philips FS (1946) The biological actions and therapeutic applications of the B-chloroethyl amines and sulfides. Science (New York, NY) 103(2675):409–436

Gnanasammandhan MK et al (2016) Near-IR photoactivation using mesoporous silica-coated NaYF4:Yb, Er/Tm upconversion nanoparticles. Nat Protoc 11(4):688–713

Goldman A et al (2015) Temporally sequenced anticancer drugs overcome adaptive resistance by targeting a vulnerable chemotherapy-induced phenotypic transition. Nat Commun 6:6139

Goodman LS et al (1946) Nitrogen mustard therapy: use of methyl-bis(beta-chloroethyl)amine hydrochloride and tris(beta-chloroethyl)amine hydrochloride for Hodgkin's disease, lymphosarcoma, leukemia and certain allied and miscellaneous disorders. JAMA 132(3):126–132

Greco F, Vicent MJ (2009) Combination therapy: opportunities and challenges for polymer-drug conjugates as anticancer nanomedicines. Adv Drug Deliv Rev 61(13):1203–1213
Hainaut P, Plymoth A (2013) Targeting the hallmarks of cancer: towards a rational approach to next-generation cancer therapy. Curr Opin Oncol 25(1):50–51
Han H, Davis ME (2013) Single-antibody, targeted nanoparticle delivery of camptothecin. Mol Pharm 10(7):2558–2567
Hanahan D, Weinberg RA (2000) The hallmarks of cancer. Cell 100(1):57–70
Hanahan D, Weinberg RA (2011) Hallmarks of cancer: the next generation. Cell 144(5):646–674
Harasym TO et al (2007) Increased preclinical efficacy of irinotecan and floxuridine coencapsulated inside liposomes is associated with tumor delivery of synergistic drug ratios. Oncol Res 16(8):361–374
Heyn RM et al (1960) The comparison of 6-mercaptopurine with the combination of 6-mercaptopurine and azaserine in the treatment of acute leukemia in children: results of a cooperative study. Blood 15:350–359
Hu C-MJ, Zhang L (2012) Nanoparticle-based combination therapy toward overcoming drug resistance in cancer. Biochem Pharmacol 83(8):1104–1111
Huang P et al (2014) Combination of small molecule prodrug and nanodrug delivery: amphiphilic drug-drug conjugate for cancer therapy. J Am Chem Soc 136(33):11748–11756
Hudis CA (2007) Trastuzumab--mechanism of action and use in clinical practice. New England J Med 357(1):39–51
Jhaveri AM, Torchilin VP (2014) Multifunctional polymeric micelles for delivery of drugs and siRNA. Front Pharmacol 5:77
Jiang T et al (2014) Gel–liposome-mediated co-delivery of anticancer membrane-associated proteins and small-molecule drugs for enhanced therapeutic efficacy. Adv Funct Mater 24(16):2295–2304
Jiang T et al (2015) Furin-mediated sequential delivery of anticancer cytokine and small-molecule drug shuttled by graphene. Adv Mater 27(6):1021–1028
Jordan VC (2003) Tamoxifen: a most unlikely pioneering medicine. Nat Rev Drug Discov 2(3):205–213
Kaelin WG (2005) The concept of synthetic lethality in the context of anticancer therapy. Nat Rev Cancer 5(9):689–698
Kala S et al (2014) Combination of dendrimer-nanovector-mediated small interfering RNA delivery to target Akt with the clinical anticancer drug paclitaxel for effective and potent anticancer activity in treating ovarian cancer. J Med Chem 57(6):2634–2642
Kalos M et al (2011) T cells with chimeric antigen receptors have potent antitumor effects and Can establish memory in patients with advanced leukemia. Sci Transl Med 3(95):95ra73–95ra73
Kaneshiro TL, Lu Z-R (2009) Targeted intracellular codelivery of chemotherapeutics and nucleic acid with a well-defined dendrimer-based nanoglobular carrier. Biomaterials 30(29):5660–5666
Karaca, M. et al., 2016. Micelle Mixtures for Co-administration of Gemcitabine and GDC-0449 to treat Pancreatic Cancer. *Mol. Pharm.*, p.acs.molpharmaceut.5b00971.
Karra N et al (2013) Antibody conjugated PLGA nanoparticles for targeted delivery of paclitaxel palmitate: efficacy and biofate in a lung cancer mouse model. Small 9(24):4221–4236
Katragadda U et al (2013) Combined delivery of paclitaxel and tanespimycin via micellar nanocarriers: pharmacokinetics, efficacy and metabolomic analysis. PLoS One 8(3):e58619
Kaye SB (1993) P glycoprotein (P-gp) and drug resistance--time for reappraisal? Br J Cancer 67(4):641–643
Keung AJ et al (2014) Using targeted chromatin regulators to engineer combinatorial and spatial transcriptional regulation. Cell 158(1):110–120
Kim E et al (2010) Prostate cancer cell death produced by the co-delivery of Bcl-xL shRNA and doxorubicin using an aptamer-conjugated polyplex. Biomaterials 31(16):4592–4599
Kim CS et al (2015) Co-delivery of protein and small molecule therapeutics using nanoparticle-stabilized nanocapsules. Bioconjug Chem 26(5):950–954

Kneuer C et al (2000) Silica nanoparticles modified with aminosilanes as carriers for plasmid DNA. Int J Pharm 196(2):257–261
Kolishetti N et al (2010) Engineering of self-assembled nanoparticle platform for precisely controlled combination drug therapy. Proc Natl Acad Sci U S A 107(42):17939–17944
Kong YW et al (2012) microRNAs in cancer management. Lancet Oncol 13(6):e249–e258
Kormann MSD et al (2011) Expression of therapeutic proteins after delivery of chemically modified mRNA in mice. Nat Biotechnol 29(2):154–157
Kulhari H, Pooja D, Prajapati SK, Chauhan AS (2011) Performance evaluation of PAMAM dendrimer based simvastatin formulations. Int J Pharm 405(1–2):203–209, http://doi.org/10.1016/j.ijpharm.2010.12.002
Lammers T et al (2009) Simultaneous delivery of doxorubicin and gemcitabine to tumors in vivo using prototypic polymeric drug carriers. Biomaterials 30(20):3466–3475
Lee JH, Nan A (2012) Combination drug delivery approaches in metastatic breast cancer. J Drug Delivery 2012(8):1–17
Lee ALZ et al (2011) Synergistic anticancer effects achieved by co-delivery of TRAIL and paclitaxel using cationic polymeric micelles. Macromol Biosci 11(2):296–307
Lee MJ, Ye AS, Gardino AK, Heijink AM, Sorger PK, MacBeath G, Yaffe MB (2012) Sequential application of anticancer drugs enhances cell death by rewiring apoptotic signaling networks. Cell 149(4):780–794, http://doi.org/10.1016/j.cell.2012.03.031
Li Y et al (2001) PEGylated PLGA nanoparticles as protein carriers: synthesis, preparation and biodistribution in rats. J Control Release 71(2):203–211
Li S-D et al (2008) Tumor-targeted delivery of siRNA by self-assembled nanoparticles. Mol Ther 16(1):163–169
Li F et al (2015) Multiple layer-by-layer lipid-polymer hybrid nanoparticles for improved FOLFIRINOX chemotherapy in pancreatic tumor models. Adv Funct Mater 25(5):788–798
Liao L et al (2014) A convergent synthetic platform for single-nanoparticle combination cancer therapy: ratiometric loading and controlled release of cisplatin, doxorubicin, and camptothecin. J Am Chem Soc 136(16):5896–5899
Longley DB, Harkin DP, Johnston PG (2003) 5-Fluorouracil: mechanisms of action and clinical strategies. Nat Rev Cancer 3(5):330–338
Luo B et al (2008) Highly parallel identification of essential genes in cancer cells. Proc Natl Acad Sci U S A 105(51):20380–20385
Luo J et al (2009) A genome-wide RNAi screen identifies multiple synthetic lethal interactions with the Ras oncogene. Cell 137(5):835–848
Machover D et al (1982) Treatment of advanced colorectal and gastric adenocarcinomas with 5-FU combined with high-dose folinic acid: a pilot study. Cancer Treat Rep 66(10):1803–1807
Mahoney KM, Rennert PD, Freeman GJ (2015) Combination cancer immunotherapy and new immunomodulatory targets. Nat Rev Drug Discov 14(8):561–584
Matsumoto Y et al (2016) Vascular bursts enhance permeability of tumour blood vessels and improve nanoparticle delivery. Nat Nanotechnol 11(6):533–538
Matsumura Y, Maeda H (1986) A new concept for macromolecular therapeutics in cancer chemotherapy: mechanism of tumoritropic accumulation of proteins and the antitumor agent smancs. Cancer Res 46(12):6387–6392
Mayer LD, Janoff AS (2007) Optimizing combination chemotherapy by controlling drug ratios. Mol Interv 7(4):216–223
Meng H, Mai WX, Zhang H, Xue M, Xia T, Lin S, Wang X, Zhao Y, Ji Z, Zink JI, Nel AE (2013) Codelivery of an optimal drug/siRNA combination using mesoporous silica nanoparticles to overcome drug resistance in breast cancer in vitro and in vivo. ACS Nano 7:994–1005
Meng H et al (2015) Use of a lipid-coated mesoporous silica nanoparticle platform for synergistic gemcitabine and paclitaxel delivery to human pancreatic cancer in mice. ACS Nano 9(4):3540–3557
Mi Y, Zhao J, Feng S-S (2013) Targeted co-delivery of docetaxel, cisplatin and herceptin by vitamin E TPGS-cisplatin prodrug nanoparticles for multimodality treatment of cancer. J Control Release 169(3):185–192

Misra R et al (2014) Reversal of multidrug resistance in vitro by co-delivery of MDR1 targeting siRNA and doxorubicin using a novel cationic poly(lactide-co-glycolide) nanoformulation. Int J Pharm 475(1–2):372–384

Morsut L et al (2016) Engineering customized cell sensing and response behaviors using synthetic notch receptors. Cell 164(4):780–791

Morton, S.W. et al., 2014. A nanoparticle-based combination chemotherapy delivery system for enhanced tumor killing by dynamic rewiring of signaling pathways. *Science signaling*, 7(325), pp.ra44–ra44. Available at: http://eutils.ncbi.nlm.nih.gov/entrez/eutils/elink.fcgi?dbfrom=pubmed&id=24825919&retmode=ref&cmd=prlinks.

Moxley JH et al (1967) Intensive combination chemotherapy and X-irradiation in Hodgkin's disease. Cancer Res 27(7):1258–1263

Nagpal K, Singh SK, Mishra DN (2010) Chitosan nanoparticles: a promising system in novel drug delivery. Chem Pharm Bull 58(11):1423–1430

Nassirpour R et al (2013) miR-221 promotes tumorigenesis in human triple negative breast cancer cells. A. Ahmad, ed. PLoS One 8(4):e62170

Ngo VN et al (2006) A loss-of-function RNA interference screen for molecular targets in cancer. Nature 441(7089):106–110

Nishiyama N, Kataoka K (2006) Current state, achievements, and future prospects of polymeric micelles as nanocarriers for drug and gene delivery. Pharmacol Ther 112(3):630–648

Park K (2013) Facing the truth about nanotechnology in drug delivery. ACS Nano 7(9):7442–7447

Park J et al (2012) Combination delivery of TGF-β inhibitor and IL-2 by nanoscale liposomal polymeric gels enhances tumour immunotherapy. Nat Mater 11(10):895–905

Patil Y et al (2009) Nanoparticle-mediated simultaneous and targeted delivery of paclitaxel and tariquidar overcomes tumor drug resistance. J Control Release 136(1):21–29

Peer D et al (2007) Nanocarriers as an emerging platform for cancer therapy. Nat Nanotechnol 2(12):751–760

Perrault SD et al (2009) Mediating tumor targeting efficiency of nanoparticles through design. Nano Lett 9(5):1909–1915

Piovan C et al (2012) Oncosuppressive role of p53-induced miR-205 in triple negative breast cancer. Mol Oncol 6(4):458–472

Platt RJ et al (2014) CRISPR-Cas9 knockin mice for genome editing and cancer modeling. Cell 159(2):440–455

Prabhakar, U. et al., 2013. Challenges and key considerations of the enhanced permeability and retention effect for nanomedicine drug delivery in oncology. In Cancer research. American Association for Cancer Research, pp. 2412–2417.

Pritchard JR et al (2013) Defining principles of combination drug mechanisms of action. Proc Natl Acad Sci U S A 110(2):E170–E179

Qhobosheane M et al (2001) Biochemically functionalized silica nanoparticles. Analyst 126(8):1274–1278

Ramsey S (2005) What do we want from our investment in cancer research? Health Aff 24:W5R101–W5R104

Ribas A et al (2013) Hepatotoxicity with combination of vemurafenib and ipilimumab. New England J Med 368(14):1365–1366

Rösler A, Vandermeulen GW, Klok HA (2001) Advanced drug delivery devices via self-assembly of amphiphilic block copolymers. Adv Drug Deliv Rev 53(1):95–108

Rothenberg M, Ling V (1989) Multidrug resistance: molecular biology clinical relevance. J Natl Cancer Inst 81(12):907–910

Rozema DB et al (2007) Dynamic PolyConjugates for targeted in vivo delivery of siRNA to hepatocytes. Proc Natl Acad Sci U S A 104(32):12982–12987

Saad M, Garbuzenko OB, Minko T (2008) Co-delivery of siRNA and an anticancer drug for treatment of multidrug-resistant cancer. Nanomedicine 3(6):761–776

Santra S et al (2001) Conjugation of biomolecules with luminophore-doped silica nanoparticles for photostable biomarkers. Anal Chem 73(20):4988–4993

Schmidt MM, Wittrup KD (2009) A modeling analysis of the effects of molecular size and binding affinity on tumor targeting. Mol Cancer Ther 8(10):2861–2871

Seino M et al (2016) Time-staggered inhibition of JNK effectively sensitizes chemoresistant ovarian cancer cells to cisplatin and paclitaxel. Oncol Rep 35(1):593–601

Sengupta S et al (2005) Temporal targeting of tumour cells and neovasculature with a nanoscale delivery system. Nature 436(7050):568–572

Shen J et al (2014) Simultaneous inhibition of tumor growth and angiogenesis for resistant hepatocellular carcinoma by co-delivery of sorafenib and survivin small hairpin RNA. Mol Pharm 11(10):3342–3351

Shi M et al (2009) Doxorubicin-conjugated immuno-nanoparticles for intracellular anticancer drug delivery. Adv Funct Mater 19:1689–1696

Shim G et al (2011) Trilysinoyl oleylamide-based cationic liposomes for systemic co-delivery of siRNA and an anticancer drug. J Control Release 155(1):60–66

Shimizu T et al (2012) The clinical effect of the dual-targeting strategy involving PI3K/AKT/mTOR and RAS/MEK/ERK pathways in patients with advanced cancer. Clin Cancer Res 18(8):2316–2325

Shin H-C et al (2011) A 3-in-1 polymeric micelle nanocontainer for poorly water-soluble drugs. Mol Pharm 8(4):1257–1265

Shin H-C et al (2012) Pharmacokinetic study of 3-in-1 poly(ethylene glycol)-block-poly(D, L-lactic acid) micelles carrying paclitaxel, 17-allylamino-17-demethoxygeldanamycin, and rapamycin. J Control Release 163(1):93–99

Skipper HE et al (1954) Observations on the anticancer activity of 6-mercaptopurine. Cancer Res 14(4):294–298

Sokolova V, Epple M (2008) Inorganic nanoparticles as carriers of nucleic acids into cells. Angew Chem Int Ed 47(8):1382–1395, http://doi.org/10.1002/anie.200703039

Song XR et al (2009) Reversion of multidrug resistance by co-encapsulation of vincristine and verapamil in PLGA nanoparticles. Eur J Pharm Sci 37(3–4):300–305

Stylianopoulos T et al (2012) Multistage nanoparticles for improved delivery into tumor tissue. In: D Nejat (ed) Nanomedicine. Academic Press, pp 109–130

Tardi P, Johnstone S et al (2009a) In vivo maintenance of synergistic cytarabine:daunorubicin ratios greatly enhances therapeutic efficacy. Leuk Res 33(1):129–139

Tardi PG, Dos Santos N et al (2009b) Drug ratio-dependent antitumor activity of irinotecan and cisplatin combinations in vitro and in vivo. Mol Cancer Ther 8(8):2266–2275

Tekade RK et al (2008) Surface-engineered dendrimers for dual drug delivery: a receptor up-regulation and enhanced cancer targeting strategy. J Drug Target 16(10):758–772

Tekade RK et al (2009) Exploring dendrimer towards dual drug delivery: pH responsive simultaneous drug-release kinetics. J Microencapsul 26(4):287–296

Thiagarajan G, Greish K, Ghandehari H (2013) Charge affects the oral toxicity of poly(amidoamine) dendrimers. Eur J Pharm Biopharm 84(2):330–334

Tobío M et al (1998) Stealth PLA-PEG nanoparticles as protein carriers for nasal administration. Pharm Res 15(2):270–275

Vaklavas C, Forero-Torres A (2012) Safety and efficacy of brentuximab vedotin in patients with Hodgkin lymphoma or systemic anaplastic large cell lymphoma. Ther Adv Hematol 3(4):209–225

van Vlerken LE et al (2010) Augmentation of therapeutic efficacy in drug-resistant tumor models using ceramide coadministration in temporal-controlled polymer-blend nanoparticle delivery systems. AAPS J 12(2):171–180

Vivero-Escoto J, Elnagheeb M (2015) Mesoporous silica nanoparticles loaded with cisplatin and phthalocyanine for combination chemotherapy and photodynamic therapy in vitro. Nanomaterials 5(4):2302–2316

Wang Y et al (2006) Co-delivery of drugs and DNA from cationic core-shell nanoparticles self-assembled from a biodegradable copolymer. Nat Mater 5(10):791–796

Wei W, Lv P-P, Chen X-M, Yue Z-G, Fu Q, Liu S-Y et al (2013) Codelivery of mTERT siRNA and paclitaxel by chitosan-based nanoparticles promoted synergistic tumor suppression. Biomaterials 34(15):3912–3923, http://doi.org/10.1016/j.biomaterials.2013.02.030
Wilhelm S et al (2016) Analysis of nanoparticle delivery to tumours. Nat Rev Mater 1(5):16014
Williams HD et al (2013) Strategies to address low drug solubility in discovery and development. Pharmacol Rev 65(1):315–499
Wong M-Y, Chiu GNC (2010) Simultaneous liposomal delivery of quercetin and vincristine for enhanced estrogen-receptor-negative breast cancer treatment. Anticancer Drugs 21(4):401–410, http://doi.org/10.1097/CAD.0b013e328336e940
Wu J et al (2007) Reversal of multidrug resistance by transferrin-conjugated liposomes co-encapsulating doxorubicin and verapamil. J Pharm Pharm Sci 10(3):350–357
Xu X et al (2013) Enhancing tumor cell response to chemotherapy through nanoparticle-mediated codelivery of siRNA and cisplatin prodrug. Proc Natl Acad Sci U S A 110(46):18638–18643
Yang J et al (2006) Preparation of poly epsilon-caprolactone nanoparticles containing magnetite for magnetic drug carrier. Int J Pharm 324(2):185–190
Zhang L, Radovic-Moreno AF, Alexis F, Gu FX, Basto PA, Bagalkot V, Jon S, Langer RS, Farokhzad OC (2007) Co-delivery of hydrophobic and hydrophilic drugs from nanoparticle–aptamer bioconjugates. ChemMedChem 2:1268–1271
Zhang N et al (2013) Trastuzumab-doxorubicin conjugate provides enhanced anti-cancer potency and reduced cardiotoxicity. J Cancer Ther 04(01):308–322
Zhang Y et al (2016) Polymeric prodrug grafted hollow mesoporous silica nanoparticles encapsulating near-infrared absorbing dye for potent combined photothermal-chemotherapy. ACS Appl Mater Interfaces 8(11):6869–6879
Zheng C et al (2013) Polypeptide cationic micelles mediated co-delivery of docetaxel and siRNA for synergistic tumor therapy. Biomaterials 34(13):3431–3438
Zhu C et al (2010) Co-delivery of siRNA and paclitaxel into cancer cells by biodegradable cationic micelles based on PDMAEMA-PCL-PDMAEMA triblock copolymers. Biomaterials 31(8):2408–2416
Zoli W et al (2001) In vitro preclinical models for a rational design of chemotherapy combinations in human tumors. Crit Rev Oncol Hematol 37(1):69–82

Part VIII
Market Situation and Commercialization of Nanotechnology

Chapter 17
The Commercialization of Medical Nanotechnology for Medical Applications

David W. Hobson

Abstract The potential for nanotechnology to advance medical science has been demonstrated and is increasing as the ability to manufacture and combine an ever expanding selection of nanotechnologies creates a wide variety of opportunities. Mechanisms for the commercialization of these "nanomedical" products have met with both success and failure but always with valuable lessons to apply toward advancing the future for the technology platforms. Nanotechnology applications in medicine are already having a significant impact on marketed products as well as increasing the number of commercially viable products in both pharmaceutical and medical device pipelines directed toward a wide variety of therapeutic applications. Commercial uses of nanotechnology in the development and formulation of new pharmaceuticals as well as incorporation into the designs of medical devices are increasing and hold much promise for commercial success. Regulatory agencies and legal firms worldwide are cautiously examining some aspects of many nanotechnology products toward demonstrating sustainable histories of safety and efficacy in these various medical applications. Approaches that have been successful in commercializing approved and currently marketed nanomedical products have met with many challenges and have identified and resolved critical issues that could have prevented or hindered market entry. The commercial path becoming increasingly better defined for new and developing products that hold much promise to significantly improve medicine in areas such as nanomaterial-enabled pharmaceuticals, targeted nanomolecular therapeutics and diagnostics, nanobrachytherapy agents and nano-enabled medical devices.

Keywords Nanotechnology • Nanomaterial • Nanoparticle nanomedicine • Nanosurgery • Commercialization • Marketing • Toxicology • Risk assessment • Valuation • Investment

---

D.W. Hobson (✉)
LoneStar PharmTox LLC, P.O.Box 4169, Bergheim, Texas 78004, USA
e-mail: dave@lonestarpharmtox.com

A. Prokop, V. Weissig (eds.), *Intracellular Delivery III*, Fundamental Biomedical Technologies 8, DOI 10.1007/978-3-319-43525-1_17

Abbreviations

AFM	Atomic force microscopy
EMA	European Medicines Agency
FDA	United States Food and Drug Administrations
MRI	Magnetic Resonance Imaging
PET	positron emission tomography
SEM	Scanning electron microscopy
STP	Scanning tunneling microscopy

17.1 Introduction

Nanotechnology is an established, advancing, and cutting edge technology platform with demonstrated potential to change the world well into the foreseeable future. Technically, it is the controlling and manipulating of matter at the atomic and/or molecular scale and deals with structures the size of 100 nm or smaller in at least one dimension. As a technology platform it is highly diverse and holds much potential as a foundation for invention and innovative advancements in many industries including many aspects of medicine and medical practice.

On the forefront of material science, nanotechnology includes the ability to manipulate atoms and create molecular structures at will that could include essentially any nuclide that is stable enough to complete the creation of the molecular structure.. This platform represents an amazing and landmark ability for mankind to interact with and manipulate essentially any stable atomic material in the known universe to create novel molecular structures.

It is no wonder then that we should wish to employ this fantastic ability toward the improvement of health and quality of life through the discovery, development, successful application and commercialization of medically, agriculturally and environmentally oriented nanotechnology.

Innovative medical applications of nanotechnologies with a profound impact on medical practice and health care have been widely anticipated and are only now becoming more commonplace in a variety of medical practice areas (Bleeker et al. 2015).

The use of nanotechnology in medical products is not an entirely recent concept. In rudimentary forms, nanotechnology has been used in medical products for a few decades, often without recognition in the form of products containing nanoparticles and/or nanoemulsions. Amazing and wonderful uses of nanomedical technology were envisioned decades ago as well. In fact Richard Feynman himself along with his friend Albert Hibbs suggested the possibility of placing a "mechanical surgeon" inside a blood vessel that would look around in the heart and correct faulty valves almost 60 years ago (Feynman 1959). What Feynman and Hibbs imagined in the past is now advancing rapidly and is close to becoming reality as we look to and engage the future of nanotechnology.

Applications of medical nanotechnology are abundant. The application of nanotechnology in medicine, referred to as nanomedicine, is offering numerous exciting possibilities in healthcare. These include opportunities in the areas of pharmaceuticals, diagnostics, surgical instruments, medical robotics, imaging technology, prosthetics, etc. and wherever our imagination might take us.

Cancer therapeutics have been of considerable medical nanotechnology research and development interest (Kolhe and Parikh 2012; Park et al. 2016; Piktel et al. 2016).

The use of nanotechnology in the development of microsurgery is essential and seeks to enable the development of surgical procedures and methods that are more precise and accurate in the targeting of surgical sites. Enhancing surgical site visualization, reducing trauma, improving outcomes and shortening recovery time are all anticipated as these new techniques will enable surgery to be done at the tissue, cellular and even genetic levels (Kohli and Elezzabi 2009; Ronchi et al. 2012). Nanotechnology also can be used in combination with traditional surgery as well as microsurgery for regenerative therapeutic effects (Tan et al. 2016).

Nanotechnology development in robotics and tissue engineering areas is of high interest. Three-dimensional (3-D) tissue printing that is leading toward the printing of surgical aids, prosthetics and perhaps even synthetic organs has been advanced extensively by advancements with and incorporation of nanotechnology (Tamjid et al. 2013; AlAli et al. 2015; de Mel 2016).

The 1996 Nobel laureate, Richard Smalley, recognized that "human health has always been determined on the nanometer scale; this is where the structural properties of the machines of life work in every one of the cells in every living thing. The practical impact of nanoscience on human health will be huge (Burgess 2012; Godin and Ferrari 2012)".

Of course for any of these visions of nanotechnology to come to practical use they must be discovered, developed and successfully commercialized. In fact, without successful commercialization, full realization of these visions for nanomedical technology cannot occur. So, cutting-edge nano material science and engineering funding and support for basic research enables the discoveries and inventions that create opportunities for future commercialization of new and better medical products. However, without commercial interest and those that are skilled and willing to invest and risk financial loss in ventures of potential significant medical and financial gain, these new generation nanomedical products and technologies cannot come to full fruition and have opportunity for commercial success. Commercialization is the nexus or process that joins the world of basic medical research invention and innovation with the business of medicine.

Fortunately, over past decades the potential for nanotechnology to advance medical science has shown promise and in at least a few instances, commercial success has been realized and demonstrated to both the medical as well as business communities. Unfortunately, there has also been failure and financial loss. This has led to increasing but cautious interest in the business sector as to the ability to develop, manufacture and combine an ever increasing selection of new and innovative nanotechnologies into useful nanomedical products. Nevertheless, the rapidly growing number of inventions and patents in nanomedical technology creates a wider variety of opportunities than were possible with earlier nanotechnology offerings and investor interest, while cautious, remains high.

Although past experiences with commercialization of "nanomedical" products has had mixed results, there have been valuable lessons learned to apply toward advancing the future and increasing the commercial success for nanomedical technology platforms.

This chapter provides a brief history of the commercialization of nanotechnology for use in medical applications as well as some of what has been learned toward increasing the potential for successful commercialization of medical nanoproducts. Nanotechnology applications in medicine are already having a significant impact on marketed products as well as increasing the number of commercially viable products in both pharmaceutical and medical device pipelines. These products are directed toward a very wide variety of therapeutic applications, leaving essentially no area of medicine untouched. There are now clear examples of uses of nanotechnology in the formulation of new pharmaceuticals as well as incorporation into new medical devices. The medical industry and regulatory agencies worldwide are demonstrating strong commitments toward establishing effective and practical approaches to demonstrate the safety and efficacy of these emerging new nanomedical products. Successful experience in commercializing now approved and currently marketed nanomedical products is available and can be examined and reviewed with hope to identify issues and to improve the commercialization process and success for many new and developing products such as targeted nanomolecular therapeutics, nanobrachytherapy agents and nano-enabled medical devices.

17.2 Historical Advancement of Medical Nanotechnology

The earliest intentional use of nanotechnology in medical products involved the use of nanoparticles in the formulation of products such as topical pharmaceuticals and sunscreen formulations (Korting et al. 1990; Bennat and Muller-Goymann 2000; Alvarez-Roman et al. 2001). This level of nanotechnology development has been referred to as "first generation" nanotechnology (Roco 2004). Examples include the use of zinc or titanium nanoparticles to block ultraviolet light in sunscreen formulations, the creation of nanoemulsions of oil in water or water in oil to enhance the solubility and stability of active pharmaceutical agents in various drug products, and the manufacture of nano-particulate carriers of active pharmaceutical agents with various nanostructures to improve target tissue delivery.

The antimicrobial properties of nanoparticles were recognized decades ago and have been successfully employed, especially silver nanoparticles, in bandages, as coatings on medical devices or in suspensions used for wound and burn treatments (Morones et al. 2005). Not too surprisingly, very soon after these initial uses of nanoparticles in medical products,, there were concerns of potential nanoparticle toxicity. This concern led to increased scrutiny and launched investigations related to the safe use of nanoparticles in sunscreens as well as other medical applications that is now an established interest and topic of research interest at toxicology meetings and in the toxicology literature contributing to the new toxicology subdiscipline of "nanotoxicology" (Borm et al. 2006; Nohynek et al 2007).

Historically, the technologic and commercial advancement of nanotechnology from the first generation of nanoparticle applications to increasingly more complex and potentially profitable "next" generation nanomedical products has occurred almost as anticipated by Roco (2004) to include four or five "generations" of nanotechnology. Roco (2004) identified and described four generations of nanotechnology development with the then current generation, as that of passive nanostructures and went on to describe time progressive, 2nd, 3rd and 4th generations as active nanostructures, systems of nano-systems and molecular nano-systems., This concept of generations has been reasonably accurate over the past few decades and is useful in understanding the levels of complexity of nanomedical technologies that have been and are now emerging from invention and basic research and moving toward commercialization. Originally, Roco (2004) anticipated a future time after the 4th generation when products from all of the previous four generations of nanotechnology would advance and be incorporated into more complex products that would not otherwise be possible. This "combinatorial" end state envisioned by Rocco (2004) has been recognized and termed a 5th generation by others and is identified as a generation where not only the technologies of previous generations converge but also where this convergence results in the potential for infinite growth of medical nanotechnology that Kennedy (2007) has termed a "singularity." The following descriptions to include the views of Rocco (2004) plus a 5th generation are provided to incorporate the types of medical nanoproducts associated with each generation and their approximate (~) years of initial emergence as useful construct for relating the time when a nanotechnology platform was recognized to its maturity to create nanomedical product opportunities for commercialization.

1st Generation (~1980s) – Passive Nanostructures

- Use of nanoparticles in pharmaceuticals as active ingredients, formulation excipients, polymers, coatings of drugs and medical devices, etc.

2nd Generation (~2000) – Active Nanosctructures

- Nanotechnology targeted pharmaceuticals, adaptive structures in formulations and medical devices such as receptor or immunologically targeted pharmaceuticals and the use of "stealth" coatings on nanoparticles to avoid immune system recognition.

3rd Generation (~2005) – Large Scale Nanosystems

- Nanotechnology-enabled (nano-enabled) medical diagnostic agents that are detected interrogated and/or imaged by nano-enabled medical devices or equipment. Therapeutic agents that incorporate nanotechnology that are activated, manipulated or detected by nano-enabled devices and/or medical equipment.

4th Generation (~2010) – Molecular Nanosystems

- The use of atomic force microscopy and other nanoengineering tools such as nanolithography (e.g. optical, X-ray and dip pen lithography) to design molecular devices and therapeutics that function "by design" based on their unique and engineered molecular structures. Molecules designed and enabled by nanoengineering for disease intervention and cancer treatment at the subcellular level would be examples of this generation.

5th Generation (~2015) – Converging Nanotechnologies and Infinite Growth

- Termed the "Singularity" by Kennedy (2007) and not defined as a distinct generation by Roco (2004), this current timeframe is exhibiting a distinct emergence of large, complex, technology platforms and processes that utilize a variety of different technologies and nanotechnologies from different applications that converge to create "convergence" nano products such as the use of nano-enabled three dimensional (3D) printing of nano-enabled solutions to result in 3D printed viable tissues and organs. Without such convergence the platform would not be possible. This represents an ultimate level of technology convergence that is increasingly including technologies that are expected to result in "self" recognition, repair and even modification within limits that are only left to imagination and the limits of technology.

This "generations" approach is useful in describing the stage-wise emergence and advancement of medical nanotechnology and can be useful in developing perspective as to the maturity, possibilities and promise of generational technology platforms. A view toward commercial development within each generation results in a stratified focus on the potential for alternate or competitive technologies with similar levels of complexity to compete. One might envision or even create companies or business entities (which has actually been done) organized by these generations such as ACME Active Nanotechnologies, ACME Medical Nanoparticules or ACME Nanosystems Inc., etc. that build portfolios of products around proprietary nanotechnology platforms that focus on and advance within a single generation of nanotechnology advancement. This single generation level of commercialization generally will produce technologies that enable medical nanotechnologies but are not focused principally on commercializing the nanomedical technologies. Companies principally focused on nanomedical technology commercialization focus on specific applications of medical nanotechnology and the problems or unmet needs they address and seek to innovate and commercialize solutions based on available nanotechnology across all relevant technology platforms/generations and further research advancements leading to new intellectual property. This view is more readily defined by the entrepreneurs and markets served and is also is more easily and effectively presented to various funding sources including investors that are typically focused on market size, opportunity, innovation, intellectual property position and return on investment.

Concern about nanoproduct safety has grown along with the development of increasingly complex nanotechnologies and now includes the potential for immunotoxicity as well as the more traditional concerns for carcinogenicity as well as reproductive, developmental, genetic and endocrine toxicity (David et al. 2016; Lorscheidt and Lamprecht 2016; Radomska et al. 2016).. Interestingly, this increased concern has resulted in new research opportunities to improve toxicological testing and develop new screening tests (Nel et al. 2013; Huo et al. 2015; Delaval et al. 2016). This has created a new and developing market for equipment capable of supporting this testing through improvements in the characterization of and quantification of nanomaterial exposures used in safety testing as well as in determining the molecular localization of nanomaterials within the test system following exposure (Wang et al. 2016). So, in effect, the advancement of nanotechnology not only creates commercialization opportunities for the nanomaterials themselves but also for equipment, processes and techniques to manufacture and test the nanotechnologies as well.

Identifying promising nanotechnology products from a commercialization view is generally more practical by identifying product focus areas and markets for medical nanotechnology and then aligning the emerging opportunity with the market and determining at what stage of commercialization (see Sect. 17.4, The Commercialization Process below) the particular product opportunity is at and how much investment may be required.

Current nanomedical product commercialization focus areas include diagnostics, imaging, pharmaceutical, therapeutic, biomaterial, tissue engineering and medical devices. There are, of course, many other possible commercialization focus areas/markets that will emerge but those described below will provide a basis for understanding the current and advancing nature of the many product opportunities that are available for emerging nanomedical technology platforms to address.

17.2.1 Diagnostics

Nanotechnology-based sensors (e.g. nanowires, nanotubes, nanoparticles, cantilevers, and micro-/nanoarrays) can enable fast and high throughput detection of disease biomarkers with higher sensitivity and lower sample consumption. Nanotechnology also offers hope for the early detection of viruses, bacteria, and circulating tumor cells, as well as for single cell analysis.

A theranostic system could be interpreted as a synergistic combination of therapeutic compounds with diagnostic agents that promise better prognosis and improved treatments for many diseases including cancer. Coincident with the development of nanoparticle systems for drug delivery, the development of theranostic systems have also largely benefited from advancements in nanoscience, since many nanoparticle-based delivery platforms represent a natural and logical choice for constructing nano-enabled theranostic systems. Such "nanotheranostic" systems have been conceived and are currently advancing toward commercialization in a number of areas (Cui and Wang 2016).

One very active area of nanotheragnostic product development is in cancer diagnosis and treatment (Roy et al. 2016). The use of nanotechnology carriers for cancer cell targeting and to improve the solubility and stability of antineoplastic therapeutics is becoming more popular as molecular targeting of cancer comes to the forefront of state of the art therapeutics. Along with this comes the ability to incorporate a ligand or label into or onto the nanostructure that allows the localization and diagnostic monitoring of the structure (Gautier et al. 2013; Shapira et al. 2011)

17.2.2 Imaging

Targeted imaging nanoprobes and theragnostics (e.g. magnetic nanoparticles, liposomes, dendrimers, quantum dots) may form platforms for less invasive and more accurate means to diagnose diseases (e.g. cancer) at their earliest stages and monitor disease progression. Other possible opportunities include reporting the *in vivo* efficacy of therapeutics, monitoring the biodistribution and kinetics of the nanocarrier in the body and locating tumors and their margins.

It has long been recognized that nanoparticles are selectively taken up by the mononuclear phagocyte system (MPS) and sequestered until they are metabolized or localized in the MPS cells indefinitely (Rolland et al. 1989). Nanoparticles serve as candidates to image the MPS, which comprises liver, spleen and the lymphatics. The majority of nanoparticles are taken up by the MPS due to the high content of macrophages present in those tissues (Baetke et al. 2015).

Nanoparticles (<5 nm) may be used as molecular diagnostics for extravascular targets. Therapeutic nanoparticle formulations are often designed to have slow renal excretion and long blood circulation times to help improve target site accumulation and to increase therapeutic efficacy. Shorter circulation times are often preferred for target-specific diagnostic nanoparticles to help lower the target-unspecific background and to improve the non-clinical translation to clinical imaging

Recently, a first-in-human clinical trial by Phillips et al. (2014) demonstrated the potential diagnostic application of 6–7 nm large hybrid silica nanoparticles (C dots) for hybrid PET-optical imaging agents for lesion detection, cancer staging, and treatment management in patients with metastatic melanoma.

17.2.3 Pharmaceuticals

Nanoscale delivery vehicles can (1) enhance the therapeutic efficacy and minimize adversities associated with available drugs; (2) enable new classes of therapeutics; and (3) encourage the re-investigation of pharmaceutically suboptimal but biologically active new molecular entities that were previously considered undevelopable.

Much recent interest has been directed toward the development of stealth, ligand-anchored, multifunctional nanocarriers which have included: radioisotope-polyethylene glycol (PEG) – human serum albumin (HA) nanoparticles; hyaluronate-targeted, PEGylated poly(lactic-co-glycolic acid; PLGA) and polycaprolactone (PCL) core-corona nanoparticles; chondroitin sulphate and HA-coupled polyethyleneimine nanoparticles, folate, methotrexate and fluorescent dye trifunctionalized multiwalled carbon nanotubes;, hyaluronate and estradiol-coupled multiwalled carbon nanotubes; and dye -coated liposomes (Jain et al. 2015a, b).

Photodynamic nanoparticles and molecularly-targeted nanoprobes strengthen the concept and possibility of personalized, theragnostic nanomedicines for the potential early detection of lesions, noninvasive imaging of disease-associated molecular signatures and molecularly targeted therapeutics (Garg et al. 2015; Yang et al. 2016).

Unlike small molecule drugs, drug-containing particles (drug-loaded nanoparticles) are taken up by the M cells in Peyer's patches in the lymphoid tissue of the small intestine after oral administration, and are transported from the gastrointestincal tract lumen to intraepithelial lymphoid cells and then into the bloodstream through the lymphatic system (Shakweh et al. 2004; Lopes et al. 2014). This special transport pathway can be employed to enhance the bioavailability of nanoparticle-encapsulated drugs, while avoiding enzymatic degradation in enterocytes, first-pass metabolism in the liver and simultaneous reduction in dose- and drug-associated toxicity. This allows for the formulation of anticancer drugs and other drugs within PLGA nanoparticles to be absorbed unchanged into the vascular system following oral administration. This has been used to increase oral dose absorption of some active pharmaceutical ingredients with a reduction in hepatotoxicty compared with oral administration of the small molecule drug in an oral vehicle. The successful utilization of folate-coupled, PEGylated PLGA nanoparticles for enhancing the oral bioavailability of the antidiabetic drug, insulin is a significant achievement of this nature, due to its high degree of enzymatic degradation in the gastrointestinal tract and poor intestinal permeability. Enhancement of oral bioavailability of the antifungal drug, amphoterecin B, using solid-lipid nanoparticles, and the anticancer drug – tamoxifen – using self-nanoemulsifying drug-delivery systems are additional examples (Jain et al. 2012; Kapse et al. 2012; do Carmo et al. 2014).

Many nanomedicine applications expand traditional pharmaceutical drug development capabilities and are something of 'old drugs in new clothes', that not only extend the patent protection of an old drug but also offer tremendous potential to improve drug targeting to reduce dose as well as the potential for adverse effects. The number of nanomedicine patents and products in the market is steadily increasing; more and more nanopharmaceuticals and devices are entering the clinical trial phase of development and, hopefully, will result in an increase in approved drugs that have been enabled and/or enhanced with the use of nanotechnology.

17.2.4 Therapeutics

Some nanomaterials have unique therapeutic properties that differ from conventional drugs, and can, therefore, be directly used to treat diseases. For example, hafnium oxide- and gold-based nanoparticles can greatly enhance X-ray therapy; gold nanoshells/nanorods, carbon nanotubes, magnetic nanoparticles can be used to induce hypothermia to kill cancer cells; and nanocrystalline silver is being used as an antimicrobial agent in wound healing applications (Chaloupka et al. 2010; Dabiri et al. 2016).

17.3 Biomaterials

Biocompatible nanomaterial commercialization is occurring with new materials to replace or improve conventional materials in medical devices and prosthetic materials (Ghanaati et al. 2014; Nuñez-Anita et al. 2014). Nanomaterials with optimized mechanical properties are developed as biomaterials that can be used in medical implants (e.g. dental restorative and bone substitute or replacement). Nanocoatings are one of the most important types of these new nanomaterial applications (Keeney et al. 2015), Nanocoatings can be designed to increase biocompatibility and/or tissue adhesion (either positively or negatively depending on the intended use) thus improve integration with the surrounding tissues. This quality is useful in the development and advancement medical implants used in cardiology (stent coating), orthopedics (coating on joint replacement implants) and dentistry (dental implant coatings) and provide a broad and growing area for commercialization (Richter et al. 2005; Subramani et al. 2009; Gurzawska et al. 2012). In addition, the natural antimicrobial properties of some nanomaterials are employed in coatings, and in wound care products and medical textiles that have already been commercialized and others that are in development (Wilkinson et al. 2011; Yamada et al. 2015).

17.3.1 Biosensors

Biosensor performance and sensitivity can be improved using nanomaterials in their construction and the use of nanomaterials has allowed the introduction of many new signal transduction technologies (Kasili and Vo-Dinh 2005). Because of their submicron dimensions, nanosensors, nanoprobes and other nanosystems have allowed simple and rapid analyses *in vivo*. Portable, lab on a chip (LOC) instruments capable of analyzing multiple components are now possible and are becoming commercially available (Vashist et al. 2012).

A wide variety of nanomaterials, such as gold nanoparticles, carbon nanotubes (CNTs), magnetic nanoparticles and quantum dots, are being applied to biosensor development because of their unique physical, chemical, mechanical, magnetic and optical properties, and because they can markedly enhance the sensitivity and specificity of detection in very small devices (Reverté et al. 2016; Rezaei et al. 2016).

The fabrication of optical sensor nanodevices with integrated Mach–Zehnder interferometric (MZI) configurations where the optical waveguides must have two main features: monomode behaviour and a high surface sensitivity has been possible for several years and commercialization of these biosensors based on total internal reflection waveguides with nanometer dimensions is now practical and nearing completion (Prieto et al. 2003).

Graphene-based nanomaterials, due to their unique physicochemical properties, versatile surface functionalization, ultra-high surface area, and good biocompatibility, have attracted considerable recent interest in numerous biomedical applications including uses in biosensors, drug delivery, bioimaging, theranostics, etc. and are being readied for commercialization (Lin et al. 2016).

17.3.2 Tissue Engineering

Nanotechnology can enable the design and fabrication of biocompatible, tissue scaffolds at the nanoscale and control the spatiotemporal release of biological factors resembling native extracellular matrix to physicochemically direct cell behaviors, and eventually lead to the creation of implantable, manufactured tissues as candidates for commercialization (Sudhakar et al. 2015).

Drug delivery systems that incorporate drug-containing nanoparticles (liposomes, dendrimers) within matrices (polymeric scaffolds and hydrogels) for tissue engineering can produce sustained delivery of drugs from three dimensional matrices and can be useful for tissue engineering applications (Dorj et al. 2014; Hsu et al. 2016; Sundar et al. 2016). The delivery of drugs into damaged cells is also a tissue engineering application made feasible by a variety of nanomaterials (Korrapati et al. 2016). Carbon nanotubes, dendrimers, and liposomes are becoming increasingly important tissue engineering scaffold materials and drug delivery systems for delivery of drugs to impaired tissues.

With the emergence of nanotechnology, researchers proved that the drawbacks of existing implants and techniques can be resolved as nanomaterials can be modified to mimic the surface properties of natural tissues. Moreover, nanomaterials exhibit superior properties when compared to conventional materials which make them suitable candidates to improve various tissue growth technologies and use in implants. Potential applications of nanotechnology to create biodegradable scaffolds for bone, vascular, and bladder tissue regeneration are also envisioned (Hsu et al. 2016; Sundar et al. 2016).

17.3.3 Medical Devices and Accessories

Wound dressings with silver nanoparticles to impart antimicrobial characteristics were among some of the initial uses of nanotechnology in wound dressings that are classified and regulated as medical devices (Pereira and Bártolo 2016).

Nanotechnology applications involving the unique electrical and magnetic properties of materials on the nanoscale are showing promise in medical devices used in neurology and cardiology, especially toward better understanding the causes and improving the treatment of cardiac arrhythmia (Leite et al. 2007; Mann et al. 2012). In addition, there are nanotechnologies that enable the development of batteries with small size and lower weight for use in implanted and portable medical devices including active implants, biosensors and other imaginative applications (Lowy and Patrut 2008; Iost et al. 2016).

Antimicrobial nanoparticle coatings have been developed and are poised for commercialization and use on medical devices, operating room equipment, various hospital surfaces, etc. (Rivero et al. 2011; Duda et al. 2015; Palza 2015).

Medical nano-sized robots (nanobots, nanites, or nanomachines) have been conceptualized for over 20 years as "smart" and very tiny (nano-sized) devices constructed to incorporate a variety of biocompatible nanometer-scale materials. Potential uses of nanobots include targeted repair of tissues and cells, targeted sensing of tissue and organ microenvironments, delivering of transfection agents, in situ assembly and service or repair of implanted devices, etc. within the host's body have been envisioned but, to date, are still a distant reality (Jacob et al. 2011). Nevertheless, these molecular machines are a definite part of the conceptual future of nanomedicine and are in the realm of fifth generation nanotechnology where the invention and discovery aspects of the nanobot commercialization process are active and advancing toward prototype development. There are significant technologic and regulatory challenges ahead but many talented inventors and entrepreneurs are currently knocking on the door to the reality of having medical nanobots soon available for commercialization (Jacob et al. 2011; Eveleth 2015).

All of the above focus areas of nanomedical product development are ripe with opportunities for commercialization with active research and development activities on a global basis to arrive at new or novel nanomedical technologies that lead to new or improved advances in healthcare.

The variety of commercialization opportunities within even a single focus area to address current and emerging medical needs is typically very wide. So, the first step toward understanding and arriving at a specific commercialization opportunity begins with identification of an innovation that may be advanced by a recent discovery that becomes essential to driving and motivating further applied research and developmental research effort toward addressing a well-identified commercial need with at least reasonable potential for entrepreneurial interest and potential for financial support.

17.4 Entrepreneurship

High-tech businesses (large or small) have long been considered the driving force behind global knowledge-based economic development (Rausch 1998). Academic institutions with nanotechnology discovery capabilities have attempted to position themselves to serve the high-tech industry through consulting, licensing, and creating university spinoff companies. In the U.S., for example, the National Nanotechnology Initiative (NNI) recognizes that moving technology from laboratories to the marketplace is the immediate challenge following discovery and has identified U.S. government resources to help foster technology transfer from government funded programs to the private sector (NNI 2016). Many academic institutions that are developing nanotechnology have a reasonable awareness of commercialization interests in their discovery capabilities and at least attempt to develop active strategies toward building an entrepreneurial culture to assist academics with the transfer their inventions to the market to achieve the maximum value (Hobson 2009; Kaur et al. 2014). Larger academic institutions and some regional business development organizations often have “technology incubators” that may have some experience with the commercialization of nanotechnology opportunities. In less experienced, smaller, academic institutions, opportunities may be delayed, lost or not achieve their full potential for a variety of reasons that are largely related to nanomedical market knowledge, access and commercialization experience. In such cases, sometimes the scientist or engineer may need to become the entrepreneur.

Fortunately, there is a developing literature specific to the commercialization of nanotechnology that presents and provides example strategies for choosing a commercialization approach, financing a startup, marketing a product, and planning an exit that can be of help (Hobson 2009; Kaur et al. 2014).

Common reasons for startup company failures are discussed and guidelines to overcome these challenges are suggested in this literature (Hobson 2009; Kaur et al. 2014). Careful market research and the examination case studies of successful and failed companies can also be very helpful.

In general, entities that are successful in the commercialization of any medical product, including nanomedical products, demonstrate excellence in identifying a market need, assembling a motivated management team, effectively managing technical and business resources, and in securing and following the advice of experienced mentors and consultants. In doing this they substantially and effectively eliminate barriers and efficiently achieve a successful market entry and a timely commercialization program exit.

Negotiating is a big part of entrepreneurship and arriving at a fair negotiated value for an emerging medical nanotechnology between two or more parties is generally not a simple or quick process. It is important to remember that developing intellectual property into commercial products through in-licensing and out-licensing is not a zero-sum game. Both the buyer and seller are looking to get something of value (often from different perspectives) and beneficial out of the deal and entrepreneurial entities that are successful work very hard to arrive at the much-sought win-win deals

Negotiation is a skill and art that has to be learned and applied at the highest level in advanced technology commercialization. This is skill is not often naturally well developed by medical professionals and academics. Even at academic institutions with prestigious business schools, this skill is not offered as part of an advanced degree program in the medical sciences. However with a reasonable effort, this skill can be learned and developed by those willing to brave the business world and spend at least some time gaining the necessary experience either through coursework, executive business degree programs or just plain hard work and engaging experienced mentors. Once learned, the medical professional or academic that can interact effectively in both worlds can often negotiate with a high degree of success as familiarity with the science and medical aspect is valuable. Excellent negotiation skills are useful and often necessary at almost all steps in the commercialization process. It is also important to learn and appreciate that although the general principles are the same, people negotiate somewhat differently in the international marketplace due to different perspectives, needs and cultural factors.. Realizing and knowing this can make the difference between success and failure when negotiating the licensure or sale of a nanomedical technology for commercialization between international parties.

Negotiation skills, while important, only one of many strengths that a successful CEO or COO for an emerging medical nanotechnology company. Lafley and Tichy (2011) provide an interesting large industry perspective on CEO selection to include succession considerations that entrepreneurs can learn from. Large, medical industry CEOs may not be entrepreneurs but calculated risk taking is sometimes required even though they are generally more oriented toward strategically identifying, controlling and reducing risk. Managing a risky situation is not something these executives seek to do. In contrast, the nanomedical technology company entrepreneur and CEO may have former large industry experience and needs to understand the essentials of how large company CEOs think for best success in licensing or acquisition deals, but managing risk is a daily activity in the entrepreneurial world of commercializing medical nanotechnology. Skilled entrepreneurs know and appreciate that their role is to create and build new opportunities and “venture” (which, by definition, is a risky or daring journey or undertaking) to advance them to commercial success. Accomplishing this with success generally requires well developed abilities to negotiate deals and manage projects effectively and these are high level skills that should not be overlooked, especially when hiring someone to serve as either the CEO or COO of a nanomedicine company who will likely one-day be required to successfully negotiate a big deal with a large medical company CEO and business team.

17.5 The Commercialization Process

The commercialization of any new medical product, including those that contain or are enabled by the use of nanotechnology can be envisioned as following a generally defined process as shown in Fig. 17.1. This process will have nuances for each product but, generally, the process includes major features common to any medical product. So, commercialization planning begins with a general concept then applies specific considerations for the particular nanoproduct. In reality, no two products will follow exactly the same commercialization path and allowances must be made for product unique variables including for example, the regulatory pathway, the nature and complexity of the manufacturing process, the market characteristics, level of funding required and source as well as the investor type and level of interest, the strength of the intellectual property portfolio, etc,.

The commercialization process for a nanomedical product often begins with a discovery or invention of something that stimulates innovative thought and activities that result in a product concept that is proven to have potential market viability and is feasible to be developed, manufactured, tested, attain regulatory approval and marketed within practical limits of current technology and the investment time frame.. After this, market acceptance and success await new and novel products that are safe, effective and either addresses an unmet medical need or significantly advances the standard of care in some medical field (Flynn and Wei 2005).

The general commercialization process for a nanomedical product is elucidated below as discrete, stage-wise activities that have as their objective the creation and marketing of new nanomedical products. Brief descriptions of each aspect are provided for additional clarification.

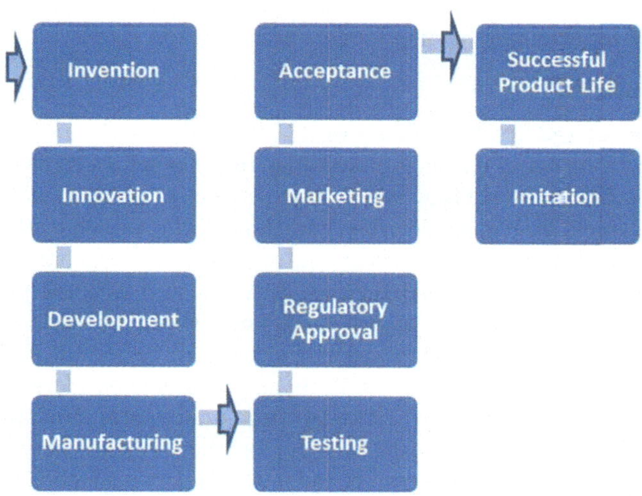

Fig. 17.1 Generalized view of the stages in the commercialization process for a medical nanotechnology product

(I) **Invention**

A new creation that may be a material, device, process, or discovery that reflects extraordinary creative ability or skill, and that makes a distinct and recognized contribution to and advancement of science.

(II) **Innovation**

Differentiated from invention, as taking the original invention, discovery or idea and applying it toward a practical need and, as a consequence, results in something new and potentially useful, that "breaks into" or "disrupts" and or advances the medical market or medical practice. Innovation is related to, may include or sometimes even require an invention, but not the same as, invention

- **Proof of Concept**
  This is a practical demonstration in principle of the feasibility and potential of practical use or application, with the purpose of verifying that the product concept or theory has the potential of being useful as proposed. A proof of concept is usually small scale demonstration and may or may not be a complete validation of the new product as would be necessary for regulatory approval and successful market introduction. This demonstration is often important to securing intellectual property and investment interest so the study design, conduct and quality are, nevertheless, important considerations.
- **Securing and Licensing of Intellectual Property**
  This involves ensuring that the nanomedical product is unique in important and significant ways such that it can be protected at least for the time until it is marketed from infringement and competition by a portfolio of patents and trade secrets. This is essential to both the valuation of the product but also for obtaining funding for commercialization.

(III) **Development**

- **Project and Strategic Planning**
  Development of a formal project plan by an experienced project management specialist that is guided by an experienced regulatory affairs specialist with input from manufacturing, non-clinical, clinical and marketing experts to establish an overall strategy and plan to document, organize, manage, prepare overall cost and time estimates as well as track progress from development to market approval. This plan should be sufficiently detailed to include all major events and may incorporate best and worst case contingencies for critical activities. Early and accurate consideration of regulatory requirements is essential for addressing critical regulatory documentation needs, especially for products where safety and efficacy testing is required, in a correct and most efficient manner. If the product is a medical device, documentation of the steps in design and development can be a regulatory necessity. This plan should be linked to strategic

planning for the company as well as to the company business plan. This is necessary for commercialization in order to secure investment and also to provide credible projections for investors and other business activities.

- **Analytical and test methods**
  Methods to analyze and test the product need to be developed, validated and documented and then made ready for use in testing prototypes as well as in the evaluation of manufacturing product release stability as well as non-clinical and clinical samples..

- **Prototype evaluation and manufacturing**
  Development and selection of a prototype nanomedical product to manufacture for use in non-clinical and clinical testing that is required for clinical trials and final product approval to market.

(IV) **Manufacturing**

- **Process development**
  This involves development of the product manufacturing process for commercial success. Often the approach used during the research phase is not the best approach for commercialization. Examples of general manufacturing approaches that are used for the production of medical nanomaterials include:

  (a) **Top Down vs Bottom Up**
  The top down approach to manufacturing nanomaterials refers to slicing or successive cutting of a bulk material to get nano sized particles. The bottom up approach refers to the build-up of a material from the bottom: atom by atom, molecule by molecule or cluster by cluster. Cost, time and safety can be factors in the selection of which approach is best for commercialization.

  (b) **Self-assembly**
  Self-assembly of molecules and molecular clusters into pre-determined and precise nanostructures with speed and efficiency at low cost may be essential to controlling costs and ensuring the market availability for some nanomedical products.

- **Batch production**
  Production of pilot batches then following by production of registration batches (to support non-clinical and clinical studies).

- **Stability**
  Stability studies are conducted as required to establish the shelf life of a formulated product or a medical device that has degradative properties or components.

(V) **Testing**

- **Chemical and/or materials testing**
  Materials performance, degradation testing, packaging, stability, shelf life, sterilization method selection, etc.
- **Non-clinical testing**
  Safety (toxicology) testing, basis for efficacy demonstration (in vitro and/or in vivo), pharmacokinetics, targeting (biomolecular, cellular, tissue, organ) efficiency, dosimetry, etc.
- **Clinical testing**
  Clinical safety and efficacy, diagnostic performance, etc. In pharmaceutical development, phase 1, 2 or 3 clinical studies.

(VI) **Regulatory Approval**
All medical products worldwide are subject to regulations focused on their intended use that are aimed at protecting the public from harm and unreasonable risk. Most regulatory agencies conduct the risk assessment of nanomedical products on a case-by-case basis. The potential for release or absorption of nanomaterials leading to a potentially harmful exposure to nanomaterials is of high interest and concern for regulators. The extent to which any potential risk identified by a well-designed, non-clinical safety testing program is offset by the clinical benefit of the nanoproduct is the basis of what is sometimes presented as "nano" risk assessment as well as regulatory approval to market the product. This aspect of commercialization should begin with the early stage development of an overall regulatory strategy that facilitates and leads to the best possible opportunity for regulatory approval for useful, safe and effective product.

(VII) **Marketing**
This involves not just the activity of marketing the product to the identified potential user communities but also can involve the aesthetics of the product (shape, color, feel, odor, etc.), the packaging, labeling, advertising, conducting of focus group discussions, etc. The importance of this activity to be involved well before the product actually is actively marketed should be obvious for commercialization success even though the technical and scientific perspectives may not always appreciate this.

(VIII) **Acceptance**
The product must gain acceptance by the user communities at least to the extent that early adoption of the product represents enough market share in profit to sustain and grow the product to the satisfaction of minimal investment expectations both in timing and magnitude of the return at the very least.

(IX) **Successful Product Life**

The lifetime of a product is dependent on too many factors to note here, but for most investments, financial success of the product is expected within just a few years (2–5) of launch depending on initial expectations based on the size of the investment, market challenges (to include competition) and how effectively the product is positioned and priced. Some products in competitive markets have a lifetime of only a few years while others may become a leading product mainstay for many years and attain name recognition as best in class. See more below with respect to market "success" from different commercialization perspectives.

(X) **Imitation (Generics)**

It has often been said in colloquial language that "Imitation is the sincerest form of flattery." Originally, this was stated by Budgell (1714) as "Imitation is a kind of artless flattery." Which may be more appropriate for the imitation of, initially, innovative technology. However received, for a company facing the loss of market share due to competition by imitation, this may be a challenge in many ways. Patent infringement and/or the unlawful taking of intellectual property may be an issue. A competitor making a lawful product that is somehow inferior to the Innovator's product can be substituted as a generic and produce clinical outcomes that cause both products to come under scrutiny and even to be removed from the market. Sometimes, illegal claims and substitutions by an imitation product can lead to lawsuits that are time consuming and costly, resulting in an unexpected expense to the Innovator that may or may not result in obtaining legal relief and/or payment of damages. So imitation must be expected for any highly successful product and prudent preparations made to best prepare for various potential challenges by imitators legal or otherwise.

Understanding this general process and learning how to plan and manage for success is the subject of numerous publications and courses of specialized project management and total quality management training. Some of these lead to professional certifications in project management such as the well-known Project Management Professional (PMP) or Six Sigma programs. Most effective business plans for a nanomedical company will include and incorporate well-developed elements of product commercialization planning as the foundation of strategic and financial planning. Often, a project management expert is engaged and is continuously actively monitoring successful commercialization projects. So, the importance of establishing a well thought out plan and continuously updating and refining it as the product progresses toward commercialization cannot be understated. Companies that fail to do this effectively very often fail altogether or fail to achieve full commercial potential (Belassi and Tukel 1996).

17.6 Risk and Hazard Assessment

There are two types of risks that must be carefully assessed and are critical to the successful commercialization of a medical nanotechnology, clinical risk and business risk. These are further described below and both will strongly affect the valuation of the product as well as its commercialization success in critical ways.

(a) **Clinical Risk**

Clinical risk assessment for medical nanotechnology follows the same principles as that for any other medical technology in the same use class (pharmaceutical, medical device, biosensor, implant, etc.). This assessment includes identification and evaluation of potential risks that may arise from the fabrication, formulation, manufacturing and packaging processes; non-clinical (*in vitro* and *in vivo*) toxicologic evaluation and clinical safety evaluation.

Clinical risk for medical nanoproducts is evaluated on a case by case basis by regulatory bodies like the U.S. FDA and EMA but is subject to the considerations and informational needs that may be found current published guidance and directive documents issued by the regulatory authority. Case by case assessment has allowed consideration of innovative nuances between products that may require differential thinking and consideration in developing appropriate guidance and regulatory paths for nanoproducts with differing compositions and features that is shown by the number and variety of FDA nanoparticle-based pharmaceutical approvals to date (Hamburg 2012; Monica 2012; Bobo et al. 2016; Landesman-Milo and Peer 2016).

In the EU, for example, a nanomedical device should comply with the essential requirements as indicated in the current medical device directive that includes the general requirement that devices must be designed and manufactured in such a way that, when used under the conditions and for the purposes intended, they will not compromise the clinical condition or the safety of patients, or the safety and health of users or, where applicable, other persons, provided that any risks which may be associated with their intended use constitute acceptable risks when weighed against the benefits to the patient and are compatible with a high level of protection of health and safety.

The safety (toxicological) evaluation requirements for emerging nanomedical products and most especially the use of nanomaterials in pharmaceuticals and medical devices are becoming established internationally. All safety evaluations require careful and complete physical and chemical characterization of the incorporated nanomaterial. While there are cases, where nanomaterials may produce similar toxicologic results to their compositional chemical/molecular substances, this cannot be presumed as experience now shows that some may be more toxic and some may not (SCENIHR 2009). The U.S. FDA and EU, in general, allow that a classical risk assessment can be performed consisting of exposure estimation, hazard identification, hazard characterization, and risk assessment. However it is known and recognized internationally that nanoparticles exhibit specific characteristics that differ from the characteristics of larger sized nanomedical products and must be identified to regulatory authorities and be characterized using data that meets regulatory data quality standards such as Good Manufacturing Practices (GMP) or Good Laboratory Practices (GLP) as required by the country regulatory authority. The potential for clinical risk is assessed using toxicological data also meeting GLP requirements and is evaluated relative to the intended clinical use of the product.

The European Commission and its Scientific Committee on Emerging and Newly Identified Health Risks (SCENIHR 2015), have published the final opinion 'Guidance on the Determination of Potential Health Effects of Nanomaterials Used in Medical Devices'. The guidance provides information on how to evaluate the risk when a nanomaterial is used in a medical device and it should be considered in conjunction with the ISO 10993–1:2009 standard 'Biological evaluation of medical devices' (ISO 2009).

ISO 10993–1:2009 categorizes devices and safety evaluation requirements for medical device products with considerations as to:

- Type of device (surface device, external communicating device and implant device),
- Location of contact (skin, mucosal membrane, breached or compromised surface, tissue/bone/dentin, or circulating blood),
- Contact time [varying from limited contact (equal or less than 24 h), prolonged contact (>24 h to 30 days), and permanent contact (>30 days].

Depending on the intended use, these categorical criteria are applied to a specific nanomedical device to arrive at the types of biological effects (e.g. cytotoxicity, genotoxicity, sensitization, systemic toxicity) that need to be considered for the biological safety evaluation of the medical device. The SCENIHR guidance also takes this approach, and it clearly distinguishes between non-invasive and invasive medical devices (SCENIHR 2015).

The clinical safety of a nanomedical product must take into consideration the use and exposure routes, including dermal, inhalation, mucosal, oral, parenteral, otic and ocular exposure. Failure to get this right and complete the necessary studies for regulatory review and approval can result in serious setbacks and additional time and expense to a commercialization program.

In general, the nanomaterial-related clinical risk of nanomedical products is mainly associated with the likelihood of the release of free nanoparticles from the product and the potential for resulting toxic effects in specific cells, tissues and target organs. Evaluations of nanomaterial safety must include a detailed characterization and identification of the that may include, but not be limited to, determination of the chemical composition/identity, particle size and size distribution (primary/secondary particles), physical form and morphology, particle and mass concentration, specific surface area, surface chemistry, surface charge, redox potential, solubility and partition properties, pH, viscosity, density and pore density, dustiness, tendency to form agglomerations or aggregates in biological conditions and whether they become coated with a protein corona to include characterization of the types of coronal proteins.

Nanotoxicology is becoming a more defined sub-discipline of toxicology that is recognized within regulatory bodies and is increasingly considered during product development and registration with respect to its relationship with clinical risk assessment. Therefore, early assessment of a commercialization opportunity by a toxicologist experienced with the safety evaluation requirements for nanomedical products should be considered in order to properly evaluate the product and help resolve any potential clinical safety issues before they become significant commercialization barriers.

(b) **Business Risk**

The business risk associated with nanomedical product commercialization is essentially the same as that when evaluating the risk to the business for any advanced medical technology. The evaluation of business risk typically involves or is associated with an investment decision. Parameters that must be assessed include, but are not limited to, significant manufacturing challenges, safety hazards and research/development/production costs for the nanomedical product, the strength of the intellectual property position, the market size and overall skills of the development and management team to run the project and business, respectively, and to collaboratively and effectively address the development and regulatory challenges and needs as they arise. So, when analysts evaluate the commercialization potential for a nanomedical product opportunity, this includes quite a number of factors other than just a valuation of the nanomedical product itself and often includes the company's overall risk which could lead to a failure of the opportunity and the management. This is why assembling an appropriately experienced and well-coordinated development and management team is essential to obtaining funding support sufficient for commercialization. Limited or no prior experience and a track record of success that is lacking often result in decisions not to fund or to only partially fund a business entity for commercialization until improvements are made, regardless of the safety, market opportunity, and intellectual property position of the technology. In some cases, where the technology exhibits a substantial enough business opportunity, the funding source may require that the management team include experts selected by the funding organization or even that the entire original management team be replaced by personnel selected by the funding organization.

Business risk analysis also values a company's ability to place their products profitably into the target market. This is would typically be evaluated by rating the risk that the offered nanomedical product would not be and stay competitive within the investment period (3–12-years) that depends on the type of product and market.

The strength of the intellectual property position and whether or not patents provide enough security is important. Often, failure to secure a competitive intellectual property position can be a sufficient barrier to prevent funding and commercialization.

Fortunately, nanomedicine is in its formative stages an there is substantial potential to develop highly competitive and market essential technologies to meet unmet needs and those truly skilled with successful commercialization of nanomedical products are few. So, due to business re-organizations, downsizing and other growth issues in the large pharmaceutical and medical devices industries and the high interest in nanomedical products, the ability to put together a skilled development and management team of able learners with reasonable experience in taking medical products to at least similar medical markets is possible. With good team selections, business risk can be effectively reduced with reasonable effort and attention that is focused by well-developed product planning and enhanced by learning from the experience of others whenever possible.

17.7 Securing Intellectual Property and Licensing (IP Strategy)

In the commercialization of medical nanotechnology it is important to understand the differences between different types of intellectual property. Patenting is often thought of as the principal means of protecting intellectual property, however, a comprehensive IP strategy for a medical nanoproduct may involve patenting, publishing and trade secrets. Valkonen (2012) notes that patenting may be advantageous to a company intellectual property strategy in the following cases:

- "Preventing others from utilizing the results of one's own product development
- Preventing others from obtaining a patent that would be harmful to oneself
- Pricing one's own products as desired by taking advantage of exclusive rights
- Increasing one's own patent base, which provides for increased freedom of action in case of patent Frenchman disputes
- Obtaining concrete clearly determine material for technological cop cooperation
- Possibly obtaining license revenue without one's own production input
- Promoting marketing by creating a favorable image."

Patenting may sometimes be challenging because of the cost and fees for maintenance of a patent portfolio may be cost prohibitive for many individuals and some small companies even though there are usually lower fees for small businesses. Also, patenting results in necessary disclosures that become publically available within months of filing the application and such disclosures may be harmful if the nature of the invention lends itself to modifications and infringements that could be too numerous to protect by expensive patent litigation.

There are alternatives to patenting that still create valuable intellectual property. These are publications to create prior art that inhibits others from patenting and the creation of trade secrets and then protecting and maintaining them as business secrets for as long as possible..

Publication has advantages when the invention is not substantially different from other prior art and uses from various innovations are not readily obvious. This is also an effective IP strategy in situations where there is little or no economic benefit and there is sufficient freedom to operate even when there's a potential for third-party use. Publication is also a means of informing about the invention in a way that gets attention by the community of technical users that would feel less likely to use or adopt or use the invention if it were patented in subject to secrecy and/or patent infringement penalties. It gives the freedom to proceed knowing that the invention cannot be claimed or patented by others because of his existence as prior art associated with inventor by publication.

Valkonen (2012) indicates that trade secrets are valuable when,

- there is a chance to keep the innovation secret,
- the product/technology has short life cycle
- the innovation reveals too much know how which cannot be discovered from the product itself
- patent infringement cannot be easily discovered

Therefore it is very important to successful commercialization of a nano medical product to develop a comprehensive and effective IP strategy. There are a number approaches toward developing an effective IP strategy and different theories as to how to best protect IP under different conditions in different situations.

An effective patent strategy also is forward-looking toward the potential even likelihood of infringement. Just as there are different types of patents there are different forms of infringement. Successful commercialization involves having a reasonable understanding of how the IP portfolio must be constructed to best prevent infringement. In this regard it is unfortunate to reveal that even patent examiners sometimes leave the government positions with substantial knowledge as to how to break or infringe on existing patents. So, keeping this in mind, an experienced patent attorney that has been involved in litigation can often be of substantial value in the development of the overall IP strategy and development of a valuable IP portfolio.

Nanotechnology patents often have unique features that make these technologies excellent candidates for patent filing however one has to remember that in emerging technology areas such as nanotechnology where many patents are being submitted each month, it may take some time complete the review process and to finally receive a patent due to the area being relatively congested and the number of government patent examiners limited. The objectives of patenting also may be different for company that simply needs to protect a marketed product as opposed to a university or research institution that wishes to protect and invention for the purposes of a lucrative technology transfer or licensing opportunity to a commercial business entity.

The big pharmaceutical company business model, which has typically depended on the continual emergence of "blockbuster" products to generate profits, is clearly broken or is at least breaking. The availability of new small molecule drugs appears to be coming to an end and the cost of commercializing a new small drug by traditional means is prohibitive. Patent expiration on numerous blockbusters in recent years is altering the drug landscape. Pharmaceutical companies are also facing other challenges that necessitate looking to more efficient development and implementation of novel R&D strategies, including those that focus on nanotechnology acquisitions and, in the case of medical devices, miniaturization. Clearly, there is enormous excitement and expectation regarding nanomedicine's potential impact. However, securing valid and defensible patent and other IP protection is more critical than ever (Bawa 2007).

Early forecasts for nanomedicine commercialization are encouraging but there are numerous bottlenecks and hurdles as well. The issuance of surprisingly broad patents for nanomedical inventions by the US Patent and Trademark Office (PTO) is problematic and may result in overlapping claims between competing products that may require litigation for resolution. Some believe that the U.S. National Nanotechnology Initiative's (NNI) widely cited definition of nanotechnology is inaccurate and irrelevant from a nanomedicine patent and IP perspective and is cause of an inadequate patent classification system that is in use by the PTO. The resulting chaotic patent landscape in various sectors of nanomedicine is thought to create uncertainty as to the validity and enforceability of numerous issued patents and it is claimed that the nanotechnology industry also suffers from overlapping patents. Changes in the PTO to fix a problematic patent process especially in the case of nanotechnology patents has been recommended and it is felt that a more robust patent system is needed to stimulate the commercialization of nanomedicine products that have good market viability, improve patient quality of life and reduce healthcare costs (Bawa et al. 2005; Bawa 2007; Bosetti and Vereeck 2012).

The success of nanomedicine is clearly dependent upon effective protection of IP rights. In order to avoid disputes with the current problematic patent situation in the U.S., it has been recommended that inference practices and strategic patenting could be used where, in the case of a dispute, parties can fall back on re-examination, cross-licensing and patent litigation. Cross-licensing agreements have also been recommended as they allow parties to access technology, create synergies and exclude third-party competitors. Ultimately, experts indicate that solving the patent problems in the nanotechnology industry is a necessary step for future success (Bosetti and Vereeck 2012).

Patenting has been shown to be a key indicator of trends in the commercialization of nanoparticles with cancer treatment showing notable progress relative to other medical sectors due to the rapid development of potentially useful nanomaterials and new treatment strategies to be investigated (Fierro et al. 2014). In 2014, Fierro et al. showed that that polymers and proteins were the main focus of the innovative efforts in patented nanoparticles systems for cancer treatment.

Bawa (2008) made an interesting observation with regard to patent strategies that apply to and should be noted by small companies with nanoproducts: "Often, larger competitors employ frivolous lawsuits to pressure smaller companies or start-ups whose patents stand in their way, or which they wish to acquire. Frequently, the cost in executive time and corporate money for the smaller company or start-up becomes so onerous that it caves in to a licensing agreement. One viable strategy to avoid being taken over is to license the patent to the large competitor, in whose interest it then becomes to protect its position by protecting and defending the patent." Bawa (2008) suggests nine considerations and strategies that bionanotechnology inventors can follow in order to protect an invention even before a patent application is drafted or filed.

1. **Avoid an early publication or any public disclosure.** The inventor should refrain from publishing a description of, publicly presenting, submitting grant proposals for, or offering the invention for sale prior to filing a patent application. Often a company releases information on a new product, or discusses details during negotiations prior to filing a patent application. All of these activities create prior art against the inventor."
2. **Consider obtaining a foreign patent.** Filing a bionanotechnology patent in a foreign country should be carefully considered and should largely depend upon commercial considerations. If there is an interest in expanding into foreign markets, then obtaining patents abroad should be seriously considered. Furthermore, even if the inventor does not plan to establish a market for the particular bionanotechnology invention in a foreign country, obtaining a patent there could be critical in securing licensing deals (and discouraging unlicensed copying or use by foreign competitors)."
3. **Beware of pre-grant publication of U.S. patent applications.** Today, as part of the application process, all U.S. patent applications are published 18 months from the earliest filing date. In effect, this implies that almost always a patent application, as filed, will eventually appear in the public domain (whether or not it is patented) and will be available to competitors."
4. **Maintain proper laboratory notebooks.** Laboratory notebooks often contain valuable and critical information that may not be readily apparent to a company or its R&D facility. Laboratory notebooks are also useful to patent practitioners to establish the date of an invention, especially in light of a competitor's challenge in court as to who invented first in what is known as 'interference proceeding'."
5. **Conduct a 'prior art' search and a 'freedom-to-operate' search.** It is highly recommended that a proper prior art search be conducted prior to filing a patent application. The purpose of this is to gauge the competition. This may also assist the inventor to design around potential prior art. Moreover, because the patent owner does not automatically have the right to practice his/her invention, it may be wise to conduct a 'freedom-to-operate search' of the issued bionanotechnology patent prior to investing in and commercializing it."
6. **Educate employees and researchers.** It is important that business and IP professionals within a company educate scientists to spot potential inventions during the R&D phase, as this may not always be apparent to them. In fact, a company should implement policies involving incentives where scientists are rewarded for reporting or submitting invention disclosures. This may be especially critical in a university setting where generating invention disclosures may be less of an incentive to researchers who are promoted or tenured based on their research grants. Scientists often overlook the fact that their inventions can be patented. Further, "patent awareness" may enable a researcher to pursue a particular research path that has a greater likelihood of leading to a patentable invention."

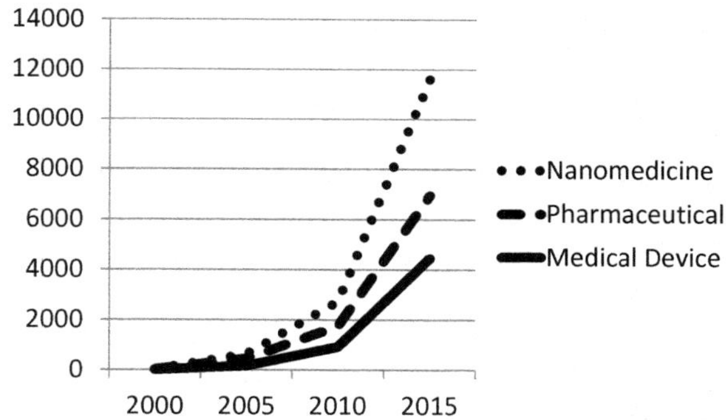

Fig. 17.2 Cumulative U.S. Nanomedical Patents 2000 to 2015. Categories shown are the cumulative number of patents in the three categories of Nanomedicine, (nano) Pharmaceuticals and (nano) Medical Devices from the years 2000 to 2015

7. **Require strong employment agreements and safeguard IP.** Companies must require all employees to sign agreements that clearly specify that all company inventions, intellectual property, and proprietary information is company property and cannot be disclosed or exploited by any employee at any time. This could become critical if a former employee joins a competitor company or research laboratory."
8. **Employ standard terminology while drafting patent applications.** The fact remains that bionanotechnology is an inherently difficult topic for discussion, in part due to the confusion surrounding its definition. Although it is well recognized in patent law that a patent applicant can be his or her own lexicographer, it is recommended that an applicant should employ standard language in bionanotechnology patent applications whose meaning is well recognized in the pharmaceutical, medical, or biotechnology fields. Furthermore, the language should be precise and the use of terms consistent throughout the claims and specification (avoid synonyms and be repetitive in the use of phrases when appropriate). This will prevent confusion at the patent office as well as prevent possible prosecution delay."
9. **Relative ease of obtaining 'broad' patents in bio-nanotechnology.** Broad patents continue to be issued by the PTO in bionanotechnology The overburdened PTO faces new challenges and problems as it attempts to handle the enormous backlog in bio-nanotechnology applications filed and the torrent of improperly reviewed patents granted."

In 2007, registered nanotechnology patents worldwide included the involvement of 35 countries in global nanotechnology patent distribution (du Toit et al. 2007). Approximately 3000 patents were issued in the U.S. between 1996 and 2007 that included the term 'nano' in the patents, with a considerable number having application in nanomedicine (du Toit et al. 2007). The number of nanotechnology patents worldwide has been steadily increasing, led by the United States with patents arising particularly universities and research centers (Antunes et al. 2012).. Patenting of medical nanotechnology has been a popular strategy as the increasing number of U.S. patents alone over the period of 2000–2015 as shown in Fig. 17.2. Clearly growth in the cumulative number of patents in each of the topic areas searched

(Nanomedicine; Nano-Pharmaceuticals; and Nanomedical Devices) and shown in Fig. 17.2 is growing geometrically with no signs of slowing. Clearly, as noted above, the enforceability of numerous, broad, U.S. patents for medical nanotechnology presents a growing problem with a clear potential for those that employ strategies to obtain "after the fact" infringing patents and seek to profit from the threat of litigation rather than advancing the clinical benefits that the original innovator hoped to commercialize with patent protection. The ownership of technology via a patent and the ability to obtain sufficient financial support to commercialize the technology can sometimes be quite different issues in a litigious and highly competitive, emerging technology field. So, inventors are encouraged to learn and apply a variety of strategies to protest their medical nanoproduct IP.

17.8 Valuation and Technology Transfer

No universal method for technology valuation exists. In fact, different methods will often be used within one organization to develop a multifactorial valuation of a new technology. Medical nanotechnology is no different except that factors relating to the incorporation of nanotechnology and intended medical use are considered. The valuation method chosen depends on the nature of technology under consideration and whether the valuation is relative to a technology buyer or a technology seller. In the end, however, what most matters is the accuracy of the estimations and assumptions about whether a product will be a business and clinical success, if it will be safe and how much users might be willing pay for it. Accurate estimation of the size of the potential market and the adoption rate for the product are critically important in technology valuation (Schmittlein and Mahajan 1982).

Successful development and commercialization of medical nanotechnology products hinges on efficient and effective technology transfer and technology trading systems. This requires the existence or, if necessary, development of a technology market or exchange for the specific type of medical nanoproduct that, in turn, requires reliable valuation methodologies.

Entire medical nanotechnology firms can also be subject to valuation for acquisitions, mergers and investment. This level of valuation usually includes valuation of the entire IP portfolio but also may include valuation of the physical assets (equipment), the experience of the personnel and even the benefits and risks of non-financial factors such as the ability of key scientific and technical staff to create and capture IP, the experience and skills/knowledge of the management team. A firm should expect to rate relatively high on mean value when they have a comprehensive understanding of the path toward commercialization and evidence significant, relevant and recent experience with successful commercial development and exhibit a demonstrated ability to successfully manage commercialization projects and know the target markets and any competition well.

Medical nanotechnology that has demonstrated income potential may be valued higher than an emerging product that has not produced income.. Most nanomedical product companies take some time to demonstrate an ability to produce a substantial income within the investment period required for valuation (typically with a few years of actual data) so most small and developing firms lack the investment history to be able to support the notion that they are an attractive investment or fund with substantial data to support any reasonable degree of confidence.

17.9 Funding and Financing

In developing a workable financing strategy, a nanomedical product company must first decide on short term milestones involved in creating a profitable company. For example, companies may need to complete a business plan and project plan, obtain funding, develop a prototype, integrate their technology with a partner's technology and then define the appropriate path to regulatory approval. Each step involves particular costs and by separating the total investment into smaller, discrete amounts, the company can approach investors with the understanding that if it meets its goals at each stage then additional funds will be sought.

Many investors will also want to understand the exit strategy for the medical nanoproduct as well as understand the time anticipated for an event, e.g. licensure, sale, market entry, that could lead to an investment return. Information that has been conservatively prepared to allow for this sort of force casting as well as an example forecast prepared by the company or innovator that is establishing a company is required and generally prepared in advance of any meetings with investors.

A comprehensive business plan that takes into account all expenses necessary to meet goals is important for three reasons. First, it can prevent a "fire sale" in which the company must sell equity quickly (usually at a discounted price) because it must raise capital to meet expenses. Second, planning prevents raising more money than the company needs and consequently prevents unnecessary dilution to existing stockholders. Third, identifying costs in the planning state enables management to prioritize and eliminate expenses that are not essential to reaching its goals.

(a) **Public Funding**

Public funding for nanotechnology has been available from government sources in most developed countries and especially the United States, and Great Britain as well as a number of European Union countries (Germany, France, The Netherlands, Belgium and Italy in particular), Japan and China. Initially, in most of these countries, funding was provided to establish "centers of excellence" and other early stage activities to help the countries become established and take leading roles in the development of new products across many sectors including medical nanotechnology. Today, funding is focused more on the development of nanotechnologies to address specific needs for technology advancement and securing new intellectual property and/or addressing specific issues of interest to the countries and their various funding sources. So, in commercializing a nanomedical product with public funding, considerable effort may be expected in identifying potential funding sources then writing well targeted grants and/or proposals to secure funding from sources with specific interests in what appears to be an increasingly competitive funding environment. Using public funding to completely commercialize a nanomedical product may take a long time and may not have a high degree of expectation for success, especially in getting into the market quickly. On the other hand, public funding may not require giving up equity. At some point though, private funding will be required to accelerate and complete commercialization.

(b) **Private Funding**

There are many sources of private funding that include, angel investors, venture capital firms, banks, private and public companies, etc. All are known to have funded nanomedical product development efforts. These investment sources are typically the best for supporting commercialization efforts, especially when the product has developed beyond proof of concept and has well thought out mature development and business strategies that include realistic and conservative financial projections that meet the requirements for the individual investors. Investors typically have minimum equity requirements for providing their funding that can occasionally be negotiated but only within limitations based on the amount of funding involved, the timing of events leading to investment returns and the amount of expected return.

Most investors and investment firms have a stream of opportunities that they are considering and don't care to spend too much time with a potential investment that does not fit their "sweet spot" meaning particular area of technology and/or market expertise and experience. Medical nanoproduct entrepreneurs do best with these funding opportunities when they do their homework on the interests and funding history of the source and are well prepared when and if they have their first meeting.

(c) **Public/Private Partnerships**

These partnerships have become a popular form of securing investment for commercialization, particularly with public sources that don't want or can't financially support the full commercialization alone. The public funding source often makes a grant award based on the ability of the product innovator to be able to secure some specified matching amount of additional funding from private sources within a specified time frame. This type of funding generally starts with a public funding source that promotes the early stage research and development of technologies for medical applications selecting promising development projects for funding.. The public funding for the project is then provided with specific, contractual, time and amount contingencies for the innovator to comply with at specified times or milestones or otherwise lose the support. Some equity may be given up when this funding is provided.

No matter the source of funding, it is extremely important to adequately research funding source interests and requirements before investing time, cash on hand and effort in pursuit of funding. Most public funds have specific information and submission requirements as well as are focused on specific areas and amounts that they can typically support so matching the product commercialization needs with the right fund in the first place as quickly as possible is prudent and can make what often can be a challenging endeavor at least a little less frustrating.

Funding characteristics for the development of medical nanotechnology already does have some history.

In 2005, Flynn et al. observed that for the development and advancement of nanotechnology. money was pouring into the development of nanotechnologies on a global level from government and corporate entities. From fiscal year 2005 to fiscal year 2008, the US government has authorized $3.7 billion for nanoscale science and engineering. In June 2004, Europe, needing to spend more to be competitive with the United States, authorized billions in funding and in effect said, "the race is on." Thirty-eight industrialized nations of the world have biotech initiatives under way with varying budgets. Nanotechnology is a global business with global applications. Flynn and Wei (2005) and others predicted that these historic and projected increases in funding and market growth will continue to fuel the emerging nanotechnology and nanomedicine markets.

As of this writing in 2016, however, there are no new nanomedicine funding opportunities listed on the U.S. National Institutes of Health website and the most recent awarded funding for nanomedicine occurred in 2006 when the NIH issued a request for applications (RFA) to fund Nanomedicine Development Centers in addition to four that were funded in 2005. The most recent progress report from the Nanomedicine Center for Mechanobiology Directing the Immune Response was a few years ago in 2012 and it would appear that this center is no longer funded and in operation. A similar fate appears to have occurred for the Nanomedicine Center for Nucleoprotein Machines, so it would appear that NIH funding for nanomedicine center activities is meager to non-existent at this time. Of course there is still ongoing development of NIH-funded nanomedical technology within other projects not specifically falling under the general funding topic area of "nanomedicine" The NIH-funded Northeastern University (NU) Center for Translational Cancer Nanomedicine (CTCN), for example, was established in September 2010 as part of Phase 2 of the National Cancer Institute's Alliance for Nanotechnology in Cancer program. The CTCN is one of nine Centers of Cancer Nanotechnology Excellence (CCNE) across the country that was awarded a 5-year $13.5 million grant from the National Cancer Institute (NCI) Alliance in an open nationwide competition. With collaborators at Beth Israel Deaconess Hospital; Harvard Medical School; Tufts University, Auburn University and Nemucore Medical Innovations, Inc.the CTCN is also utilizing the support and facilities of the NU-based Center for High-rate Nanomanufacturing.

It would appear that in the U.S., government funding for nanomedicine has shifted toward addressing specific clinical goals rather than making "general" grants to establish nanomedicine oriented centers of excellence as was done in the past. This pattern places more emphasis on the invention and innovative development of working concepts and prototypes that can be directed toward providing solutions to problems and focusing on issues of interest to the funding entities that may be local, state or national, public or private.

European funding for research in nanomedicine has been supported by the European Commission through its Framework Programmes for Research and Development as well as the Horizon 2020 program. In contrast to the NIH program in the U.S., the EU program generally funds projects rather than centers and appears to be viable and showing recent funding awards for projects such as "Nanotherapeutics for antibiotic resistant emerging bacterial pathogens," "Tissue engineering of the right heart outflow tract by a biofunctionalized bioresorbable polymeric valved tube," "Reinforced Bioresorbable Biomaterials for Therapeutic Drug Eluting Stents, Nanotherapeutics to Treat Antibiotic Resistant Gram-Negative Pneumonia Infections", "Novel combination of biopolymers and manufacturing technologies for production of a peripheral nerve implant containing an internal aligned channels array," and many more.

Nanomedicine for diagnosis and treatment of disease, as a discipline has now been around for decades, with the first nanotherapeutic products being approved in the mid-1990s. Several nanomedicine centers were established during the period of 2004–2006. Many of these centers were set up to facilitate and accelerate the speed of translation of the research from the laboratory to the market (Venkatraman 2014). There has been progress made in the last 15 years, and some translation efforts have become commercialization successes (see definitions of "Success" below). There has not been a true "blockbuster" nanopharmaceutical product yet but the opportunity is clearly present as applications and potential uses of nanotechnology in medicine including pharmaceutical development are growing rapidly.

With all this activity and interest, why does the application of nanotechnology take so long to become commercially viable in medical applications? In part this may be due to the reality of bringing up a new and emerging technology platform to safe and effective use as well as establishing genuine market acceptance in the highly regulated and competitive world of medical products versus the "hype: or in this case "nanohype" if you will of selling (or perhaps over selling) the promise to the funding entities and investment community (Vashist et al. 2012).

In addition, there is some consumer concern about the general safety of nanomaterials. Some nanomaterials that have been used in nanomedical products (e.g. titanium dioxide nanoparticles in sunscreens and silver nanoparticles in antimicrobials) have generated uncertainty over the environmental consequences of the use of nanotechnology and some have uncertainty as to the occupational health impact (in handling during manufacturing). These have been thought to be contributing factors to investor and market concerns (Venkatraman 2014). In addition, these concerns and the realization that nanomaterials have some characteristics that require somewhat more characterization and elaborate analysis during non-clinical safety testing that may lead to somewhat longer times in the pre-clinical development phase as well as higher overall development costs (Tinkle et al. 2014).

The scaling up of nanomedical products can also be more challenging, costly and time consuming than that for "traditional" medical products in the same categories (Venkatraman 2014).

Decisions to invest and finance projects with long commercialization times are always subject to much scrutiny and cost-to-benefit analysis. So, it is wise to have as many of these concerns (non-clinical safety, environmental and occupational safety, scale up, etc.) in mind when developing the nanomedical product concept and designing the plan for commercialization. The more these issues have been addressed and resolved satisfactorily and supported by actual results and real solutions the less uncertainty there will be in the investment decisions. Careful consideration of the cost of goods (to include intellectual property support, development and marketing costs) for the nanomedical product, the time to market, size of the market and benefit to the market (in the first year of introduction and sales projections over a conservative product life of 7–10 years) can help to move the cost-to-benefit considerations in favor of support.

It is important to realize that, fundamentally, researchers and research institutions are generally focused on publications and the creation of intellectual property with hopes for significant financial interest and support. In contrast, commercialization involves market and financially focused investors that require clearly defined development paths that realistically address and complete all the steps necessary to commercialize with an expected minimal rate of return. It takes talented entrepreneurs to bring these two different but dependent interests together.

Successful commercialization of a nanomedical product begins with the innovator and those wishing to move the product forward into the clinic. Then there are those who must work to identify and address potential barriers and show investors and funding entities a realistic and workable path toward arriving at a cost-efficient, profitable product with significantly improved efficacy for the intended use.

17.10 Regulatory Approval

Obtaining regulatory approval is obviously an issue and potential barrier for commercialization of a nanomedical product and can be a concern for the global regulatory picture for review and approval of innovative nanomedical products (Wacker et al. 2016). As indicated above, while certainly a challenge, this does not necessarily have to be an onerous issue. Recognizing that regulators have an important job to do toward protection of the public (which generally includes everyone including the company personnel, investors, etc. unless the company is foreign to the country where approval is sought). Respecting and understanding this role to include gaining a familiarity with all regulations and guidances that are applicable to each proposed nanoproduct is essential. There are reputable and experienced consultants and firms in most countries that maintain an up to date knowledge of these requirements and can provide assistance toward planning for commercialization at the earliest possible stage. Engaging this expertise is certainly more cost and time affective than attempting to "do it yourself" in most all cases. They can help avoid costly and time consuming issues and also will often help to negotiate cost efficient pricing with other contract organizations that are needed to commercialize.

The U.S. FDA as well as regulators in other countries need to know if a medical product contains nanotechnology even if the nanotechnology is not associated with the active pharmaceutical ingredient or essential to the function of a medical device (Hamburg 2012; Monica 2012). The FDA, for example, has issued a guidance document that addresses and explains this need clearly entitled, "Guidance for Industry: Considering Whether an FDA-Regulated Product Involves the Application of Nanotechnology" (FDA 2014). So, it is wise to not let anyone, including a trade association or inexperienced advisor lead you to believe that not mentioning or "hiding" the nanotechnology in a medical product is a good strategy when seeking regulatory approval of a product that contains nanotechnology. The regulatory agency view of the company's commitment to safety can be as important to the agency concern for product safety and quality of the non-clinical and clinical data that has been submitted for review to demonstrate product safety.

17.11 Market Acceptance/Adoption

Cancer is a market of opportunity for nanotherapeutics and is the second worldwide cause of death, exceeded only by cardiovascular diseases (Pérez-Herrero and Fernández-Medarde 2015). Because it is characterized by uncontrolled cell proliferation with an absence of cell death that typically generates an abnormal cell mass or tumor, there is much interest and opportunity in the development of intracellularly targeted cancer therapeutics with various applications of nanotechnology.

Surgery and radiotherapy are the primary treatment used for local and non-metastatic cancers, anti-cancer drugs (chemotherapy, hormone and biological therapies) are the choice currently used in metastatic cancers (Pérez-Herrero and Fernández-Medarde 2015). Chemotherapeutic targeting is typically based on inhibiting the division of rapidly growing cells, which is a characteristic of the cancerous cells, but unfortunately, this approach also affects normal, rapidly proliferating, cells such as the spermatocytes, oocytes, basal keratinocytes, hair follicles, bone marrow progenitor cells and cells lining the gastrointestinal tract thereby resulting in adverse effects (hair loss, diminished sperm and white blood cell count, etc.) and other effects associated on proliferative normal cells that are characteristic side effects of chemotherapy. This destruction of normal cells along with cancer cells, the toxicity of conventional chemotherapeutic drugs, as well as the development of multidrug resistance, are supporting the demand for intracellularly-targeted treatments with improved efficacy and minimal toxicity based on more selective and effective interaction with cellular targets and less overall systemic exposure (mg/kg dose). Medical nanotechnology is well positioned to advance this market and is showing real promise.

There has been an increase in FDA-approved targeted cancer drugs in recent years that block biologic transduction pathways and/or specific cancer proteins to induce the death of cancer cells by means of apoptosis and stimulation of the immune system, or specifically deliver chemotherapeutic agents to cancer cells, minimizing the undesirable side effects (Pérez-Herrero and Fernández-Medarde 2015). Intracellularly targeted, nanotechnology-enabled, therapies can include cell-selective, alteration of cancer cell specific cell signaling by means of vascular nanoparticle uptake, monoclonal antibody binding or the action of cellular toxins or metabolic inhibitors. These antibodies, cytotoxic agents and metabolic inhibitors can be bound to, or otherwise associated with different nanoparticles and nanostructures that act as cytotoxic drug carriers or actively targeted cancer cell anti-proliferative agents by acting on a number of known mechanisms and/or molecular targets. In some cases, the cytotoxic agent is a drug active ingredient (i.e. doxorubicin, taxol, etc.) that has previously received FDA approval. With high specificity targeting the exposure to the cytotoxic agent is generally reduced such that the concern for safety and the potential for adverse effects is diminished such that the commercialization path may lessened somewhat due to the need for extensive safety testing typically required for new pharmaceuticals.. The safety of the nanomedical drug carriers does need to be confirmed on a case by case basis, but once the safety of a specific carrier is established at a sufficiently high exposure concentration, concern for the safety is diminished and the same carrier can be used as a carrier for other anticancer agents with less overall concern.

There are several clinical trials in progress to evaluate the anti-cancer efficacy of polymer-protein and polymer-drug conjugates, liposomal formulations, including immunoliposomes, polymeric micelles and polymeric nanoparticles (Pérez-Herrero and Fernández-Medarde 2015). In contrast, there are no FDA approvals or current clinical trials in process with carbon nanotube or dendrimers carriers due to their unresolved toxicity. The interest and market potential that these types of nanomedical formulations have is evidenced by the fact that 90 % of all scientific papers based on cancer therapeutics with polymeric nanoparticles have been published since 2009 (Pérez-Herrero and Fernández-Medarde 2015).

The impact of market acceptance/adoption on nanomedical product valuation is critical. How much of the product is sold or used and at what rate demand for the product develops after market introduction are very important factors in determining the market value of a marketed nanomedical product. However, a nanomedical product's success depends not just on the number of people who try it once, but on the number of repeat users. This is referred to as the adoption process, in which a product goes from being new in the marketplace to being an established product (or, in some cases where a product is at the end of its lifecycle, obsolete).

Naturally, product adoption and market acceptance have a large effect on the revenue generated. In many cases, product acceptance begins slowly and the adoption rate increases as clinicians and patients become aware of the product and begin to use it with some routine. "Early adopters" are users that are first to use the product and begin to demonstrate the product's market and revenue potential. When the new nanomedical product approaches full market penetration and saturation, the rate of adoption and acceptance slows and it should be expected that there always users that are "late adopters" or that will never adopt the product.

At some point other, competing products that may have improved safety and/or efficacy will typically enter an especially lucrative market and reduce market share for the initial nanomedical product. In some cases, novel or more advanced nanotechnologies emerges that result in improvements in safety and/or efficacy that will replace the initial product completely.

Estimating the potential market value of a product using sales projections (the income approach) must consider more than just the total market size, but also include a realistic projection of the rate at which the product will be adopted and whether or not the product is likely to achieve complete market saturation. With older products, one must always consider the likelihood of reduced acceptance that comes at the end of the product lifecycle and emergence of competing/improved products. With popular pharmaceuticals that have shown long-term safety and efficacy, there is sometimes additional market potential and product life in the over the counter market but at this time, we are far from that point with any prescriptive nanomedical pharmaceutical product.

17.12 Success

Just as there is no single or optimal path to success for the commercialization for all nanomedical technologies even in the same treatment class, there are also different ways in which success is measured depending upon one's particular point of view and expectation.

Success is generally measured in terms of time to market and some degree or indicator of market success. Timewise, this is the time from invention to market or innovation to market when there is no invention but rather an innovative combination of technologies developed to a marketable product. In the case of an imitation, it follows that success would be measured relative to the time of creation of an imitation with commercial potential and its arrival in the market.

Clinically, market success is a guarded conclusion but is allowed when a product meets or exceeds its therapeutic expectations and maintains a favorable risk to benefit ratio throughout the patient population over its product lifecycle.

From a business perspective, success is usually measured with respect to whether a product enters the market when expected or before and meets or exceeds its net profit expectations thereby resulting in the greatest returns to the company and investors.

So, great success may be viewed differently from different points of view. Clinically as entering the market with efficiency of a minimal time frame and then exceeding clinical expectations, or businesswise as getting to market in less time than anticipated and exceeding net profit expectations giving investors more return than anticipated. Taking these views together, it is clear that, ideally, a completely successful product would be one that enters the market with efficiency in minimal time, quickly exceeds clinical expectations and then, having a high level of market acceptance, produces much greater net profit than anticipated. This would be the very best outcome a successful nanomedical commercialization program could expect and there is no nanomedical product to date that has yet achieved this, but with so much opportunity, this is certain to occur soon and savvy investors are cautiously aware of this…

Therefore, when initiating a commercialization process, it should always be assumed that it will be a "journey" with unexpected events, iterative learning, continuous improvement and adjusted planning as small successes and setbacks may occur along the way. While, on rare occasion, a charismatic individual seemingly may "will" a product to market success, know without question that behind the scenes (even if there is little or no recognition by the charismatic leadership) that there were a lot of activities in process involving talented individuals that were absolutely necessary to make the "dream come true.." Thomas A. Edison, the famous inventor, is quoted as having said (c. 1903) "genius is one percent inspiration and ninety-nine percent perspiration" and was known to work himself and his laboratory employees very long and hard, but they also invented many now everyday items and successfully commercialized many of inventions (Rosanoff 1932). In his statement, Edison was referring what some saw as his seeming "at will" ability to produce many useful new inventions, but in reality there was a lot of behind the scenes effort involved and the same will always be true for any successful commercialization of a medical product (nanomedical or otherwise). This might look "easy" or "at will" from the outsider perspective but hard work is involved in most every commercial success and should always be expected and planned for as well as making sure to appropriately reward those that contributed to the success.

17.13 Lessons Learned

There are important lessons to be learned working with, observing and/or studying several companies that achieved success with their medical product commercialization programs (nanomedical or otherwise) but were savvy enough to recognize early enough in the program when to make a change that proved critical to their overall success, including a temporary or permanent work stoppage as necessary.

Venditto and Szoka (2013). review and describe the difficult and challenging commercialization path for clinical development of nanoparticle-based drugs. They identify that "inventors create the technology necessary for drug development including cell targeting markers, protein engineering technology, drug discovery, etc. Then innovators remix the invented technology into a drug formulation and navigate the approval process. Lastly, imitators exploit the technology of the newly approved drug to ride the wave of success that preceded them." In nanomedicine, it can readily be seen that invention, innovation and imitation is operating continuously to arrive at new project and product concepts to be funded and attempt to take to market. This has been applied to the commercialization steps noted in this article in an attempt to provide as complete a picture of the process as possible.

Other lessons learned in the commercialization of nanomedical products include the following:

- Strategically define and develop product opportunities: Strategic development of the commercialization plan is always better than leaving things to chance and working more "opportunistically." All strategically-planned commercialization programs, of course, have opportunistic decisions to make but these should be done within a well-developed commercialization strategy.
- Take time to understand the target market as well as the risks and benefits of the commercialization opportunity: This means that one must carefully evaluate your product opportunity against the anticipated market against its ability to be disruptive to current and emerging products as well as attempt to identify and define any potential risks against conservative estimates of benefit.
- Successful leaders of commercialization programs are not born: Experience and knowledge count as does the ability to learn continuously from one's own mistakes and those of others. Even talented first time entrepreneurs must continuously develop their knowledge and test their skills while learning quickly from experience. When selecting executive officers for a commercialization opportunity or emerging company, this should always be in mind.
- Leadership by Example: The best commercialization programs generally have leadership that leads by example and not intimidation or an atmosphere of servitude. In these programs, everyone involved is engaged, prepared and motivated to identify and solve problems as they occur and not wait and see if someone else will do the work. The upper management is clearly observed "walking their talk" and are prepared to clean floors and take out the trash should it become necessary to the success of the project and business.
- Always think safety: It can never be too early to start thinking realistically and constructively about the potential for any risks to humans or the environment from not only the medical use of the product but also during the product lifecycle that includes manufacturing, distribution and disposal as well. Failing to recognize and incorporate safety early in the commercialization plan can result in funding delays, loss of the opportunity and possibly even catastrophic consequences.
- There is no "easy" button: Commercialization of a any product and most especially an emerging technology based nanomedical product is never easy but those who recognize and accept the challenges then plan strategically are best positioned to experience the satisfaction and benefits of seeing a new, safe and effective product enter the medical marketplace.
- Understand, monitor and carefully manage financial "burn rate" and conservatively established "flameout date": Successful commercialization programs know their finances and have leaders that continuously work to get the most efficiency both in funds and time to market in their planning.
- Know who to market the emerging product to and who to avoid: The nanomedical product market is hot, especially for products directed toward an unmet need or may be disruptive in improving the safety and/or efficacy of products cur-

rently offered in a current lucrative market. Investors and companies wishing to acquire new products to market have their own interests in mind which likely do not have the needs of an emerging company at the forefront. When fundraising, licensing and selling a new nanomedical product it is always important and never unnecessary to carefully protect intellectual property at all levels and during each step of the commercialization activity. Also, be aware that many if not most government regulatory officials do not take kindly to being marketed as to the business opportunity or comparative cost merits of products under consideration for approval. Doing this not only wastes time but also sets the wrong tone for what must be objective evaluations of risk and clinical benefit of the nanomedical product.

17.14 Future of Medical Nanotechnology Commercialization

Molecular manufacturing and nanoproduct assembly on a commercial scale has long been a prime concern for the future advancement of medical nanotechnology (Merkle 1999). Wonderful medical advances are anticipated from the medical nanotechnology field and steady progress is occurring while enabling technologies emerge and advance as well as international regulatory frameworks with which to effectively evaluate risks and benefits of products submitted for approval and registration. The potential ability to attack illness and repair injury at the molecular level is already here at the discovery level (Ma et al. 2016; Santoso and Yang 2016; Liu et al. 2015). Further development toward commercialization of these new generation medical products will take time and funding but we hold to the dream that, through managing these opportunities to success, one day soon, cancer, diabetes and antibiotic resistant infections as well as neurodegenerative and cardiovascular disease will be treatable and perhaps even cured. Until then, we must work diligently and effectively at all levels and dimensions of medical nanotechnology commercialization to invent, innovate and develop the opportunities that will make this dream become practical reality.

17.15 Summary

Nanotechnology has been for more than two decades and will continue into the foreseeable future to be a broad, expanding and major enabling technology platform for a variety of evolutionary and revolutionary changes in medicine. The field of "nanomedicine" is no longer in its infancy and is becoming established and positioned to produce many commercial products. The impact of nanotechnology on drug delivery has already helped to improve the efficacy of commercially-available therapeutics and has extended the commercial life of older pharmaceuticals as well as has enabled entirely new therapeutics. In tissue engineering, nanotechnology has also supported new approaches that could facilitate reconstruction of complex tissue architectures and three-dimensional tissue and organ constructions that in future may make organ donation less of a problem. Nanomedicine arrived and the future looks bright for many decades to come with continually growing families and types of innovative nanoscale materials and nanotechnology-enabled combinatorial technologies. With the successful commercialization of nanomedical products, it seems very clear that the new field of nanomedicine will not only improve and significantly advance conventional therapies, but will also serve as a bridge to future therapeutics that will be revolutionary, amazing and quite profitable.

References

AlAli AB, Griffin MF, Butler PE (2015) Three-dimensional printing: surgical applications. Eplasty 15:e37, 352–67

Alvarez-Roman R, Barre G, Guy RH, Fessi H (2001) Biodegradable polymer nanocapsules containing a sunscreen agent: preparation and photoprotection. Eur J Pharm Biopharm 52:191–195

Antunes AM, Alencar MS, da Silva CH, Nunes J, Mendes FM (2012) Trends in nanotechnology patents applied to the health sector. Recent Pat Nanotechnol 6(1):29–43

Baetke SC, Lammers T, Kiessling F (2015) Applications of nanoparticles for diagnosis and therapy of cancer. Br J Radiol 88(1054):20150207. doi: 10.1259/bjr.20150207. Epub 2015 Jun 12

Bawa R (2007) Patents and nanomedicine. Nanomedicine 2(3):351–374

Bawa R (2008) Patenting inventions in bionanotechnology: a guide for scientists and lawyers. Chapter 27. In: Reisner DE (ed) Bionanotechnology: global prospects. CRC Press, Boca Raton, pp 309–338

Bawa R, Bawa SR, Maebius SB, Flynn T, Wei C (2005) Protecting new ideas and inventions in nanomedicine with patents. Nanomedicine 1(2):150–158

Belassi W, Tukel OI (1996) A new framework for determining critical success/failure factors in projects. Int J Proj Manag 14(3):141–151

Bennat C, Muller-Goymann CC (2000) Skin penetration and stabilisation of formulations containing microfine titanium dioxide and a physical UV filter. Int J Cosmet Sci 22:271–283

Bleeker EAJ, Evertz S, Geertsma RE, Peijenburg WJGM, Westra J, Wijnhoven SWP (2015) Assessing health and environmental risks of nanoparticles: current state of affairs in policy, science and areas of application, RIVM Report 2014–0157. National Institute for Public Health and the Environment, Bilthoven, 146 pp

Bobo D, Robinson KJ, Islam J, Thurecht KJ (2016) Corrie SR (2016) Nanoparticle-based medicines: a review of FDA-approved materials and clinical trials to date. Pharm Res 14

Borm PJA, Robbins D, Haubold S, Kuhlbusch T, Fissan H, Donaldson K, Schins RPF, Stone V, Kreyling W, Lademann J, Krutmann J, Warheit D, Oberdorster E (2006) The potential risks of nanomaterials: a review conducted out for ECETOC. Part Fibre Toxicol 3:11. doi:10.1186/1743-8977-3-11

Bosetti R, Vereeck L (2012) The impact of effective patents on future innovations in nanomedicine. Pharm Pat Anal 1(1):37–43

Budgell E (1714) The Spectator No. 605, October 1714

Burgess R (2012) Understanding nanomedicine: an introductory textbook. Pan Stanford, Hoboken, 524 pp

Chaloupka K, Malam Y, Seifalian AM (2010) Nanosilver as a new generation of nanoproduct in biomedical applications. Trends Biotechnol 28(11):580–588

Cui H, Wang J (2016) Progress in the development of nanotheranostic systems. Theranostics 6(7):915–917

Dabiri G, Damstetter E, Phillips T (2016) Choosing a wound dressing based on common wound characteristics. Adv Wound Care 5(1):32–41

David CA, Owen A, Liptrott NJ (2016) Determining the relationship between nanoparticle characteristics and immunotoxicity: key challenges and approaches. Nanomedicine 11(11):1447–1464

de Mel A (2016) Three-dimensional printing and the surgeon. Br J Surg 103(7):786–788

Delaval M, Wohlleben W, Landsiedel R, Baeza-Squiban A, Boland S. (2016) Assessment of the oxidative potential of nanoparticles by the cytochrome c assay: assay improvement and development of a high-throughput method to predict the toxicity of nanoparticles. Arch Toxicol [Epub ahead of print]

do Carmo FA, Sathler PC, Zancan P, Rodrigues CR, Castro HC, de Sousa VP, Sola-Penna M, Cabral LM (2014) Therapeutic nanosystems for oral administration of insulin. Curr Pharm Biotechnol 15(7):620–628

Dorj B, Won JE, Purevdorj O, Patel KD, Kim JH, Lee EJ, Kim HW (2014) Novel therapeutic design of microporous-structured biopolymer scaffolds for drug loading and delivery. Acta Biomater 10(3):1238–1250

du Toit LC, Pillay V, Choonara YE, Pillay S, Harilall SL (2007) Patenting of nanopharmaceuticals in drug delivery: no small issue. Recent Pat Drug Deliv Formul 1(2):131–142

Duda F, Bradel S, Bleich A, Abendroth P, Heemeier T, Ehlert N, Behrens P, Esser KH, Lenarz T, Brandes G, Prenzler NK (2015) Biocompatibility of silver containing silica films on Bioverit® II middle ear prostheses in rabbits. J Biomater Appl 30(1):17–29

Eveleth R (2015) Why there aren't yet nanobot doctors. The Atlantic, Aug 6, 2015. http://www.theatlantic.com/technology/archive/2015/08/nanobot-treatment-doctors-cancer/400613/. Accessed 24 June 2016

FDA (2014) Guidance for industry: considering whether an FDA-regulated product involves the application of nanotechnology. U.S. Department of Health and Human Services, Food and Drug Administration, Office of the Commissioner. June 2014, 14 pp. http://www.fda.gov/downloads/RegulatoryInformation/Guidances/UCM401695.pdf. Accessed 24 June 2016

Feynman R (1959) There's plenty of room at the bottom; an invitation to enter a new field of physics. From a transcript of a talk given on December 29th 1959 at the annual meeting of the American Physical Society at the California Institute of Technology

Fierro IM, de Menezes Alencar MS, Lins Mendes FM, de Souza Mendes C, Nunes BF, de Souza Antunes AM (2014) Nanoparticles applied to antineoplastic agents: a patent landscape. Pharm Pat Anal 3(6):613–623

Flynn T, Wei C (2005) The pathway to commercialization for nanomedicine. Nanomed Nanotechnol Biol Med 1:47–51

Garg T, Jain NK, Rath G, Goyal AK (2015) Nanotechnology-based photodynamic therapy: concepts, advances, and perspectives. Crit Rev Ther Drug Carrier Syst 32(5):389–439

Gautier J, Allard-Vannier E, Hervé-Aubert K, Soucé M, Chourpa I (2013) Design strategies of hybrid metallic nanoparticles for theragnostic applications. Nanotechnology 24(43):432002

Ghanaati S, Lorenz J, Obreja K, Choukroun J, Landes C, Sader RA (2014) Nanocrystalline hydroxyapatite-based material already contributes to implant stability after 3 months: a clinical and radiologic 3-year follow-up investigation. J Oral Implantol 40(1):103–109

Godin B, Ferrari M (2012) Cardiovascular nano medicine: a posse ad esse. Methodist Debakey Cardiovasc J 8(1):2–5

Gurzawska K, Svava R, Jørgensen NR, Gotfredsen K (2012) Nanocoating of titanium implant surfaces with organic molecules. Polysaccharides including glycosaminoglycans. J Biomed Nanotechnol 8(6):1012–1024

Hamburg MA (2012) FDA's approach to regulation of products of nanotechnology. Science 336(6079):299–300

Hobson DW (2009) Commercialization of nanotechnology. Wiley Interdiscip Rev Nanomed Nanobiotechnol 1(2):189–202
Hsu HJ, Bugno J, Lee SR, Hong S. (2016) Dendrimer-based nanocarriers: a versatile platform for drug delivery. Wiley Interdiscip Rev Nanomed Nanobiotechnol [Epub ahead of print]
Huo L, Chen R, Shi X, Bai R, Wang P, Chang Y, Chen C (2015) High-content screening for assessing nanomaterial toxicity. J Nanosci Nanotechnol 15(2):1143–1149
Iost RM, Crespilho FN, Kern K, Balasubramanian K (2016) A primary battery-on-a-chip using monolayer graphene. Nanotechnology 27(29):29LT01 [Epub ahead of print]
ISO (2009) Biological evaluation of medical devices – Part 1: evaluation and testing within a risk management process. International Standards Organisatoin. ISO 10993–1:2009, 21pp
Jacob T, Hemavathy K, Jacob J, Hingorani A, Marks N, Ascher E (2011) A nanotechnology-based delivery system: nanobots. Novel vehicles for molecular medicine. J Cardiovasc Surg 52(2):159–167
Jain S, Valvi PU, Swarnakar NK, Thanki K (2012) Gelatin coated hybrid lipid nanoparticles for oral delivery of amphotericin B. Mol Pharm 9(9):2542–2553
Jain S, Bhankur N, Swarnakar NK, Thanki K (2015a) Phytantriol based "stealth" lyotropic liquid crystalline nanoparticles for improved antitumor efficacy and reduced toxicity of docetaxel. Pharm Res 32(10):3282–3292
Jain S, Spandana G, Agrawal AK, Kushwah V, Thanki K (2015b) Enhanced antitumor efficacy and reduced toxicity of docetaxel loaded estradiol functionalized stealth polymeric nanoparticles. Mol Pharm 12(11):3871–3884
Kapse SV, Gaikwad RV, Samad A, Devarajan PV (2012) Self nanoprecipitating preconcentrate of tamoxifen citrate for enhanced bioavailability. Int J Pharm 429(1–2):104–112
Kasili PM, Vo-Dinh T (2005) Optical nanobiosensor for monitoring an apoptotic signaling process in a single living cell following photodynamic therapy. J Nanosci Nanotechnol 5(12):2057–2062
Kaur IP, Kakkar V, Deol PK, Yadav M, Singh M, Sharma I (2014) Issues and concerns in nanotech product development and its commercialization. J Control Release 193:51–62
Keeney M, Jiang XY, Yamane M, Lee M, Goodman S, Yang F (2015) Nanocoating for biomolecule delivery using layer-by-layer self-assembly. J Mater Chem B Mater Biol Med 3(45):8757–8770
Kennedy J. (2007) Nanotechnology: the future is coming sooner than you think. A Joint Economic Committee ctudy. Joint Economic Committee United States Congress. March 2007. 20 pp. http://www.thenewatlantis.com/docLib/20120213_TheFutureisComingSoonerThanYouThink.pdf. Accessed 24 June 2016
Kohli V, Elezzabi AY (2009) Prospects and developments in cell and embryo laser nanosurgery. Wiley Interdiscip Rev Nanomed Nanobiotechnol 1(1):11–25
Kolhe S, Parikh K (2012) Application of nanotechnology in cancer: a review. Int J Bioinform Res Appl 8(1–2):112–125
Korrapati PS, Karthikeyan K, Satish A, Krishnaswamy VR, Venugopal JR, Ramakrishna S (2016) Recent advancements in nanotechnological strategies in selection, design and delivery of biomolecules for skin regeneration. Mater Sci Eng C Mater Biol Appl 67:747–765
Korting HC, Zienecke H, Schafer-Korting M, Braun-Falco O (1990) Liposome encapsulation improves efficacy of betamethasone dipropionate in atopic eczema but not in psoriasis vulgaris. Eur J Clin Pharmacol 39(4):349–351
Lafley AG, Tichy N (2011) The art and science of finding the right CEO. Harvard Business Review. Oct 2011, https://hbr.org/2011/10/the-art-and-science-of-finding-the-right-ceo. Accessed 24 June 2016
Landesman-Milo D, Peer D (2016) Transforming nanomedicines from lab scale production to novel clinical modality. Bioconjug Chem 27(4):855–862
Leite EA, Grabe-Guimarães A, Guimarães HN, Machado-Coelho GL, Barratt G, Mosqueira VC (2007) Cardiotoxicity reduction induced by halofantrine entrapped in nanocapsule devices. Life Sci 80(14):1327–1334

Lin J, Chen X, Huang P (2016) Graphene-based nanomaterials for bioimaging. Adv Drug Deliv Rev. [Epub ahead of print]

Liu J, Wei T, Zhao J, Huang Y, Deng H, Kumar A, Wang C, Liang Z, Ma X, Liang XJ (2015) Multifunctional aptamer-based nanoparticles for targeted drug delivery to circumvent cancer resistance. Biomaterials 91:44–56

Lopes MA, Abrahim BA, Cabral LM, Rodrigues CR, Seiça RM, de Baptista Veiga FJ, Ribeiro AJ (2014) Intestinal absorption of insulin nanoparticles: contribution of M cells. Nanomedicine 10(6):1139–1151

Lorscheidt S, Lamprecht A (2016) Safety assessment of nanoparticles for drug delivery by means of classic in vitro assays and beyond. Expert Opin Drug Deliv. [Epub ahead of print]

Lowy DA, Patrut A (2008) Nonobatteries: decreasing size power sources for growing technologies. Recent Pat Nanotechnol 2(3):208–219

Ma L, Liu TW, Wallig MA, Dobrucki IT, Dobrucki LW, Nelson ER, Swanson KS, Smith AM. (2016) Efficient targeting of adipose tissue macrophages in obesity with polysaccharide nanocarriers. ACS Nano. [Epub ahead of print]

Mann EE, Thompson LC, Shannahan JH, Wingard CJ (2012) Changes in cardiopulmonary function induced by nanoparticles. Wiley Interdiscip Rev Nanomed Nanobiotechnol 4(6):691–702

Merkle RC (1999) Biotechnology as a route to nanotechnology. Trends Biotechnol 17(7):271–274

Monica JC Jr (2012) FDA's evolving approach to nanotechnology. Food Drug Law J 67(4):405–411

Morones JR, Elechiguerra JL, Camacho A, Holt K, Juan B, Kouri JB, Ramırez JT, Yacaman MJ (2005) The bactericidal effect of silver nanoparticles. Nanotechnology 16:2346–2353

Nel A, Xia T, Meng H, Wang X, Lin S, Ji Z, Zhang H (2013) Nanomaterial toxicity testing in the 21st century: use of a predictive toxicological approach and high-throughput screening. Acc Chem Res 46(3):607–621

NNI (2016) Tech transfer and commercialization. Webpage..http://www.nano.gov/techtransfer. Accessed 24 June 2016

Nohynek GJ, Lademann J, Ribaud C, Roberts MS, Grey goo on the gkin? (2007) Nanotechnology, cosmetic and sunscreen safety. Crit Rev Toxicol 37:251–277

Nuñez-Anita RE, Acosta-Torres LS, Vilar-Pineda J, Martínez-Espinosa JC, de la Fuente-Hernández J, Castaño VM (2014) Toxicology of antimicrobial nanoparticles for prosthetic devices. Int J Nanomedicine 9:3999–4006

Palza H (2015) Antimicrobial polymers with metal nanoparticles. Int J Mol Sci 16(1):2099–2116

Park JS, Kim DH, Kim HN, Wang CJ, Kwak MK, Hur E, Suh KY, An SS, Levchenko A (2016). Directed migration of cancer cells guided by the graded texture of the underlying matrix. Nat Mater Published online 14 March 2016

Pereira RF, Bártolo PJ (2016) Traditional therapies for skin wound healing. Adv Wound Care 5(5):208–229

Pérez-Herrero E, Fernández-Medarde A (2015) Advanced targeted therapies in cancer: drug nanocarriers, the future of chemotherapy. Eur J Pharm Biopharm 93:52–79

Phillips E, Penate-Medina O, Zanzonico PB, Carvajal RD, Mohan P, Ye Y et al (2014) Clinical translation of an ultrasmall inorganic optical-PET imaging nanoparticle probe. Sci Transl Med 6(260):260ra149

Piktel E, Niemirowicz K, Wątek M, Wollny T, Deptuła P, Bucki (2016) Recent insights in nanotechnology-based drugs and formulations designed for effective anti-cancer therapy. J Nanobiotechnol 14(1):39

Prieto F, Sep lveda B, Calle A, Llobera A, Domínguez C, Abad A, Montoya A, Lechuga LM (2003) An integrated optical interferometric nanodevice based on silicon technology for biosensor applications. Nanotechnology 14:907–912

Radomska A, Leszczyszyn J, Radomski MW (2016) The nanopharmacology and nanotoxicology of nanomaterials: new opportunities and challenges. Adv Clin Exp Med 25(1):151–162

Rausch LM (1998) High-tech industries drive global economic activity. U.S. National Science Foundation. Division of Science Resource Studies: Issue Brief. Directorate for Social, Behavioral and Economic Sciences. NSF 98–319 July 20, 1998. http://www.nsf.gov/statistics/issuebrf/sib98319.htm. Accessed 24 June 2016

Reverté L, Prieto-Simón B, Campàs M (2016) New advances in electrochemical biosensors for the detection of toxins: nanomaterials, magnetic beads and microfluidics systems. A review. Anal Chim Acta 908:8–21

Rezaei B, Ghani M, Shoushtari AM, Rabiee M (2016) Electrochemical biosensors based on nanofibres for cardiac biomarker detection: a comprehensive review. Biosens Bioelectron 78:513–523

Richter GM, Stampfl U, Stampfl S, Rehnitz C, Holler S, Schnabel P, Grunze M (2005) A new polymer concept for coating of vascular stents using PTFEP (poly(bis(trifluoroethoxy)phosphazene) to reduce thrombogenicity and late in-stent stenosis. Invest Radiol 40(4):210–218

Rivero PJ, Urrutia A, Goicoechea J, Zamarreño CR, Arregui FJ, Matías IR (2011) An antibacterial coating based on a polymer/sol–gel hybrid matrix loaded with silver nanoparticles. Nanoscale Res Lett 6(1):305–312

Roco M (2004) Nanoscale science and engineering: unifying and transforming tools. AIChE J 50(5):890–897

Rolland A, Collet B, Le Verge R, Toujas L (1989) Blood clearance and organ distribution of intravenously administered polymethacrylic nanoparticles in mice. J Pharm Sci 78(6):481–484

Ronchi P, Terjung S, Pepperkok R (2012) At the cutting edge: applications and perspectives of laser nanosurgery in cell biology. Biol Chem Apr 393(4):235–248

Rosanoff MA (1932). Edison in his laboratory. Harper's Monthly, September 1932; 402–17

Roy E, Patra S, Madhuri R, Sharma PK (2016) Stimuli-responsive poly(N-isopropyl acrylamide)-co-tyrosine-gadolinium: iron oxide nanoparticle-based nanotheranostic for cancer diagnosis and treatment. Colloids Surf B: Biointerfaces 142:248–258

Santoso MR, Yang PC (2016) Magnetic nanoparticles for targeting and imaging of stem cells in myocardial infarction. Stem Cells Int 2016:4198790. doi:10.1155/2016/4198790, Epub 2016 Apr 3

SCENIHR (2009) Risk assessment of products of nanotechnologies. European Commission, Scientific Committee on Emerging and Newly Identified Health Risks. European Commission 2009, 71 pp. http://ec.europa.eu/health/ph_risk/committees/04_scenihr/docs/scenihr_o_023.pdf. Accessed 24 June 2016

SCENIHR (2015) Guidance on the determination of potential health effects of nanomaterials used in medical devices. Commission, Scientific Committee on Emerging and Newly Identified Health Risks. European Commission 2015, 77 pp. http://ec.europa.eu/health/scientific_committees/emerging/docs/scenihr_o_045.pdf. Accessed 24 June 2016

Schmittlein DC, Mahajan V (1982) Maximum likelihood estimation for an innovation diffusion model of new product acceptance. Mark Sci 1(1):57–78

Shakweh M, Ponchel G, Fattal E (2004) Particle uptake by Peyer's patches: a pathway for drug and vaccine delivery. Expert Opin Drug Deliv 1(1):141–163

Shapira A, Livney YD, Broxterman HJ, Assaraf YG (2011) Nanomedicine for targeted cancer therapy: towards the overcoming of drug resistance. Drug Resist Updat 14(3):150–163

Subramani K, Jung RE, Molenberg A, Hammerle CH (2009) Biofilm on dental implants: a review of the literature. Int J Oral Maxillofac Implants 24(4):616–626

Sudhakar CK, Upadhya N, Verma A, Jain A, Charyulu RN, Jain S (2015) Chapter 1 – Nanomedicine and tissue engineering. In: Grohens Y, Ninan N, Thomas S (eds) Nanotechnology applications for tissue engineering. Elsevier, Amsterdam, pp 1–19

Sundar DS, Antoniraj MG, Kumar CS, Mohapatra SS, Houreld NN, Ruckmani K (2016) Recent trends of biocompatible and biodegradable nanoparticles in drug delivery: a review. Curr Med Chem. [Epub ahead of print]

Tamjid E, Simchi A, Dunlop JW, Fratzl P, Bagheri R, Vossoughi M (2013) Tissue growth into three-dimensional composite scaffolds with controlled micro-features and nanotopographical surfaces. J Biomed Mater Res A 101(10):2796–2807
Tan A, Chawla R, Mahdibeiraghdar S, Jeyaraj R, Rajadas J, Hamblin MR, Seifalian AM (2016) Nanotechnology and regenerative therapeutics in plastic surgery: the next frontier. J Plast Reconstr Aesthet Surg 69(1):1–13
Tinkle S, McNeil SE, Mühlebach S, Bawa R, Borchard G, Barenholz YC, Tamarkin L, Desai N (2014, April) Nanomedicines: addressing the scientific and regulatory gap. Ann N Y Acad Sci 1313:35–56
Valkonen P (2012) How to set up an effective IP strategy and manage a nanotechnology-based patent portfolio. Chapter 5. In: Helwegen W, Escoffier L (eds) Nanotechnology commercialization for managers and scientists. Pan Stanford Publishing, Singapore, pp 103–130
Vashist SK, Venkatesh AG, Mitsakakis K, Czilwik G, Roth G, von Stetten F, Zengerle R (2012) Nanotechnology-based biosensors and diagnostics: technology push versus industrial/healthcare requirements. BioNanoSci 2:115–126
Venditto VJ, Szoka FC Jr (2013) Cancer nanomedicines: so many papers so few drugs! Adv Drug Deliv Rev 65(1):80–88
Venkatraman S (2014) Has nanomedicine lived up to its promise? Nanotechnology 25(37):372501
Wacker MG, Proykova A, Santos GM (2016) Dealing with nanosafety around the globe-regulation vs. innovation. Int J Pharm 509(1–2):95–106
Wang Y, Li C, Yao C, Ding L, Lei Z, Wu M (2016) Techniques for investigating molecular toxicology of nanomaterials. J Biomed Nanotechnol 12(6):1115–1135
Wilkinson LJ, White RJ, Chipman JK (2011) Silver and nanoparticles of silver in wound dressings: a review of efficacy and safety. J Wound Care 20(11):543–549
Yamada M, Foote M, Prow TW (2015) Therapeutic gold, silver, and platinum nanoparticles. Wiley Interdiscip Rev Nanomed Nanobiotechnol 7(3):428–445
Yang W, Fu LM, Wen X, Liu Y, Tian Y, Liu YC, Han RC, Gao ZY, Wang TE, Sha YL, Jiang YQ, Wang Y, Zhang JP (2016) Nanoprobes for two-photon excitation time-resolved imaging of living animals: in situ analysis of tumor-targeting dynamics of nanocarriers. Biomaterials 100:152–161

Index

A

Abraxane, 50, 57, 61, 62, 84, 96, 102, 127, 190
Active targeting, 4, 10–13, 24, 69, 168, 260, 267, 285, 287, 326, 328–332, 346, 348, 350, 394
Animal model, 4, 13, 30, 150, 157, 158, 188, 191, 265, 294, 334, 335, 347, 383
Anti-angiogenic therapy, 201–221
Antibody, 11, 38, 53, 73, 128, 151, 165, 203, 260, 310, 326, 346, 380, 439
Anti-cancer nanodrug, 132–143, 166, 168, 186, 187
Anticancer therapy, 42, 49–62, 262

B

Bacterial-derived minicell, 170, 174, 183–184, 186
Bioavailability, 73, 81, 127–129, 131, 134, 139, 140, 256, 261, 288, 413
Biomedical imaging, 273–297
Blood half-life, 92, 131

C

CAFs. *See* Cancer-associated fibroblast (CAFs)
Cancer, 6, 39, 50, 69, 126, 150, 166, 202, 231, 256, 285, 302, 324, 347, 371, 377, 410
Cancer-associated fibroblast (CAFs), 14, 74–79, 106, 154, 155, 308, 312
Cancer treatment, 8, 43, 54, 132, 187, 191, 205, 291, 295, 377–383, 410, 429
Cell specific, 15–16, 30, 37–44, 177, 184, 185, 218, 439
Cell tracking, 129, 132, 292
Chemoadjuvant, 55, 57
Chemotherapy, 7, 25, 43, 54, 55, 57, 62, 72, 73, 79, 134, 140, 142, 170, 176, 203–205, 210, 221, 233, 258, 336, 377–379, 382, 384, 391, 393, 438
Chondroitinase, 61
Clinical translation, 7, 62, 127, 165–167, 179
Collagenase, 14, 58–61, 106, 312
Commercialization, 3–30, 405–444
Cross-species extrapolation, 370–372

D

DDS. *See* Drug delivery system (DDS)
Disposition, 17, 342–348, 350, 355, 357, 358, 382, 392
Doxil, 9, 18, 21, 59, 61, 62, 84, 96, 127, 128, 142, 175, 184, 186, 189, 190, 203
Drug ADME, 372
Drug delivery, 4, 38, 79, 127, 152, 165, 204, 248, 256, 282, 317, 342, 366, 379, 411
Drug delivery system (DDS), 11–13, 26, 38, 84, 100, 104, 142, 152, 204, 206, 214, 215, 221, 248, 256, 257, 282, 323–336, 341–358, 415

E

ECM. *See* Extracellular matrix (ECM)
Endothelial cells, 10, 71, 84, 87, 88, 90, 99–101, 150–153, 177, 180, 191, 203–207, 210–221, 260, 261, 305, 310, 312–314, 345, 346

A. Prokop, V. Weissig (eds.), *Intracellular Delivery III*, Fundamental Biomedical Technologies 8, DOI 10.1007/978-3-319-43525-1

Enhanced permeability and retention (EPR) effect, 10–13, 71, 84, 87, 142, 150, 152, 156, 166–168, 185, 188, 203, 204, 206, 256, 259, 260, 285, 287, 292, 308, 324–326, 328, 332, 335, 347, 348, 394
Enhanced permeability effect, 10–12, 150, 167, 203, 256, 308, 324, 347, 348, 394
Enlarged spleen, 137
Extracellular matrix (ECM), 8–9, 14, 49–62, 71, 73, 74, 77–79, 81, 84, 105, 106, 154, 244, 257, 259, 304, 306–308, 310, 312, 316, 348, 357, 415
Extracellular matrix components, 8, 9, 79, 81, 154, 257, 312

F

Formulation, 11, 12, 41, 55, 57, 59, 99, 101, 105, 127, 128, 175–178, 186–188, 265, 275–281, 283, 285, 286, 288, 289, 292, 295, 326, 342, 343, 347, 349, 350, 355, 358, 366, 367, 387, 391, 393, 408, 409, 413, 424, 441

H

Hepatotoxicity, 138, 394
High-energy method, 277, 280, 281
Hyaluronidase, 9, 50, 53–58, 106, 312

I

Imaging, 4, 40, 59, 80, 126, 175, 211, 242, 282, 302, 326, 350, 407
Immune cell labeling, 129, 132
Intralipid®, 10, 125–143
Investment, 7, 19, 29, 410, 411, 419–423, 426, 432–434, 436, 437

L

Liposome, 10, 38, 75, 129, 167, 203, 237, 262, 287, 303, 342, 381, 412
Low-energy method, 276, 280–281

M

Magnetic resonance imaging (MRI), 12, 13, 24, 53, 129, 131–132, 134, 213, 216, 219, 220, 291–293, 295, 296, 303, 305, 315, 317
Market, 6, 29, 30, 43, 50, 54, 107, 408, 410, 411, 413, 417–423, 426–430, 432–443
Marketing, 417, 419, 420, 422, 427, 437
Mechanism of action, 379
Micelles, 11, 14, 20, 22, 23, 25, 41, 84, 85, 95, 96, 98, 99, 140, 167, 178, 186, 191, 239, 261–264, 278, 283, 286, 303, 314, 324–326, 332, 342, 343, 383–385, 393, 439
Microenvironment, 14, 24, 40, 67–107, 150, 221, 256–260, 262, 265, 267, 301–318, 377, 380, 385–387, 416
Microscopic mass spectrometry, 323–336
Mitochondrial uptake, 239–241
Mitochondria targeting, 21, 234, 236, 237, 239, 240, 243, 247, 248
Molecular imaging, 24, 302, 315–317
Molecular targeting, 16, 17, 28, 302, 309, 311, 313, 316, 335, 379, 380, 412, 439
MRI. *See* Magnetic resonance imaging (MRI)

N

NanoCargo, 229–248
Nanocarriers, 16–22, 24, 49–62, 67–107, 127, 129, 134, 155, 165–168, 174, 175, 179, 181, 184, 186, 189, 191, 219, 221, 282, 285, 287, 292, 317, 375–395, 412, 413
Nano-drug delivery systems, 341–358
Nano-emulsion, 273–297
Nanomaterial, 7, 92, 129, 240, 308, 309, 313, 314, 317, 383, 411, 414, 415, 421, 422, 424, 425, 429, 436
Nanomedicine, 6, 74, 132, 150, 165, 275, 302, 386, 407
Nanoparticle delivery, 11, 142, 301–318, 395
Nanoparticle nanomedicine, 413
Nanosurgery, 407
Nanotechnology, 7, 82, 126, 165, 234, 282, 385, 406
Nephrotoxicity, 137, 138

O

Off-target accumulation, 143
Oncolytic virus, 54, 61, 62

P

Particulate nanocarrier, 166, 167, 173–175, 181, 186, 189
Passive targeting, 10, 11, 15, 71, 102, 168, 260, 285, 287, 324, 332–335, 394
Patenting, 28–30, 427–429, 431
Pathophysiology, 150, 168

PBPK. *See* Physiologically-based pharmacokinetic modelling (PBPK)
Pharmacodynamics, 15, 127, 256, 326, 341–358, 382
Pharmacokinetic modelling, 364–367
Pharmacokinetic/pharmacodynamics profile, 27, 28, 127, 132, 135, 140, 142
Pharmacokinetics, 6, 14, 15, 25–28, 38, 40, 44, 127, 128, 142, 150, 156, 157, 166, 167, 177, 178, 188, 256, 283, 287, 292, 297, 302, 313, 326, 341–358, 370, 382, 385, 394, 422
Physiologically-based pharmacokinetic modelling (PBPK), 363–372
Platinum-containing nanodrug, 134
Polymeric nanoparticle, 87, 90, 95, 167, 181–183, 186, 237, 238, 262, 283, 284, 393, 439
Polymersome, 62, 388
Precision drugs, 37–44

R

Rational drug combinations, 379, 380, 395
Redox, 231, 255–268, 384, 425
RES. *See* Reticuloendothelial system (RES)
Reticuloendothelial system (RES), 18, 106, 125–143, 260, 282, 283, 285, 302, 324
Retrovector, 185
Risk assessment, 422, 424, 426

S

Silica nanoparticles, 9, 264, 316, 384, 412
Stimuli-responsive nanoparticles, 262, 394
Stromal cell, 14, 78, 150, 153–156, 304, 308, 314
Superparamagnetic iron-oxide nanoparticle, 129, 267
Surface functionalization, 87, 244, 283, 284, 415
Synergy, 107, 140, 381, 383, 386, 387
Synthetic lethality, 380
Systems pharmacology, 6, 107

T

TAMs. *See* Tumor-associated macrophages (TAMs)
Targeted delivery, 26, 165, 191, 214, 260, 267, 324, 329, 342, 350–355
Targeting, 4, 38, 58, 69, 127, 165, 202, 233, 260, 282, 302, 324, 342, 379, 407
Targeting efficiency, 12, 81, 342, 343, 347, 349, 352, 355, 358
Targeting ligand, 11, 15, 94, 166, 168, 173, 174, 176, 179–182, 185–189, 191, 206, 260, 262, 265, 395
Technological methods, 350, 427
Toxicity profile, 179, 189, 386, 392
Toxicology, 13, 27, 127, 140, 409, 411, 422, 424–426
Toxic side effects, 125–143
Tumor, 6, 39, 50, 69, 129, 150, 166, 202, 235, 256, 282, 302, 324, 347, 377, 411
Tumor-associated macrophages (TAMs), 14, 72–74, 76, 101, 155, 156, 308
Tumor endothelial cells, 99, 203–207, 210–221, 260
Tumor microenvironment, 24, 67–107, 221, 256–260, 262, 265, 267, 301–318, 385, 386

V

Valuation, 420, 424, 426, 432–433, 439
Vasculature, 11, 14, 42, 69, 71, 75, 84, 87, 93, 95–101, 150–153, 156–158, 167, 181, 185–187, 191, 201–221, 257–259–260, 285, 304–306, 308, 309, 312–316

The manufacturer's authorised representative in the EU is Springer
Nature Customer Service Centre GmbH, Europaplatz 3, 69115 Heidelberg,
Germany. If you have any concerns regarding our products, please
contact ProductSafety@springernature.com

Printed and bound by CPI Group (UK) Ltd, Croydon, CR0 4YY
30/09/2026
02225571-0017